Handbook of Experimental Pharmacology

Volume 246

More information about this series at http://www.springer.com/series/164

Mohamed Chahine
Editor

Voltage-gated Sodium Channels: Structure, Function and Channelopathies

Springer

Editor
Mohamed Chahine
CERVO Brain Research Center
Institut Universitaire en Santé Mentale
Quebec, Canada

Department of Medicine
Université laval
Quebec, Canada

ISSN 0171-2004 ISSN 1865-0325 (electronic)
Handbook of Experimental Pharmacology
ISBN 978-3-030-07982-6 ISBN 978-3-319-90284-5 (eBook)
https://doi.org/10.1007/978-3-319-90284-5

Printed on acid-free paper

This Springer imprint is published by the registered company Springer International Publishing AG part of Springer Nature.
The registered company address is: Gewerbestrasse 11, 6330 Cham, Switzerland

Preface

This is an update and expansion of a previous HEP volume and well-received book entitled *Voltage-Gated Sodium Channels*, edited by Peter Ruben, which details the current state of knowledge of voltage-gated Na^+ channels, their pharmacology, and related diseases. The book chapters authored by internationally recognized experts cover a broad array of topics, including the structural basis of Na^+ channel function, methodological advances in the study of Na^+ channels, the pathophysiology of Na^+ channels, and drug and toxin interactions with these channels.

Na^+ channels play a fundamental role in the initiation and propagation of electrical signals in many excitable cells. Mammalian Na^+ channels are composed of one α-subunit (260 kDa), which forms the core of the channel and is responsible for voltage-dependent gating and ion permeation. The α-subunit is composed of four homologous domains (DI-DIV), each with six α-helical transmembrane-spanning segments (S1–S6). In a state-of-the-art review, Nishino and Okamura present an overview of the molecular and evolutionary aspects of Na^+ channel α-subunits and discuss their contribution to evolutionary changes in animals. The α-subunit forms complexes with auxiliary β-subunits that regulate their trafficking and gating properties. At least four distinct β-subunit subtypes (β1–β4) have been identified. β-subunits are relatively small proteins (33–37 kDa) composed of a single transmembrane α-helix, a short intracellular C-terminus, and a large extracellular N-terminus incorporating an immunoglobulin-like fold similar to that in cell adhesion molecules. The review by Molinarolo et al. summarizes the evolutionary history of β-subunits. Mutations in the genes encoding β-subunits are linked to a variety of diseases, including epilepsy, sudden infant death syndrome, sudden unexpected death in epilepsy, cancer, neuropathic pain, and some major neurodegenerative disorders. Bouza and Isom review the physiopathology of these important regulatory subunits, which also partner with a number of proteins. Balse and Eichel provide an overview of the partners that have been characterized to date, with a focus on the cardiac Na^+ channel. The human genome harbors nine genes encoding voltage-gated Na^+ channels. Mutations in these genes cause dysfunctions of these channels and underlie Na^+ channelopathies. Recent structure–function studies have given us a better understanding of the structural elements of voltage-gated Na^+ channels that are involved in gating and various channel states and drug and toxin binding sites as well as of the topological arrangement of the channel in the membrane.

Na^+ channel α-subunits are also modulated via posttranslational modifications, including phosphorylation, ubiquitination, palmitoylation, nitrosylation, glycosylation, and SUMOylation. Mercier et al. review the biosynthesis and transport of Na^+ channels as well as the mechanisms involved in their anterograde/retrograde trafficking and subcellular targeting.

Aromolaran et al. review the molecular and structural outcomes of posttranslational modulation (expression, gating, trafficking) of $Na_V1.5$, the cardiac Na^+ channel, through the activation of protein kinase A (PKA) and protein kinase C (PKC), while Pei et al. review all aspects of the posttranslational modification of Na^+ channels. They also discuss the underlying mechanisms of posttranslational modifications that may open the way to the development of new drugs.

Na^+ channels are also regulated by the extracellular pH. Peters et al. review the mechanisms of proton block of the Na^+ channel and the impact this mechanism has on disease states.

Gamal El-Din et al. explore the structural and functional aspects of prokaryotic and eukaryotic Na^+ channels. Payandeh and Hackos reviewed several aspects of structure–function relationships of $Na_V1.7$ Na^+ channel, an important peripheral channel implicated in pain and a promising target for subtype-selective $Na_V1.7$ channel modulators. Na^+ channel dysfunctions cause cardiac arrhythmias, skeletal muscle disorders such as paramyotonia congenita, congenital pain, and neurological disorders.

Na^+ channel mutations are associated with many human channelopathies, including inherited syndromes such as skeletal muscle disorders, which are reviewed by Cannon, and arrhythmic cardiac disorders, which are reviewed by Savio-Galimberti. Mutations in several peripheral Na^+ channels have been shown to underlie congenital pain syndrome (gain-of-function) or lack of pain (loss-of-function); Lampert et al. reviewed the role of these channels in pain. The impact of cell damage on Na^+ channels is modeled and reviewed by Joos et al.

Mutations in the voltage sensor domain of Na^+ channels have been implicated in the generation of leak currents known as gating pore currents or omega currents, which in turn have been implicated in several cardiac and neuromuscular disorders. Groome et al. review the gating pore concept as well as the biophysical properties of gating pores and their involvement in a number of human disorders.

Although the concept of state-dependent drug binding is well accepted, the molecular mechanism underlying this phenomenon is not well understood. The prevailing view is that conformational changes in the local anesthetic drug binding site associated with the voltage-dependent activation and inactivation of Na^+ channels enhance drug binding and stabilize channels in non-conducting states. O'Leary and Chahine review the mechanisms of Na^+ channel gating and the models used to describe drug binding and Na^+ channel inhibition. Farinato et al. present a few other pharmacological aspects of Na^+ channels and summarize the concept of the benzothiazolamine scaffold as an interesting tool to build new Na_v channel blockers with promising pharmacological and clinical properties. Zhorov et al. summarize the structural models of ligand binding to Na^+ channels.

Natural toxins have long been used as high-affinity probes to study the molecular structure of Na^+ channels. By binding to a precise site, they modify one or more functional properties of Na^+ channels. They can thus be used to probe various components of the channel and study their functional properties. The three-dimensional structures of some toxins are now known, making them powerful tools for probing different regions of the channel or even providing insights into the three-dimensional structures of individual components. Ji summarizes some of the current views on Na^+ channel–toxin interactions.

This book will be of interest to both fundamental and clinical researchers seeking to understand the molecular basis of voltage-gated Na^+ channels as well as those interested in identifying targets for various pharmacotherapies.

Quebec, Canada Mohamed Chahine

Contents

Part I

Evolution of Voltage-Gated Sodium Channels

Evolutionary History of Voltage-Gated Sodium Channels

Atsuo Nishino and Yasushi Okamura

Contents

Abstract

Every cell within living organisms actively maintains an intracellular Na^+ concentration that is 10–12 times lower than the extracellular concentration. The cells then utilize this transmembrane Na^+ concentration gradient as a driving force to produce electrical signals, sometimes in the form of action potentials. The protein family comprising voltage-gated sodium channels (Na_Vs) is essential for such signaling and enables cells to change their status in a regenerative manner and to rapidly communicate with one another. Na_Vs were first predicted in squid and were later identified through molecular biology in the electric eel. Since then, these proteins have been discovered in organisms ranging from bacteria to humans. Recent research has succeeded in decoding the amino acid sequences of a wide variety of Na_V family members, as well as the three-dimensional structures of some. These studies and others have uncovered several of the major steps in the functional and structural transition of Na_V proteins that has occurred along the course of the evolutionary history of organisms. Here we

A. Nishino (✉)
Department of Biology, Faculty of Agriculture and Life Science, Hirosaki University, Hirosaki, Aomori, Japan
e-mail: anishino@hirosaki-u.ac.jp

Y. Okamura
Integrative Physiology, Graduate School of Medicine, Osaka University, Suita, Osaka, Japan

M. Chahine (ed.), *Voltage-gated Sodium Channels: Structure, Function and Channelopathies*, Handbook of Experimental Pharmacology 246,
https://doi.org/10.1007/164_2017_70

present an overview of the molecular evolutionary innovations that established present-day Na_V α subunits and discuss their contribution to the evolutionary changes in animal bodies.

Keywords
Bilaterian · Ca_V · Chordate · Gene duplication · Myelination · Na_V · Vertebrate

1 Introduction

Hodgkin and Huxley in the 1940s–1950s suggested the presence of pores that enabled selective permeation of Na^+ and K^+ ions that was dependent on membrane depolarization and shaped the action potentials (APs). The Na^+-based APs were first recorded from the giant axons of the squid *Loligo* (e.g., Hodgkin and Huxley 1945, 1952). Since then, we have come to recognize voltage-gated Na^+ channels (Na_Vs) as a core element of the nerve impulse that supports essentially all brain function. William A. Catterall and colleagues succeeded in purifying the biochemical component of Na_Vs from rat brains (e.g., Beneski and Catterall 1980; Hartshorne and Catterall 1981; Hartshorne et al. 1985). Noda et al. (1984) isolated a cDNA encoding the pore-forming α subunit of the Na_V from the electric eel, *Electrophorus*. Thereafter, cDNAs for other Na_V α subunits have been cloned from mammals, other vertebrates, and invertebrates. In the fruit fly, *Drosophila melanogaster*, Na_V mutants called *para* were isolated, and the later studies have clearly linked the genotypes of the Na_V α subunit to cellular/whole body-level phenotypes (e.g., Loughney et al. 1989). Then the gene expression of Na_V in response to neural inductive signaling was shown to underlie development of membrane excitability in neurons in a simple chordate model, the ascidian *Halocynthia roretzi* (Okamura et al. 1994). In recent years, Na^+ channels derived from marine bacteria were utilized to increase our understanding of the structural and biophysical basis of Na^+ ion selectivity and voltage dependence (e.g., Payandeh and Minor 2015; Catterall and Zheng 2015). And very recently, genes from the cockroach *Periplaneta* and from *Electrophorus* were utilized to finally resolve the 3D structures of metazoan Na_Vs (Shen et al. 2017; Yan et al. 2017). This brief history of Na^+ channel research reflects the long, but cooperative, struggle to uncover the essence of nature (membrane excitability here) through the use of appropriate biological materials at appropriate times, seemingly making manifest August Krogh's Principle (Krebs 1975).

The aforementioned analyses were performed using Na_Vs from a variety of animal species and demonstrated that the amino acid sequences of the channel proteins have changed to varying degrees and have incorporated innovations in accordance with the evolution of the animals harboring the channels. Given the well-known statement by T. Dobzhansky that nothing in biology makes sense except in the light of evolution, the diversity of Na_Vs provides us with rich insight. Like other gene families, the Na_V gene family grew through gene duplication, sequence changes, and natural selection. These processes now enable animals to utilize similar but independent Na_Vs at different times during development and/or in specific cell types. In this review, we will outline mammalian, vertebrate, chordate, and metazoan Na_V α subunit diversity and the cell type-specific usage of different isoforms, and we will introduce some of their molecular innovations that correlate with the evolution of animal lifestyles.

2 Structural Outlines of Voltage-Gated Sodium Channels

Metazoan Na_Vs consist of four serially homologous sections, domains I-IV (Fig. 1). Each domain contains six α-helices constituting transmembrane regions S1-S6 linked to each other by extracellular or intracellular loops (Fig. 1a). As a result,

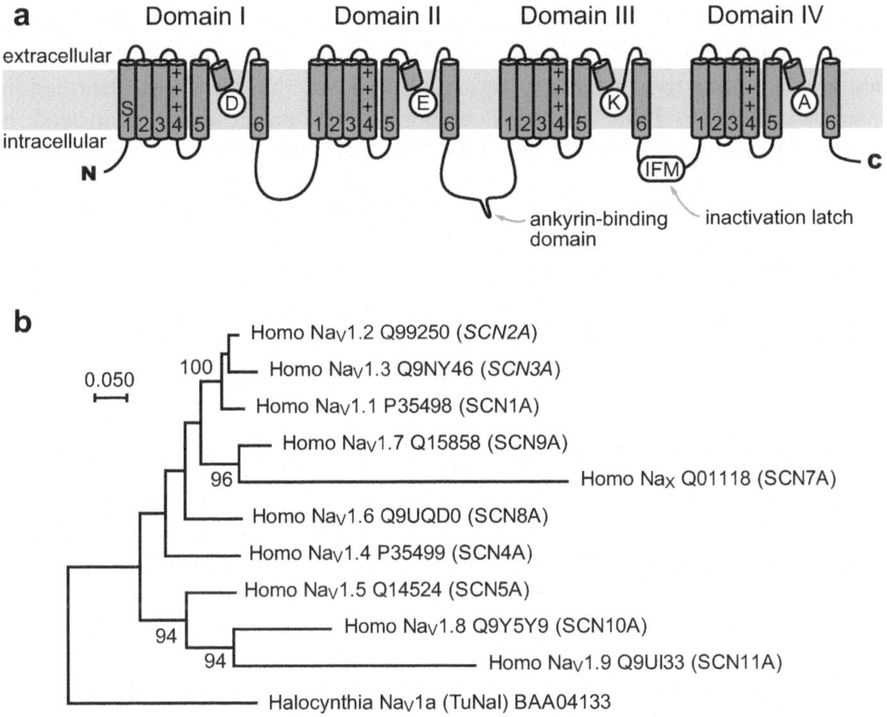

Fig. 1 Mammalian Na_V1 channel α subunits. (**a**) A schematic image of mammalian Na_V1 channels. Diagnostic characteristics of Na_V1s are indicated. Na_V1s have 24 transmembrane segments and are composed of serially homologous domains I–IV, each of which contains segments S1–S6. S4 segments harbor evenly spaced positively charged residues (+). Amino acids at the inner vertices of the loops between S5 and S6 pore-forming segments (P-loops) constitute an ion-selectivity filter. In the case of Na_V1s, the pore signature is Asp/Glu/Lys/Ala (D/E/K/A). The loop between domains II and III contains an ankyrin-binding motif sequence that is reportedly important to localize the Na_V1 channels to the axon initial segment and nodes of Ranvier. The sequence between domains III and IV functions as the inactivation latch (inactivation ball), the core of which is represented by the well-conserved Ile-Phe-Met (I-F-M) triplet. (**b**) A molecular phylogenetic tree of mammalian Na_V1 α subunits constructed by the maximum-likelihood method based on gap-free 1,372 amino acid positions using MEGA7. The sequence from the ascidian *Halocynthia* is used as the out-group. The bootstrap values over 80 are shown. The four groups ($Na_V1.1/1.2/1.3/1.7$, $Na_V1.4$, $Na_V1.6$, $Na_V1.5/1.8/1.9$) are recognized (see text and Table 1)

the full-length metazoan Na_V polypeptide contains a total of 24 transmembrane (TM) regions spanning about 2,000 amino acids. As was proposed for voltage-gated potassium channels (K_Vs), the S1-S4 segments in each domain function as a voltage sensor (e.g., Catterall 2000). In particular, the S4 segment, in which every third amino acid is a positive charged Arg or Lys residue, is considered essential for sensing membrane voltage. In the resting state, S4 is positioned closer to the intracellular side of the membrane. Upon membrane depolarization, S4 moves in an extracellular direction, and this change of conformation stimulates the channel gate to open (Catterall 2000; Shen et al. 2017; Yan et al. 2017). The S5 and S6 segments and the loop between them (P-loop) from each domain occupy a quarter of the central portion of the protein such that the four domains together form the wall, gate, and ion-selectivity filter of the channel (Catterall 2000; Shen et al. 2017).

Mammals, including humans, express nine Na_V isoforms (Na_V1.1 to Na_V1.9). In addition, a protein most similar to Na_V1.7, called Na_X, has also been identified in mammals (Fig. 1b, Table 1). Na_X does not exhibit ion permeability. Instead, it appears to function as a sensor of extracellular Na^+ ion and is now thought to be involved in ionic homeostasis in the body (Hiyama et al. 2002, 2004; Hiyama and Noda 2016). The cDNA sequences, predicted amino acid sequences, locations of the respective encoding genes (*SCN1A-11A*) on the chromosomes, spatiotemporal expression patterns, sensitivity to toxins [e.g., tetrodotoxin (TTX)], single-channel conductances, characteristics of inactivation, and biological functions of these isoforms have all been comprehensively analyzed (some characteristics of mammalian Na_Vs are listed in Table 1) (Goldin 2001; Catterall et al. 2005). For instance, mammalian Na_V1.1, 1.2, and 1.3 are encoded by *SCN1A*, *2A*, and *3A*, respectively, and these genes are tandemly arrayed within the genome (on the "q" arm of chromosome 2 in the case of humans) close to the *HoxD* gene cluster (Table 1). The primary structures of these isoforms are mutually similar, suggesting they emerged through relatively recent tandem gene duplications (Fig. 1b). This tandem cluster also includes the gene for Na_V1.7 (*SCN9A*), which is expressed in the peripheral nervous system (PNS), including the dorsal root ganglia (DRG)

Table 1 Mammalian voltage-gated sodium channel α subunits[a]

Channel name	Gene symbol	Adj. Hox cluster	Approx. TTX IC_{50}	Tissue to function
Na_V1.1	*SCN1A*	*HoxD*	10 nM	CNS
Na_V1.2	*SCN2A*	*HoxD*	10 nM	CNS
Na_V1.3	*SCN3A*	*HoxD*	10 nM	CNS
Na_V1.4	*SCN4A*	*HoxB*	10 nM	Skeletal muscle
Na_V1.5	*SCN5A*	*HoxA*	1–10 μM	Heart
Na_V1.6	*SCN8A*	*HoxC*	<10 nM	CNS
Na_V1.7	*SCN9A*	*HoxD*	10 nM	PNS
Na_V1.8	*SCN10A*	*HoxA*	>10 μM	PNS
Na_V1.9	*SCN11A*	*HoxA*	1 μM	PNS
Na_X	*SCN7A*	*HoxD*	–	CNS and others

[a]This table is modified from Goldin (2001)

(Sangameswaran et al. 1997; Toledo-Aral et al. 1997). The gene encoding Na_X (*SCN7A*) is also found in this cluster. $Na_V1.7$ and Na_X constitute a sister clade in the molecular phylogenetic tree, and the duplication of these two genes occurred in parallel with the splits of $Na_V1.1$, 1.2, and 1.3 (Fig. 1b) (Catterall et al. 2005). They commonly show high sensitivity to TTX, with IC_{50}s around 10 nM (summarized in Goldin 2001). In addition to the channels already mentioned, the central nervous system (CNS) expresses $Na_V1.6$, which exhibits characteristic persistent current and a large resurgent current, while the PNS expresses $Na_V1.8$ and 1.9. The heart expresses $Na_V1.5$, encoded by *SCN5A*, which clusters with *SCN10A* encoding $Na_V1.8$ and *SCN11A* encoding $Na_V1.9$, close to the *HoxA* cluster (Table 1). $Na_V1.8$ and 1.9 are also presumed to have emerged through tandem gene duplication that occurred after the two rounds of whole genome duplication in the ancestor of jawed vertebrates (see below) and share relatively slow activation kinetics (Fig. 1b) (Lai et al. 2004). Along with $Na_V1.5$, $Na_V1.8$ and 1.9 are relatively insensitive to TTX, with IC_{50} in the range of 1–10 μM or more (Gellens et al. 1992; Akopian et al. 1996; Tate et al. 1998). Skeletal muscles use $Na_V1.4$ for APs, and its gene, *SCN4A*, is not clustered with other Na_V genes and neighbors the *HoxB* cluster (Table 1). $Na_V1.4$ in skeletal muscle is as sensitive to TTX as Na_Vs in the CNS (NaV1.1–1.3, and 1.6).

All nine Na_Vs possess an activation gate that opens in response to membrane depolarization and an ion-selectivity filter that enables selective Na^+ permeation. The extracellular fluid around living cells generally contains a high concentration of Na^+, while the intracellular fluid has a lower Na^+ concentration. Na_V gating allows Na^+ to flow into the cell and depolarize the membrane, though the channel soon shuts through the process of inactivation. The inactivation function is one that the Na_V α subunits themselves possess and enables immediate repolarization of the membrane to sharpen the AP (Hodgkin and Huxley 1952). All known Na_Vs show some degree of inactivation, which is known to require the linker sequence between domains III and IV (Stühmer et al. 1989), a region called the "inactivation ball" or "inactivation latch" (Fig. 1a). The latch is modeled to fit into the open pore and block ion permeation (ball-and-chain model). Within the sequence of this linker, three consecutive hydrophobic amino acids, Ile-Phe-Met, are well shared by vertebrate Na_Vs and are totally conserved in all the human Na_V isoforms. This triplet, especially the central Phe residue, is essential for inactivation, and thus hydrophobic interaction between the latch and the open pore would be important (West et al. 1992; see also a recent revision to this model proposed from the structural study by Yan et al. 2017).

The P-loop between S5 and S6 in each of the four domains protrudes into the central canal of Na_Vs, and the inner vertices of the loops are thought to constitute the ion-selectivity filter (Heinemann et al. 1992; Shen et al. 2017). The residues for Na^+ ion selectivity in all mammalian Na_Vs are Asp from domain I, Glu from domain II, Lys from domain III, and Ala from domain IV (asymmetric D/E/K/A signature) (Fig. 1a). By contrast, most Ca_Vs, which have a similar 24-TM conformation, show a symmetric E/E/E/E pore signature (e.g., Hinemann et al. 1992). The 3–4 acidic residues behind the D/E/K/A signature residues (E/E/D/D in all mammalian Na_Vs) form the "outer ring," which is also significant for positively charged ion permeability (Catterall 2000). The region around the signature residue in domain I is a definitive binding target for TTX (Noda et al. 1989). This selectivity filter is located midway through the canal, and the outer ring is a bit extracellular to the signature. On the intracellular side is a central cavity enclosed by charged residues, with the predicted activation gate composed of the S6 segments from domains I-IV aligned along the channel fenestration (Catterall 2000; Shen et al. 2017).

These structural features, including the 24-TM segments in the four serially homologous domains, the voltage-sensor domains, D/E/K/A selectivity filter, central cavity, activation gate, and inactivation latch are shared by all mammalian Na_V α subunits (e.g., Catterall 2000; Catterall et al. 2005).

3 Historical Origin of Voltage-Gated Sodium Channels and Their Related Proteins

The discovery of Na_Vs in bacteria (BacNa_Vs) confirmed that Na^+-permeable channel proteins gated in response to changes in membrane potential had already evolved in prokaryotes (Ren et al. 2001; Payandeh and Minor 2015). However, these BacNa_Vs are composed of homotetrameric 6-TM subunits that contain compact S1-S6 segments (e.g., Payandeh and Minor 2015; Catterall and Zheng 2015). Because a BacNa_V can be changed into Ca^{2+}-selective channel through simple mutation(s), it is thought that prokaryotes can also possess voltage-gated Ca^{2+} channels (Yue et al. 2002). In addition, the K_VAP channel, the first voltage-gated ion channel whose structure was resolved, is a K^+ channel from prokaryotic archaea (Jiang et al. 2003). It is thus evident that prokaryotes likely made use of a wide variety of voltage-gated, ion-specific 6-TM channels long before the emergence of eukaryotes. The 6-TM segments can be functionally divided into a voltage-sensing unit composed of segments S1-S4 and an ion channel pore domain corresponding to segments S5-S6, including the P-loop in between. Members of the prokaryotic KcsA-related channel and eukaryotic Kir channel families are composed of homotetrameric 2-TM helices, which are thought to represent units of the tetraradial pore domains (e.g., Bichet et al. 2003). The K2P channel family members show a tandem repeat of two pore domains, dimerization of which produces a channel pore of pseudotetrameric symmetry (e.g., Honoré 2007). Recent research has revealed that the voltage-sensor domain can also function independently of the pore domain. For instance, the recently identified voltage-sensing phosphatase (VSP) possesses S1-S4-like segments coupled to phosphatase domain that catalyze membrane phospholipid dephosphorylation in response to changes in membrane potential (Murata et al. 2005). It is also known that the primary structure of a voltage-dependent proton channel, H_V1/voltage-sensor-domain-only protein (VSOP), is comparable to segments S1-S4 of 6-TM voltage-gated ion channel subunits (Ramsey et al. 2006; Sasaki et al. 2006).

All of the prokaryotic voltage-gated ion channels identified so far are composed of homotetrameric 6-TM subunits. Eukaryotic voltage-gated K^+ channels and Ca^{2+}-permeable Catsper channels are still tetrameric compositions of 6-TM subunits (Liebeskind et al. 2013). In addition to these, eukaryotes express larger voltage-gated ion channel proteins composed of 12-TM or 24-TM segments. The 12-TM channels are two-pore channels (TPCs), which are cation channels functioning within intracellular organelles such as endosomes and lysosomes (Calcraft et al. 2009). The 24-TM ion channel family includes not only Na_Vs but also L-, T-, and N/P/Q/R-type Ca_Vs (also called Ca_V1, Ca_V2, and Ca_V3, respectively) and cation leak channels (NALCN). It is thought that eukaryotic 24-TM channels harboring four-time serially homologous domains are derived from two rounds of tandem duplication of an original 6-TM factor. In fact, the amino acid sequences of domain I of the Na_V and Ca_V1, 2, and 3 channels are more similar to domain III than to II or IV, while the sequence in domain II is more like that of IV than domain I or III (Strong et al. 1993; Liebeskind et al. 2013). It is also known that the amino acid sequences of these four domains are closer to each other, and to those of the Catsper isoforms, than to the sequences of $BacNa_V$s or K_Vs. These relationships suggest eukaryotic 24-TM channels were not derived from duplication of a $BacNa_V$, but arose instead through two sequential rounds of duplication of a gene encoding a Catsper-like 6-TM protein; one duplication gave rise to a gene encoding 12-TM segments containing domain I and II, and a second tandem duplication established the present conformation composed of domains I-IV (Liebeskind et al. 2013). This process was significant in that the ion channel molecules got to be formed by a single polypeptide stretch, not by four identical subunits, which would facilitate accumulation of "asymmetric" mutations independently within domains I to IV to make the channel "pseudotetrameric." This implies that this process would potentiate future molecular evolutionary fine-tuning for specific functions – e.g., Na^+ selectivity, fast inactivation, anchoring, etc. (see below).

In addition to the channels mentioned to far, another 24-TM subfamily, Na_V2, has been identified in invertebrates, and its members have amino acid sequences similar to Na_V1, but exhibits a D/E/E/A pore signature (Salkoff et al. 1987; Sato and Matsumoto 1992; Nagahora et al. 2000; Zhou et al. 2004; Zakon 2012; Gur Barzilai et al. 2012; Moran et al. 2015). Later analyses proved that the members of this family are permeable to Ca^{2+}, and it has recently been proposed that this family be renamed Ca_V4 (Zhou et al. 2004; Gosselin-Badaroudine et al. 2016). It has also been reported that a mutant $Na_V1.2$ channel of rat giving a D/E/E/A pore signature is permeable not only to Na^+ but also to Ca^{2+} and K^+ and that the presence of a physiological concentration of Ca^{2+} in the extracellular fluid blocks permeation of Na^+ through the mutant channel (Heinemann et al. 1992). This nonselective permeation of cations is also observed in a cnidarian Na_V family channel having the D/E/E/A pore (called $NvNa_V2.1$), although it does not show the blockade by extracellular Ca^{2+} (Gur Barzilai et al. 2012). Given its sequence similarity to Na_V1 and its functional characteristics, we will refer to this subfamily as $Na_V2(Ca_V4)$ here.

The major diversification events of these families of 24-TM channels of animals, namely, evolutionary splits of the Na_V, Ca_V, and NALCN families, predate the origin of metazoan animals (Fig. 2a). The split of the $Na_V1/Na_V2(Ca_V4)$ clade from the Ca_V1–3 molecular clades occurred before the divergence of animals and

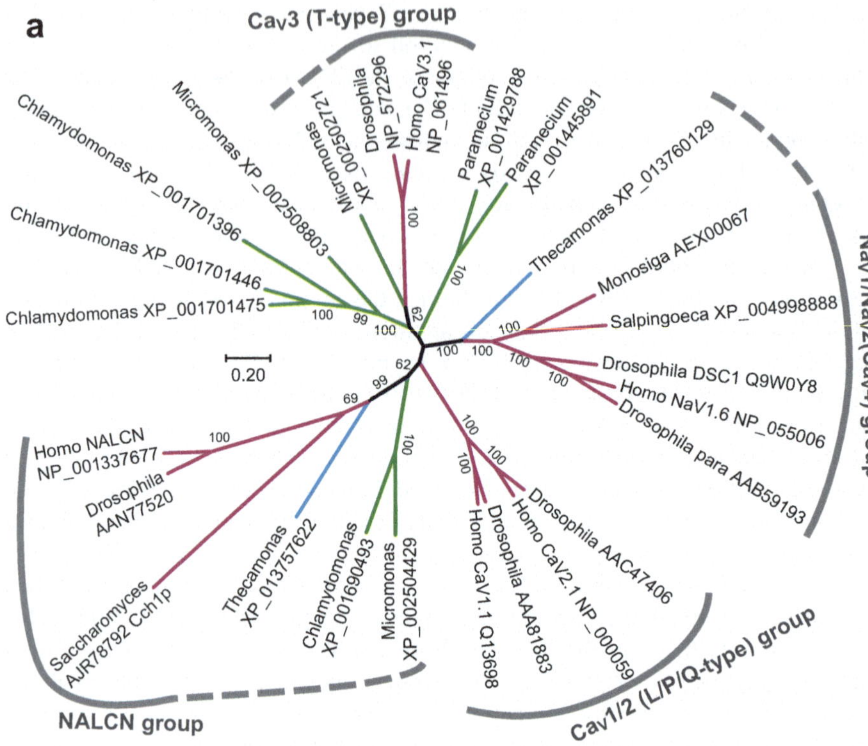

b

				Domain I	Domain II	Domain III	Domain IV
NALCN group	Chlre	XP_001690493		TLTGW	NLENW	TLNGY	TGDNW
	Thetr	XP_013757662		TLEGW	AGDAW	SLEGW	TGEDW
	Sacce	AJR78792		SANTF	STENW	SLEGW	FGEGW
	Drome	AAN77520		SQEGW	TQEAW	SFKGW	TGEDW
	Homsa	NP_001337677	NALCN	SQEGW	TQEGW	SLKGW	TGEDW
Cav1/2 group	Drome	AAC47406		TMEGW	TGEDW	TGEGW	TGEAW
	Homsa	NP_000059	CaV2.1	TMEGW	TGEDW	TGEGW	TGEAW
	Drome	AAA81883		TLEGW	TGEDW	TFEGW	TGEAW
	Homsa	Q13698	CaV1.1	TMEGW	TGEDW	TFEGW	TGEAW
Nav1/2 group	Thetr	XP_013760129		TMDEW	IGE-W	TFEGW	TASGW
	Monbr	AEX00067		TLDFW	CGE-W	TFEGW	TGAGW
	Salro	XP_004998888		LLDFW	CCE-W	TFEGW	TAAGW
	Drome	Q9W0Y8	DSC1	TLDYW	CGE-W	TFEGW	TSAGW
	Homsa	NP_055006	NaV1.6	TQDYW	CGE-W	TFKGW	TSAGW
	Drome	AAB59193	para	TQDFW	CGE-W	TFKGW	TSAGW
Cav3 etc.	Parte	XP_001429788		TTEGW	TNESW	TTEGW	TGEDW
	Micco	XP_002502721		TLEGW	TGEDW	TTEGW	TGEAW
	Drome	NP_572296		SLEGW	TQEDW	SRDGW	TGDNW
	Homsa	NP_061496	CaV3.1	TLEGW	TQEDW	SKDGW	TGDNW
	Micco	XP_002508803		SMEGW	TGENW	TLEGW	TGEAW
	Chlre	XP_001701475		TQEGW	TMENW	TLELW	TGESW
	Chlre	XP_001701446		TLSGW	TGENW	TMSLW	TGESW

Fig. 2 Molecular phylogeny of eukaryotic 24-TM channels. (**a**) An unrooted maximum-likelihood tree of 24-TM channels found in green algae (*Chlamydomonas reinhardtii*, *Micromonas commoda*), the ciliate (*Paramecium tetraurelia*), the fungi (*Saccharomyces cerevisiae*), the apusozoan flagellate (*Thecamonas trahens*), choanoflagellates (*Monosiga brevicollis*, *Salpingoeca rosetta*), and metazoans (*Drosophila melanogaster*, *Homo sapiens*). Accession numbers of the

choanoflagellates, the protozoan group most closely related to the metazoan clade (Zakon 2012; Liebeskind et al. 2012; Moran et al. 2015). NALCN-like sequences are also found in fungi, and the history of this family can be traced back to the origin of opisthokonta, the monophyletic group containing animals, choanoflagellates, and fungi (Torruella et al. 2011; Liebeskind et al. 2012). Figure 2a shows here that in fact the representative unicellular eukaryotes distantly related with each other, such as the yeast *Saccharomyces*, the ciliate *Paramecium*, and green algae *Chlamydomonas* and *Micromonas*, possess different types of the 24-TM channel proteins. For instance, *Saccharomyces*, *Chlamydomonas*, and *Micromonas* express a 24-TM channel related to NALCN of animals, while *Paramecium*, *Chlamydomonas*, and *Micromonas* use other types of 24-TM channels closer to Na_Vs or Ca_Vs of animals (Fig. 2a). This suggests that the gene duplication of NALCN, Na_V, and Ca_V (even between $Ca_V1/2$ and Ca_V3) families preceded the split of Unikonta (including animals, fungi, and amoebozoans) and Bikonta (plants, algae, etc.), namely, had occurred close to or before the origin of eukaryotes (Fig. 2a) (see also the phylogenetic views of eukaryotic lineages in Roger and Simpson 2008; Rogozin et al. 2009; Cavalier-Smith 2010). Despite the deep origin of the voltage-gated 24-TM Ca^{2+}/Na^{+} channels, it is also known that many eukaryotic groups lack the genes of them, probably because of secondary loss. Brunet and Arendt (2015) argued that the losses of 24-TM Na_V/Ca_V channels that had happened in varied eukaryote lineages are tightly correlated with the absence of flagella in those organisms. The Ca_V channels of *Paramecium* and *Chlamydomonas* are localized in the membrane of their cilia/flagella, in which the channels function to generate APs to change flagellar/ciliary beating waveforms (Machemer and Ogura 1979; Fujiu et al. 2009). The flagellar localization of the 24-TM Ca_V/Na_V channels may be a preadaptation for the later emergence of neurons (Brunet and Arendt 2015).

While the phylogenetic relationships among the protein sequences from various eukaryotes are based on overall sequence similarities, the ion-selectivity filter signatures (D/E/K/A, E/E/E/E, E/E/D/D, and E/E/K/E in mammalian Na_V1, $Ca_V1/2$, Ca_V3, and NALCN, respectively) were established relatively recently (Fig. 2b) (Liebeskind et al. 2011, 2012). This means that the specific ion selectivity of each family may have changed over the course of evolution. For instance, the D/E/K/A pore signature of Na_V1 is thought to have emerged at the origin of

Fig. 2 (continued) sequences are indicated in the tree. The groups of $Na_V1/Na_V2(Ca_V4)$, $Ca_V1/2$, Ca_V3, and NALCN and their extended clades are indicated with *gray* solid and *dashed lines*. The bootstrap values over 60 are shown. The tree was constructed using the WAG model on MEGA7 from gap-free 847 amino acid positions aligned by MUSCLE program. (**b**) The pore signatures (at the putative ion-selectivity filter in the P-loops) of the eukaryotic 24-TM channels. The signature sequences were obtained from the alignment used for constructing the tree shown in (**a**). The species codes (first column) made of the first three and two letters from the genus and species name, respectively, and accession numbers (second column) are shown. The amino acids at the pore signatures are highlighted with colors [Asp (D), *magenta*; Glu (E), *red*; Lys (K), *blue*; Ala (A), *yellow*; other polar amino acids, *green*]

Bilateria (the clade of animals with bilateral body plans). The evidence is that all Na_V-type 24-TM genes found so far in choanoflagellates (the closest relatives to metazoans) (*Monosiga* and *Salpingoeca* in Fig. 2) and in ctenophores and placozoans (non-bilaterian metazoans), encode the D/E/E/A filter motif, not D/E/K/A, which implies the $Na_V2(Ca_V4)$-type pore signature is original (Liebeskind et al. 2011; Zakon 2012; Gur Barzilai et al. 2012; Moran et al. 2015). Cnidarians, another major non-bilaterian metazoan group, express several isoforms of Na_V-like 24-TM polypeptides with D/E/E/A (called $Na_V2.1$, 2.3, and 2.4), D/E/E/T ($Na_V2.2$), and D/K/E/A ($Na_V2.5$) pore signatures, and a phylogenetic analysis suggested that the latter two of the isoforms were derived from the first (Fig. 3) (Gur Barzilai et al. 2012; Moran et al. 2015). The cnidarian D/K/E/A channels ($Na_V2.5$) are presumably Na^+-selective, representing convergent evolution of Na^+-selective channels that was independent from the origin of the D/E/K/A Na_V1 subfamily in bilaterians (see below) (Anderson et al. 1993; Gur Barzilai et al. 2012). This molecular parallel evolution may be related to independent acquisitions of large bodies moving fast in cnidarians (e.g., jellyfish) and in bilaterians. On the other hand, the pore signature of cnidarian NALCN-like genes is commonly E/E/E/E, which likely ensures Ca^{2+} selectivity, while those in all known bilaterians have E/E/K/E at the selectivity filter, making the channels to be cation nonselective (Lu et al. 2007; Liebeskind et al. 2012).

From the evidence summarized above, it can be predicted that at their origin animals possessed two or three Ca_V subfamily genes, one $Na_V2(Ca_V4)/Na_V1$ subfamily gene and one NALCN-like gene (Moran and Zakon 2014; Moran et al. 2015). Interestingly, none of the channels are Na^+-selective, but are Ca^{2+}-preferential. Animals may have evolved as organisms lacking voltage-gated Na^+-selective channels, and the true metazoan Na^+ channels, Na_V1s with the D/E/K/A signature, emerged in a bilaterian ancestor via gene duplication and diversification that at the same time gave rise to the $Na_V2(Ca_V4)$ subfamily.

4 Evolution of Bilaterians and Voltage-Gated Sodium Channel Proteins

Bilateria is composed of animals having bilateral body plans, which are classified into three superphyla, Deuterostomia, Lophotrochozoa, and Ecdysozoa. These animal groups share a basic repertoire of 24-TM ion channel paralogues of three subtypes of Ca_V (Ca_V1–3), two of $Na_V1/Na_V2(Ca_V4)$, and one of NALCN leak channel. The molecular phylogenetic analysis suggests that the Na_V1 subfamily was diverged from the $Na_V2(Ca_V4)$ subfamily and originated in the last common ancestor of bilaterians (Fig. 3) (Liebeskind et al. 2011; Moran et al. 2015). The origin of Na_V1 in fact correlated with the development of the D/E/K/A pore signature and therefore represents the occurrence of "true" Na^+ selectivity in 24-TM channels (Fig. 3) (Liebeskind et al. 2011). This channel enabled bilateral animals to specifically utilize Na^+, the most abundant cation within the environment in which they had adapted. Combined with the function of the Na^+ pump (Na,K-ATPase), Na_V1 would have conferred a fast electric responsiveness to excitable cells by means of the large

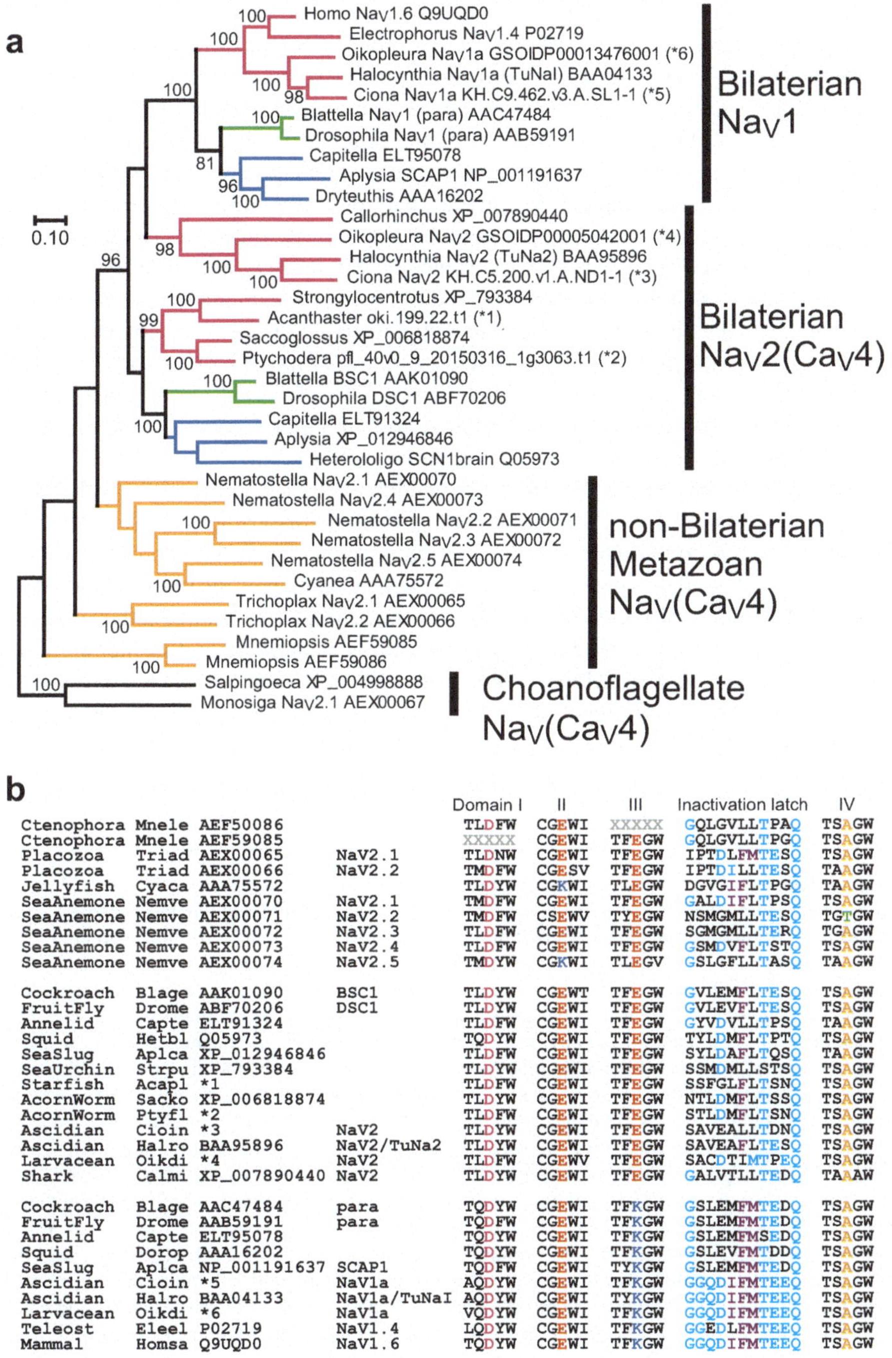

Fig. 3 Molecular phylogeny of metazoan $Na_V1/Na_V2(Ca_V4)$ channels. (**a**) A maximum-likelihood tree of Na_V-related channels from the ctenophore (*Mnemiopsis leidyi*), the placozoan (*Trichoplax adhaerens*), the jellyfish (*Cyanea capillata*), the sea anemone (*Nematostella vectensis*), squids

driving force across the membrane. Na_V1 would also be important in that it could mediate steep depolarization without direct stimulation of the intracellular processes triggered by Ca^{2+}, a major second messenger in the wide variety of unicellular and multicellular organisms. It is also possible the inactivation latch in the loop between domains III and IV developed at the same time (Fig. 3b), which would have facilitated fast recovery to the resting state of membrane potential and thus enabled cells to minimize the changes in intracellular ionic conditions. Importantly, the molecular development of Na_V1 would be related not only to fast generation/propagation of APs but also to the occurrence of refractory periods after depolarization. Repetitive excitation of neurons is facilitated by these characteristics of Na_V1, which enables the nerves to encode neural signals based on the AP frequencies. Fast neural transmission would have supported evolution of larger bodies, and the ability to inactivate itself can ensure unidirectional flow of signals within the neural network. This would have offered segregation of input (sensory) systems from output (motor) systems, permitting development of the CNS.

The evolutionary emergence and divergence of bilaterians represents the geographical event called the "Cambrian explosion." It is inspiring to consider that the origin of Na_V1 is concordant with the evolution of bilaterians. Paleontologists have suggested that predator-prey relationships were established during the Cambrian era, which stimulated increases in body size and complexity and also the sophistication of sensory organs, leading to acceleration of movement and further elaboration of the CNS (Conway-Morris 1986; Gould 1990; Parker 2003). One hypothesis has proposed that development of eyes was a key step in this so-called evolutionary big bang (Parker 2003). Given this context, it would be reasonable to predict that

---

Fig. 3 (continued) (*Doryteuthis opalescens*, *Heterololigo bleekeri*), the sea slug (*Aplysia californica*), the annelid worm (*Capitella teleta*), the fruit fly (*Drosophila melanogaster*), the cockroach (*Blattella germanica*), acorn worms (*Ptychodera flava*, *Saccoglossus kowalevskii*), the starfish (*Acanthaster planci*), the sea urchin (*Strongylocentrotus purpuratus*), ascidians (*Ciona intestinalis*, *Halocynthia roretzi*), the larvacean (*Oikopleura dioica*), the elephant fish chimaera (*Callorhinchus milii*), the electric eel (*Electrophorus electricus*), and humans (*Homo sapiens*). Monophyly of bilaterian Na_V1s is evident, while that of $Na_V2(Ca_V4)$ channels is not clear here. The sequences from non-bilaterian metazoans, ecdysozoans, lophotrochozoans, and deuterostomes are labeled with *yellow*, *blue*, *green*, and *magenta* branches, respectively. NCBI accession numbers and other ID codes are indicated in the tree. The sequences of *O. dioica* are obtained from OikoBase (http://oikoarrays.biology.uiowa.edu/Oiko/); those of *C. intestinalis* are from Ghost database (http://ghost.zool.kyoto-u.ac.jp/cgi-bin/gb2/gbrowse/kh/); those of *A. planci* and *P. flava* are from the OIST genome browsers (http://marinegenomics.oist.jp/gallery/gallery/index). The bootstrap values over 80 are shown. The tree was constructed using the WAG model on MEGA7 from gap-free 812 amino acid positions aligned by MUSCLE program. (**b**) The pore signatures and the region corresponding to the inactivation latch in metazoan Na_V-related channels. The sequences shown are mostly identical to those in (**a**). The species codes and color codes are same as in Fig. 2. Sequences corresponding to core triplet of the inactivation latch (I-F-M) are also marked by *purple*. *Cyan* residues indicate the amino acids identical to those around the inactivation latch of the mammalian $Na_V1.6$. *1-*6: the genemodel IDs are shown in (**a**) to identify the sequences in the genome browser of each organism the molecular phylogenetic origin of Na_V1 provided a physiological basis for the bilaterian ancestor that potentiated the explosive evolution.

Na_V1 originated through gene duplication and diversification that gave rise to the $Na_V2(Ca_V4)$ as well (Fig. 3a). Consequently, most bilaterians express two subtypes of Na_V family components, Na_V1 and $Na_V2(Ca_V4)$. For example, the genome of *Drosophila melanogaster* harbors *para*, which encodes a Na_V1-type channel, and *DSC1*, which encodes a $Na_V2(Ca_V4)$ channel (Salkoff et al. 1987; Ramaswami and Tanouye 1989; Loughney et al. 1989; Hong and Ganetzky 1994; Kulkarni et al. 2002; Zhang et al. 2013). Major invertebrate lineages, such as mollusks, annelids, arthropods, and chordates all express $Na_V2(Ca_V4)$ proteins in addition to Na_V1. $Na_V2(Ca_V4)$ family proteins contain voltage sensors – i.e., S4 segments containing evenly spaced, positively charged residues, and their pore signature is generally D/E/E/A. When exogenously expressed in *Xenopus* oocytes, $Na_V2(Ca_V4)$ family proteins from the cockroach and honeybee (called BSC1 and $AmCa_V4$, respectively) are more permeable to the divalent cations Ca^{2+} and Ba^{2+} than to Na^+ (Zhou et al. 2004; Gosselin-Badaroudine et al. 2016). These channels are reportedly insensitive to TTX, exhibit relatively slow activation and inactivation, and can be blocked by Cd^{2+} or Zn^{2+}. Phenotypic analyses of *DSC1* gene mutants in *Drosophila* have suggested its functions in olfaction or odor-responsive behavior and also in stabilizing the performance of neural circuits under stresses (Kulkarni et al. 2002; Zhang et al. 2013).

While bilateral animals share orthologues of Na_V1 and $Na_V2(Ca_V4)$, it has also been revealed that some animal lineages have lost either the Na_V1 or $Na_V2(Ca_V4)$ subtype, or both. It is well known, for example, that the genome of *Caenorhabditis elegans* contains neither Na_V1 nor $Na_V2(Ca_V4)$ (e.g., Okamura et al. 2005). Whether these paralogues are present or absent is thought to reflect the physical characteristics and lifestyle of each animal group, including body size, locomotion speed, and/or complexity of neural processing. Although the nematode lacks any Na_V-class channels, it is not true that this species is "primitive." The nematode lost this protein because it was not essential for its interstitial life. In fact, nematodes have flourished around the globe without it.

The Na_V1 family channels are absent from echinoderms and hemichordates, the group collectively called Ambulacraria, which constitutes the sister clade of the phylum Chordata (Fig. 3) (see also Widmark et al. 2011; Gur Barzilai et al. 2012). This indicates that echinoderms and hemichordates (ambulacrarians) secondarily lost the fast Na_V1, while the $Na_V2(Ca_V4)$ with the D/E/E/A pore signature typical for that subfamily remains. This suggests these animal groups are incapable of fast sodium spikes. These animals are small during the larval stage (about 0.1–5.0 mm), and the adults (about 1–10 cm or more sometimes) are generally slow moving. They are not fast predators but protective; echinoderm adults are covered with calcite skeletons, and sometimes also with spines, while hemichordates are generally buried in the seafloor. They develop an ectodermal nerve net over the entire body, and their CNS is relatively rudimentary (Hyman 1955; Holland 2003, 2016; Nomaksteinsky et al. 2009). Their evolutionary status may be regarded like an atavism – i.e., reminiscent of the status of animals before the origin of bilaterians. In other words, they live a "slow life" that is a consequence of the loss of Na_V1. On the contrary, the last common ancestor of bilaterians had a Na_V1-type channel and lived a "quick life." Therefore, the long-standing controversy around how the less centralized nervous system seen in echinoderms or hemichordates was integrated into the well-centralized nervous system of vertebrates may not be valid (for reviews, e.g., Holland 2003, 2016). It is noteworthy that recent analyses of the CNS of a polychaete annelid (ragworm) provided surprising evidence of its deep anatomical similarity to the vertebrate CNS (Tessmar-Raible et al. 2007; Tomer et al. 2010; Vergara et al. 2017). A well-centralized nervous system in the last common ancestor of bilaterians would be consistent with the evolution of Na_V1.

5 Voltage-Gated Sodium Channels in Chordates

As the bilaterians diversified, the chordate lineage led to vertebrates and two other animal groups, the amphioxi (cephalochordates, lancelets) and tunicates (urochordates). Recent comparative genomic analyses showed that the amphioxi diverged first within these three groups, and tunicates and vertebrates form a sister group (Delsuc et al. 2006; Putnam et al. 2008). Amphioxi and tunicates provided "observation windows" for researchers to investigate past situations before establishment of vertebrate bodies.

Amphioxi inhabit sandy shores and live as filter feeders. They are able to swim out of the sand and dive back into it very quickly. While they develop segmented somites, they lack developed eyes or an expanded brain (Willey 1894). Their repertoire of Na_V-related α subunits also appears primitive. An amphioxus likely possesses five genes encoding $Na_V1/Na_V2(Ca_V4)$ channel α subunits. At least three of the genes are classified as Na_V1 family, and the other two are in the $Na_V2(Ca_V4)$ clade (Fig. 4a). The three Na_V1 channels of amphioxi are paralogous with each other and share the D/E/K/A pore signature. On the other hand, the differences in their amino acid sequences are considerable, implying differing functions of the isoforms and situation-dependent differential utilizations (tissue- or life stage-specific expression, etc.). A clearer sign of functional specificity is seen in the $Na_V2(Ca_V4)$ proteins; the one (depicted as Na_V2a in Fig. 4) contains a D/D/Q/A, not the D/E/E/A, pore signature at the ion-selectivity filter. Although we do not know the ion permeability of amphioxus Na_V2a, it is conceivable the D/D/Q/A signature invests the excitable membranes of this animal with a regulatory option. Examination of these organisms enables us to monitor an "evolutionary experiment" carried out through gene duplication before the emergence of vertebrates (Ohno 1970).

Similar traces are also found in the tunicate lineage, which possess four types of $Na_V1/Na_V2(Ca_V4)$ proteins. Ascidians, constituting a representative tunicate class, abundantly distribute along the shores around the world. Their adult form is sessile, while the larvae are in the form of tiny tadpoles that swim in the sea. Ascidians have been utilized as a research model, because of their kinship to vertebrates, but also because of their abundance and the availability of mature gametes among other

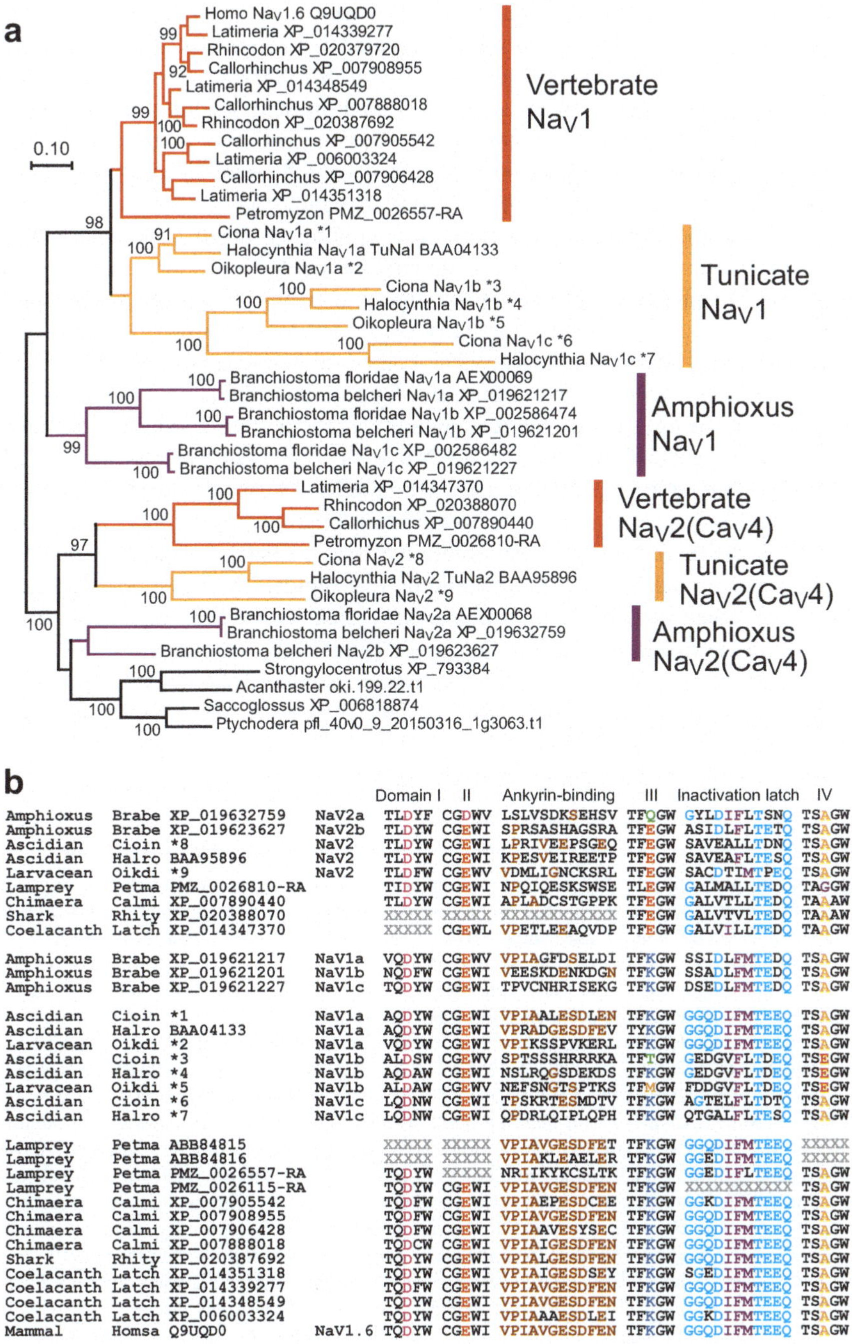

Fig. 4 Molecular phylogeny of chordate Na_V1 and $Na_V2(Ca_V4)$ channels. (**a**) A maximum-likelihood tree of Na_V-related channels from acorn worms (*Ptychodera flava*, *Saccoglossus kowalevskii*), the starfish (*Acanthaster planci*), the sea urchin (*Strongylocentrotus purpuratus*), amphioxi (*Branchiostoma belcheri* and *B. floridae*), ascidians (*Ciona intestinalis*, *Halocynthia*

features. Ascidians have contributed to ion channel studies through the "mosaicism" of their embryogenesis. Historical studies revealed that neuron-like Na^+ spikes could be evoked in the neural cell-lineage blastomere of embryos whose cleavage was arrested using an inhibitor of cytokinesis (Takahashi and Yoshii 1981; Takahashi and Okamura 1998). Developmental expression of this Na^+ current in the neural cell-lineage blastomere is dependent on a fibroblast growth factor-like inductive signal from a neighboring endomesodermal blastomere, which represents neural induction. This process of differentiation in membrane excitability is firmly correlated with the gene expression of a Na_V1 channel, originally called TuNaI (referred to as Na_V1a here) (Okado and Takahashi 1988; Okamura et al. 1994; Takahashi and Okamura 1998). This Na_V1a α subunit is actually expressed in all known neuronal types (Okamura et al. 1994; Okada et al. 1997). Later studies carried out before and after the genomic sequencing of several species of ascidians revealed that ascidians have four genes encoding Na^+ channel α subunits (Nagahora et al. 2000; Okamura et al. 2005; Brozovic et al. 2016). One encodes Na_V1a (TuNaI) containing the typical D/E/K/A pore signature, an inactivation latch with the I-F-M triplet between domains III and IV, and a sequence similar to the ankyrin-binding motif found in the loop between domains II and III of vertebrate Na_V1s (see below). Another encodes a $Na_V2(Ca_V4)$ subfamily protein containing the typical D/E/E/A pore signature (Na_V2, previously called TuNa2), but lacking clear consensus sequences for the inactivation latch and ankyrin-binding motif (Fig. 4) (Nagahora et al. 2000). This gene encoding $Na_V2(Ca_V4)$ is also expressed in some, but not all, neurons in ascidians (Nagahora et al. 2000). The nested patterns of Na_V1 and $Na_V2(Ca_V4)$ gene expression are reminiscent of the patterns of *para* and *DSC1* expression in *Drosophila* embryos (Hong and Ganetzky 1994).

Fig. 4 (continued) roretzi), the larvacean (*Oikopleura dioica*), the lamprey (*Petromyzon marinus*), the elephant fish chimaera (*Callorhinchus milii*), the whale shark (*Rhincodon typus*), the coelacanth (*Latimeria chalumnae*), and humans (*Homo sapiens*). The clades of Na_V1 and $Na_V2(Ca_V4)$ are clearly divided. Vertebrates possess $Na_V2(Ca_V4)$. The sequences from amphioxi, tunicates (ascidians and larvaceans), and vertebrates are labeled with *purple*, *yellow*, and *red* branches, respectively. NCBI accession numbers and other ID codes are mostly indicated in the tree. *1-*9 indicate the genemodel IDs in the genome browser of each organism: *1, KH.C9.462.v3.A.SL1–1; *2, GSOIDP00013476001; *3, KH.C1.1161.v1.A.ND1–1; *4, Harore.CG.MTP2014.S1.g14830; *5, GSOIDP00011229001; *6, KH.C10.502.v2.A.SL1–1; *7, Harore.CG.MTP2014.S25.g02359; *8, KH.C5.200.v1.A.ND1–1; *9, GSOIDP00005042001. The *O. dioica* sequences are obtained from OikoBase (http://oikoarrays.biology.uiowa.edu/Oiko/); the *C. intestinalis* sequences are from Ghost database (http://ghost.zool.kyoto-u.ac.jp/cgi-bin/gb2/gbrowse/kh/); the *H. roretzi* sequences are from Aniseed database (https://www.aniseed.cnrs.fr/); the *P. marinus* sequences are from the UCSC genome browser gateway (http://genome-asia.ucsc.edu/cgi-bin/hgGateway); and the *A. planci* and *P. flava* sequences are from the OIST genome browsers (http://marinegenomics.oist.jp/gallery/gallery/index). The bootstrap values over 90 are shown. The tree was constructed using the WAG model on MEGA7 from gap-free 526 amino acid positions aligned by MUSCLE program. (**b**) The pore signatures and the regions corresponding to ankyrin-binding motif and the inactivation latch in chordate Na_V-related channels. The species codes and color codes are as used in Figs. 2 and 3. Amino acids identical to the mammalian $Na_V1.6$ ankyrin-binding motif are indicated by *brown*. *1-*9: the genemodel IDs identical to those in (**a**)

The other two channels in ascidians, called here Na_V1b and Na_V1c (previously named Na_V3 and 4, respectively, in Okamura et al. 2005), are categorized in the Na_V1 family, and the tunicate Na_V1a, b, and c constitute a clade different from that including the vertebrate Na_V1s (Fig. 4a). While the ion-selectivity filter signature of the tunicate Na_V1c is the same as that in typical Na_V1-type channels (D/E/K/A), Na_V1b exhibits a D/E/(K or T or M)/E pore signature (Fig. 4b) (see also Widmark et al. 2011). We do not know the ion permeability of either channel. Temporal expression patterns estimated from the counts of expressed sequenced tags (ESTs) in the ascidian *Ciona intestinalis* (Satou et al. 2003) suggests the tunicate Na_V1a (TuNaI) is expressed in the larval and adult nervous systems which is consistent with in situ analyses of this gene expression pattern (Okamura et al. 1994, 2005; Okada et al. 1997). Na_V2 is estimated to be expressed in the nervous systems of larvae and juveniles, and possibly on the juvenile endostyle, which is an organ putatively homologous to the vertebrates' thyroid gland. The EST counts suggest expression of Na_V1b occurs during the larval stage. Our preliminary examination of the spatial expression pattern in *C. intestinalis* showed that the Na_V1b gene is expressed in neurons in the CNS and PNS and in some of the muscle cells in the larva. The EST counts predict a small amount of Na_V1c is expressed during the larval stage, though our preliminary in situ hybridization analysis did not detect any clear signal during the larval stage. Also, intriguing is the detection of EST counts for Na_V1b and Na_V1c in mature hermaphroditic adults producing gametes, not in young immature adults. In the eggs of ascidians, steep membrane depolarization is evoked in response to fertilization, which is known to be mediated by an unknown voltage-gated Na^+ channel that is somewhat permeable to Ca^{2+} along with Na^+ (Okamoto et al. 1977; Fukushima 1981; Okamura and Shidara 1987). Thus Na_V1b and/or Na_V1c may be involved in this process.

Single-channel recordings from cleavage-arrested neuronal blastomeres of the ascidian *Halocynthia roretzi* support this view. The electrophysiology has uncovered three types of voltage-gated sodium currents that turn over with time after fertilization to matured stage (Okamura and Shidara 1990a, b). The "Type A" Na^+ current shows only one decay phase during voltage-dependent inactivation, suggesting that a single type of Na_V is responsible for this current. Type A currents are seen in every blastomere within early embryos, and its expression level appears highest at the gastrula stage. This type of Na^+ current is identical to that in the fertilization potential of *Halocynthia* eggs represented by a Na^+-dependent AP (Fukushima 1981). The Na_V current in *Halocynthia* eggs is insensitive to TTX but is highly sensitive to scorpion toxin and local anesthetics (Okamoto et al. 1977). These data suggest that the tunicate Na_V1b, which has an atypical pore signature and is thus resistant to TTX, is involved in the Type A current. On the other hand, voltage-dependent inactivation of the "Type C" current shows two different, fast and slow, phases of decay (Okamura and Shidara 1987), which is reminiscent of the Nav1.6 channel in mammalian neurons. The Type C is the most predominant in differentiated neuronal blastomeres and is suppressed by microinjection of antisense DNA targeting the *Halocynthia* Na_V1a gene *TuNaI* (Okamura et al. 1994). During the short period between the disappearance of Type A and appearance of Type C currents during the developmental course of neural-type membrane excitability, an unusual voltage-gated Na^+ current, "Type B," is transiently expressed (Okamura and Shidara 1990a). This current shows persistent gating behavior with multiple short openings (burst activity). At present, the relationship between the classically characterized diversity of voltage-gated Na^+ currents and the ascidian Na_V isoforms remains unclear, though it appears that Na_V1a (TuNa1) carries Type C current. It would be interesting to know whether the tunicate Na_V1b or Na_V1c carries the Type A current and what underlies the Type B current.

The tree topology shown in Fig. 4 suggests that the ancestor of tunicate Na_V1 paralogues became the seed from which there was further molecular evolution of the Na_V1 channels in modern vertebrates. The Na_V1a of tunicates shares an ankyrin-binding motif sequence with the vertebrate Na_V1s (Fig. 1) (Hill et al. 2008), which consists of a dozen amino acids residing in the loop between domains II and III (Fig. 4b), while the Na_V1b and Na_V1c proteins almost lost it. Similarly, the I-F-M inactivation latch that is conserved in the tunicate Na_V1a has been lost in the paralogues, Na_V1b and Na_V1c (Fig. 4b). Another ankyrin-binding motif similar to that in vertebrate Na_V1s is also found in vertebrate KCNQ2/3 ($K_V7.2/7.3$) K^+ channels, and these motifs are crucial for ankyrin-G binding and for anchoring of Na_V1s and KCNQ2/3s at the axon initial segment (AIS) and nodes of Ranvier in myelinated neurons (Garrido et al. 2003; Lamaillet et al. 2003; Pan et al. 2006; Hill et al. 2008). The sequences of the motif in ascidians' Na_V1a varies somewhat, but ~70% of the amino acids are conserved (Fig. 4b). Even the Na_V1 channel in amphioxi, depicted as *Branchiostoma* Na_V1a in Fig. 4, has ~40% identity, though we find no other traces in invertebrate Na_V1s (Fig. 4b) (Hill et al. 2008). Myelination and resultant saltatory conduction are regarded as a feature of jawed vertebrates (gnathostomes) (Zalc et al. 2008; Zalc 2016). Despite the absence of nodes of Ranvier in amphioxi and tunicates, the ankyrin-binding motif emerged in these animals and may have initiated interaction with ankyrins within neurons. This may be a key property that the ancestral gene of the tunicate Na_V1 paralogues retained, and the reason it was selected as the seed for further evolution in the vertebrate lineage.

The situations seen in amphioxi and tunicates inform us that the isoforms occurring in these so-called "protochordate" organisms through gene duplication differentially evolved, leading to changes, even into the pore signature (as seen in Na_V1b of tunicates and Na_V2b of amphioxi). The gene duplications can confer specific regulatory options to each of duplicated isoforms as indicated so far (Ohno 1970). What occurred in these organisms is a prelude to what has occurred in the vertebrate lineage: another story of gene duplication and functional differentiation.

6 Evolution of Na_V1 Channels in Vertebrates

The monophyletic vertebrate lineage has given rise to agnathans (hagfish and lampreys), cartilaginous fish (sharks, skates, and rays), ray-finned fish (bichirs, sturgeons, gars, bowfin, and teleosts), lobe-finned fish (paraphyletic group of coelacanths and lungfish), and tetrapods (amphibians, reptiles, birds, and mammals) (e.g., Amemiya et al. 2013). During this process of radiation, these organisms inhabited and adapted to various environments in salt and freshwater and in wet and dry terrestrial areas. An ability for predation has been especially well developed in this animal lineage, and several ideas have been proposed in that regard (e.g., Gans and Northcutt 1983). Myelination and saltatory conduction along neuronal axons, as well as the developmental capacities derived from neural crest and placode cells that especially enhanced sensory systems, referred to as "new head," have enabled vertebrates to become larger and predatory (e.g., Gans and Northcutt 1983; Zalc 2016). The presence of active predators in turn stimulated greater ability to efficiently recognize the predators so as to escape (Parker 2003). Increased complexity of sensory inputs, higher ordered neural processing, and high-speed regulation of locomotion would have strongly supported the radiation of vertebrates under water and on land. Na_V1 function was definitely essential to those evolutionary steps.

The lamprey *Petromyzon marinus* shows multiple types of predicted transcripts encoding Na_V1 channel α subunits. Two Na_V1 isoforms have previously been identified (Hill et al. 2008; Zakon 2012), but four types, at least, may be there (Fig. 4b). Cartilaginous fish, including the elephant fish (chimaera) *Callorhinchus milii*, and the whale shark *Rhincodon typus* also appear to harbor four to five (or more) predicted transcripts for Na_V1 isoforms (Fig. 4; some sequences were too short and thus omitted here). Our molecular phylogenetic analysis suggests the Na_V1 isoforms in cartilaginous fish well reflect an original state (one gene in each of NaV1.4, NaV1.5/1.8/1.9, NaV1.6, and NaV1.1/1.2/1.3/1.7 groups), before the extensive duplication that occurred in amniotes and in teleosts (Figs. 4 and 5) (see below) (Widmark et al. 2011). It remains difficult, however, to know precise full-length sequence data for these transcripts, and there is not yet sufficient data available to draw firm conclusions.

Three lamprey Na_V1 α subunits for which longer sequence information was found in the database have the D/E/K/A pore signature. On the other hand, at least two of them contain the I-F-M inactivation ball in the loop between domains III and IV, while the third has an I-F-L triplet. The ankyrin-binding motif is conserved to varying degrees (50–100%) in their domain II-III loops (Fig. 4b). The variation in sequence motifs may represent differences of the molecular functions among them, which implies efficient "evolution by gene duplication" working in this gene family from the beginning of the vertebrate lineage (Ohno 1970).

The ankyrin-binding motif may reportedly work to locate and accumulate Na_V1s at the AIS in the neurons of lampreys (Hill et al. 2008). True myelin sheaths have not been found along the axons of lampreys, though molecular traces of myelination have been detected (Smith et al. 2013). On the other hand, the proximal portion of neuronal axons in lampreys is thinner than elsewhere along the axon. This narrow initial segment decreases local capacitance and conductance and supports the occurrence of steep APs. Localization of Na_V1 at high density in the AIS makes sense for efficient induction of APs (Hill et al. 2008; Kole and Stuart 2012), and full establishment of Na_V1 localization at the AIS via its ankyrin-binding motif may have facilitated the increase in body size of the vertebrate

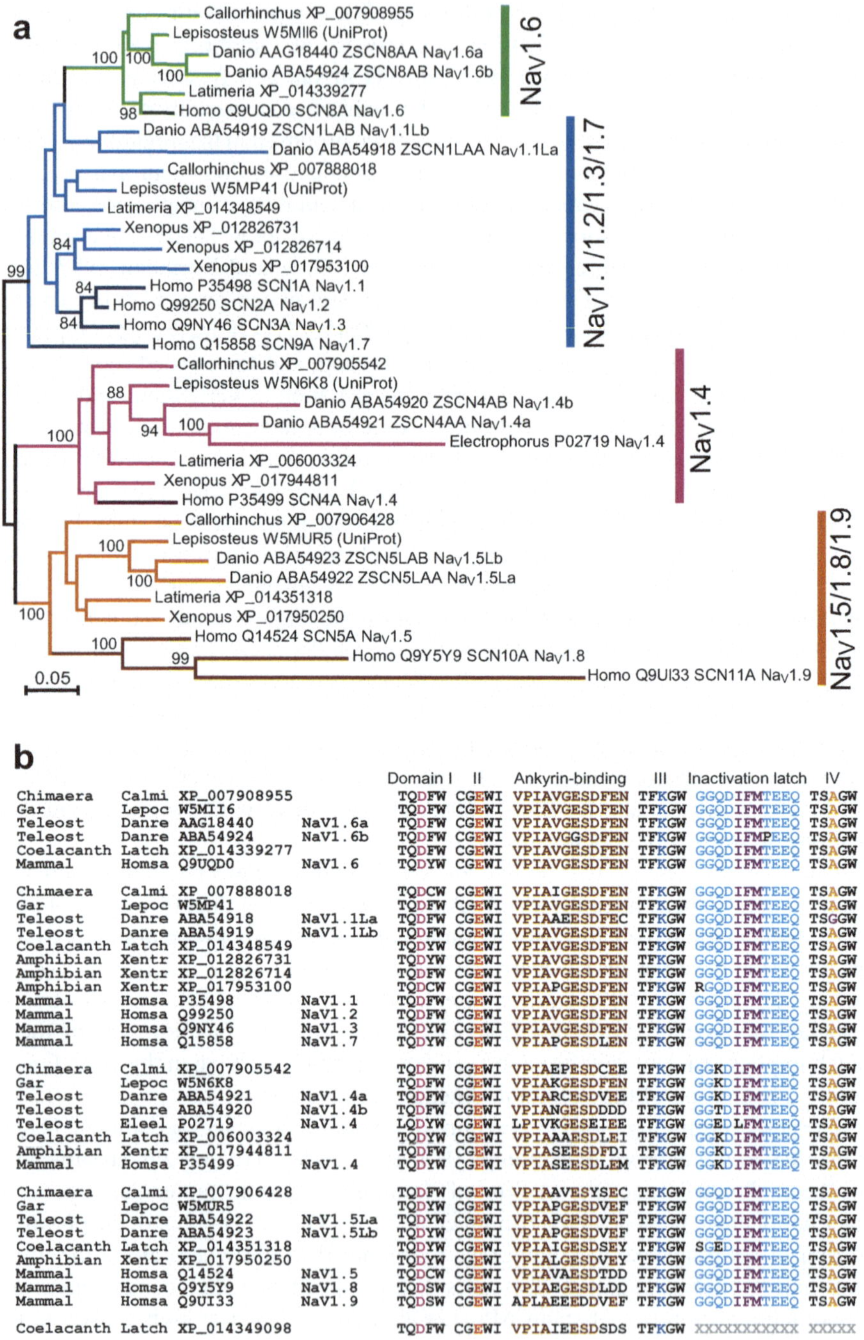

Fig. 5 Molecular phylogeny of vertebrate Na_V1s. (**a**) A maximum-likelihood tree of Na_V1 α subunits from the elephant fish chimaera (*Callorhinchus milii*), the spotted gar (*Lepisosteus oculatus*), the teleost zebrafish (*Danio rerio*), the coelacanth (*Latimeria chalumnae*),

ancestor. This AIS machinery may be a preadaptation for the evolution of nodes of Ranvier and saltatory conduction in gnathostomes (jawed vertebrates). KCNQ2/3 (Kv7.2/7.3) channels emerged in gnathostomes and, together with Na_V1s, were directed toward the AIS and nodes of Ranvier via their ankyrin-binding motifs, concomitantly with the evolutionary appearance of true myelination (Hill et al. 2008). This innovation would not only change neuronal morphology, but would greatly accelerate the rates of signal relay and processing among sensory organs, neurons themselves, and effectors, which would in turn contribute to the evolution of jaws and the predatory lifestyle.

Interestingly, the genomes of the lamprey, cartilaginous fish, and the coelacanth *Latimeria* possess the gene for $Na_V2(Ca_V4)$ channel (Fig. 4). The pore signature is D/E/E/G in the *Petromyzon* $Na_V2(Ca_V4)$, but in cartilaginous fish and the coelacanth, the pore signature is the common D/E/E/A (Fig. 4b). It had been thought that $Na_V2(Ca_V4)$ was present only in invertebrate animals, but it is now recognized that this channel remains in the vertebrate lineage. On the other hand, we did not find $Na_V2(Ca_V4)$ in the genomes of ray-finned fish or amphibians, which suggests the gene encoding this channel was independently lost from the lineages of ray-finned fish and tetrapods. Considering that both "gene loss" events appear to have occurred as the animals left the sea for inland environments, changes in the ionic conditions may have decreased the selection pressure to keep the $Na_V2(Ca_V4)$ gene. However, since the operational principles of this type of channel are still unclear, this issue remains to be determined.

7 Independent Gene Duplications of Na_V1 in Teleosts and Amniotes

Recent comparative genome analyses indicate that the last common ancestor of ray-finned fish and tetrapods possessed at least four types of Na_V1, each of which was linked to *HoxA-D* clusters (Widmark et al. 2011; Zakon et al. 2011). The genes encoding $Na_V1.1$, 1.2, 1.3, and 1.7 in amniotes, as well as the Na_X shared by therian mammals, emerged from the ancestral gene linked with *HoxD* cluster through a series of gene duplications, probably in the sequence of [(1.3 (1.2, 1.1)) (1.7, X)] (Fig. 5) (Widmark et al. 2011; Zakon et al. 2011). On the other hand, the gene

Fig. 5 (continued) the amphibian frog (*Xenopus tropicalis*), and humans (*Homo sapiens*). $Na_V1.4$ (*magenta*), $Na_V1.6$ (*green*), and $Na_V1.5/1.8/1.9$ (*yellow*) are monophyletic, while NaV1.1/1.2/1.3/1.7 sequences (*blue*) does not form a clade in this tree. The branches of mammalian Na_V1s are marked with darker colors. NCBI accession numbers are indicated in the tree. The *L. oculatus* sequences are obtained from Ensembl (http://www.ensembl.org/Lepisosteus_oculatus/Info/Index), and thus sequence IDs are of UniProt. The bootstrap values over 80 are shown. The tree was constructed using the WAG model on MEGA7 from gap-free 1,183 amino acid positions aligned by MUSCLE program. (**b**) The pore signatures and the regions corresponding to ankyrin-binding motif and the inactivation latch in vertebrate Na_V1s. The species codes and color codes are as used in Figs. 2, 3, and 4 encoding $Na_V1.6$, a neural Na_V1 linked to *HoxC*, apparently experienced no gene duplications and was single throughout amniote evolution (Fig. 5). In mammals, $Na_V1.1$, 1.2, 1.3, and 1.6 are mainly expressed in CNS neurons and contribute to rapid APs, while $Na_V1.1$ and 1.6 are also expressed to some extent in PNS neurons, and $Na_V1.7$ is selectively expressed in PNS neurons (Table 1) (Goldin 2001). This situation implies that the amniote ancestor utilized $Na_V1.7$ (more precisely the mother gene of 1.7 and 1.1/1.2/1.3) in PNS and $Na_V1.6$ in CNS to independently control the excitability of PNS and CNS neurons, respectively. Later, $Na_V1.1$–1.3, derived from a $Na_V1.7$-like ancestral protein, were recruited for the operation of the CNS in amniotes. The diversity of $Na_V1.1$, 1.2, 1.3, and 1.6 (i.e., the CNS subtypes) enabled independent control of their subcellular distributions in individual neurons, since expression of these subtypes in CNS is not only cell type specific but also cellular compartment specific (Hu et al. 2009; Lorincz and Nusser 2010; Zakon et al. 2011). This diversification of these CNS subtypes would have fine-tuned membrane excitabilities among CNS neurons and led to higher level brain performance in ancestral amniotes, which would contribute to forebrain expansion in amniotes (Zakon et al. 2011).

Like $Na_V1.6$, the amniote $Na_V1.4$ linked to *HoxB* was not duplicated and remained single. This subtype has presumably remained to work in the skeletal muscle of amniotes as well as anamniotes (Widmark et al. 2011; Zakon et al. 2011). Other subtypes, $Na_V1.5$, 1.8, and 1.9, are clustered with *HoxA* and were derived via sequential gene duplication, probably in the sequence [1.5 (1.8, 1.9)] (Fig. 5) (Widmark et al. 2011; Zakon et al. 2011). The mammalian (likely all the amniote) $Na_V1.5$ is expressed in the heart (Table 1) (Rogart et al. 1989), as is the $Na_V1.5$ (more precisely the ancestor gene for $Na_V1.5$ and 1.8/1.9) in shark and lungfish (Zakon et al. 2011). The genomes of the lizard *Anolis* and chick *Gallus* contain the genes encoding $Na_V1.8$ and 1.9, as well as 1.5, while that of the frog *Xenopus* does not, suggesting that these innovations occurred with the evolution of amniotes, as in the case of $Na_V1.1/1.2/1.3/1.7$ diversification (Zakon et al. 2011; see also Fig. 5).

It is known that $Na_V1.8$ and $Na_V1.9$ are specialized to work in the nociceptive DRG neurons, a neural crest derivative, in mammals (and likely in all amniotes) (Table 1) (e.g., Lai et al. 2004). The amniote genes encoding $Na_V1.8$ and 1.9 were derived from a $Na_V1.5$-like mother gene and predominantly function for nociception. Unlike other Na_V1s, $Na_V1.8$ and 1.9 show slow activation and inactivation (reviewed in Lai et al. 2004), despite a conserved I-F-M inactivation latch in both isoforms (Fig. 5b). The slow kinetics of $Na_V1.8/1.9$ may favor generation of graded membrane potential changes, which could be linearly related to nociceptive input. In fact, these channels, especially $Na_V1.9$, are expressed in small unmyelinated neurons (Lai et al. 2004), which is consistent with the fact that the ankyrin-binding motif is not well conserved in mammalian $Na_V1.9$ (Fig. 5b). This suggests critical aspects of nociceptive properties were innovations in amniotes. Another possible benefit of using multiple Na_V1s for nociception is that the slow $Na_V1.8$ and 1.9, fast 1.7, and possibly fast 1.5 may make possible special tuning of the propagation speed of nociceptive signals (Lai et al. 2004). This would enable, for example, adjustment for larger body size or constant/high body temperature in ancestral amniotes. However, answering this question must await testing whether Na_V1s expressed in nociceptive peripheral neurons do in fact function in nociceptive neurons in amniotes other than mammals and also await characterization of nociceptive properties in anamniotes.

This expansion of the Na_V1 gene family in amniotes, and the additional event needed to derive $Na_V1.7$ and Na_X in the therian ancestor, was mediated through tandem gene duplication. This kind of tandem gene duplication was not detected in their neighboring genes nor in the amniote Ca_V genes. Thus "evolution by gene duplication" in this case apparently occurred specifically to Na_V1 family genes of amniotes (Zakon et al. 2011). This diversification pattern is in contrast to what happened in the lineage of ray-finned fish, where another round of whole genome duplication took place in the ancestor of teleosts (e.g., Novak et al. 2006b; Braasch et al. 2016). Indeed, in the zebrafish genome, at least eight genes for Na_V1 isoforms have been identified – *zscn1Laa/ab*, *5Laa/ab*, *8aa/ab*, and *4aa/ab* associated with *HoxDa/b*, *HoxAa/b*, *HoxCa/b*, and *HoxBa/b* clusters, respectively, although the *HoxDb* cluster was lost from teleosts (Novak et al. 2006b; Widmark et al. 2011; Zakon et al. 2011, Won et al. 2012). The mode of gene duplication that took place in teleosts was not specific to genes encoding Na_V1s; it was global and thus distinct from what happened in amniotes. The gene expression pattern analyses of the zebrafish Na_V1 isoforms revealed considerable variation in the expressed isoforms among the CNS, PNS, heart, and skeletal muscles (Novak et al. 2006a; Won et al. 2012). For example, while the zebrafish heart expresses both of the two paralogous $Na_V1.5$ genes (*zscn5Laa* and *ab*), PNS cells from the DRG express only one of the paralogues (*5Laa*) (Novak et al. 2006a; Won et al. 2012). Electrophysiological studies of zebrafish DRG neurons confirmed that a type of Na^+ current is slowly inactivating and recorded from smaller neurons (presumably mediated by the $Na_V1.5$ encoded in *zscn5Laa*), and another Na^+ current is rapidly inactivating and recorded from larger neurons (Won et al. 2012). Given that other fast Na_V1s are also expressed in the DRG (Won et al. 2012), functional diversification of peripheral neurons and differential utilization of fast and slow Na_V1 isoforms may have been established in parallel in teleosts and mammals, although the degree of functional diversification in the DRG neurons (and also in the kinetics of the expressed Na_V1s) appears greater in those of mammals.

A remarkable feature of the amniotes' body is an elaborate forebrain with a large number of neurons. Ancestral amniotes must have developed multiple sensory systems related to their large body size and to terrestrial habitation, which enabled them to utilize complex environmental stimuli for complex motor behaviors (Zakon et al. 2011). "Specific" expansion of the genes encoding Na_V1 isoforms in amniotes, namely, the evolution of $Na_V1.1$–1.3 and $1.8/1.9$, provided functional options that supported radiation of amniotes. This may have relieved the limitation on body size and facilitated modal innovations in the sensory system and high-speed signal processing in the CNS, which constituted a morphological and physiological framework preadapted for the evolution of humankind.

8 Concluding Remarks

Along the history of voltage-gated Na^+ channels, one can see repetitive duplication and diversification of functional units. To establish the 24-TM channel family, two rounds of tandem duplication of a Catsper-like 6-TM unit were required to build a polypeptide composed of serially homologous quadruple domains. The unification of the 24-TM segments made possible asymmetric mutation within a single pseudotetraradial channel such that this structure became a unit for further duplication and diversification that gave rise to various types of voltage-gated channels. Na_V channels were established under this molecular evolutionary trend, and the asymmetric pore signatures of Na_V1 (D/E/K/A) and $Na_V2(Ca_V4)$ (D/E/E/A), as well as the asymmetric domains that include the inactivation latch and the ankyrin-binding domain emerged. In the lineage leading to Na_V channels, some of the genes having emerged via duplication were evolutionary dead ends (e.g., Na_V2 paralogues in cnidarians including $Na_V2.5$ with the D/K/E/A pore signature or Na_V1b and 1c in tunicates), while others became the seeds for future duplication and diversification. The repertoire of Na_V-related channels shrank in some animal lineages but grew in others. Loss of the Na_V1 gene in ambulacrarians is an example of the former, while special expansion of Na_V1 in amniotes is an example of the latter; and each is in good accord with their specific evolutionary steps for adaptation. The presence/absence of Na_V and the multiplication of their isoforms led to fundamental modal shifts in the membrane excitability in animal cells. These fundamental changes expanded/restricted the possibilities for physiological and morphological adaptation of animal bodies, extending the limit on body size, increasing the speed of sensory signal processing and locomotion, facilitating predatory life or habitation on land with an elaborate brain. The evolutionary history of Na_Vs reflects well the major steps in the broader evolution of organisms on earth, with the former possibly serving as a motive force for the latter. The involvement of $Na_V2(Ca_V4)$ channels in animal evolution was also touched on here. Several ion channel groups appear to have lost their Ca^{2+} permeability in parallel during the course of vertebrate evolution (Schredelseker et al. 2010; Nishino et al. 2011; Hirai et al. 2017). Further studies of the types of Ca^{2+}-permeable/impermeable Na_V-related channels may provide another key to understanding unknown aspects of cellular and organismal adaptations to their environment.

Acknowledgments Drs. Patrick Lemaire, Hiroki Nishida, and Hitoshi Sawada allowed us to utilize genome datasets from ascidians, *Halocynthia roretzi* and *H. aurantium*, before the publication of the data.

References

Akopian AN, Sivilotti L, Wood JN (1996) A tetrodotoxin-resistant voltage-gated sodium channel expressed by sensory neurons. Nature 379:257–262

Amemiya CT, Alföldi J, Lee AP, Fan S, Philippe H, Maccallum I, Braasch I, Manousaki T, Schneider I, Rohner N, Organ C, Chalopin D, Smith JJ, Robinson M, Dorrington RA,

Gerdol M, Aken B, Biscotti MA, Barucca M, Baurain D, Berlin AM, Blatch GL, Buonocore F, Burmester T, Campbell MS, Canapa A, Cannon JP, Christoffels A, De Moro G, Edkins AL, Fan L, Fausto AM, Feiner N, Forconi M, Gamieldien J, Gnerre S, Gnirke A, Goldstone JV, Haerty W, Hahn ME, Hesse U, Hoffmann S, Johnson J, Karchner SI, Kuraku S, Lara M, Levin JZ, Litman GW, Mauceli E, Miyake T, Mueller MG, Nelson DR, Nitsche A, Olmo E, Ota T, Pallavicini A, Panji S, Picone B, Ponting CP, Prohaska SJ, Przybylski D, Saha NR, Ravi V, Ribeiro FJ, Sauka-Spengler T, Scapigliati G, Searle SM, Sharpe T, Simakov O, Stadler PF, Stegeman JJ, Sumiyama K, Tabbaa D, Tafer H, Turner-Maier J, van Heusden P, White S, Williams L, Yandell M, Brinkmann H, Volff JN, Tabin CJ, Shubin N, Schartl M, Jaffe DB, Postlethwait JH, Venkatesh B, Di Palma F, Lander ES, Meyer A, Lindblad-Toh K (2013) The African coelacanth genome provides insights into tetrapod evolution. Nature 496:311–316

Anderson PAV, Holman MA, Greenberg RM (1993) Deduced amino acid sequence of a putative sodium channel from the scyphozoan jellyfish *Cyanea capillata*. Proc Natl Acad Sci U S A 90:7419–7423

Beneski DA, Catterall WA (1980) Covalent labeling of protein components of the sodium channel with a photoactivable derivative of scorpion toxin. Proc Natl Acad Sci U S A 77:639–643

Bichet D, Haass FA, Jan LY (2003) Merging functional studies with structures of inward-rectifier K^+ channels. Nat Rev Neurosci 4:957–967

Braasch I, Gehrke AR, Smith JJ, Kawasaki K, Manousaki T, Pasquier J, Amores A, Desvignes T, Batzel P, Catchen J, Berlin AM, Campbell MS, Barrell D, Martin KJ, Mulley JF, Ravi V, Lee AP, Nakamura T, Chalopin D, Fan S, Wcisel D, Cañestro C, Sydes J, Beaudry FE, Sun Y, Hertel J, Beam MJ, Fasold M, Ishiyama M, Johnson J, Kehr S, Lara M, Letaw JH, Litman GW, Litman RT, Mikami M, Ota T, Saha NR, Williams L, Stadler PF, Wang H, Taylor JS, Fontenot Q, Ferrara A, Searle SM, Aken B, Yandell M, Schneider I, Yoder JA, Volff JN, Meyer A, Amemiya CT, Venkatesh B, Holland PW, Guiguen Y, Bobe J, Shubin NH, Di Palma F, Alföldi J, Lindblad-Toh K, Postlethwait JH (2016) The spotted gar genome illuminates vertebrate evolution and facilitates human-teleost comparisons. Nat Genet 48:427–437

Brozovic M, Martin C, Dantec C, Dauga D, Mendez M, Simion P, Percher M, Laporte B, Scornavacca C, Di Gregorio A, Fujiwara S, Gineste M, Lowe EK, Piette J, Racioppi C, Ristoratore F, Sasakura Y, Takatori N, Brown TC, Delsuc F, Douzery E, Gissi C, McDougall A, Nishida H, Sawada H, Swalla BJ, Yasuo H, Lemaire P (2016) ANISEED 2015: a digital framework for the comparative developmental biology of ascidians. Nucleic Acids Res 44:D808–D818

Brunet T, Arendt D (2015) From damage response to action potentials: early evolution of neural and contractile modules in stem eukaryotes. Philos Trans R Soc Lond B Biol Sci 371:20150043

Calcraft PJ, Arredouani A, Ruas M, Pan Z, Cheng X, , Hao X, Tang J, Rietdorf K, Teboul L, Chuang K-T, Lin P, Xiao R, Wang C, Zhu Y, Lin Y, Wyatt CN, Parrington J, Ma J, Evans AM, Galione A, Zhu MX (2009) NAADP mobilizes calcium from acidic organelles through two-pore channels. Nature 459:596–600

Catterall WA (2000) From ionic currents to molecular mechanisms: the structure and function of voltage-gated sodium channels. Neuron 26:13–25

Catterall WA, Zheng N (2015) Deciphering voltage-gated Na^+ and Ca^{2+} channels by studying prokaryotic ancestors. Trends Biochem Sci 40:526–534

Catterall WA, Goldin AL, Waxman SG (2005) International union of pharmacology. XLVII. Nomenclature and structure-function relationships of voltage-gated sodium channels. Pharmacol Rev 57:397–409

Cavalier-Smith T (2010) Kingdoms Protozoa and Chromista and the eozoan root of the eukaryotic tree. Biol Lett 6:342–345

Conway-Morris S (1986) The community structure of the Middle Cambrian Phyllopod Bed (Burgess Shale). Palaeontology 29:423–467

Delsuc F, Brinkmann H, Chourrout D, Philippe H (2006) Tunicates and not cephalochordates are the closest living relatives of vertebrates. Nature 439:965–968

Fujiu K, Nakayama Y, Yanagisawa A, Sokabe M, Yoshimura K (2009) *Chlamydomonas CAV2* encodes a voltage-dependent calcium channel required for the flagellar waveform conversion. Curr Biol 19:133–139

Fukushima Y (1981) Identification and kinetic properties of the current through a single Na^+ channel. Proc Natl Acad Sci U S A 78:1274–1277

Gans C, Northcutt RG (1983) Neural crest and the origin of vertebrates: a new head. Science 220:268–273

Garrido JJ, Giraud P, Carlier E, Fernandes F, Moussif A, Fache MP, Debanne D, Dargent B (2003) A targeting motif involved in sodium channel clustering at the axon initial segment. Science 300:2091–2094

Gellens ME, George AL Jr, Chen L, Chahine M, Horn R (1992) Primary structure and functional expression of the human cardiac tetrodotoxin-insensitive voltage-dependent sodium channel. Proc Natl Acad Sci U S A 89:554–558

Goldin AL (2001) Resurgence of sodium channel research. Annu Rev Physiol 63:871–894

Gosselin-Badaroudine P, Moreau A, Simard L, Cens T, Rousset M, Collet C, Charnet P, Chahine M (2016) Biophysical characterization of the honeybee DSC1 orthologue reveals a novel voltage-dependent Ca^{2+} channel subfamily: Ca_V4. J Gen Physiol 148:133–145

Gould SJ (1990) Wonderful life: the Burgess Shale and the nature of history. WW Norton & Co., New York

Gur Barzilai M, Reitzel AM, Kraus JE, Gordon D, Technau U, Gurevitz M, Moran Y (2012) Convergent evolution of sodium ion selectivity in metazoan neuronal signaling. Cell Rep 2:242–248

Hartshorne RP, Catterall WA (1981) Purification of the saxitoxin receptor of the sodium channel from rat brain. Proc Natl Acad Sci U S A 78:4620–4624

Hartshorne RP, Keller BU, Talvenheimo JA, Catterall WA, Montal M (1985) Functional reconstitution of the purified brain sodium channel in planar lipid bilayers. Proc Natl Acad Sci U S A 82:240–244

Heinemann SH, Terlau H, Stühmer W, Imoto K, Numa S (1992) Calcium channel characteristics conferred on the sodium channel by single mutations. Nature 356:441–443

Hill AS, Nishino A, Nakajo K, Zhang G, Fineman JR, Selzer ME, Okamura Y, Cooper EC (2008) Ion channel clustering at the axon initial segment and node of Ranvier evolved sequentially in early chordates. PLoS Genet 4:e1000317

Hirai S, Hotta K, Kubo Y, Nishino A, Okabe S, Okamura Y, Okado H (2017) AMPA glutamate receptors are required for sensory-organ formation and morphogenesis in the basal chordate. Proc Natl Acad Sci U S A 114:3939–3944

Hiyama TY, Noda M (2016) Sodium sensing in the subfornical organ and body-fluid homeostasis. Neurosci Res 113:1–11

Hiyama TY, Watanabe E, Ono K, Inenaga K, Tamkun MM, Yoshida S, Noda M (2002) Na_X channel involved in CNS sodium-level sensing. Nat Neurosci 5:511–512

Hiyama TY, Watanabe E, Okado H, Noda M (2004) The subfornical organ is the primary locus of sodium-level sensing by Na_X sodium channels for the control of salt-intake behavior. J Neurosci 24:9276–9281

Hodgkin AL, Huxley AF (1945) Resting and action potentials in single nerve fibers. J Physiol 104:176–195

Hodgkin AL, Huxley AF (1952) Currents carried by sodium and potassium ions through the membrane of the giant axon of *Loligo*. J Physiol 116:449–472

Holland ND (2003) Early central nervous system evolution: an era of skin brains? Nat Rev Neurosci 4:617–627

Holland ND (2016) Nervous systems and scenarios for the invertebrate-to-vertebrate transition. Phil Trans R Soc Lond B 371:20150047

Hong CS, Ganetzky B (1994) Spatial and temporal expression patterns of two sodium channel genes in *Drosophila*. J Neurosci 14:5160–5169

Honoré E (2007) The neuronal background K2P channels: focus on TREK1. Nat Rev Neurosci 8:251–261

Hu W, Tian C, Li T, Yang M, Hou H, Shu Y (2009) Distinct contributions of $Na_V1.6$ and $Na_V1.2$ in action potential initiation and backpropagation. Nat Neurosci 12:996–1002

Hyman LH (1955) The invertebrates: Echinodermata. The coelomic Bilateria. McGraw-Hill, New York

Jiang Y, Lee A, Chen J, Ruta V, Cadene M, Chait BT, MacKinnon R (2003) X-ray structure of a voltage-dependent K^+ channel. Nature 423:33–41

Kole MH, Stuart GJ (2012) Signal processing in the axon initial segment. Neuron 73:235–247

Krebs HA (1975) The August Krogh principle: "For many problems there is an animal on which it can be most conveniently studied". J Exp Zool 194:221–225

Kulkarni NH, Yamamoto AH, Robinson KO, Mackay TFC, Anholt RR (2002) The DSC1 channel, encoded by the *smi60E* locus, contributes to odor-guided behavior in *Drosophila melanogaster*. Genetics 161:1507–1516

Lai J, Porreca F, Hunter JC, Gold MS (2004) Voltage-gated sodium channels and hyperalgesia. Annu Rev Pharmacol Toxicol 44:371–397

Lamaillet G, Walker B, Lambert S (2003) Identification of a conserved ankyrin-binding motif in the family of sodium channel alpha subunits. J Biol Chem 278:27333–27339

Liebeskind BJ, Hillis DM, Zakon HH (2011) Evolution of sodium channels predates the origin of nervous systems in animals. Proc Natl Acad Sci U S A 108:9154–9159

Liebeskind BJ, Hillis DM, Zakon HH (2012) Phylogeny units animal sodium leak channels with fungal calcium channels in an ancient, voltage-insensitive clade. Mol Biol Evol 29:3613–3616

Liebeskind BJ, Hillis DM, Zakon HH (2013) Independent acquisition of sodium selectivity in bacterial and animal sodium channels. Curr Biol 23:R948–R949

Lorincz A, Nusser Z (2010) Molecular identity of dendritic voltage-gated sodium channels. Science 328:906–909

Loughney K, Kreber R, Ganetzky B (1989) Molecular analysis of the *para* locus, a sodium channel gene in *Drosophila*. Cell 58:1143–1154

Lu B, Su Y, Das S, Liu J, Xia J, Ren D (2007) The neuronal channel NALCN contributes resting sodium permeability and is required for normal respiratory rhythm. Cell 129:371–383

Machemer H, Ogura A (1979) Ionic conductances of membranes in ciliated and deciliated *Paramecium*. J Physiol 296:49–60

Moran Y, Zakon HH (2014) The evolution of the four subunits of voltage-gated calcium channels: ancient roots, increasing complexity, and multiple losses. Genome Biol Evol 6:2210–2217

Moran Y, Liebeskind BJ, Zakon HH (2015) Evolution of voltage-gated ion channels at the emergence of Metazoa. J Exp Biol 218:515–525

Murata Y, Iwasaki H, Sasaki M, Inaba K, Okamura Y (2005) Phosphoinositide phosphatase activity coupled to an intrinsic voltage sensor. Nature 435:1239–1243

Nagahora H, Okada T, Yahagi N, Chong JA, Mandel G, Okamura Y (2000) Diversity of voltage-gated sodium channels in the ascidian larval nervous system. Biochem Biophys Res Commun 275:558–564

Nishino A, Baba SA, Okamura Y (2011) A mechanism for graded motor control encoded in the channel properties of the muscle ACh receptor. Proc Natl Acad Sci U S A 108:2599–2604

Noda M, Shimizu S, Tanabe T, Takai T, Kayano T, Ikeda T, Takahashi H, Nakayama H, Kanaoka Y, Minamino N, Kangawa K, Matsuo H, Raftery MA, Hirose T, Inayama S, Hayashida H, Miyata T, Numa S (1984) Primary structure of electrophorus electricus sodium channel deduced from cDNA sequence. Nature 312:121–127

Noda M, Suzuki H, Numa S, Stühmer W (1989) A single point mutation confers tetrodotoxin and saxitoxin insensitivity on the sodium channel II. FEBS Lett 259:213–216

Nomaksteinsky M, Röttinger E, Dufour HD, Chettouh Z, Lowe CJ, Martindale MQ, Brunet JF (2009) Centralization of the deuterostome nervous system predates chordates. Curr Biol 19:1264–1269

Novak AE, Taylor AD, Pineda RH, Lasda EL, Wright MA, Ribera AB (2006a) Embryonic and larval expression of zebrafish voltage-gated sodium channel α-subunit genes. Dev Dyn 235:1962–1973

Novak AE, Jost MC, Lu Y, Taylor AD, Zakon HH, Ribera AB (2006b) Gene duplications and evolution of vertebrate voltage-gated sodium channels. J Mol Evol 63:208–221

Ohno S (1970) Evolution by gene duplication. Springer, New York

Okada T, Hirano H, Takahashi K, Okamura Y (1997) Distinct neuronal lineages of the ascidian embryo revealed by expression of a sodium channel gene. Dev Biol 190:257–272

Okado H, Takahashi K (1988) A simple "neural induction" model with two interacting cleavage-arrested ascidian blastomeres. Proc Natl Acad Sci U S A 85:6197–6201

Okamoto H, Takahashi K, Yamashita N (1977) One-to-one binding of a purified scorpion toxin to Na channels. Nature 266:465–468

Okamura Y, Shidara M (1987) Kinetic differences between Na channels in the egg and the neutrally differentiated blastomere in the tunicate. Proc Natl Acad Sci U S A 84:8702–8706

Okamura Y, Shidara M (1990a) Changes in sodium channels during neural differentiation in the isolated blastomere of the ascidian embryo. J Physiol 431:39–74

Okamura Y, Shidara M (1990b) Inactivation kinetics of the sodium channel in the egg and the isolated, neutrally differentiated blastomere of the ascidian. J Physiol 431:75–102

Okamura Y, Ono F, Okagaki R, Chong JA, Mandel G (1994) Neural expression of a sodium channel gene requires cell-specific interactions. Neuron 13:937–948

Okamura Y, Nishino A, Murata Y, Nakajo K, Iwasaki H, Ohtsuka Y, Tanaka-Kunishima M, Takahashi N, Hara Y, Yoshida T, Nishida M, Okado H, Watari H, Meinertzhagen IA, Satoh N, Takahashi K, Satou Y, Okada Y, Mori Y (2005) Comprehensive analysis of the ascidian genome reveals novel insights into the molecular evolution of ion channel genes. Physiol Genomics 22:269–282

Pan Z, Kao T, Horvath Z, Lemos J, Sul JY, Cranstoun SD, Bennett V, Scherer SS, Cooper EC (2006) A common ankyrin-G-based mechanism retains KCNQ and Na_V channels at electrically active domains of the axon. J Neurosci 26:2599–2613

Parker A (2003) In the blink of an eye. Basic Books, New York

Payandeh J, Minor DL Jr (2015) Bacterial voltage-gated sodium channels ($BacNa_V$s) from the soil, sea, and salt lakes enlighten molecular mechanisms of electrical signaling and pharmacology in the brain and heart. J Mol Biol 427:3–30

Putnam NH, Butts T, Ferrier DE, Furlong RF, Hellsten U, Kawashima T, Robinson-Rechavi M, Shoguchi E, Terry A, Yu JK, Benito-Gutiérrez EL, Dubchak I, Garcia-Fernàndez J, Gibson-Brown JJ, Grigoriev IV, Horton AC, de Jong PJ, Jurka J, Kapitonov VV, Kohara Y, Kuroki Y, Lindquist E, Lucas S, Osoegawa K, Pennacchio LA, Salamov AA, Satou Y, Sauka-Spengler T, Schmutz J, Shin-I T, Toyoda A, Bronner-Fraser M, Fujiyama A, Holland LZ, Holland PW, Satoh N, Rokhsar DS (2008) The amphioxus genome and the evolution of the chordate karyotype. Nature 453:1064–1071

Ramaswami M, Tanouye MA (1989) Two sodium-channel genes in *Drosophila*: implications for channel diversity. Proc Natl Acad Sci U S A 86:2079–2082

Ramsey IS, Moran MM, Chong JA, Clapham DE (2006) A voltage-gated proton-selective channel lacking the pore domain. Nature 440:1213–1216

Ren D, Navarro B, Xu H, Yue L, Shi Q, Clapham DE (2001) A prokaryotic voltage-gated sodium channel. Science 294:2372–2375

Rogart RB, Cribbs LL, Muglia LK, Kephart DD, Kaiser MW (1989) Molecular cloning of a putative tetrodotoxin-resistant rat heart Na^+ channel isoform. Proc Natl Acad Sci U S A 86:8170–8174

Roger AJ, Simpson AGB (2008) Evolution: revisiting the root of the eukaryote tree. Curr Biol 19: R165–R167

Rogozin IB, Basu MK, Csürös M, Koonin EV (2009) Analysis of rare genomic changes does not support the unikont-bikont phylogeny and suggests cyanobacterial symbiosis as the point of primary radiation of eukaryotes. Genome Biol Evol 1:99–113

Salkoff L, Butler A, Wei A, Scavarda N, Giffen K, Ifune C, Goodman R, Mandel G (1987) Genomic organization and deduced amino acid sequence of a putative sodium channel gene in *Drosophila*. Science 237:744–749

Sangameswaran L, Fish LM, Koch BD, Rabert DK, Delgado SG, Ilnicka M, Jakeman LB, Novakovic S, Wong K, Sze P, Tzoumaka E, Stewart GR, Herman RC, Chan H, Eglen RM, Hunter JC (1997) A novel tetrodotoxin-sensitive, voltage-gated sodium channel expressed in rat and human dorsal root ganglia. J Biol Chem 272:14805–14809

Sasaki M, Takagi M, Okamura Y (2006) A voltage sensor-domain protein is a voltage-gated proton channel. Science 312:589–592

Sato C, Matsumoto G (1992) Primary structure of squid sodium channel deduced from the complementary DNA sequence. Biochem Biophys Res Commun 186:61–68

Satou Y, Kawashima T, Kohara Y, Satoh N (2003) Large scale EST analyses in *Ciona intestinalis*: its application as northern blot analyses. Dev Genes Evol 213:314–318

Schredelseker J, Shrivastav M, Dayal A, Grabner M (2010) Non-Ca^{2+}-conducting Ca^{2+} channels in fish skeletal muscle excitation-contraction coupling. Proc Natl Acad Sci U S A 107:5658–5663

Shen H, Zhou Q, Pan X, Li Z, Wu J, Yan N (2017) Structure of a eukaryotic voltage-gated sodium channel at near-atomic resolution. Science 355:eaal4326

Smith JJ, Kuraku S, Holt C, Sauka-Spengler T, Jiang N, Campbell MS, Yandell MD, Manousaki T, Meyer A, Bloom OE, Morgan JR, Buxbaum JD, Sachidanandam R, Sims C, Garruss AS, Cook M, Krumlauf R, Wiedemann LM, Sower SA, Decatur WA, Hall JA, Amemiya CT, Saha NR, Buckley KM, Rast JP, Das S, Hirano M, McCurley N, Guo P, Rohner N, Tabin CJ, Piccinelli P, Elgar G, Ruffier M, Aken BL, Searle SM, Muffato M, Pignatelli M, Herrero J, Jones M, Brown CT, Chung-Davidson YW, Nanlohy KG, Libants SV, Yeh CY, McCauley DW, Langeland JA, Pancer Z, Fritzsch B, de Jong PJ, Zhu B, Fulton LL, Theising B, Flicek P, Bronner ME, Warren WC, Clifton SW, Wilson RK, Li W (2013) Sequencing of the sea lamprey (*Petromyzon marinus*) genome provides insights into vertebrate evolution. Nat Genet 45(415–421):421e1–421e2

Strong M, Chandy KG, Gutman GA (1993) Molecular evolution of voltage-sensitive ion channel genes: on the origins of electrical excitability. Mol Biol Evol 10:221–242

Stühmer W, Conti F, Suzuki H, Wang X, Noda M, Yahagi N, Kubo H, Numa S (1989) Structure parts involved in activation and inactivation of the sodium channel. Nature 339:597–603

Takahashi K, Okamura Y (1998) Ion channels and early development of neural cells. Physiol Rev 78:307–337

Takahashi K, Yoshii M (1981) Development of sodium, calcium and potassium channels in the cleavage-arrested embryo of an ascidian. J Physiol 315:515–529

Tate S, Benn S, Hick C, Trezise D (1998) Two sodium channels contribute to the TTX-R sodium current in primary sensory neurons. Nat Neurosci 1:653–655

Tessmar-Raible K, Raible F, Christodoulou F, Guy K, Rembold M, Hausen H, Arendt D (2007) Conserved sensory-neurosecretory cell types in annelid and fish forebrain: insights into hypothalamus evolution. Cell 129:1389–1400

Toledo-Aral JJ, Moss BL, He ZJ, Koszowski AG, Whisenand T, Levinson SR, Wolf JJ, Silos-Santiago I, Halegoua S, Mandel G (1997) Identification of PN1, a predominant voltage-dependent sodium channel expressed principally in peripheral neurons. Proc Natl Acad Sci U S A 94:1527–1532

Tomer R, Denes AS, Tessmar-Raible K, Arendt D (2010) Profiling by image registration reveals common origin of annelid mushroom bodies and vertebrate pallium. Cell 142:800–809

Torruella G, Derelle R, Paps J, Lang BF, Roger AJ, Shalchian-Tabrizi K, Ruiz-Trillo I (2011) Phylogenetic relationships within the Opisthokonta based on phylogenomic analyses of conserved single copy protein domains. Mol Biol Evol 29:531–544

Vergara HM, Bertucci PY, Hantz P, Tosches MA, Achim K, Vopalensky P, Arendt D (2017) Whole-organism cellular gene-expression atlas reveals conserved cell types in the ventral nerve cord of *Platynereis dumerilii*. Proc Natl Acad Sci U S A 114:5878–5885

West JW, Patton DE, Scheuer T, Wang Y, Goldin AL, Catterall WA (1992) A cluster of hydrophobic amino acid residues required for fast Na^+-channel inactivation. Proc Natl Acad Sci U S A 89:10910–10914

Widmark J, Sundström G, Ocampo Daza D, Larhammar D (2011) Differential evolution of voltage-gated sodium channels in tetrapods and teleost fishes. Mol Biol Evol 28:859–871

Willey A (1894) Amphioxus and the ancestry of the vertebrates. Macmillan, New York

Won Y-J, Ono F, Ikeda SR (2012) Characterization of Na^+ and Ca^{2+} channels in zebrafish dorsal root ganglion neurons. PLoS One 7:e42602

Yan Z, Zhou Q, Wang L, Wu J, Zhao Y, Huang G, Peng W, Shen H, Lei J, Yan N (2017) Structure of the $Na_V1.4$-β1 complex from electric eel. Cell 170:470–482.e11

Yue L, Navarro B, Ren D, Ramos A, Clapham DE (2002) The cation selectivity filter of the bacterial sodium channel, NaChBac. J Gen Physiol 120:845–853

Zakon HH (2012) Adaptive evolution of voltage-gated sodium channels: the first 800 million years. Proc Natl Acad Sci U S A 109(Suppl 1):10619–10625

Zakon HH, Jost MC, Lu Y (2011) Expansion of voltage-dependent Na^+ channel gene family in early tetrapods coincided with the emergence of terrestriality and increased brain complexity. Mol Biol Evol 28:1415–1424

Zalc B (2016) The acquisition of myelin: an evolutionary perspective. Brain Res 1641:4–10

Zalc B, Goujet D, Colman D (2008) The origin of the myelination program in vertebrates. Curr Biol 18:R511–R512

Zhang T, Wang Z, Wang L, Luo N, Jiang L, Liu Z, Wu C-F, Dong K (2013) Role of the DSC1 channel in regulating neuronal excitability in *Drosophila melanogaster*: extending nervous system stability under stress. PLoS Genet 9:e1003327

Zhou W, Chung I, Liu Z, Goldin A, Dong K (2004) A voltage-gated calcium-selective channel encoded by a sodium channel-like gene. Neuron 42:101–112

Mining Protein Evolution for Insights into Mechanisms of Voltage-Dependent Sodium Channel Auxiliary Subunits

Steven Molinarolo, Daniele Granata, Vincenzo Carnevale, and Christopher A. Ahern

Contents

Abstract

Voltage-gated sodium channel (VGSC) beta (β) subunits have been called the "overachieving" auxiliary ion channel subunit. Indeed, these subunits regulate the trafficking of the sodium channel complex at the plasma membrane and simultaneously tune the voltage-dependent properties of the pore-forming alpha-subunit. It is now known that VGSC β-subunits are capable of similar modulation of multiple isoforms of related voltage-gated potassium channels, suggesting that their abilities extend into the broader voltage-gated channels. The gene family for these single transmembrane immunoglobulin beta-fold proteins extends well beyond the traditional VGSC β1–β4 subunit designation, with deep roots into the cell adhesion protein family and myelin-related proteins – where inherited mutations result in a myriad of electrical signaling disorders. Yet, very little is known about how VGSC β-subunits support protein trafficking

S. Molinarolo · C. A. Ahern (✉)
Department of Molecular Physiology and Biophysics, Iowa Neuroscience Institute, University of Iowa, Iowa City, IA, USA
e-mail: christopher-ahern@uiowa.edu

D. Granata · V. Carnevale (✉)
Institute for Computational Molecular Science, College of Science and Technology, Temple University, Philadelphia, PA, USA
e-mail: vincenzo.carnevale@temple.edu

M. Chahine (ed.), *Voltage-gated Sodium Channels: Structure, Function and Channelopathies*, Handbook of Experimental Pharmacology 246,
https://doi.org/10.1007/164_2017_75

pathways, the basis for their modulation of voltage-dependent gating, and, ultimately, their role in shaping neuronal excitability. An evolutionary approach can be useful in yielding new clues to such functions as it provides an unbiased assessment of protein residues, folds, and functions. An approach is described here which indicates the greater emergence of the modern β-subunits roughly 400 million years ago in the early neurons of Bilateria and bony fish, and the unexpected presence of distant homologues in bacteriophages. Recent structural breakthroughs containing α and β eukaryotic sodium channels containing subunits suggest a novel role for a highly conserved polar contact that occurs within the transmembrane segments. Overall, a mixture of approaches will ultimately advance our understanding of the mechanism for β-subunit interactions with voltage-sensor containing ion channels and membrane proteins.

Keywords

Bacterial ion channels · Evolution of membrane proteins and electrical signaling · Ion channel auxiliary subunits · NaV · Sodium channel · Voltage-gated ion channels

1 Sodium Channel Basics

Voltage-gated sodium channels (VGSC) are critical facilitators of electrical conduction in cardiovascular and neuronal tissues, as well as traditionally non-excitable cell types. The initial characterization of the VGSC protein complex identified two ~40 kDa proteins associated with the channel in a 1:1:1 ratio, termed the VGSC β auxiliary subunits. These β-subunits immunoprecipitated with the channel and were determined necessary for generating physiological current reproduction when the α-subunit is expressed heterologously (Hartshorne and Catterall 1981, 1984; Hartshorne et al. 1985). Since their initial identification, β-subunits have been shown to modulate VGSC gating and kinetics in addition to altering VGSC expression and pharmacology and further implications in disease phenotypes and nonconduction adhesion roles.

Four different β-subunits have been identified: β1, β2, β3, and β4 along with an alternative splice variant of β1 termed β1b (Isom et al. 1992, 1995a; Morgan et al. 2000; Yu et al. 2003; Qin et al. 2003). Membrane bound like the VGSC α-subunit, the single membrane-spanning segment of each β-subunit connects a large extracellular N-terminal domain to a short intracellular sequence. The exception to this topology is β1b whereby a skipped splice site prior to the β1 transmembrane exon results in a soluble protein. The large extracellular domain of the β-subunits is dominated by an immunoglobulin (Ig) fold of the V-set type that is highly glycosylated comprising nearly 30% of the proteins' molecular weight (Messner and Catterall 1985; Isom and Catterall 1996; Roberts and Barchi 1987). This Ig fold of β-subunits, which has similarity to the adhesion molecule contactin, targets binding for protein-protein interactions (Xiao et al. 1999). These β-subunits thus support a myriad of physiological roles, some of which are achieved through their tissue specific modulation of sodium channel expression and voltage-dependent gating. Other cellular effects appear to arise independently of their roles in tuning sodium channel electrical signaling behavior, these alternative roles rely heavily on the adhesion aspects of the β-subunit architecture.

Consistent with their primary effects on electrical signaling rhythms, β-subunits fine-tune the responses of the pore-forming α-subunit to transmembrane voltage. For instance, it was shown early on that co-expression of the neuronal channel Nav1.2 with β1 significantly accelerates the fast-inactivation process (Isom et al. 1992; Patton et al. 1994). Other VGSCs including Nav1.1–4, 1.6, and 1.7 display accelerated inactivation kinetics when β1 is co-expressed; adversely, Nav1.5 and Nav1.8 already have comparatively fast inactivation and have minimal if any inactivation modulation by β1 (Patton et al. 1994; Smith and Goldin 1998; Bennett et al. 1993; Wallner et al. 1993; Dietrich et al. 1998; Shcherbatko et al. 1999; Qu et al. 1995; Zhao et al. 2011). However, one complicating factor is that such effects on inactivation, and other channel gating modulatory effects, have varied between groups reporting results and between expression systems utilized, a possible result due to the presence of endogenous β-subunits in the expression systems (Isom et al. 1995b; Isom 2001; Chioni et al. 2009). Further, these so-called VGSC β-subunits have been recently shown to modulate multiple voltage-gated potassium channels (Deschênes and Tomaselli 2002; Marionneau et al. 2012; Nguyen et al. 2012a, b). Another striking effect the β-subunits have on VGSCs is to significantly increase current density; this effect is trafficking related resulting in channel upregulation at the plasma membrane (Kazarinova-Noyes et al. 2001). This increased presence of the VGSC at the plasma membrane allows for quicker conduction which combined with an hastened inactivation rate produces a faster return to baseline after action potential firing allowing for more rapid subsequent electrical stimulation.

β-subunits are principally located in excitable tissues (muscle and neurons) where their interaction can tune the VGSC conduction in a tissue-specific manner. While the β-subunits primary role appears to be electrically related, the do not have a direct role in the movement of ions across the plasma membrane and can be found in non-excitable tissues such as glial astrocytes, Schwann cells, and in the kidney where the β-subunit function is theorized to be adhesion focused (Oh and Waxman 1994; Isom 2002). The β-subunits are differently regulated by tissues and within a tissue, for instance, the heart has β1, β2, and β3 at the transverse-tubules and β1, β2, and β4 at the intercalated disks; β1 is found at both locations with the Y181 phosphorylation modified β1 trafficking to the ICD, while unphosphorylated proceeds at the T-tubules (Maier et al. 2004; Malhotra et al. 2004). Further evidence of subcellular regulation is the increased density of β-subunits at the nodes of Ranvier where the β-subunits can interact with VGSC to alter gating, this nodal regulation has been linked to β2 interaction with tenascin-R and tenascin-C; however, β-subunit association with nodes also requires both ankyrin and VSGCs to localize (Xiao et al. 1999; Chen et al. 2012; Srinivasan et al. 1998). β-subunit interaction with VGSC is not limited to electrophysiological modulation; expression of β1 or β3 in HEK cells alters trafficking and glycosylation content depending on the β-subunit associated (Laedermann et al. 2013). Co-immunoprecipitation analysis have identified ankyrin-G, contactin, NrCAM, NF155, and NF186 as interaction partners with β-subunits with differing specificity for each subunit and the subunit post-translational modifications (Kazarinova-Noyes et al. 2001; Malhotra et al. 2000, 2004; Ratcliffe et al. 2001; McEwen and Isom 2004). β-subunits not only have a similar Ig domain to myelin protein zero, the major constituent of the PNS myelin sheath the β-subunits also have roles in neuronal development: β-subunit axonal guidance is experimentally observed by an increased neurite length when grown on a β1 supporting monolayer; β2 nor β4 have similar effects in the supporting cells; however, increased neuronal β4 expression promotes neurite elongation (Shapiro et al. 1996; Davis et al. 2004; Zhou et al. 2012). β1 neurite pathfinding effects require the adhesion molecule contactin and Fyn kinase, suggesting lipid raft localization; the extracellular domain of β1 is sufficient to mediate this effect as β1B has similar properties (Brackenbury et al. 2008; Patino et al. 2011). Lipid raft localization is where the β-subunits are targeted by proteases including the gamma secretase complex and BACE1, that release the extracellular Ig domain and intracellular domain of the β-subunits, thus expanding the β-subunit interactions from the membrane to surrounding cells and intracellular locations (Patino et al. 2011; Kim et al. 2005, 2007; Wong et al. 2005).

2 VGSC and Human Disease

Epileptic seizures are linked to elevated BACE1 activity in early- and late-onset Alzheimer's disease, this electrical disruption is partially contributed from increased Nav1.1 surface expression, a result of the BACE1 mediated β2 intracellular domain release (Kim et al. 2007). β-subunits have been implicated more directly with other electrical disorders in humans, principally the epileptic disorders (SIDS, GEFS+, or Dravet syndrome) and cardiac disorders (Brugada syndrome, long QT syndrome, atrial and ventricular fibrillation) with known clinical mutations of β1 (R85C/H, E87Q, I106F, C121W, R125C/L, D153N, W179X, and a splice site mutation) (Scheffer et al. 2007; Xu et al. 2007; Watanabe et al. 2008, 2009; Ogiwara et al. 2012; Wallace et al. 1998; Patino et al. 2009; Fendri-Kriaa et al. 2011; Audenaert et al. 2003), β1B (H162P, P213T, R214Q, and G257R) (Patino et al. 2011; Hu et al. 2012; Yuan et al. 2014; Riuró et al. 2014), β2 (R28Q/W and D211G) (Watanabe et al. 2009; Riuró et al. 2013), β3 (R6K, L10P, V36M, V54G, V110I, A130V, and M161T) (Tan et al. 2010; Hu et al. 2009; Valdivia et al. 2009; Ishikawa et al. 2013; Wang et al. 2010), and β4 (V162G, I166L, L179F, and S206L) (Tan et al. 2010; Li et al. 2013; Medeiros-Domingo et al. 2007). These mutations are found throughout the β-subunits domains, with effects ranging from nonfunctional proteins to altered interaction with VGSCs. Null mice for the four β-subunits have been produced with differing effects. β1 has a critical role in neuronal and cardiac function, this is exhibited by β1 null effects including developmental abnormalities, perturbed axonal pathfinding, spontaneous seizures, and extended QT intervals (Brackenbury et al. 2008, 2013; Chen et al. 2004; Lopez-Santiago et al. 2007). However, other β-subunit null mice have less drastic effects. For instance, β2 are prone to seizures, β3 have arrhythmic tendencies, and β4 have balance and motor defects, none of which are as serious as the β1 effects, suggesting a less critical role or adequate compensation by other proteins (Chen et al. 2002; Hakim et al. 2008; Ransdell et al. 2017). β-subunits alter the channel pharmacology, natural drugs including the toxins STX, and μ-conotoxins have altered binding when expressed with different β-subunits. In particular, the cone snail venom μO-conotoxin MrVIB has significantly improved block on Nav1.8 with any of the four β-subunit co-expressed compared to the channel alone and differing degrees of block depending on the interacting β-subunit (Schmidt et al. 1985; Wilson et al. 2011, 2015; Zhang et al. 2013). These pharmacological effects are extended to clinical drugs which act through VGSCs. The antiarrhythmic Lidocaine has over a two-fold decrease in affinity for Nav1.5 when expressed with β1, and the antiepileptic drug carbamazepine is incapable of blocking repetitive action potential firing without β1 (Makielski et al. 1996; Uebachs et al. 2012). The adhesion properties of β-subunits make them a strong candidate for effects in cancer metastases, β-subunits have been impacted in tumor migration and invasiveness and further association to angiogenesis has been observed in tumors originating from breast, cervical, glioblastoma, NSCL, and prostate cancers (Chioni et al. 2009; Diss et al. 2008; Aronica et al. 2003; Nelson et al. 2014; Brackenbury 2012).

3 β-Subunit Homology from the Perspective of Primary Sequence

At the sequence level, the β-subunits are diverse. The human VGSC α-subunits range from 53 to 88% identity; within the β-subunits, β1 and β3 are the most similar with nearly 50% identity, subsequently the second most similar are β2 and β4 with only 27% identity, and the other comparisons come in at less than 25% identity. β2 and β4 were determined via alanine scanning to bind α-channels via a disulfide bridge at Cys55 and Cys58, respectively, that separate under reducing conditions (Chen et al. 2012; Buffington and Rasband 2013). Similar scanning of Nav1.2 determined Cys910 of Nav1.2 D_{II}S5–S6 to be the β-subunit intermolecular disulfide bonding pair (Das et al. 2016). However, not all VGSC have this conserved cysteine for covalent attachment, such as Nav1.5 the cardiac isoform which does not covalently bind with either β2 or β4.

Unlike β2 and β4, the β1-subunit is not covalently bound; if theorizing a 1:1:1 ratio of the α:β1:β2 and that β1 has a direct interaction with α, then β1 must be interacting at a differing location than β2. The extracellular domains of the β-subunits are important in their interaction, nodal localization is interrupted by the Ig domain mutant of β2 C55A that breaks the disulfide bond; the extracellular Ig loop of β1 has similar importance on VGSC interaction, evidence by a GPI-linked β1 extracellular domain can recapitulate the β1 modulatory effects (Chen et al. 2012; Makita et al. 1996a; McCormick et al. 1998). Studies making chimeras using either Nav1.2 or Nav1.4 which have significant modulation by β1 and Nav1.5 that is not modulated by β1 have identified extracellular locations of the α required for modulation (Makita et al. 1996a, b; Qu et al. 1999). Further evidence for extracellular interaction has been recently identified with voltage-clamp fluorometry, this data depicts β1 and β3 interacting with the voltage-sensing domains of Nav1.5 and suggests a close proximity of the non-covalently bound β-subunit to the channel (Sharkey et al. 1984; Zhu et al. 2017).

4 Evolutionary History of Beta-Subunits

All VGSC β-subunits are type I integral membrane proteins, i.e., they show a single transmembrane with a cytosolically located C-terminus. The extracellular V-set Ig domain is a Greek-key beta-sandwich structure resembling the antibody variable domain, which is connected to the transmembrane alpha helical segment through a neck domain (Namadurai et al. 2015). V-set domains are members of a large class of domains, namely, immunoglobulin-like domains, ubiquitously present in all kingdoms of life (Bork et al. 1994). This fold is characterized by a two-layer sandwich of seven to nine antiparallel β-strands arranged in two planes of β-sheets.

The origin of immunoglobulin-like folds has long been debated. The traditionally favored hypothesis posits that this fold is a thermodynamically favored "platonic form" toward which proteins of a given length would spontaneously converge (Lesk and Chothia 1982). The observed distribution of immunoglobulin-like domain would then be the result of convergent evolution. A second hypothesis (that is gaining acceptance), instead, places emphasis on the biological function shared by a large portion of this superfamily: many members, though not all, are involved in the processes of cell adhesion, an observation suggesting a possible origin for the evolutionary conservation of the structure. However, despite the reasonability of this scenario, sequence conservation is poor across kingdoms and therefore evolutionary relatedness is difficult to establish with an acceptable degree of statistical confidence (Namadurai et al. 2015).

In contrast to other members of this superfamily, V-set domains are found almost exclusively in metazoan Bilateria, with a few occurrences in cnidarians. A notable exception to this monophyletic distribution in present in viruses: some poxyviruses contain V-set domains in their hemagglutinin and glycoprotein genes, a possible result of horizontal gene transfer events (Dermody et al. 2009). Whether these horizontal gene transfer events took place after the emergence of animals or before is still an open question. Indeed, the second largest group of viruses showing immunoglobulin-like domains are dsDNA bacteriophages. This observation raises the hypothesis that bacteriophages have played a crucial role in enabling extensive horizontal gene transfer events in bacteria and, possibly, in early eukaryotes (Fraser et al. 2007).

The evolutionary history of V-set domains constrains the possible evolutionary emergence events of the auxiliary β-subunits. In effect, demonstrating that VGSC β-subunits could not have appeared in their current form prior to the emergence of animals. This raises immediately a question: were β-subunits a response to the merging nervous systems in animals? Intriguingly, some occurrences of the V-set domain are found in the phylum of sponges, indicating that structural templates for β-subunits might have been available before nerve cells appeared. However, homologues of β-subunits found by scouring the entire UniProt database suggest that membrane-bound V-set domains might have been co-opted into the Nav auxiliary subunits role more recently than the emergence of excitable cells, suggesting their emergence was timed as animal nervous systems cultivated complexity. Notably, all genes containing detectable homology with human β-subunits are in vertebrata (Fig. 1). In particular, when pinpointed on the tree of life, these genes are found in bony fish ogranisms (Osteichthyes), which are in turn divided into the ray-finned fish (Actinopterygii) and lobe-finned fish (Sarcopterygii), but not in cartilaginous fish ogramisms (Chondrichthyes) like sharks. Fossil records help locate the splitting between Osteichthyes and Chondrichthyes at about 420 million years ago; thus, the present form of VGSC β-subunits, appeared in organisms already possessing a fully developed nervous system.

The evolution of the β-subunits can be best appreciated by analyzing the functionally homogenous families of β1 and β3. Since both families are present in ray-finned fish (Actinopterygii) and in lobe-finned fish (Sarcopterygii, ancestors

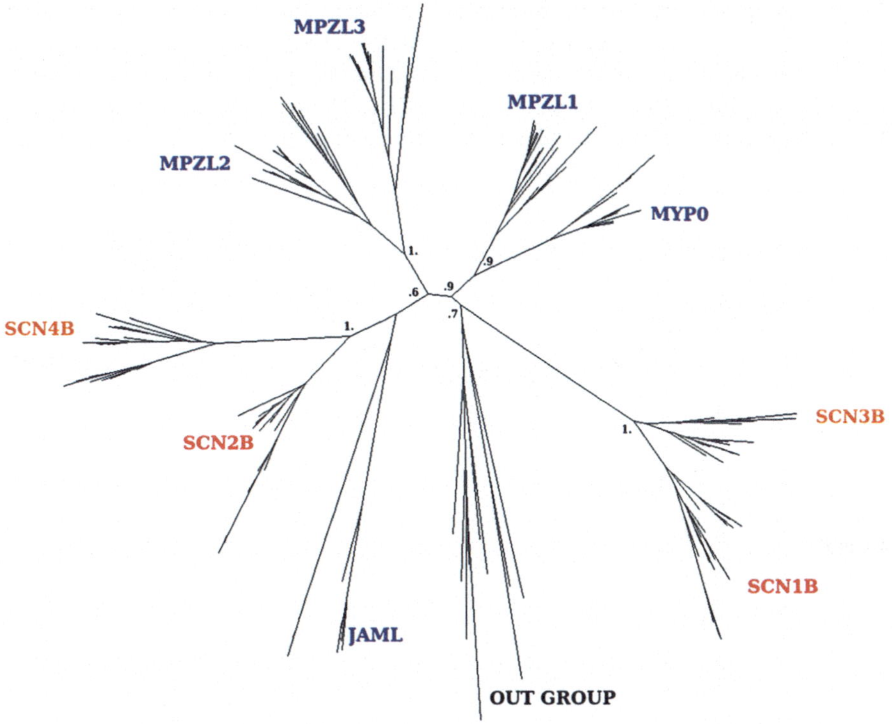

Fig. 1 Phylogenetic structure of the genes homologous to SCN1B. The dendrogram shows the major branching and their statistical significance (support). Numbers greater than 0.7 indicate a large degree of confidence in the tree structure. Groups are labeled according to the annotated genes contained in each group. Note that several myelin-associated proteins are identified as SCN1B homologues

of, among others, Reptilia and Mammalia), we can conclude that the common ancestor was present at the emergence of Euteleostomi in Silurian age, ca 420 Mya (Fig. 2). Consistently, the gene from Latimeria, which is considered the oldest representative of Sarcopterygii, is found close to the node separating the two main branches. Notably, the β1 and β3 branches show very similar organizations, completely consistent with the phylogenetic tree of Vertebrata. Since no homologous sequence from sharks has been detected, it is unlikely that the origin of these families can be further pinpointed.

An interesting insight that can be gained from this database-wide search for β-subunits homologues, is an appreciation that the overall size of the β-subunit superfamily extends well beyond the traditional β1–β4 gene nomenclature. The dendrogram shown in Fig. 1 highlights the major branches, or clusters, in which these homologous genes are organized. Labels have been added to all the groups containing sequences with experimentally validated annotation (i.e., belonging to the Swiss-Prot database). A first notable feature that emerges from this representation is that β-subunits are formed by two major groups: the first one consists of the β1 and β3 family, while the second contains the β2 and β4 families. Importantly, this phylogenetic organization of the four families is consistent with the observed functional differences (Namadurai et al. 2015). The second interesting feature conveyed by the three is the presence of several intervening groups of genes labeled as JAML, MYP0, MPZL1, MPZL2, and MPZL1. The position in the tree suggests that these groups are more similar to the members of the β2 and β4 groups than

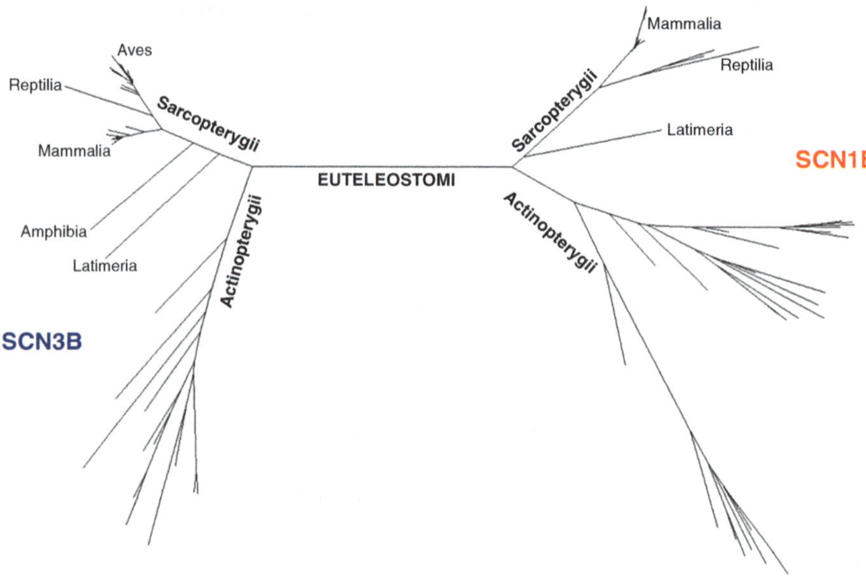

Fig. 2 Phylogenetic relationships between SCN1B and SCN3B genes. The dendrogram is restricted to the branches containing the annotated SCN1B and SCN3B genes. Note that the structure is the same for the two major branches and is consistent with known cladograms

those of the β1 and β3 one. Is it then possible that some of the genes present in these phylogenetic branches encode for a thus far unknown auxiliary subunit? Intriguing insights into a possible involvement of ion channel regulation by these genes come from the identified clinical variants. For instance, the gene group labeled MYP0 contains the human gene encoding for myelin protein zero, a glycoprotein that is a crucial structural component of the myelin sheath. Mutations in this protein are associated with the diseases Charcot-Marie-Tooth and Dejerine-Sottas. In most of their forms, these diseases are associated with demyelination. However, some forms such as Charcot-Marie-Tooth type II disease is associated with mutations in MYP0 that are not demyelinating neuropathies, despite the fact that they result in altered motor action potentials (Chapon et al. 1999). Thus, it is tempting to speculate that, besides their known cellular function, genes like MYP0 can have a more functional interaction with neuronal VGSCs.

5 Structural Features and Regions of Sequence Conservation

A recently determined structure of Nav1.4 from the electric eel, obtained through cryo-electron microscopy, provides some unique insights into possible mechanisms of action of the regulatory subunits (Yan et al. 2017). Presence of an interacting partner for Nav1.4 in the electron density was identified as β1. An interesting feature revealed by this structural model is the fact that the α- and β-subunits interact through both their transmembrane and extracellular domains (Fig. 3a). This was somewhat unanticipated given that previous structure-function studies suggested a predominate role of the β-subunit extracellular Ig domain such the transmembrane segment was predicted as a more generic interaction interface. Interestingly, β1 establishes extensive interactions with structural elements from three distinct domains from the sodium channel α-subunit: D_I, D_{III}, and D_{IV}. In particular, the extracellular part of β1 shows contacts with loop 5 from D_I and loop 6 from D_{IV}, in addition to a salt bridge with R1028 located on the S1–S2 loop of D_{III}. Besides these interactions between the two solvent exposed regions, β1 and Nav1.4 show a remarkably large number of residue-residue contacts in their hydrophobic transmembrane sections (Fig. 3b). The protein-protein interaction interface involves voltage sensor segments S0 and S2 of D_{III} and is mostly composed by interweaving bulky hydrophobic side chains.

A particularly intriguing set of residue-residue interactions between β1 and Nav1.4 are those involving polar side chains (Fig. 3c). As noted previously (Namadurai et al. 2015), the polar residues of β1 and β3 within the transmembrane segments act as potential oligomerization regions between β-subunits. Intriguingly, these polar interaction side chains are involved in seemingly specific recognition between the α- and the β-subunits. Particularly relevant is the hydrogen bond between Q174 from β1 and Y1043 from S2 of D_{III}: in spite of their marked polar character, these two side chains are located in the middle of the membrane, i.e., in a region completely inaccessible to water molecules and typically devoid of

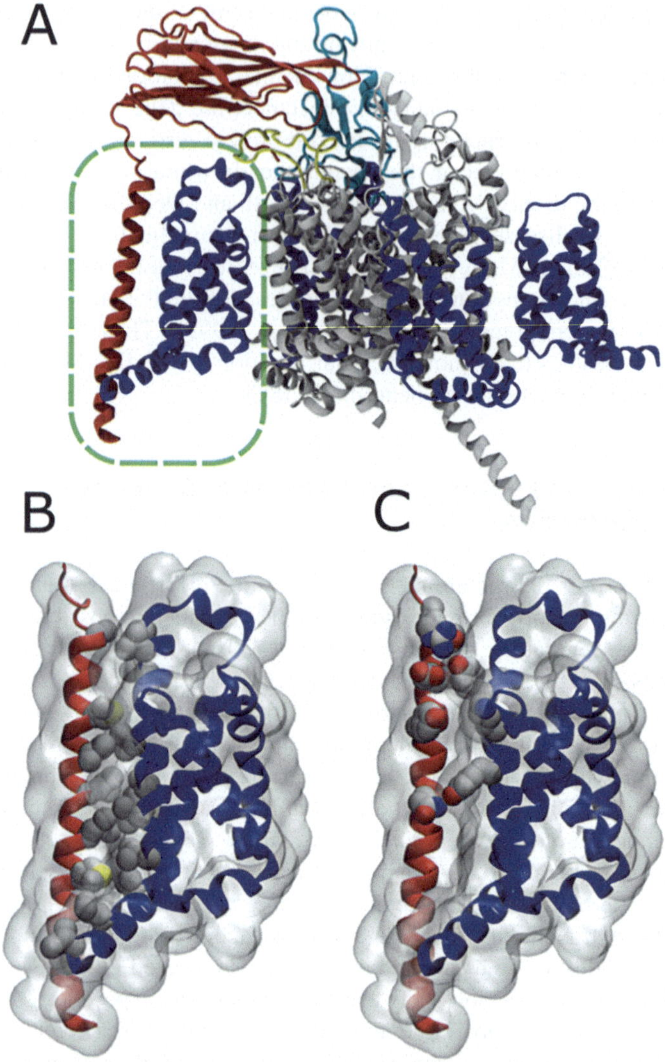

Fig. 3 Structure of the eel β1/Nav1.4 complex as determined by cryo-electron microscopy. (**a**) Cartoon representation of the complex showing β1 (red), the voltage-sensor domains (blue), the pore domain (white), the L5 loop (cyan), and the L6 loop (yellow). The green dashed line highlights the transmembrane section of the protein-protein complex constituted by the voltage-sensor domain of D_{III} and the transmembrane part of β1. (**b**, **c**) Close-up of the transmembrane region of the β1/Nav1.4 interaction surface. The gray shading shows the molecular surface of the two interacting partners; a space filling representation is used to highlight the nonpolar (**b**) and polar (**c**) side chains at the protein-protein interface

hydrogen bond donors or acceptors. The net free energy gain for establishing such interaction in this environment is typically large (between 2 and 5 kcal/mol); therefore, glutamine (and to a lesser extent asparagines) residue is often responsible for multimerization of transmembrane helices (Choma et al. 2000). In the context of the α/β complex, this residue-residue interaction may provide a structural determinant of the affinity between β1 and Nav1.4 and possibly determines the selectivity against other class of β-subunits like β2 and β4. In addition to the Q174-Y1043 contact, there are two other polar interactions involving the two proteins occurring between side chains located in the interfacial region of the lipid bilayer. The first one is established between Y167 from β1 and K1039 from S2, possibly established with an intervening water molecule or lipid head group. The second polar interaction involves an entire cluster of polar side chains from β1 (R155, S159, and E163) and the phenyl group of Y1036 on S2, whose hydroxyl group seemingly donates an H-bond to the carboxylate of E163 and accepts two H-bonds, one from R155 and the other from S159.

In light of the specific contacts highlighted by the cryo-EM structure, it is interesting to analyze the sequence conservation pattern of distinct β-subunits within multiple species. In particular, from a database-wide search for β-subunits homologues across all the sequenced organisms, one can characterize the degree of sequence conservation at each structural position and from this infer which of the contacts observed in the β1/Nav1.4 complex are crucial and have been thus preserved across evolution. The first interesting insights emerging from this analysis is that the transmembrane domain is one of the most conserved regions of the β-subunits for both the β1 and β3 and the β2 and β4 groups (Fig. 4a, b). The second interesting observation is that the polar amino acids involved in the interaction with S2 of Nav1.4 D_{III} appear to be strongly conserved (Fig. 4a). Importantly, amino acids corresponding to Q174 in the electric eel β1 are invariably polar: at this position, only glutamine or threonine side chains are observed in other species. An even stronger sequence conservation is observed at the positions corresponding to S159, E163, and Y167 of the electric eel β1. Importantly, the remaining conserved amino acids are also found to establish hydrophobic interactions in the β1/Nav1.4; these are M166, I170, and L177. As a result, a polar/nonpolar alternate pattern of conservation is apparent in the β1 and β3 family. Intriguingly, this pattern is absent in the β2 or β4 with the exception of a conserved serine amino acid, the side chains are mostly nonpolar. While it is impossible to draw strong conclusions from this information, it is tempting to speculate that the different functional role of β2 and β4 with respect to β1 and β3 might result from these differences in sequence, which, ultimately might result in a different binding mode to the α-subunit.

In conclusion, the overall understanding of voltage-gated sodium channel auxiliary subunits has benefited from a variety of biochemical, electrophysiological, and structural characterizations. An evolutionary analysis suggests homologues exist in viral bacteriophage proteins, with the modern manifestation of the beta-subunit gene family emerging within the existing nervous system of bony fish ~420 myo. Thus, it is clear that the VGSC β-subunit family has deeper roots in the foundation of neurobiology and, possibly, playing a supporting role in the formation of the

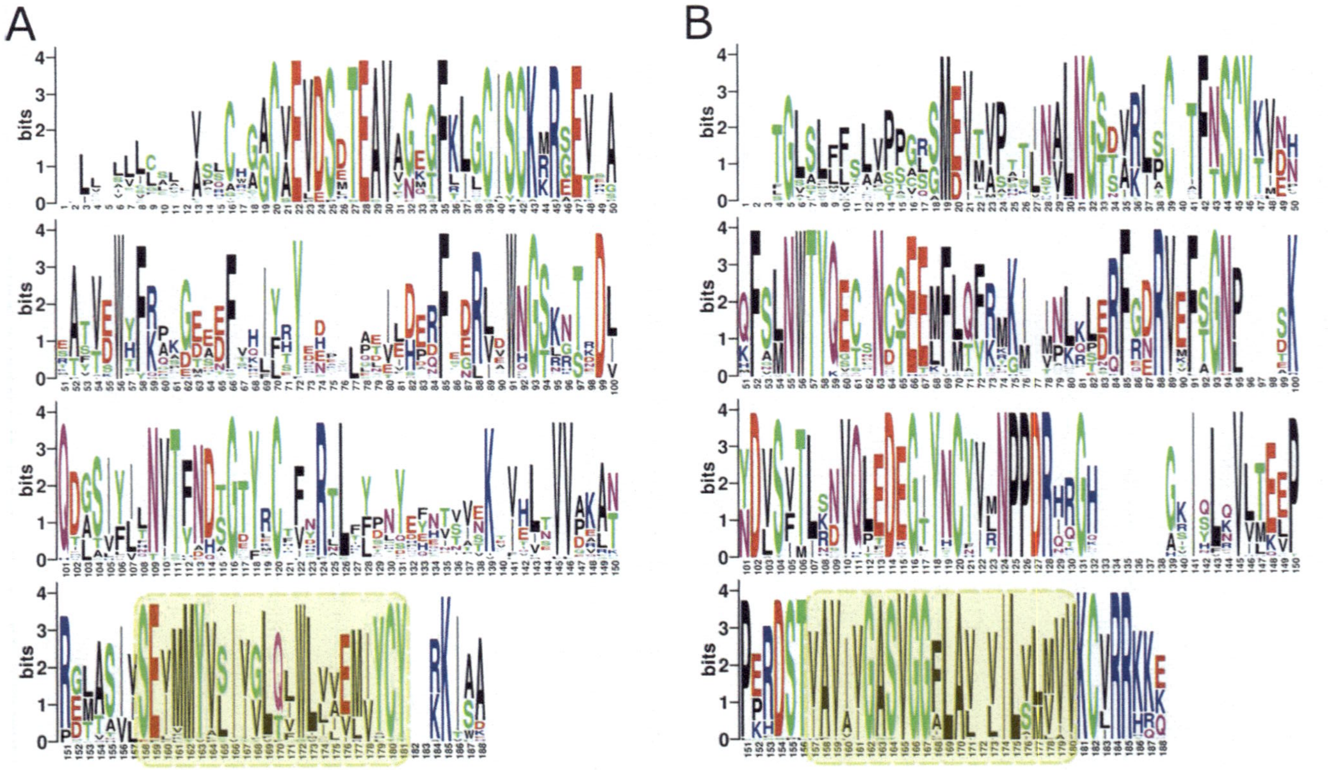

Fig. 4 Sequence conservation profile for the SCN1B/SCN3B (**a**) and SCN2B/SCNB4 (**b**) families. Shown are sequence logos: a concise graphical representation of sequence multiple sequence alignments. Amino acid one letter code is used to show the residues present in each position of the sequence. The height of each font is proportional to the frequency of occurrence of each amino acid, while the height of the entire column indicates the overall sequence conservation at a given position. The yellow shaded boxes highlight the transmembrane regions

nervous system in animals. Given their widespread expression, and multiple modulatory targets within excitable cells, they are associated with a plethora of human diseases. The advent of recent structural breakthroughs are beginning to provide a framework for how this class of diverse auxiliary subunit might regulate the voltage-dependent properties of ion channels and may herald a new era of efforts to guide structure-based therapeutics.

References

Aronica E et al (2003) Expression and regulation of voltage-gated sodium channel β1 subunit protein in human gliosis-associated pathologies. Acta Neuropathol 105(5):515–523

Audenaert D et al (2003) A deletion in SCN1B is associated with febrile seizures and early-onset absence epilepsy. Neurology 61:854–856

Bennett PB Jr, Makita N, George AL Jr (1993) A molecular basis for gating mode transitions in human skeletal muscle Na+ channels. FEBS Lett 326(1–3):21–24

Bork P, Holm L, Sander C (1994) The immunoglobulin fold. Structural classification, sequence patterns and common core. J Mol Biol 242(4):309–320

Brackenbury WJ (2012) Voltage-gated sodium channels and metastatic disease. Channels 6 (5):352–361

Brackenbury WJ et al (2008) Voltage-gated Na+ channel β1 subunit-mediated neurite outgrowth requires Fyn kinase and contributes to postnatal CNS development in vivo. J Neurosci 28 (12):3246–3256

Brackenbury WJ et al (2013) Abnormal neuronal patterning occurs during early postnatal brain development of Scn1b-null mice and precedes hyperexcitability. Proc Natl Acad Sci 110 (3):1089–1094

Buffington SA, Rasband MN (2013) Na+ channel-dependent recruitment of Navβ4 to axon initial segments and nodes of Ranvier. J Neurosci 33(14):6191–6202

Chapon F et al (1999) Axonal phenotype of Charcot-Marie-Tooth disease associated with a mutation in the myelin protein zero gene. J Neurol Neurosurg Psychiatry 66(6):779–782

Chen C et al (2002) Reduced sodium channel density, altered voltage dependence of inactivation, and increased susceptibility to seizures in mice lacking sodium channel β2-subunits. Proc Natl Acad Sci 99(26):17072–17077

Chen C et al (2004) Mice lacking sodium channel β1 subunits display defects in neuronal excitability, sodium channel expression, and nodal architecture. J Neurosci 24(16):4030–4042

Chen C et al (2012) Identification of the cysteine residue responsible for disulfide linkage of Na+ channel α and β2 subunits. J Biol Chem 287(46):39061–39069

Chioni A-M et al (2009) A novel adhesion molecule in human breast cancer cells: voltage-gated Na+ channel β1 subunit. Int J Biochem Cell Biol 41(5):1216–1227

Choma C et al (2000) Asparagine-mediated self-association of a model transmembrane helix. Nat Struct Biol 7(2):161–166

Das S et al (2016) Binary architecture of the Nav1.2-β2 signaling complex. eLife 5:e10960

Davis TH, Chen C, Isom LL (2004) Sodium channel β1 subunits promote neurite outgrowth in cerebellar granule neurons. J Biol Chem 279(49):51424–51432

Dermody TS et al (2009) Immunoglobulin superfamily virus receptors and the evolution of adaptive immunity. PLoS Pathog 5(11):e1000481

Deschênes I, Tomaselli GF (2002) Modulation of Kv4.3 current by accessory subunits. FEBS Lett 528(1–3):183–188

Dietrich PS et al (1998) Functional analysis of a voltage-gated sodium channel and its splice variant from rat dorsal root ganglia. J Neurochem 70:2262–2272

Diss J et al (2008) [Beta]-subunits of voltage-gated sodium channels in human prostate cancer: quantitative in vitro and in vivo analyses of mRNA expression. Prostate Cancer Prostatic Dis 11(4):325
Fendri-Kriaa N et al (2011) New mutation c.374C>T and a putative disease-associated haplotype within SCN1B gene in Tunisian families with febrile seizures. Eur J Neurol 18(5):695–702
Fraser JS, Maxwell KL, Davidson AR (2007) Immunoglobulin-like domains on bacteriophage: weapons of modest damage? Curr Opin Microbiol 10(4):382–387
Hakim P et al (2008) Scn3b knockout mice exhibit abnormal ventricular electrophysiological properties. Prog Biophys Mol Biol 98(2):251–266
Hartshorne RP, Catterall WA (1981) Purification of the saxitoxin receptor of the sodium channel from rat brain. Proc Natl Acad Sci 78(7):4620–4624
Hartshorne RP, Catterall WA (1984) The sodium channel from rat brain. Purification and subunit composition. J Biol Chem 259(3):1667–1675
Hartshorne RP et al (1985) Functional reconstitution of the purified brain sodium channel in planar lipid bilayers. Proc Natl Acad Sci U S A 82:240–244
Hu D et al (2009) A mutation in the β3 subunit of the cardiac sodium channel associated with Brugada ECG phenotype. Circ Cardiovasc Genet 2:270–278
Hu D et al (2012) A novel rare variant in SCN1Bb linked to Brugada syndrome and SIDS by combined modulation of Nav1.5 and Kv4.3 channel currents. Heart Rhythm 9(5):760–769
Ishikawa T et al (2013) Novel SCN3B mutation associated with Brugada syndrome affects intracellular trafficking and function of Nav1. 5. Circ J 77(4):959–967
Isom LL (2001) Sodium channel β subunits: anything but auxiliary. Neuroscientist 7(1):42–54
Isom LL (2002) The role of sodium channels in cell adhesion. Front Biosci 7(1):12–23
Isom LL, Catterall WA (1996) Na+ channel subunits and Ig domains. Nature 383(6598):307–308
Isom LL et al (1992) Primary structure and functional expression of the β subunit of the rat brain sodium channel. Science 256(5058):839–842
Isom LL et al (1995a) Structure and function of the β2 subunit of brain sodium channels, a transmembrane glycoprotein with a CAM motif. Cell 83(3):433–442
Isom LL et al (1995b) Functional co-expression of the 1 and type IIA subunits of sodium channels in a mammalian cell line. J Biol Chem 270(7):3306–3312
Kazarinova-Noyes K et al (2001) Contactin associates with Na+ channels and increases their functional expression. J Neurosci 21(19):7517–7525
Kim DY et al (2005) Presenilin/gamma-secretase-mediated cleavage of the voltage-gated sodium channel beta2-subunit regulates cell adhesion and migration. J Biol Chem 280:23251–23261
Kim DY et al (2007) BACE1 regulates voltage-gated sodium channels and neuronal activity. Nat Cell Biol 9(7):755–764
Laedermann CJ et al (2013) β1-and β3-voltage-gated sodium channel subunits modulate cell surface expression and glycosylation of Nav1. 7 in HEK293 cells. Front Cell Neurosci 7:137
Lesk AM, Chothia C (1982) Evolution of proteins formed by beta-sheets. II. The core of the immunoglobulin domains. J Mol Biol 160(2):325–342
Li RG et al (2013) Mutations of the SCN4B-encoded sodium channel β4 subunit in familial atrial fibrillation. Int J Mol Med 32(1):144–150
Lopez-Santiago LF et al (2007) Sodium channel SCN1B null mice exhibit prolonged QT and RR intervals. J Mol Cell Cardiol 43(5):636–647
Maier SKG et al (2004) Distinct subcellular localization of different sodium channel α and β subunits in single ventricular myocytes from mouse heart. Circulation 109(11):1421–1427
Makielski JC et al (1996) Coexpression of beta 1 with cardiac sodium channel alpha subunits in oocytes decreases lidocaine block. Mol Pharmacol 49(1):30–39
Makita N, Bennett PB, George AL (1996a) Multiple domains contribute to the distinct inactivation properties of human heart and skeletal muscle Na+ channels. Circ Res 78(2):244–252
Makita N, Bennett PB Jr, George AL Jr (1996b) Molecular determinants of B1 subunit-induced gating modulation in voltage-dependent Na+ channels. J Neurosci 16(22):7117–7127

Malhotra JD et al (2000) Sodium channel β subunits mediate homophilic cell adhesion and recruit ankyrin to points of cell-cell contact. J Biol Chem 275(15):11383–11388
Malhotra JD et al (2004) Tyrosine-phosphorylated and nonphosphorylated sodium channel β1 subunits are differentially localized in cardiac myocytes. J Biol Chem 279(39):40748–40754
Marionneau C et al (2012) The sodium channel accessory subunit Navbeta1 regulates neuronal excitability through modulation of repolarizing voltage-gated K(+) channels. J Neurosci 32 (17):5716–5727
McCormick KA et al (1998) Molecular determinants of Na+ channel function in the extracellular domain of the β1 subunit. J Biol Chem 273(7):3954–3962
McEwen DP, Isom LL (2004) Heterophilic interactions of sodium channel β1 subunits with axonal and glial cell adhesion molecules. J Biol Chem 279(50):52744–52752
Medeiros-Domingo A et al (2007) SCN4B-encoded sodium channel β4 subunit in congenital long-QT syndrome. Circulation 116(2):134–142
Messner DJ, Catterall WA (1985) The sodium channel from rat brain. Separation and characterization of subunits. J Biol Chem 260(19):10597–10604
Morgan K et al (2000) β3: an additional auxiliary subunit of the voltage-sensitive sodium channel that modulates channel gating with distinct kinetics. Proc Natl Acad Sci 97(5):2308–2313
Namadurai S et al (2015) A new look at sodium channel β subunits. Open Biol 5(1):140192
Nelson M et al (2014) The sodium channel β1 subunit mediates outgrowth of neurite-like processes on breast cancer cells and promotes tumour growth and metastasis. Int J Cancer 135(10):2338–2351
Nguyen HM et al (2012a) Modulation of voltage-gated K+ channels by the sodium channel β1 subunit. Proc Natl Acad Sci 109(45):18577–18582
Nguyen HM et al (2012b) Modulation of Kv1 voltage-gated potassium channels by sodium channel beta subunits. Biophys J 102(3, Suppl 1):687a
Ogiwara I et al (2012) A homozygous mutation of voltage-gated sodium channel β1 gene SCN1B in a patient with Dravet syndrome. Epilepsia 53(12):e200–e203
Oh Y, Waxman SG (1994) The beta 1 subunit mRNA of the rat brain Na+ channel is expressed in glial cells. Proc Natl Acad Sci U S A 91:9985–9989
Patino GA et al (2009) A functional null mutation of SCN1B in a patient with Dravet syndrome. J Neurosci 29(34):10764–10778
Patino GA et al (2011) Voltage-gated Na(+) channel β1B: a secreted cell adhesion molecule involved in human epilepsy. J Neurosci 31(41):14577–14591
Patton DE et al (1994) The adult rat brain β1 subunit modifies activation and inactivation gating of multiple sodium channel alpha subunits. J Biol Chem 269(26):17649–17655
Qin N et al (2003) Molecular cloning and functional expression of the human sodium channel β1B subunit, a novel splicing variant of the β1 subunit. Eur J Biochem 270(23):4762–4770
Qu Y et al (1995) Modulation of cardiac Na+ channel expression in Xenopus oocytes by β1 subunits. J Biol Chem 270(43):25696–25701
Qu Y et al (1999) Functional roles of the extracellular segments of the sodium channel α subunit in voltage-dependent gating and modulation by β1 subunits. J Biol Chem 274(46):32647–32654
Ransdell JL et al (2017) Loss of Navβ4-mediated regulation of sodium currents in adult Purkinje neurons disrupts firing and impairs motor coordination and balance. Cell Rep 19(3):532–544
Ratcliffe CF et al (2001) Sodium channel β1 and β3 subunits associate with neurofascin through their extracellular immunoglobulin-like domain. J Cell Biol 154(2):427–434
Riuró H et al (2013) A missense mutation in the sodium channel β2 subunit reveals SCN2B as a new candidate gene for Brugada syndrome. Hum Mutat 34(7):961–966
Riuró H et al (2014) A missense mutation in the sodium channel β1b subunit reveals SCN1B as a susceptibility gene underlying long QT syndrome. Heart Rhythm 11(7):1202–1209
Roberts RH, Barchi RL (1987) The voltage-sensitive sodium channel from rabbit skeletal muscle. Chemical characterization of subunits. J Biol Chem 262(5):2298–2303
Scheffer IE et al (2007) Temporal lobe epilepsy and GEFS+ phenotypes associated with SCN1B mutations. Brain 130(1):100–109

Schmidt J, Rossie S, Catterall WA (1985) A large intracellular pool of inactive Na channel alpha subunits in developing rat brain. Proc Natl Acad Sci 82(14):4847–4851

Shapiro L et al (1996) Crystal structure of the extracellular domain from P0, the major structural protein of peripheral nerve myelin. Neuron 17:435–449

Sharkey RG, Beneski DA, Catterall WA (1984) Differential labeling of the α and β1 subunits of the sodium channel by photoreactive derivatives of scorpion toxin. Biochemistry 23 (25):6078–6086

Shcherbatko A et al (1999) Voltage-dependent sodium channel function is regulated through membrane mechanics. Biophys J 77(4):1945–1959

Smith RD, Goldin AL (1998) Functional analysis of the rat I sodium channel in Xenopus oocytes. J Neurosci 18(3):811–820

Srinivasan J, Schachner M, Catterall WA (1998) Interaction of voltage-gated sodium channels with the extracellular matrix molecules tenascin-C and tenascin-R. Proc Natl Acad Sci U S A 95(26):15753–15757

Tan B-H et al (2010) Sudden infant death syndrome–associated mutations in the sodium channel beta subunits. Heart Rhythm 7(6):771–778

Uebachs M et al (2012) Loss of β1 accessory Na+ channel subunits causes failure of carbamazepine, but not of lacosamide, in blocking high-frequency firing via differential effects on persistent Na+ currents. Epilepsia 53(11):1959–1967

Valdivia CR et al (2009) Loss-of-function mutation of the SCN3B-encoded sodium channel β3 subunit associated with a case of idiopathic ventricular fibrillation. Cardiovasc Res 86 (3):392–400

Wallace RH et al (1998) Febrile seizures and generalized epilepsy associated with a mutation in the Na+-channel sz1 subunit gene SCN1B. Nat Genet 19(4):366–370

Wallner M et al (1993) Modulation of the skeletal muscle sodium channel α-subunit by the β1-subunit. FEBS Lett 336(3):535–539

Wang P et al (2010) Functional dominant-negative mutation of sodium channel subunit gene SCN3B associated with atrial fibrillation in a Chinese GeneID population. Biochem Biophys Res Commun 398(1):98–104

Watanabe H et al (2008) Sodium channel β1 subunit mutations associated with Brugada syndrome and cardiac conduction disease in humans. J Clin Invest 118(6):2260–2268

Watanabe H et al (2009) Mutations in sodium channel β1- and β2-subunits associated with atrial fibrillation. Circ Arrhythm Electrophysiol 2(3):268–275

Wilson MJ et al (2011) Navβ subunits modulate the inhibition of Nav1.8 by the analgesic gating modifier μO-conotoxin MrVIB. J Pharmacol Exp Ther 338(2):687–693

Wilson MJ et al (2015) A-and β-subunit composition of voltage-gated sodium channels investigated with μ-conotoxins and the recently discovered μO §-conotoxin GVIIJ. J Neurophysiol 113 (7):2289–2301

Wong H-K et al (2005) β subunits of voltage-gated sodium channels are novel substrates of β-site amyloid precursor protein-cleaving enzyme (BACE1) and γ-secretase. J Biol Chem 280 (24):23009–23017

Xiao Z-C et al (1999) Tenascin-R is a functional modulator of sodium channel β subunits. J Biol Chem 274(37):26511–26517

Xu R et al (2007) Generalized epilepsy with febrile seizures plus–associated sodium channel β1 subunit mutations severely reduce β subunit–mediated modulation of sodium channel function. Neuroscience 148(1):164–174

Yan Z et al (2017) Structure of the Nav1.4-β1 complex from electric eel. Cell 170(3):470–482

Yu FH et al (2003) Sodium channel β4, a new disulfide-linked auxiliary subunit with similarity to beta2. J Neurosci 23(20):7577–7585

Yuan L et al (2014) Investigations of the Navβ1b sodium channel subunit in human ventricle; functional characterization of the H162P Brugada syndrome mutant. Am J Phys Heart Circ Phys 306(8):H1204–H1212

Zhang MM et al (2013) Co-expression of Navβ subunits alters the kinetics of inhibition of voltage-gated sodium channels by pore-blocking μ-conotoxins. Br J Pharmacol 168(7):1597–1610

Zhao J, O'Leary ME, Chahine M (2011) Regulation of Nav1.6 and Nav1.8 peripheral nerve Na+ channels by auxiliary β-subunits. J Neurophysiol 106(2):608–619

Zhou T-t et al (2012) Glycosylation of the sodium channel β4 subunit is developmentally regulated and involves in neuritic degeneration. Int J Biol Sci 8(5):630

Zhu W et al (2017) Mechanisms of noncovalent β subunit regulation of Nav channel gating. J Gen Physiol 149:813–831

Part II

The Structural Basis of Sodium Channel Function

Structural and Functional Analysis of Sodium Channels Viewed from an Evolutionary Perspective

Tamer M. Gamal El-Din, Michael J. Lenaeus, and William A. Catterall

Contents

Abstract

Voltage-gated sodium channels initiate and propagate action potentials in excitable cells. They respond to membrane depolarization through opening, followed by fast inactivation that terminates the sodium current. This ON-OFF behavior of voltage-gated sodium channels underlays the coding of information and its transmission from one location in the nervous system to another. In this review, we explore and compare structural and functional data from prokaryotic and eukaryotic channels to infer the effects of evolution on sodium channel structure and function.

The online version of this article (https://doi.org/10.1007/164_2017_61) contains supplementary material, which is available to authorized users.

T. M. Gamal El-Din (✉) · M. J. Lenaeus · W. A. Catterall
Department of Pharmacology, University of Washington, Seattle, WA 98195-7280, USA
e-mail: tmgamal@uw.edu

M. Chahine (ed.), *Voltage-gated Sodium Channels: Structure, Function and Channelopathies*, Handbook of Experimental Pharmacology 246,
https://doi.org/10.1007/164_2017_61

Keywords
Activation mechanisms · Bacterial sodium channels · Eukaryotic sodium channels · Evolution of sodium channels · Gating mechanisms · Inactivation mechanisms · Selectivity of sodium channels

1 Introduction

The eukaryotic voltage-gated sodium (Na^+) channel is composed of a complex containing a pore-forming α-subunit and up to two β-subunits (Ahern et al. 2016; Catterall 2012; Catterall and Zheng 2015). The pore-forming α-subunit is sufficient to form a functional sodium channel, while the β-subunits are known to modulate the kinetics and voltage dependence of Na^+ channel activation and inactivation (Isom et al. 1995; Qu et al. 1995). Nine different isoforms of voltage-gated sodium (Na_V) channel α-subunits and four Na_V channel β-subunits have been discovered (Catterall et al. 2005). The central nervous system (CNS) expresses four main isoforms: $Na_v1.1$, $Na_v1.2$, $Na_v1.3$, and $Na_v1.6$. Peripheral nervous system (PNS) has three different isoforms $Na_v1.7$, $Na_v1.8$, and $Na_v1.9$. These three subtypes of Na_V channels are mainly associated with autonomic regulation and pain sensation. In skeletal muscles, $Na_v1.4$ is the principal Na_V channel, while $Na_v1.5$ is the primary Na_V channel in cardiac muscle. In each of these channels, the α-subunit is composed of ~2,000 amino acid residues that form four homologous domains (traditionally numbered as I–IV), each with six transmembrane helices (Fig. 1a, numbered S1–S6) (Ahern et al. 2016; Catterall 2012). The fourth transmembrane helix (S4) in each domain has 4–8 positive residues (R or K) that sense changes in voltage, while transmembrane helices S5 and S6 from each domain form the pore-lining residues of the channel (Bezanilla 2000; Catterall 2010). Eukaryotic Na_V channels are subject to broad posttranslational modifications including phosphorylation, glycosylation, palmitoylation, ubiquitination, and methylation (Catterall 1986a). This reflects the diversity of eukaryotic Na_V channel expression and the role played by posttranslational modification in fine-tuning its function. Recently published structures of Na_V channels from American cockroach (Na_vPas) and electric eel electroplax ($Na_v1.4$) reveal the subunit architecture and structural modules of eukaryotic Na_V channels (Shen et al. 2017; Yan et al. 2017).

Compared to the complexity of mammalian Na_V channels, bacterial Na_V channels are composed of four identical subunits of ~250–270 residues. Every subunit has a voltage sensor domain (VSD) and a pore domain (PD) (Payandeh et al. 2011; Ren et al. 2001). They share the major biophysical features with

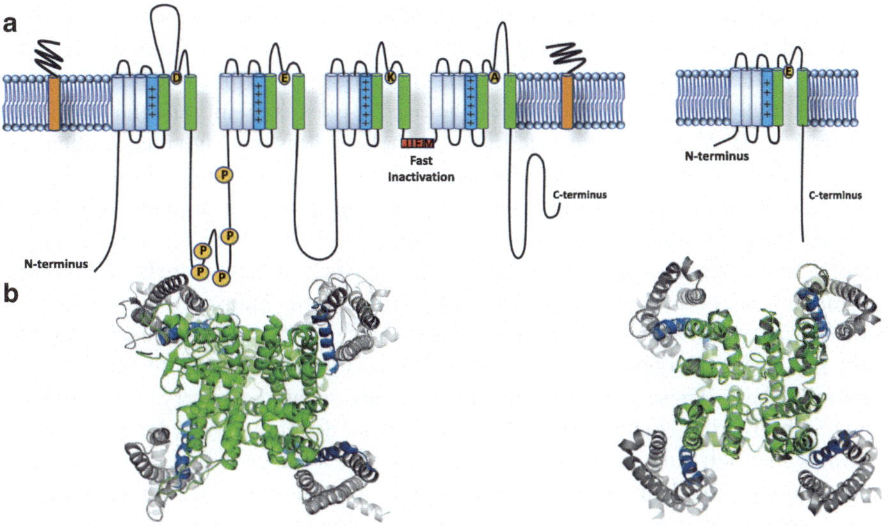

Fig. 1 Overall structures of eukaryotic and prokaryotic sodium channels. (**a**) Topology of a eukaryotic voltage-gated sodium channel showing four homologous domains. Each domain consists of six segments. Voltage-sensing segments (S1–S4) are shown in *gray* with S4 segment depicted in *marine blue*. The pore domain (S5–S6) is shown in *light green*. *Right*, topology of a single domain of a bacterial sodium channel. (**b**) *Top* view of the overall model of the eukaryotic voltage-gated sodium channel NavPas (*left*) (5X0M) and bacterial sodium channel NavAb (*right*) (PDB 3RVZ). Key structural and functional features of Na_vPas (*left*) and Na_vAb (*right*) channels are color labeled including the voltage sensor domain (S1–S3, *green*; S4, *marine blue*), pore domain (*light green*)

eukaryotic counterparts (Catterall and Zheng 2015). In contrast to eukaryotic channels, bacterial Na_V channels lack auxiliary subunits and posttranslational modifications. Crystallization of full-length bacterial Na_V channels like Na_vAb, Na_vRh, and Na_vMs and analysis of their structures at high resolution (Payandeh et al. 2011, 2012; Sula et al. 2017; Zhang et al. 2012) make them invaluable models for studying the structural basis of ion conduction, activation, inactivation, and drug interaction. In this review, we use an evolutionary perspective to analyze the differences and similarities between bacterial and eukaryotic Na_V channels and how they might have evolved over billions of years.

2 The Voltage-Sensing Module

2.1 The Excitable Membrane and the Voltage Sensor

Na_V channels are integral membrane proteins that rely on the potential generated by selective ion conductance and unequal concentrations of ions across the plasma membrane (Hille 2001). This separation of ions across the lipid bilayer plus high resting membrane permeability to K^+ results in a membrane potential that ranges from −40 to −95 mV depending on the cell type (measured from inside with respect to the outside). Since the plasma membrane is a lipid bilayer of ~30 Å thickness, this produces an electric field across the plasma membrane that can be as high as 3×10^7 V/m. Voltage-gated ion channels exploit this high electric field by coupling protein conformational changes to physiological changes in the electric field strength. From a structural standpoint, the voltage sensor domains of eukaryotic and bacterial Na_V channels are very similar in that there are four voltage-sensing units per channel and each voltage-sensing domain consisting of four transmembrane helices numbered S1–S4 by convention. Transmembrane helix S4 is the key player in sensing the voltage across the cell membrane by placing four to eight positively charged R or K residues across the membrane (Bezanilla 2000; Catterall 2010). These gating charge residues are located at every third position along S4 high order channels (Fig. 2). An important difference between prokaryotic channels and their eukaryotic counterparts is that the number of charged S4 residues is fixed among the four subunits in prokaryotic, homotetrameric channels but can vary a great deal among the subunits and homologous domains of higher-order channels along S4 (Fig. 2). Crystal structures show that the S1, S2, and S3 helices of the voltage sensor module surround the S4 segment and form a gating canal, which facilitates the movement of S4 (Payandeh et al. 2011, 2012; Zhang et al. 2012).

The S1, S2, and S3 helices contain polar or negative residues that create the gating canal that catalyzes the movement of S4 from its inward resting-state position to progressively more activated states upon depolarization (Fig. 2a, b). The positively charged residues in the S4 segments were proposed to form ion pairs with polar/negatively charged amino acid residues present in these surrounding helices (Catterall 1986a, b; Guy and Seetharamulu 1986). S1 has an N residue (N25 in Na_vAb numbering) that is highly conserved in bacterial and eukaryotic sodium channels (Fig. 1S). Another less conserved residue is E32 (Na_vAb numbering), which is present in some bacterial Na_V channels and in two or three domains of eukaryotic Na_V channels. The S2 segment has two negative (or one negative and one polar) residues separated by nine, mostly hydrophobic, residues. On the N-terminal side, either D, E, or N is present, and on the C-terminal, E is always present. In-between these two negative residues, a set of hydrophobic residues (including F56 in NavAb numbering) serve as a hydrophobic *plug* that prevents ionic leak through VSD. This hydrophobic constriction site (HCS, Fig. 2) is conserved among voltage sensors from many different proteins, including K_v, Na_v, Ca_v, proton channels, and voltage-dependent phosphatase (VSP) enzymes (Li et al. 2014; Payandeh et al. 2011; Tao et al. 2010; Wu et al. 2016). In the case of segment S3, E80 (Na_vAb numbering) is a very conserved residue in bacterial and eukaryotic Na_V channels. The high sequence identity of residues within VSD indicates that gating has been conserved through evolution (Figs. 1S and 2a).

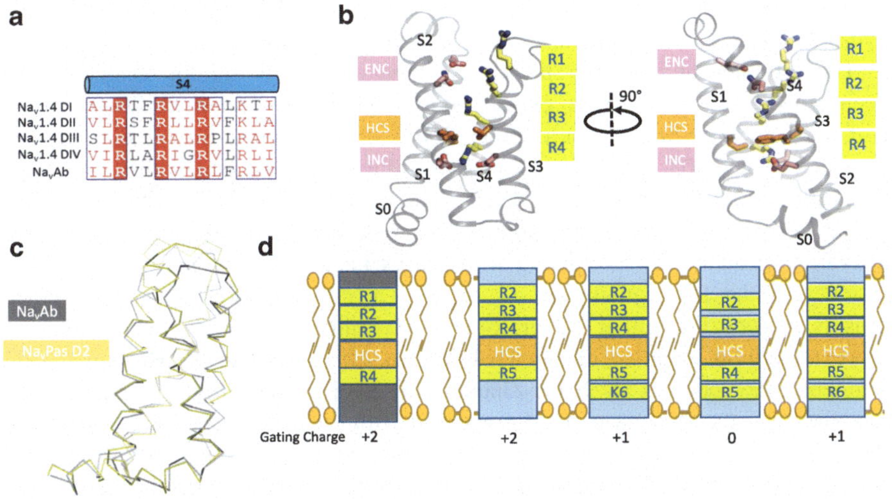

Fig. 2 The prokaryotic voltage sensor shares structural features with eukaryotic channels. (**a**) Sequence alignment of S4 segment of Na$_v$Ab with Na$_v$1.4 showing the conservation of residues discussed in the text. (**b**) Structural model of Na$_v$Ab in orthogonal views, highlighting residues and structural regions discussed in the text. Helices are shown as ribbon models, while important side chains are shown as sticks. R1–R4 refer to voltage-sensing arginine residues; ENC, HCS, and INC refer to extracellular negative cluster, hydrophobic constriction site, and intracellular negative cluster, respectively. (**c**) Structure-based alignment of the voltage sensor of Na$_v$Ab and Domain II of the recently solved Na$_v$Pas structure. Alignment was performed using the entire voltage sensor sequence of each respective protein. (**d**) Cartoon model highlighting the variability of gating charge positions in Na$_v$Pas when compared to Na$_v$Ab. The *gray* model represents the S4 helix of Na$_v$Ab, while the *blue* helices represent those of Na$_v$Pas. Gating charges are shown as *yellow boxes*, the HCS is shown as an *orange box*, and lipids are drawn in a cartoon format. We have assigned a gating charge position to each S4 helix, shown in the line below the diagram and meant to highlight the different position of gating charges in DI–DIV on Na$_v$Pas

2.2 A Conserved Mechanism of Activation

There is a consensus among voltage-gated ion channel biophysicists that the electromechanical coupling between the voltage sensor module and the activation gate of the pore domain follows a sliding-helix mechanism (Vargas et al. 2012). The S4 segment serves as the main voltage sensor by virtue of its high concentration of positively charged residues (Catterall 1986a). Mutagenesis studies of these residues resulted in reduction of the steepness of voltage-dependent gating and a shift of its voltage dependence (Logothetis et al. 1992; Papazian et al. 1991; Stuhmer et al. 1989). The S1–S3 segments form the structure of the gating pore through which S4 moves from its resting to the activated state (Catterall 2010). The shape of the voltage-sensing module looks like an hour glass, with a large extracellular aqueous cleft (~10 Å) and a smaller intracellular cleft (Fig. 2b). The hydrophobic constriction site (HCS, Fig. 2b) is present between these two aqueous clefts.

Very highly conserved residues (I in S1, F in S2, and V or I in S3) form this hydrophobic seal, which prevents ionic leak through voltage-sensing module.

The activated and resting states of voltage sensors of sodium channels have been probed with gating pore current studies (Gamal El-Din et al. 2010; Gamal El-Din et al. 2014; Sokolov et al. 2005, 2008), disulfide cross-linking experiments (DeCaen et al. 2008, 2009, 2011), and substituted cysteine accessibility methods (Larsson et al. 1996; Yang et al. 1996). These studies showed that the R3 or R4 gating charges seal the gating pore at the hydrophobic constriction site during activation, probably by interacting with nearby negatively charged or hydrophilic residues in the S2 segment (E, D, or N). Crystal structures of many voltage sensors showed that either R3 or R4 has proximity to the outermost residue in the extracellular negative cluster (N49 in Na_vAb or D48 in Na_vRh) on the S2 segments (Payandeh et al. 2011; Zhang et al. 2012). Interestingly, computational models using the Rosetta method combined with disulfide cross-linking revealed the existence of three different activated states of the bacterial sodium channel NaChBac (Yarov-Yarovoy et al. 2012). In activated state I, gating charges R1 and R2 interact with the external negative cluster, while R3 is located between N49 and F56. In activated state II, R1 is solvated in the extracellular aqueous cleft, while R2 interacts with E43 and R3 interacts with N49. In activated state III, R1 is still solvated, R2 and R3 interact with E43 and N49, respectively, and R4 passes through the hydrophobic constriction site and is located just on the extracellular side of F56.

Lack of a crystal structure of a Na_V channel in the resting state has placed primary emphasis on experimental and computational methods to assist in seeing the unseen. Fluorometric studies have shown that the electric field is focused around the part of the S4 segment that interacts with the hydrophobic constriction site (Ahern and Horn 2005; Asamoah et al. 2003). The substituted cysteine accessibility method (SCAM), in which the gating charge of interest is mutated to cysteine and perfusion of cysteine-reactive agent (MTSET, MTSES, or MTSEA) probes its aqueous accessibility, showed that R1 is not accessible from either side in mammalian or bacterial Na_V channels (Blanchet and Chahine 2007; Yang and Horn 1995), which means that it is buried within the gating canal. R2, on the other hand, was accessible from both sides in the resting state, indicating that it can move through the hydrophobic constriction site in a pair of structurally similar resting states. Mutation of R1, R2, or both to neutral, shorter, and uncharged residues causes a leak current, the gating pore current, through the mutated voltage-sensing module. R1 and R2 gating pore currents are active in the resting state and shut off in the activated state (Gamal El-Din et al. 2010, 2014; Sokolov et al. 2005; Starace and Bezanilla 2001, 2004).

Based on the available data regarding resting and activated states of the voltage sensor, we can now consider the current evidence regarding the gating transition between these states. When the potential difference across the plasma membrane is reduced during depolarization, the voltage-sensing modules undergo a conformation transition from the resting state to the activated state. During this transition, sequential formation and breakage of salt bridges and hydrogen-bonding interactions between arginine residues (R1–R4) on S4 and the negatively charged and polar residues on neighboring segments S1–S3 mediate low-energy passage of the S4 gating charges through the hydrophobic constriction site, as posited in the sliding helix or helical screw models (Catterall 1986a, b; Guy and Seetharamulu 1986). This transition of gating charges from resting to activated state is equivalent to moving ~8–14 elementary charges across the membrane electric field (Bezanilla 2000; Catterall 2010; Gamal El-Din et al. 2008; Hirschberg et al. 1995). Gating pore current and disulfide-locking experiments tracked this journey of the S4 segment from resting to activated states (Gamal El-Din et al. 2010, 2014; DeCaen et al. 2008, 2009, 2011). There is agreement that the S4 helix slides ~10 Å outward through the gating pore formed by the S1, S2, and S3 segments, accompanied by ~30° rotation and a sideways tilt at the pivot point in the hydrophobic constriction site (Vargas et al. 2012; Yarov-Yarovoy et al. 2012). All the resting-state models show the positive gating charge residues along S4 are in position to form salt bridges with acidic residues in the S1–S3 helices or interact with the aqueous regions of the lipid head groups. These conclusions agree with the experimental data on sodium channels obtained from the disulfide cross-linking experiments (DeCaen et al. 2008, 2009, 2011; Yarov-Yarovoy et al. 2012) and measurements of gating pore currents (Gamal El-Din et al. 2014).

Even though the overall mechanism of S4 activation is very similar between the bacterial and eukaryotic Na_V channels, the details of this transition differ in several ways. First, the position of the resting state of each voltage sensor relative to the activation gate is different (Gosselin-Badaroudine et al. 2012). Second, the voltage sensors of eukaryotic Na_V channels are in a different axial position in the plasma membrane relative to the prokaryotic voltage sensor and often in different positions relative to each other (Shen et al. 2017; Yan et al. 2017). These differential positions may enable these channels to activate faster and tune individual voltage sensors for specific aspects of sodium channel function (Fig. 2d). The result of these different resting and activated states may be stabilization of the channel in various conformations over a specific range of membrane potentials as required for the physiological niche for each Na_V channel.

3 The Selectivity Filter: From Symmetry to Asymmetry

It is intuitive that ion channels could attract cations with a cluster of negative charges (Hille 2001), and, indeed, the structure of the selectivity filter of the prokaryotic Na_V channels showed just such a mechanism (Payandeh et al. 2011). These channels are considered the ancestors of eukaryotic Na_V and Ca_V channels, and each contains a quartet of negatively charged E residues arranged around the selectivity filter, confirming this idea (Payandeh et al. 2011; Ren et al. 2001). The selectivity filter itself is formed of four identical segments, located in the "turn" in the "P1 helix-turn-P2 helix" structure from each domain. The selectivity filter of bacterial Na_V channels has the signature sequence motif TLX_1SWX_2, where X_1 is the site of high-field-strength anionic site in most channels. The only exception is

Na_vMs, where X_2 is the high-field-strength site instead of X_1. The selectivity filter in Na_vAb has the sequence (TL*E*SWSM). It has been suggested that the carboxylate groups of the four E residues (one from each domain) form an extracellular symmetric acidic coordination site ($Site_{HFS}$), while the backbone carbonyls of the four L and T residues form the central and inner coordination sites ($Site_{CEN}$ and $Site_{IN}$), respectively. The crystal structure of Na_vMs showed three Na^+ at almost the same sites proposed for Na_VAb (Naylor et al. 2016). Ordered water molecules were observed in the Na_VAb and Na_VMs structures in the lower part of the selectivity filter. Molecular dynamics studies elucidated the catalytic mechanism of Na^+ conduction (Chakrabarti et al. 2013; Ulmschneider et al. 2013). Remarkably, highly degenerate but energetically favorable dunking motions, in which one to three of the carboxylate side chains at $Site_{HFS}$ bend at a single torsion angle and move inward accompanying partially, dehydrated Na^+ (Chakrabarti et al. 2013).

While mammalian Na_V channels are virtually impermeable to Ca^{2+} under physiological conditions, bacterial Na_V channels show a significant permeability for Ca^{2+}, as demonstrated by permeability ratios of P_{Ca}/P_{Na} of 0.15, 0.14, 0.27, and 0.2 for NaChBac, Na_vRh, Na_vAb, and Na_vMs, respectively (Naylor et al. 2016; Tang et al. 2014; Yue et al. 2002; Zhang et al. 2012). This indicates that ancient bacterial Na_V channels might have served as a conductance pathway for both Na^+ and Ca^{2+}. Over billions of years of evolution, the homomeric structure of eukaryotic Na_V channels was converted into a single polypeptide chain by covalently linking the C- and N-termini of two neighboring subunits. This resulted in loss of the fourfold symmetry of the pore and selectivity filter, as well as the signature motif at $Site_{HFS}$. This stepwise evolution is reflected in the many different signature motifs at $Site_{HFS}$ in invertebrate sodium channels: EEEE, EEKE, EEAA, DEAA, etc. (Stephens et al. 2015). The signature motif EEEE has remained the same for Ca_v channels, the position of the carbon backbone stayed the same, but the geometry of the locations of the E residues side chains changed (Wu et al. 2016) (Fig. 3). On the other hand, the evolution of bacterial to eukaryotic Na_V channels involved changes in the chemical signature of residues that select Na^+ to give much higher selectivity. K appeared in Domain III as the residue conferring higher selectivity for Na^+ and preventing the permeability of Ca^{2+} (Favre et al. 1996). Studies that target this residue in Na_V channels by mutating it to E rescued Ca_V conductance (Heinemann et al. 1992). Overall, the structure of the carbon backbone of the high-field-strength site of the cockroach NavPas channel resembles that of the quartet of E residues in Na_vAb, though the exact placement of the side chains is uncertain (Fig. 3b).

Two amino acids have been conserved throughout the evolution process, W179 and T175 (Na_VAb numbering). T175 forms a hydrogen bond with W179 of a neighboring subunit. This bond is a key interaction site to connect adjacent subunits at the selectivity filter and to stabilize the overall structure of the selectivity filter (Payandeh et al. 2011).

The apparent similarity of Na^+ and Ca^{2+} permeation through eukaryotic ion channels indicate that they have evolved from the same ancestor. In both cases, Na^+ and Ca^{2+} have to be partially hydrated in order to move through the relatively

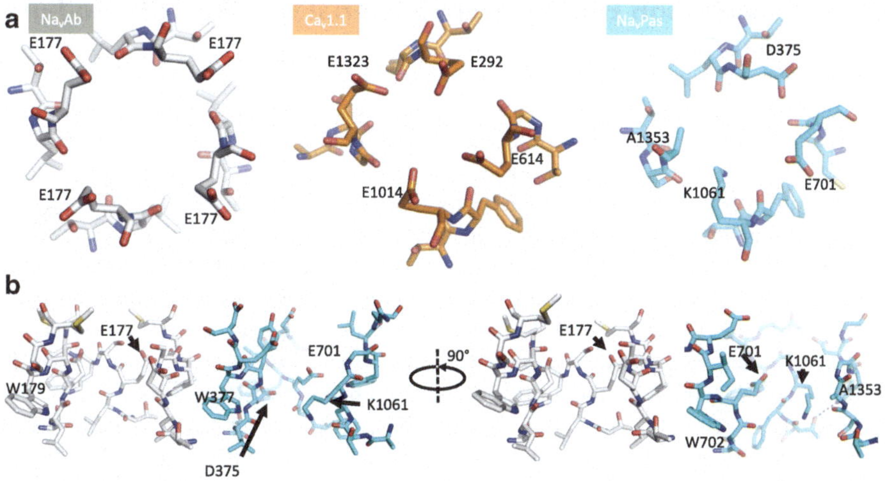

Fig. 3 The selectivity filters of Na_vAb, $Ca_v1.1$, and Na_vPas. (**a**) Comparing the selectivity filters from a view above the plane of the membrane. Each selectivity filter is shown in stick format, with the residue equivalent to HFS2 (E177 in Na_vAb) highlighted. Na_vAb is shown in *gray*, $Ca_v1.1$ in *orange*, and Na_vPas in *light blue*. (**b**) Side view of Na_vAb and Na_vPas selectivity filters highlighting the position of residues important for selectivity (E177 in Na_vAb, D375, E701, K1061, A1353 in Na_vPas) and overall structure of the selectivity filter (W179 in Na_vAb and W377 and W702 in Na_vPas are highlighted)

wide selectivity filter (Hille 1971; Naylor et al. 2016; Payandeh et al. 2011; Tang et al. 2014; Wu et al. 2016) (Fig. 3). This is fundamentally different from K^+ channels where K^+ ions pass completely dehydrated through the selectivity filter (Zhou et al. 2001). Moreover, in both Na_V and Ca_V channels, there is an extracellular cluster of negatively charged residues that recruit cations to the selectivity filter. In contrast, K_V channels use only interactions with backbone carbonyls to select and conduct K^+ (Naylor et al. 2016; Wu et al. 2016). Another major difference between permeation in eukaryotic versus bacterial Na_V and Ca_V channels on one hand and K_V channels on the other hand is flexibility versus rigidity of the selectivity filter. The finding that the size of the selectivity filter of Na_V and Ca_V channels is much wider than K_V channels indicates that the side chains of the surrounding residues have more degrees of freedom. Classic studies of Ca_V channels suggested that carboxylate side chains of the EEEE locus move independently to form a high-affinity binding site for incoming Ca^{2+} (Sather and McCleskey 2003; Yang et al. 1993). Molecular dynamics studies of Na_vAb showed that the E side chains of $Site_{HFS}$ undergo a "dunking" motion with each permeating Na^+ (Chakrabarti et al. 2013). In contrast, the relative tightness of the K_V channel selectivity filter requires backbone carbonyls to select the right ions, while the side chains of the selectivity filter residues point away from the lumen of the pore (Morais-Cabral et al. 2001; Zhou et al. 2001).

4 Inactivation Evolved from Bacterial to Eukaryotic Sodium Channels

In contrast to the conserved mechanism of activation between bacterial Na_V channels and their eukaryotic counterparts, inactivation mechanisms have evolved over billions of years to become faster and more efficient as required for the rapid information processing in excitable cells of multicellular organisms. Eukaryotic Na_V channels have two main types of inactivation: fast inactivation that occurs on a time scale of 1–5 ms (Hodgkin and Huxley 1952a, c, d) and slow inactivation that occurs on a time scale of hundreds of milliseconds to seconds (Adelman and Palti 1969; Palti and Adelman 1969; Rudy 1975, 1981). Mechanistically, fast inactivation takes place when a short peptide located in the segment between domains III and IV (IFM) binds to the intracellular mouth of the pore, which stops ion conduction (Catterall 2012). Slow inactivation, on the other hand, is believed to result from collapse of the pore of the Na_V channel (Payandeh et al. 2012; Zhang et al. 2012).

4.1 Slow Inactivation of Eukaryotic Sodium Channels

When Na_V channels are exposed to prolonged depolarizations, either by applying one long positive pulse or a series of high-frequency repetitive ones, they stop Na^+ conduction by entering the slow-inactivated state (Ulbricht 2005; Vilin and Ruben 2001). Just as entry into this inactivated state takes much longer than fast inactivation, recovery from it also proceeds very slowly with time constants in the range of hundreds of milliseconds to minutes, compared to 20 - 30 ms for fast inactivation. During this time, excitable membranes become refractory to further stimulation. Slow inactivation seems to be an ancient mechanism that Na_V channels used to protect cells against high stress conditions that generate highly repetitive stimuli.

Studies of slow inactivation of eukaryotic Na_V channels using site-directed mutagenesis methods implicated amino acid residues in the selectivity filter and surrounding S5 and S6 segments in conformational changes that accompany slow inactivation (Balser et al. 1996; Benitah et al. 1996, 1997; Vilin and Ruben 2001). Mutation of W402 located in the selectivity filter of Domain I of $Na_v1.4$ dramatically reduces slow inactivation (Balser et al. 1996). Four charged residues within the selectivity filter of $rNa_v1.4$ (E403/DI, E758/DII, D1214/DIII, and D1532/DIV) move toward each other during establishment of slow inactivation (Xiong et al. 2003), as indicated by disulfide bond formation between substituted cysteine residues. Similarly, F1236C was accessible only from the outside during short depolarizing pulses, which did not induce slow inactivation but became inaccessible during long depolarizing pulses (6 s), which induce slow inactivation (Ong et al. 2000).

4.2 Studies of Slow Inactivation of Bacterial Sodium Channels

Bacterial Na_V channels lack the equivalent linker between Domain III and IV that is responsible for fast inactivation in eukaryotic Na_V channels and thus have only slow inactivation (Pavlov et al. 2005). The crystal structure of Na_vAb showed that its selectivity filter is rigidly anchored by a hydrogen bond (~ 3.0 Å) between T175 (equivalent to T in domains I, III, and IV in Na_V1.4) and W179 (equivalent to W402 in Na_V1.4) in neighboring subunits [(Payandeh et al. 2011, 2012); Fig. 4]. However, the crystal structure of an inactivated state of Na_vAb/WT showed that two of the key T175–W179 interactions have become very weak, forming hydrogen bonds with a length ~3.8 Å compared to ~3 Å in case of pre-open state (Payandeh et al. 2012). This suggests that the hydrogen bond to W at this outer side of the selectivity filter is keeping the pore open for permeation of hydrated Na^+. In the inactivated structure of Na_vAb, S180 has flipped into a different conformation and interacts with the carboxylate side chain of E177 of a neighboring subunit (Payandeh et al. 2012). This movement led to the asymmetric collapse of two of the four pore-lining S6 segments of Na_vAb. In order to have a highly conductive channel, the geometry at the selectivity filter must be perfectly sized to coordinate and conduct a hydrated square-planar Na^+, as in the Na_VAb-I217C structure (Chakrabarti et al. 2013; Payandeh et al. 2011). Therefore, the change in size and shape of the selectivity filter in the slow-inactivated state of Na_VAb/WT (Payandeh et al. 2012) is likely to make it poorly conductive.

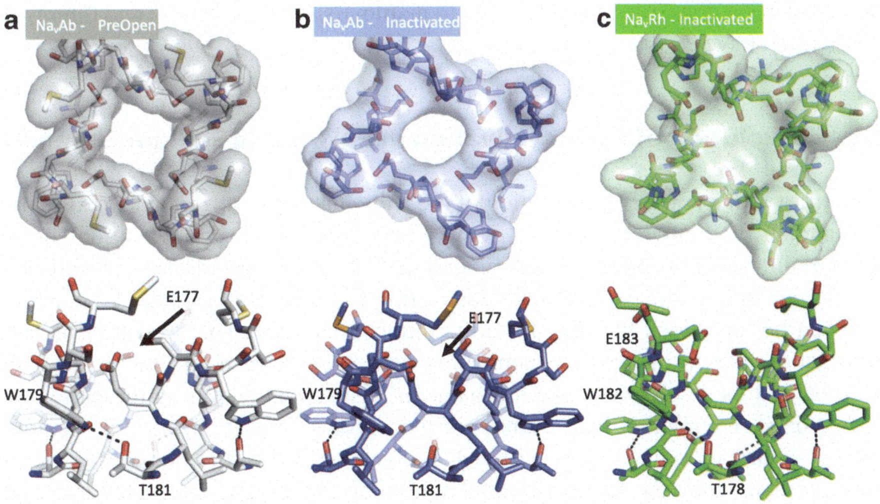

Fig. 4 The selectivity filter changes conformation as sodium channels inactivate. (**a–c**) Space filling (*top row*) and stick models (*bottom row*) are shown for pre-open Na_vAb (*gray*, pdb 3RVY), inactivated Na_vAb (*purple*, pdb 4EKW AB tetramer), and NavRh (*green*, pdb 4dxw) highlighting conformational changes as the channel moves into the inactivated state. Residues and hydrogen bonds discussed in the text are highlighted in the stick figures

Structure-function studies suggested that both activation and inactivation of bacterial Na_V channels occur via twisting and bending movements of the S6 pore-lining helix at a glycine hinge, such as G219 in NaChBac (Zhao et al. 2004b). The same mechanism has been proposed for the bacterial KcsA and MthK channels (Cuello et al. 1998; Jiang et al. 2002a, b; Perozo et al. 1999). In NaChBac, mutation of the neighboring T residue (T220) also greatly slows inactivation during the pulse (Lee et al. 2012; Zhao et al. 2004a, b). Crystal structures of Na_vMs show a kink at T219 that is equivalent to the glycine hinge in NaChBac (McCusker et al. 2012). Recently, the same kink in Na_vAb has been observed, indicating a similar mechanism of twisting and bending movements of the S6 pore-lining helix (Lenaeus et al. 2017). Mutation of T206 at the bending point in the Na_VAb S6 segment causes a negative shift of the voltage dependence of activation and is also required for early voltage-dependent inactivation (Gamal El-Din et al. 2017). These results suggest that this segment of the S6 segment undergoes two conformational changes, one that initiates opening of the activation gate and a second that begins the process of inactivation.

The amino acid residues that move during slow inactivation of mammalian Na_V channels are also intimately involved in inactivation of bacterial Na_V channels. For example, residues E403 in Domain I, E758 in Domain II, D1214 and F1236 in Domain III, and D1532 in Domain IV of Na_V1.4 move when the channel is subject to high-frequency depolarizing pulses. E403 and E758 are equivalent to S180, F1236 is equivalent to I176, and D1532 is equivalent to M181 in Na_vAb, which all move during the transition from the pre-open to the slow-inactivated state (Payandeh et al. 2011, 2012). Evidently, the molecular machinery for slow inactivation has been conserved from bacteria to mammals, even as the rate and timing of inactivation have evolved.

4.3 Evolution of Fast Inactivation in Eukaryotic Sodium Channels

In contrast to the conserved evolution of the slow inactivation from bacterial to the mammalian sodium channel, current inactivation during application of a depolarizing stimulus has evolved tremendously over billions of years by addition of a fast inactivation using the linker connecting Domain III to Domain IV to act as a hinged lid and block the pore from the intracellular side. The need for efficient information processing in the mammalian nervous system might be the driving force behind the evolution of the mechanisms of inactivation during the pulse.

A typical mammalian Na_V channel in the nerve or muscle opens very rapidly upon depolarization and closes within 1–2 ms, even if the stimulus is still on. In their seminal work on action potential initiation and propagation, Hodgkin and Huxley (1952b, c, d) described the Na^+ channel conductance as being proportional to m^3h, where m and h are gating particles responsible for activation and inactivation, respectively (Hille 2001). This was the first indication that the fast inactivation process is voltage dependent, which predicted that there is a differential functionality of voltage sensor modules in eukaryotic Na_V channels.

Currently, there is a consensus that S4 segments of domains I, II, and III are mainly responsible for activation, while Domain IVS4 is primarily responsible for initiating and maintaining fast inactivation (Capes et al. 2013; Chanda and Bezanilla 2002; Kuhn and Greeff 1999; Rogers et al. 1996; Sheets et al. 1999).

The structural motif that is responsible for translating the electrical signal sensed by the S4 segment in Domain IVS4, and thus initiating fast inactivation, is the intracellular linker between domains III and IV (Vassilev et al. 1988, 1989). Cutting this motif (Stuhmer et al. 1989) or mutating the critical hydrophobic residues (IFM) abolishes inactivation during the pulse (Kellenberger et al. 1997; Patton et al. 1992; West et al. 1992). The inactivation gate NMR structure shows that it has an alpha-helix backbone capped at its N-terminus by two turns, with the hydrophobic IFM residues located in this turn motif. Structural studies confirmed the physiological results that the IFM triad and adjacent threonine residue are essential components of the latch (Rohl et al. 1999). This structure is seen exactly in the corresponding segment in the cryo-EM structure of $Na_V1.4$ (Yan et al. 2017).

5 Modulation of Sodium Channels by Their C-Terminal Tail

In eukaryotic Na_V channels, the intracellular extensions of S6 segments begin the large C-terminal segments following each homologous domain. They are ~250 residues in length and they constitute the major targets for channel regulation. The C-terminal domains regulate fast inactivation (Cormier et al. 2002; Deschenes et al. 2001; Mantegazza et al. 2001; Nguyen and Goldin 2010) as well as slow inactivation that is modulated by interaction of calmodulin with a conserved site in the proximal segment (Gabelli et al. 2016). The newly published cryo-EM structure of a $Na_V1.4$ channels reveals extensive interactions between the C-terminal domain and Domain III–IV linker, which may explain the ability of the C-terminal tail to modulate fast inactivation (Shen et al. 2017) (Fig. 5).

Most bacterial Na_V channels have a short (~40 residues) C-terminal tail. The crystal structure of the tail was resolved for Na_vMs, Na_vAe, Na_vSulP, and Na_VAb (Bagneris et al. 2013; Irie et al. 2012; Lenaeus et al. 2017; Shaya et al. 2014). There is a striking similarity among these structures, in which the C-terminal tails from four subunits form a four-helix bundle (the neck) proximal to the activation gate, which terminates in a four-stranded coiled coil. While there is a consensus about the similarity of C-terminal crystal structure, there are differences in its function. Originally, it was proposed that C-terminal is essential for tetramer formation (Powl et al. 2010; Shaya et al. 2014), but subsequent reports argued that it is only important for stabilizing subunit-subunit interactions (Mio et al. 2010). Indeed, many bacterial Na_V channels are functional even after deleting the entire C-terminal (Arrigoni et al. 2016; Irie et al. 2012; Lenaeus et al. 2017). Arrigoni et al. showed that the proximal part of the C-terminal (the neck domain) is subject to temperature-dependent unfolding transitions during gating, while the coiled-coil part stays intact (Arrigoni et al. 2016). They found that only the neck has a major effect on channel activation profile, with a minor role for the coiled coil. Their

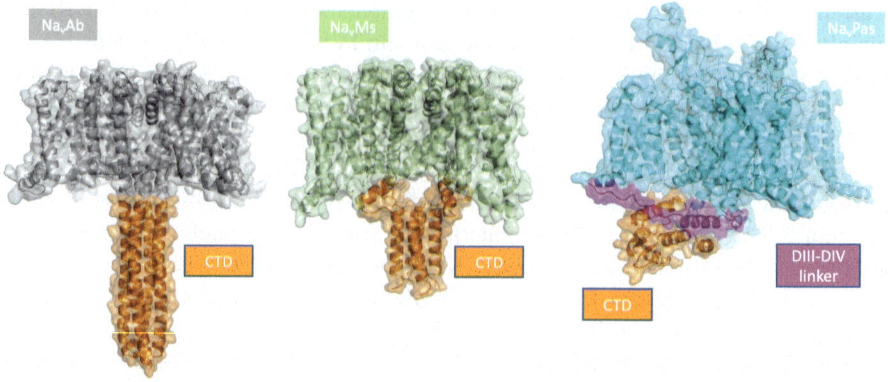

Fig. 5 The interaction between the C-terminal domain and the pore domain is different in prokaryotic channels and Na$_V$Pas. Cartoon models of prokaryotic channels Na$_V$Ab (closed state, *gray*), Na$_V$Ms (open state, *light green*), and Na$_V$Pas (uncertain state, *blue*). In each case, a cartoon model is shown with overlying space-filling figure to highlight the interactions between C-terminal and membrane-spanning domains. Each channel shows the pore and voltage-sensing domains in a single color, then the associated C-terminal domain in *orange* to allow for contrast. The Na$_V$Pas model also shows the DIII–DIV linker associated with inactivation in *pink*. The implications related to inactivation are discussed in the text

experiments demonstrated little effect of the C-terminal tail on inactivation. In contrast, C-terminal truncations modulate early voltage-dependent inactivation of Na$_V$Ab during single depolarizations (Gamal El-Din et al. 2017). Progressive deletions from the C-terminus cause graded increases in the rate of the early voltage-dependent inactivation process, without major effects on the rate of activation. This reflects a loss of voltage dependence, as if truncation of the C-terminal tail removes a voltage-dependent brake on the early phase of the inactivation process. A similar effect has been shown previously in the KcsA channel where truncation of the C-terminal tail enhanced inactivation kinetics (Cuello et al. 1998). In contrast, deleting the C-terminal tail of Na$_V$SulP slowed its early phase of inactivation during single depolarizations, and point mutations that disrupt the four-helix bundle also caused that effect (Irie et al. 2012). These results suggest that the C-terminal tail can modulate the voltage dependence and kinetics of the early phase of voltage-dependent inactivation of different Na$_V$ channels in different directions, depending on the molecular context.

The fine-tuning of inactivation during the pulse via the C-terminal tail has been preserved over billions of years of evolution, as truncation or mutation of C-terminal residues of eukaryotic Na$_V$ channels affects inactivation kinetics. Truncation of 129 residues of the distal C-terminus of Na$_V$1.3 accelerates fast inactivation by approximately fourfold (Nguyen and Goldin 2010). The C-terminus has a strong influence on kinetics and voltage dependence of inactivation of brain (Na$_V$1.2), cardiac (Na$_V$1.5), and skeletal muscle (Na$_V$1.4) channels and is primarily responsible for their differing rates of channel inactivation (Deschenes et al. 2001; Mantegazza et al. 2001).

The interplay between the early and late phases of inactivation in bacterial Na_V channels is quite evident. This explains the recent results showing that truncation of the C-terminal tail of Na_VAb/WT also removes the late, use-dependent phase of inactivation (Gamal El-Din et al. 2013, 2017). Cutting as few as ten residues from the distal end of the tail was both sufficient to abolish use-dependent inactivation at low frequency (0.2 Hz); however, at higher frequency (1 Hz), ~ 20–30% use-dependent inactivation occurs during the first few pulses, and then the peak current stays constant for up to 50 pulses. Overall, the movement of S6 segments seems to be finely tuned by the C-terminal tail in eukaryotic or bacterial Na_V channels.

6 Conclusion

As this brief review shows, sodium channels are ancient membrane signaling proteins with many functional properties conserved from bacteria to human. The simple bacterial sodium channels have provided an invaluable resource for high-resolution structural analysis of sodium channel function. The recent addition of cryo-EM structures of eukaryotic sodium channels at near-atomic resolution shows striking conservation of the core structures of sodium channels over this full length of evolutionary time. Looking ahead, the combination of structural studies of bacterial and eukaryotic sodium channels seems likely to reveal the mechanisms of their core functions in atomic detail and their complex regulation in eukaryotes with increasing resolution. It is an exciting time in this field of research!

References

Adelman WJ Jr, Palti Y (1969) The effects of external potassium and long duration voltage conditioning on the amplitude of sodium currents in the giant axon of the squid, Loligo pealei. J Gen Physiol 54:589–606

Ahern CA, Horn R (2005) Focused electric field across the voltage sensor of potassium channels. Neuron 48:25–29

Ahern CA, Payandeh J, Bosmans F, Chanda B (2016) The hitchhiker's guide to the voltage-gated sodium channel galaxy. J Gen Physiol 147:1–24

Arrigoni C, Rohaim A, Shaya D, Findeisen F, Stein RA, Nurva SR, Mishra S, McHaourab HS, Minor DL Jr (2016) Unfolding of a temperature-sensitive domain controls voltage-gated channel activation. Cell 164:922–936

Asamoah OK, Wuskell JP, Loew LM, Bezanilla F (2003) A fluorometric approach to local electric field measurements in a voltage-gated ion channel. Neuron 37:85–97

Bagneris C, Decaen PG, Hall BA, Naylor CE, Clapham DE, Kay CW, Wallace BA (2013) Role of the C-terminal domain in the structure and function of tetrameric sodium channels. Nat Commun 4:2465

Balser JR, Nuss HB, Chiamvimonvat N, Pérez-García MT, Marban E, Tomaselli GF (1996) External pore residue mediates slow inactivation in mu-1 rat skeletal muscle sodium channels. J Physiol 494:431–442

Benitah JP, Tomaselli GF, Marban E (1996) Adjacent pore-lining residues within sodium channels identified by paired cysteine mutagenesis. Proc Natl Acad Sci U S A 93:7392–7396

Benitah JP, Ranjan R, Yamagishi T, Janecki M, Tomaselli GF, Marban E (1997) Molecular motions within the pore of voltage-dependent sodium channels. Biophys J 73:603–613
Bezanilla F (2000) The voltage sensor in voltage-dependent ion channels. Physiol Rev 80:555–592
Blanchet J, Chahine M (2007) Accessibility of four arginine residues on the S4 segment of the Bacillus halodurans sodium channel. J Membr Biol 215:169–180
Capes DL, Goldschen-Ohm MP, Arcisio-Miranda M, Bezanilla F, Chanda B (2013) Domain IV voltage-sensor movement is both sufficient and rate limiting for fast inactivation in sodium channels. J Gen Physiol 142:101–112
Catterall WA (1986a) Molecular properties of voltage-sensitive sodium channels. Annu Rev Biochem 55:953–985
Catterall WA (1986b) Voltage-dependent gating of sodium channels: correlating structure and function. Trends Neurosci 9:7–10
Catterall WA (2010) Ion channel voltage sensors: structure, function, and pathophysiology. Neuron 67:915–928
Catterall WA (2012) Voltage-gated sodium channels at 60: structure, function and pathophysiology. J Physiol 590:2577–2589
Catterall WA, Zheng N (2015) Deciphering voltage-gated Na^+ and Ca^{2+} channels by studying prokaryotic ancestors. Trends Biochem Sci 40:526–534
Catterall WA, Goldin AL, Waxman SG (2005) International Union of Pharmacology. XLVII. Nomenclature and structure-function relationships of voltage-gated sodium channels. Pharmacol Rev 57:397–409
Chakrabarti N, Ing C, Payandeh J, Zheng N, Catterall WA, Pomes R (2013) Catalysis of Na^+ permeation in the bacterial sodium channel Na_VAb. Proc Natl Acad Sci U S A 110:11331–11336
Chanda B, Bezanilla F (2002) Tracking voltage-dependent conformational changes in skeletal muscle sodium channel during activation. J Gen Physiol 120:629–645
Cormier JW, Rivolta I, Tateyama M, Yang AS, Kass RS (2002) Secondary structure of the human cardiac Na^+ channel C terminus: evidence for a role of helical structures in modulation of channel inactivation. J Biol Chem 277:9233–9241
Cuello LG, Romero JG, Cortes DM, Perozo E (1998) pH-dependent gating in the Streptomyces lividans K^+ channel. Biochemistry 37:3229–3236
DeCaen PG, Yarov-Yarovoy V, Zhao Y, Scheuer T, Catterall WA (2008) Disulfide locking a sodium channel voltage sensor reveals ion pair formation during activation. Proc Natl Acad Sci U S A 105:15142–15147
DeCaen PG, Yarov-Yarovoy V, Sharp EM, Scheuer T, Catterall WA (2009) Sequential formation of ion pairs during activation of a sodium channel voltage sensor. Proc Natl Acad Sci U S A 106:22498–22503
DeCaen PG, Yarov-Yarovoy V, Scheuer T, Catterall WA (2011) Gating charge interactions with the S1 segment during activation of a Na^+ channel voltage sensor. Proc Natl Acad Sci U S A 108:18825–18830
Deschenes I, Trottier E, Chahine M (2001) Implication of the C-terminal region of the α-subunit of voltage-gated sodium channels in fast inactivation. J Membr Biol 183:103–114
Favre I, Moczydlowski E, Schild L (1996) On the structural basis for ionic selectivity among Na^+, K^+, and $Ca2^+$ in the voltage-gated sodium channel. Biophys J 71(6):3110–3125
Gabelli SB, Yoder JB, Tomaselli GF, Amzel LM (2016) Calmodulin and $Ca(2^+)$ control of voltage gated $Na(^+)$ channels. Channels (Austin) 10:45–54
Gamal El-Din TM, Grogler D, Lehmann C, Heldstab H, Greeff NG (2008) More gating charges are needed to open a Shaker K^+ channel than are needed to open an rBIIA Na^+ channel. Biophys J 95:1165–1175
Gamal El-Din TM, Heldstab H, Lehmann C, Greeff NG (2010) Double gaps along Shaker S4 demonstrate omega currents at three different closed states. Channels (Austin) 4(2):93–100
Gamal El-Din TM, Martinez GQ, Payandeh J, Scheuer T, Catterall WA (2013) A gating charge interaction required for late slow inactivation in the bacterial sodium channel Na_VAb. J Gen Physiol 142:181–190

Gamal El-Din TM, Scheuer T, Catterall WA (2014) Tracking S4 movement by gating pore currents in the bacterial sodium channel NaChBac. J Gen Physiol 144:147–157

Gamal El-Din TM, Lenaeus MJ, Ramanadane K, Zheng N, Catterall WA (2017) Control of slow, use dependent inactivation of NaVAb by its C-terminal tail. Biophys J 112(3, Suppl 1):105a. https://doi.org/10.1016/j.bpj.2016.11.597

Gosselin-Badaroudine P, Delemotte L, Moreau A, Klein ML, Chahine M (2012) Gating pore currents and the resting state of Nav1.4 voltage sensor domains. Proc Natl Acad Sci U S A 109:19250–19255

Guy HR, Seetharamulu P (1986) Molecular model of the action potential sodium channel. Proc Natl Acad Sci U S A 83(2):508–512

Heinemann SH, Terlau H, Stuhmer W, Imoto K, Numa S (1992) Calcium channel characteristics conferred on the sodium channel by single mutations. Nature 356:441–443

Hille B (1971) The hydration of sodium ions crossing the nerve membrane. Proc Natl Acad Sci U S A 68:280–282

Hille B (2001) Ionic channels of excitable membranes. Sinauer Associates, Sunderland, MA

Hirschberg B, Rovner A, Lieberman M, Patlak J (1995) Transfer of twelve charges is needed to open skeletal muscle Na+ channels. J Gen Physiol 106:1053–1068

Hodgkin AL, Huxley AF (1952a) Currents carried by sodium and potassium ions through the membrane of the giant axon of Loligo. J Physiol 116:449–472

Hodgkin AL, Huxley AF (1952b) The dual effect of membrane potential on sodium conductance in the giant axon of Loligo. J Physiol 116:497–506

Hodgkin AL, Huxley AF (1952c) A quantitative description of membrane current and its application to conduction and excitation in nerve. J Physiol 117:500–544

Hodgkin AL, Huxley AF (1952d) The components of membrane conductance in the giant axon of Loligo. J Physiol 116:473–496

Irie K, Shimomura T, Fujiyoshi Y (2012) The C-terminal helical bundle of the tetrameric prokaryotic sodium channel accelerates the inactivation rate. Nat Commun 3:793

Isom LL, Ragsdale DS, De Jongh KS, Westenbroek RE, Reber BF, Scheuer T, Catterall WA (1995) Structure and function of the beta 2 subunit of brain sodium channels, a transmembrane glycoprotein with a CAM motif. Cell 83:433–442

Jiang Y, Lee A, Chen J, Cadene M, Chait BT, MacKinnon R (2002a) Crystal structure and mechanism of a calcium-gated potassium channel. Nature 417:515–522

Jiang Y, Lee A, Chen J, Cadene M, Chait BT, MacKinnon R (2002b) The open pore conformation of potassium channels. Nature 417:523–526

Kellenberger S, West JW, Scheuer T, Catterall WA (1997) Molecular analysis of the putative inactivation particle in the inactivation gate of brain type IIA Na^+ channels. J Gen Physiol 109:589–605

Kuhn FJ, Greeff NG (1999) Movement of voltage sensor S4 in domain 4 is tightly coupled to sodium channel fast inactivation and gating charge immobilization. J Gen Physiol 114:167–183

Larsson HP, Baker OS, Dhillon DS, Isacoff EY (1996) Transmembrane movement of the *Shaker* potassium channel S4. Neuron 16:387–397

Lee S, Goodchild SJ, Ahern CA (2012) Local anesthetic inhibition of a bacterial sodium channel. J Gen Physiol 139:507–516

Lenaeus MJ, El-Din TMG, Ing C, Ramanadane K, Pomes R, Zheng N, Catterall WA (2017) Structures of closed and open states of a voltage-gated sodium channel. Proc Natl Acad Sci U S A 114:E3051–E3060

Li Q, Wanderling S, Paduch M, Medovoy D, Singharoy A, McGreevy R, Villalba-Galea CA, Hulse RE, Roux B, Schulten K et al (2014) Structural mechanism of voltage-dependent gating in an isolated voltage-sensing domain. Nat Struct Mol Biol 21:244–252

Logothetis DE, Movahedi S, Satler C, Lindpaintner K, Nadal-Ginard B (1992) Incremental reductions of positive charge within the S4 region of a voltage-gated K+ channel result in corresponding decreases in gating charge. Neuron 8:531–540

Mantegazza M, Yu FH, Catterall WA, Scheuer T (2001) Role of the C-terminal domain in inactivation of brain and cardiac sodium channels. Proc Natl Acad Sci U S A 98:15348–15353

McCusker EC, Bagneris C, Naylor CE, Cole AR, D'Avanzo N, Nichols CG, Wallace BA (2012) Structure of a bacterial voltage-gated sodium channel pore reveals mechanisms of opening and closing. Nat Commun 3:1102

Mio K, Mio M, Arisaka F, Sato M, Sato C (2010) The C-terminal coiled-coil of the bacterial voltage-gated sodium channel NaChBac is not essential for tetramer formation, but stabilizes subunit-to-subunit interactions. Prog Biophys Mol Biol 103:111–121

Morais-Cabral JH, Zhou Y, MacKinnon R (2001) Energetic optimization of ion conduction rate by the K+ selectivity filter. Nature 414:37–42

Naylor CE, Bagneris C, DeCaen PG, Sula A, Scaglione A, Clapham DE, Wallace BA (2016) Molecular basis of ion permeability in a voltage-gated sodium channel. EMBO J 35:820–830

Nguyen HM, Goldin AL (2010) Sodium channel carboxy terminal residue regulates fast inactivation. J Biol Chem 285(12):9077–9089

Ong BH, Tomaselli GF, Balser JR (2000) A structural rearrangement in the sodium channel pore linked to slow inactivation and use dependence. J Gen Physiol 116:653–662

Palti Y, Adelman WJ Jr (1969) Measurement of axonal membrane conductances and capacity by means of a varying potential control voltage clamp. J Membr Biol 1:431–458

Papazian DM, Timpe LC, Jan YN, Jan LY (1991) Alteration of voltage-dependence of Shaker potassium channel by mutations in the S4 sequence. Nature 349:305–310

Patton DE, West JW, Catterall WA, Goldin AL (1992) Amino acid residues required for fast Na+-channel inactivation: charge neutralizations and deletions in the III-IV linker. Proc Natl Acad Sci U S A 89:10905–10909

Pavlov E, Bladen C, Winkfein R, Diao C, Dhaliwal P, French RJ (2005) The pore, not cytoplasmic domains, underlies inactivation in a prokaryotic sodium channel. Biophys J 89:232–242

Payandeh J, Scheuer T, Zheng N, Catterall WA (2011) The crystal structure of a voltage-gated sodium channel. Nature 475:353–358

Payandeh J, Gamal El-Din TM, Scheuer T, Zheng N, Catterall WA (2012) Crystal structure of a voltage-gated sodium channel in two potentially inactivated states. Nature 486:135–139

Perozo E, Cortes DM, Cuello LG (1999) Structural rearrangements underlying K^+-channel activation gating. Science 285:73–78

Powl AM, O'Reilly AO, Miles AJ, Wallace BA (2010) Synchrotron radiation circular dichroism spectroscopy-defined structure of the C-terminal domain of NaChBac and its role in channel assembly. Proc Natl Acad Sci U S A 107:14064–14069

Qu Y, Isom LL, Westenbroek RE, Rogers JC, Tanada TN, McCormick KA, Scheuer T, Catterall WA (1995) Modulation of cardiac Na+ channel expression in Xenopus oocytes by beta 1 subunits. J Biol Chem 270:25696–25701

Ren D, Navarro B, Xu H, Yue L, Shi Q, Clapham DE (2001) A prokaryotic voltage-gated sodium channel. Science 294:2372–2375

Rogers JC, Qu Y, Tanada TN, Scheuer T, Catterall WA (1996) Molecular determinants of high affinity binding of alpha-scorpion toxin and sea anemone toxin in the S3-S4 extracellular loop in domain IV of the Na+ channel alpha subunit. J Biol Chem 271:15950–15962

Rohl CA, Boeckman FA, Baker C, Scheuer T, Catterall WA, Klevit RE (1999) Solution structure of the sodium channel inactivation gate. Biochemistry 38:855–861

Rudy B (1975) Proceedings: slow recovery of the inactivation of sodium conductance in Myxicola giant axons. J Physiol 249:22–24

Rudy B (1981) Inactivation in Myxicola giant axons responsible for slow and accumulative adaptation phenomena. J Physiol 312:531–549

Sather WA, McCleskey EW (2003) Permeation and selectivity in calcium channels. Annu Rev Physiol 65:133–159

Shaya D, Findeisen F, Abderemane-Ali F, Arrigoni C, Wong S, Nurva SR, Loussouarn G, Minor DL Jr (2014) Structure of a prokaryotic sodium channel pore reveals essential gating elements and an outer ion binding site common to eukaryotic channels. J Mol Biol 426:467–483

Sheets MF, Kyle JW, Kallen RG, Hanck DA (1999) The Na channel voltage sensor associated with inactivation is localized to the external charged residues of domain IV, S4. Biophys J 77:747–757

Shen HZ, Zhou Q, Pan XJ, Li ZQ, Wu JP, Yan N (2017) Structure of a eukaryotic voltage-gated sodium channel at near-atomic resolution. Science 355(6328). pii: eaal4326

Sokolov S, Scheuer T, Catterall WA (2005) Ion permeation through a voltage-sensitive gating pore in brain sodium channels having voltage sensor mutations. Neuron 47:183–189

Sokolov S, Scheuer T, Catterall WA (2008) Depolarization-activated gating pore current conducted by mutant sodium channels in potassium-sensitive normokalemic periodic paralysis. Proc Natl Acad Sci U S A 105:19980–19985

Starace DM, Bezanilla F (2001) Histidine scanning mutagenesis of basic residues of the S4 segment of the *Shaker* K^+ channel. J Gen Physiol 117:469–490

Starace DM, Bezanilla F (2004) A proton pore in a potassium channel voltage sensor reveals a focused electric field. Nature 427:548–553

Stephens RF, Guan W, Zhorov BS, Spafford JD (2015) Selectivity filters and cysteine-rich extracellular loops in voltage-gated sodium, calcium, and NALCN channels. Front Physiol 6:153

Stuhmer W, Conti F, Suzuki H, Wang X, Noda M, Yahadi N, Kubo H, Numa S (1989) Structural parts involved in activation and inactivation of the sodium channel. Nature 339:597–603

Sula A, Booker J, Ng LC, Naylor CE, DeCaen PG, Wallace BA (2017) The complete structure of an activated open sodium channel. Nat Commun 8:14205

Tang L, Gamal El-Din TM, Payandeh J, Martinez GQ, Heard TM, Scheuer T, Zheng N, Catterall WA (2014) Structural basis for Ca^{2+} selectivity of a voltage-gated calcium channel. Nature 505:56–61

Tao X, Lee A, Limapichat W, Dougherty DA, MacKinnon R (2010) A gating charge transfer center in voltage sensors. Science 328:67–73

Ulbricht W (2005) Sodium channel inactivation: molecular determinants and modulation. Physiol Rev 85:1271–1301

Ulmschneider MB, Bagneris C, McCusker EC, Decaen PG, Delling M, Clapham DE, Ulmschneider JP, Wallace BA (2013) Molecular dynamics of ion transport through the open conformation of a bacterial voltage-gated sodium channel. Proc Natl Acad Sci U S A 110:6364–6369

Vargas E, Yarov-Yarovoy V, Khalili-Araghi F, Catterall WA, Klein ML, Tarek M, Lindahl E, Schulten K, Perozo E, Bezanilla F et al (2012) An emerging consensus on voltage-dependent gating from computational modeling and molecular dynamics simulations. J Gen Physiol 140:587–594

Vassilev PM, Scheuer T, Catterall WA (1988) Identification of an intracellular peptide segment involved in sodium channel inactivation. Science 241:1658–1661

Vassilev P, Scheuer T, Catterall WA (1989) Inhibition of inactivation of single sodium channels by a site-directed antibody. Proc Natl Acad Sci U S A 86:8147–8151

Vilin YY, Ruben PC (2001) Slow inactivation in voltage-gated sodium channels: molecular substrates and contributions to channelopathies. Cell Biochem Biophys 35:171–190

West JW, Patton DE, Scheuer T, Wang Y, Goldin AL, Catterall WA (1992) A cluster of hydrophobic amino acid residues required for fast Na^+-channel inactivation. Proc Natl Acad Sci U S A 89:10910–10914

Wu J, Yan Z, Li Z, Qian X, Lu S, Dong M, Zhou Q, Yan N (2016) Structure of the voltage-gated calcium channel $Ca_V1.1$ at 3.6 a resolution. Nature 537:191–196

Xiong W, Li RA, Tian YL, Tomaselli GF (2003) Molecular motions of the outer ring of charge of the sodium channel: do they couple to slow inactivation? J Gen Physiol 122:323–332

Yan Z, Zhou Q, Wang L, Wu J, Zhao Y, Huang G, Peng W, Shen H, Lei J, Yan N (2017) Structure of the Nav1.4-beta1 complex from electric eel. Cell 170(3):470–482.e11

Yang N, Horn R (1995) Evidence for voltage-dependent S4 movement in sodium channels. Neuron 15:213–218

Yang J, Ellinor PT, Sather WA, Zhang JF, Tsien RW (1993) Molecular determinants of Ca^{2+} selectivity and ion permeation in L-type Ca^{2+} channels. Nature 366:158–161

Yang N, George AL Jr, Horn R (1996) Molecular basis of charge movement in voltage-gated sodium channels. Neuron 16:113–122

Yarov-Yarovoy V, Decaen PG, Westenbroek RE, Pan CY, Scheuer T, Baker D, Catterall WA (2012) Structural basis for gating charge movement in the voltage sensor of a sodium channel. Proc Natl Acad Sci U S A 109:E93–E102

Yue L, Navarro B, Ren D, Ramos A, Clapham DE (2002) The cation selectivity filter of the bacterial sodium channel, NaChBac. J Gen Physiol 120:845–853

Zhang X, Ren W, DeCaen P, Yan C, Tao X, Tang L, Wang J, Hasegawa K, Kumasaka T, He J et al (2012) Crystal structure of an orthologue of the NaChBac voltage-gated sodium channel. Nature 486:130–134

Zhao Y, Scheuer T, Catterall WA (2004a) Reversed voltage-dependent gating of a bacterial sodium channel with proline substitutions in the S6 transmembrane segment. Proc Natl Acad Sci U S A 101:17873–17878

Zhao Y, Yarov-Yarovoy V, Scheuer T, Catterall WA (2004b) A gating hinge in Na^+ channels; a molecular switch for electrical signaling. Neuron 41:859–865

Zhou Y, Morais-Cabral JH, Kaufman A, MacKinnon R (2001) Chemistry of ion coordination and hydration revealed by a K^+ channel- Fab complex at 2.0 a resolution. Nature 414:43–48

The Cardiac Sodium Channel and Its Protein Partners

Elise Balse and Catherine Eichel

Contents

Abstract

Activation of the electrical signal and its transmission as a depolarizing wave in the whole heart requires highly organized myocyte architecture and cell-cell contacts. In addition, complex trafficking and anchoring intracellular machineries regulate

E. Balse (✉)
Sorbonne Universités – UPMC Univ Paris 06 – Inserm – UMR_S 1166 – Unité de recherche sur les maladies cardiovasculaires, le métabolisme et la nutrition – Faculté de Médecine – Site Pitié-Salpêtrière, 91 boulevard de l'Hôpital, 75013 Paris, France
e-mail: elise.balse@upmc.fr

C. Eichel
Department of Neurosciences, University of Wisconsin School of Medicine and Public Health, WIMRII, 1111 Highland Avenue, Madison, WI 53705, USA

M. Chahine (ed.), *Voltage-gated Sodium Channels: Structure, Function and Channelopathies*, Handbook of Experimental Pharmacology 246,
https://doi.org/10.1007/164_2017_45

the proper surface expression of channels and their targeting to distinct membrane domains. An increasing list of proteins, lipids, and second messengers can contribute to the normal targeting of ion channels in cardiac myocytes. However, their precise roles in the electrophysiology of the heart are far from been extensively understood. Nowadays, much effort in the field focuses on understanding the mechanisms that regulate ion channel targeting to sarcolemma microdomains and their organization into macromolecular complexes. The purpose of the present section is to provide an overview of the characterized partners of the main cardiac sodium channel, $Na_V1.5$, involved in regulating the functional expression of this channel both in terms of trafficking and targeting into microdomains.

Keywords
Ankyrin G · CASK · Caveolin · Connexin 43 · Desmoglein 2 · Desmoplakin · Dystrophin-syntrophin complex · Myocyte membrane domains · $Na_V1.5$ protein partners · Plakoglobin · Plakophilin 2 · SAP97

1 Introduction

The electrical activity generated by cardiac myocytes is central to cardiac physiology from heartbeats to the secretion of natriuretic peptides. Many cardiac ionic currents and their molecular correlates have been identified and their roles in shaping action potentials (AP) described. Progress in genetic and molecular biology has provided precise information on the structure–function relationships of most channels, pumps, and exchangers expressed in cardiac myocytes. These studies provided substantial evidence that rhythm disorders are associated with modifications in ion channel function that could originate from alterations in ion channels, associated regulatory proteins, trafficking, or the cellular environment. For instance, in inherited channelopathies such as congenital long QT syndromes (LQTS), Brugada syndrome (BrS), conduction disorders, and familial forms of atrial fibrillation, these defects are carried by hundreds of naturally occurring mutations in genes encoding ion channels (Wilde and Brugada 2011). In addition, recent studies have shown that gain- or loss-of-function of ion channels may also arise from mutations carried by regulatory and anchoring partners as well as altered trafficking of ion channels (Rook et al. 2012). Finally, in acquired arrhythmias such atrial fibrillation, ion channel dysfunctions are often the consequence of structural remodeling of the myocardium, interstitial fibrosis, and gap junction disorganization (Schotten et al. 2011).

As in other excitable tissues, activation of the electrical signal and its transmission as a depolarizing wave in the whole heart requires the myocyte architecture and cell–cell contacts to be highly organized. Trafficking and anchoring mechanisms tightly regulate both the level of expression and distinct domain localization of ion channels at the cell membrane. The list of proteins, lipids, and second messengers able to contribute to the normal targeting of ion channels in cardiac myocytes is rapidly increasing. However, their precise roles in the electrophysiology of the heart are not well understood. Currently, much effort in the field focuses on understanding the mechanisms that regulate ion channel targeting to sarcolemma microdomains and their organization into macromolecular complexes. The purpose of the present section is to provide an overview of the characterized partners of the main cardiac sodium channel, $Na_V1.5$. It should be noted that beta-subunit regulation of sodium channels is outside the scope of the present section.

2 Specialized Membrane Domains in Cardiac Sarcolemma

Cardiac myocytes, similar to neurons or epithelial cells, are structurally polarized cells, possessing highly organized membrane domains. These are characterized by the presence of specific sets of proteins. In cardiac myocytes, ion channels are localized in three major functional and structural entities: (1) intercalated discs responsible for the electro-mechanical coupling, (2) T-tubules, where excitation–contraction coupling takes place, and (3) focal adhesions complexes (costameres) that link adjacent myocytes in the myocardium through interactions with the extracellular matrix (Fig. 1). While the transmission of the action potential (AP) between myocytes occurs at the intercalated disc (ID), the AP is conducted along the lateral sarcolemma where the T-tubules and focal adhesions complexes are localized. This specific organization allows the anisotropic propagation of the AP between adjacent myocytes and the harmonious depolarization of the whole myocardium.

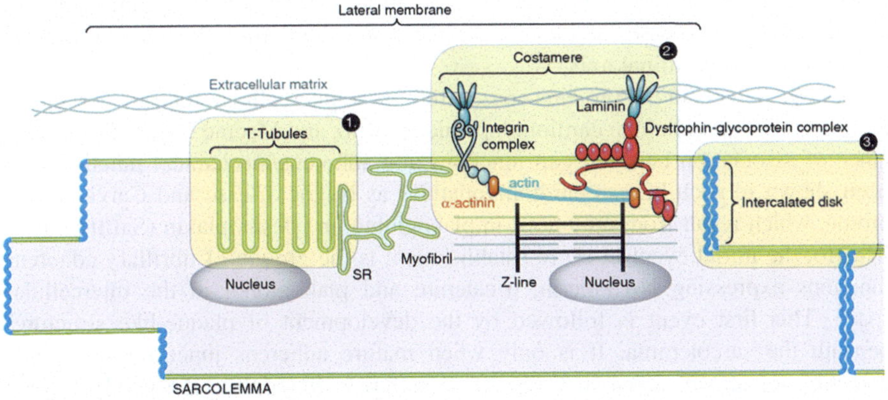

Fig. 1 Schematic representation of the three well-identified specialized domains for channel expression in cardiac myocyte (from Balse et al. 2012). (*1*) The "calcium synapse" composed of T-tubule and terminal cisternae of the sarcoplasmic reticulum, (*2*) the costamere at the lateral membrane, and (*3*) the intercalated disc that concentrates desmosomes, fascia adherens, and gap junctions

2.1 Intercalated Disk

The intercalated disc (ID) ensures the rapid and coordinated propagation of the AP along the length of the cardiac muscle conferring to the myocardium a syncytium-like function. Three types of intercellular adhesion structures are present within the ID: gap junctions, adherens junctions, and desmosomes (Clark et al. 2002).

Gap junctions constitute an intercellular adhesion structure of the ID mainly responsible for the AP transmission between adjacent cells. They are composed of two connexons (or hemichannels) formed from six connexins spanning the lipid bilayer on two adjacent cardiomyocytes (Yeager 1998). These junctions allow non-selective diffusion of ~1,000-Da molecules (Kumar and Gilula 1996; Loewenstein 1981) including ions and small molecules like second messengers, resulting in electro-metabolic coupling. Membrane regions containing gap junction channels are rigid and susceptible to collapse in response to mechanical stresses. Compensating for this, gap junctions in cardiomyocytes are always located in close proximity to other intercellular adhesion structures of the ID, notably the adherens junctions.

Adherens junctions and desmosomes are mechanical junctions responsible for intercellular adhesion. Composed of adhesion molecules that cross the sarcolemma, they bind both intracellular proteins of the cytoskeleton and proteins from the neighboring cell in the extracellular space. While adherens junctions are anchoring sites for actin to connect cells to sarcomeres, desmosomes bind intermediate filaments joining cells together. In mammals, a mixed-type junctional structure termed "hybrid adhering junction" or "area composita" exists in the cardiac ID, suggesting that it evolved to strengthen mechanical coupling (Borrmann et al. 2006; Franke et al. 2006). In the working myocardium both adherens junctions and desmosomes are directly involved in sensing and regulating mechanical stresses acting in the longitudinal axis.

Genetic defects in the proteins constituting these intercellular adhesion structures have been linked to human cardiomyopathies (Saffitz and Macrae 2010). The importance of association between gap junctions and adherens/mechanical junctions has been shown in such human cardiomyopathies as Naxos disease and Carvajal syndrome, which result from mutations in plakoglobin and desmoplakin (Saffitz 2005). In vitro, the initial event in ID re-establishment is the growth of fibrillary adherens junctions expressing N-cadherin, α-catenin, and plakoglobin in the intercellular space. This first event is followed by the development of plaque-like structures beneath the sarcolemma. It is only when mature adherens junctions develop a detectible connexin signal that gap junctions arise. Invariably, the newly formed gap junction is associated with adhesive junctions. These observations suggest that formation of adhesion junctions is a prerequisite for gap junction formation within the ID (Geisler et al. 2010). Finally, several lines of evidence indicate that altered mechanical coupling has a large impact on electrical coupling, whereas impaired electrical coupling does not affect mechanical coupling (Noorman et al. 2009).

2.2 Lateral Membrane

Under physiological conditions, cardiomyocytes are not electrically coupled in the transverse axis but the AP propagates along the lateral sarcolemma towards the ID. Two major structures can be distinguished in the lateral membrane (LM) of cardiomyocytes, the costamere and the T-tubule system. The costamere links cardiomyocytes to the extracellular matrix (ECM) and guarantees the maintenance of the three-dimensional organization of the myocardium. The T-tubule system is linked to sarcoplasmic reticulum extensions and initiates EC-coupling.

2.2.1 Costamere

A major consequence of muscle contraction is cell deformation due to shortening. During this process, the contractile machinery of sarcomeres must remain connected to both the sarcolemma and the ECM to properly coordinate contraction within the three-dimensional organization of cardiac muscle layers. Costameres are sarcolemmal transverse rib-like structures that overlie the Z lines of the sarcomere (Pardo et al. 1983a, b). By constituting physical links between Z-discs, sarcolemma, and the ECM, they sense and transmit bidirectional mechanical forces generated both intrinsically and externally (Danowski et al. 1992; Mansour et al. 2004; McCain and Parker 2011; Samarel 2005; Sharp et al. 1997). Two distinct macromolecular complexes play major regulatory roles at the costamere level: the integrin complex and the dystrophin–glycoprotein complex (DGC).

Integrins bind extracellular laminin and link to the actin cytoskeleton via interactions with talin, vinculin, and the nebulin-related protein N-RAP, as well as sarcomeric actinin (Bershadsky et al. 2003). In addition to their function as adhesion molecules, integrins are mechano-transducers in cardiomyocytes as well as in non-muscle cells (Miranti and Brugge 2002; Ross 2004; Boycott et al. 2013). The cytoplasmic domain of integrin also provides critical anchors for a number of multimolecular complexes including non-receptor type tyrosine kinases such as focal adhesion kinase (FAK), src-family tyrosin kinases, integrin-linked protein kinase (ILK), and signal mediators such as small GTPases (rac, rho, Cdc42) (Ervasti 2003; Miranti and Brugge 2002; Ross 2004; Samarel 2005). The integrin complex is essential for maintenance of the normal physiological functioning of the heart as revealed by β1-integrin and vinculin KO mice which develop dilated cardiomyopathy (Shai et al. 2002) and stress-induced cardiomyopathy respectively (Zemljic-Harpf et al. 2004).

The DGC is a large complex of proteins attached to the extracellular laminin through α dystroglycan associated with the transmembrane proteins β dystroglycan, sarcoglycans, and β dystroglycan. On the cytoplasmic side, dystrophin binds to syntrophin, dystrobrevin, and nitric oxide synthase (NOS), and attaches the entire macromolecular complex to the actin cytoskeleton. Studies using animal models of Duchenne and Becker muscular dystrophies showed physiological roles for the DGC (Allamand and Campbell 2000; Durbeej and Campbell 2002). The DGC is essential in stabilizing the sarcolemma upon physical stresses. Animals deficient in dystrophin exhibit membrane fragility and loss of membrane integrity (Cohn and

Campbell 2000; Lapidos et al. 2004). In addition to its mechanical and structural functions, the DGC is involved in cellular communication through interactions with NOS and Grb2 signaling molecules (Brenman et al. 1995; Grady et al. 1999).

2.2.2 T-Tubule System

T-tubules are characteristic narrow, tubular invaginations of the lateral sarcolemma observed in adult mammalian ventricular cardiomyocytes (Orchard and Brette 2008; Tohse et al. 2004). These digitations are in close proximity to the adjacent membrane of the sarcoplasmic reticulum (SR). The short inter-membrane space between the sarcolemma and the SR constitutes a dyadic cleft for the calcium communication that drives EC-coupling. The T-tubule system also exists in atrial cardiomyocytes but the transverse-axial tubular network is more rudimentary in these cells (Orchard and Brette 2008; Walden et al. 2009). It has been proposed that this lose organization in the atrial cells results in an absence of EC-coupling in the center of the atrial myocyte under basal conditions and, consequently, a less robust contraction then occurs in ventricular cells (Bootman et al. 2007; Mackenzie et al. 2004).

2.3 Localization of $Na_V1.5$ Channels in Membrane Microdomains of Cardiac Myocytes

The ID is specialized in the rapid propagation of the AP along cardiac fibers as it concentrates a number of ion channels. This fast propagation at the ID has been linked to the differential location of $Na_V1.5$ channels in the membrane subdomains described above, resulting in a sodium current (I_{Na}) of higher amplitude at the ID than the current recorded in the midsection (LM) by macro-patch (Verkerk et al. 2007; Lin et al. 2011; Shy et al. 2014). Furthermore, the dystrophin-deficient mdx mouse model of Duchenne muscular dystrophy revealed that the LM pool of $Na_V1.5$ is strongly reduced and I_{Na} is decreased by ~30% (Petitprez et al. 2011; Gavillet et al. 2006). Finally, detubulation experiments using formamide treatment showed that I_{Na} generated by $Na_V1.5$ channel at the T-tubule account for 20% of the total current (Orchard and Brette 2008). From these observations, a scheme can be made concerning the relative contribution of the different populations of $Na_V1.5$ in the propagation of the AP: the ID pool having a higher weight than the LM pool.

Membrane microdomain distribution of $Na_V1.5$ channels has been further elucidated by scanning ion conductance microscopy (SICm) experiments. SICm identified three topographic entities at the LM: T-tubules, crests, and z-grooves corresponding to T-tubule openings (Lab et al. 2013). Super-resolution scanning patch-clamp recordings showed that $Na_V1.5$ channels do not distribute homogenously along the LM. The channels aggregate to form clusters, the largest of which is located at crest regions (likely costamere regions of the LM) and the lowest at the T-tubule (Bhargava et al. 2013). This differential distribution of $Na_V1.5$ in LM microdomains suggests distinct roles for crests and T-tubules in electrical and contractile activities of cardiomyocytes. For instance, $Na_V1.5$ located at the crest

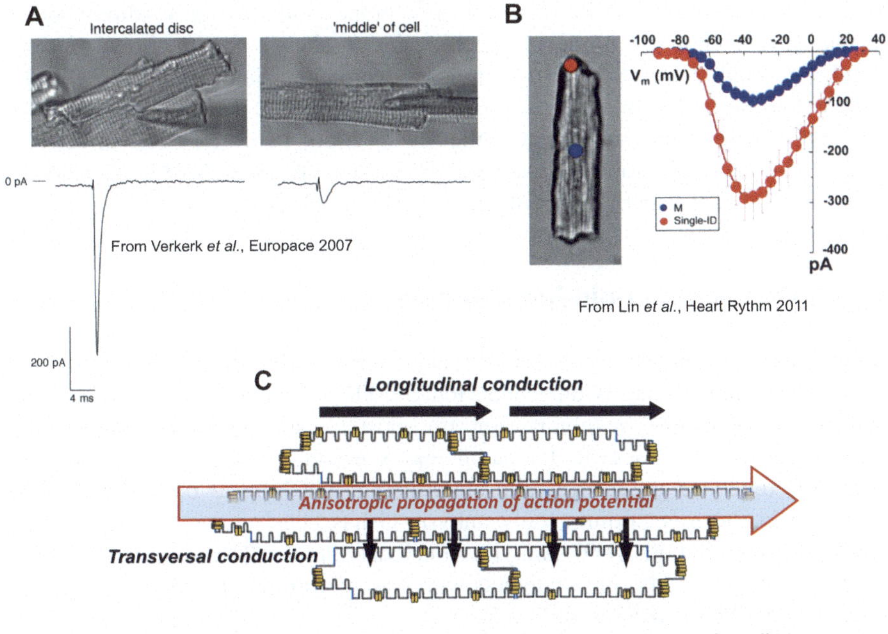

Fig. 2 Differential distribution of $Na_V1.5$ channels in membrane domains of cardiac myocytes. (**a**, **b**) Examples of sodium current amplitude recorded at the ID and midsection (corresponding to lateral membrane) showing that $Na_V1.5$ channels are concentrated at the ID. (**c**) Proposed scheme explaining that the differential distribution of $Na_V1.5$ channels favors the anisotropic conduction of the electrical impulse

could play a preferential role in the propagation of the AP along the LM whereas those located in T-tubules could be involved in EC-coupling. Altogether, these observations suggest that the differential distribution of $Na_V1.5$ channels might have profound consequence on the establishment and maintenance of the anisotropic propagation of the AP. Such an organization would facilitate longitudinal propagation of the AP over transverse propagation, and therefore determine conduction velocity (CV) in the myocardium (Spach 1999).

The above considerations explain why much research is being focused on mechanisms that regulate $Na_V1.5$ channel expression into cardiomyocytes specialized membrane domains. The current scheme is that ion channels are organized in distinct subpopulations with specific roles in the generation and the propagation of the cardiac action potential (Fig. 2).

3 $NA_V1.5$ Partners and Their Function in the Regulation of the Sodium Current

In the past decade, important observations have been made regarding the regulation of the functional expression of $Na_V1.5$ in respect to its association with specific partners. Indeed, recent studies suggest the existence of distinct populations of

$Na_V1.5$ channels located in different membrane domains of cardiomyocytes depending on their association with specific partners such as gap junctional proteins (Agullo-Pascual et al. 2014a, b), desmosomal proteins (Sato et al. 2009, 2011), actin cytoskeleton-binding proteins (Mohler et al. 2003; Makara et al. 2014), dystrophin–syntrophin complex (Gavillet et al. 2006; Petitprez et al. 2011; Shy et al. 2014), and Membrane-Associated GUanylate Kinase (MAGUK) proteins (Milstein et al. 2012; Petitprez et al. 2011; Eichel et al. 2016).

3.1 Cytoskeleton-Binding Proteins

In mammals, the ankyrin family of proteins is encoded by three genes (ANK1–3) expressed in many tissues (Cunha and Mohler 2006). The main function of ankyrin proteins is to anchor membrane ion transporters, such as ion channels and ion-transporting ATPases, to the actin and spectrin cytoskeleton. While both ankyrin-B (encoded by ANK2) and ankyrin-G (ANK3) have been found to be expressed in the myocardium, only ankyrin-G has been shown to interact with $Na_V1.5$ (Hashemi et al. 2009). Ankyrin-G is located at the ID and in the T-tubules of adult myocytes. Ankyrin-G interacts with $Na_V1.5$ through its ankyrin G binding domain (the VPIAVAESD motif on loop 2).

The first evidence for a role of ankyrin-G in the targeting and anchoring of the $Na_V1.5$ channel came from the identification of a mutation (E1053K) in a patient with BrS occurring in the ankyrin-binding motif of $Na_V1.5$ (Mohler et al. 2004). The mutation E1053K abolishes binding of $Na_V1.5$ to ankyrin-G and prevents its accumulation at cell surface in ventricular cardiomyocytes, without altering protein folding or Golgi-mediated trafficking. Although I_{Na} density is not modified, the mutation leads to modifications in gating properties of the channel. In vitro, reduction of ankyrin G expression in cardiac myocytes leads to lowered $Na_V1.5$ protein expression, retention in perinuclear regions, and reduced I_{Na} attributed to the loss of direct $Na_V1.5$-ankyrin-G interaction (the ANK repeat 14–15 β-hairpin loop tips of the ankyrin-G membrane-binding domain) (Lowe et al. 2008). In addition, ankyrin G silencing in cardiac myocytes reduces electrical coupling through decreased connexin 43 (cx43) expression and adhesion strength by lowering plakophilin 2 (PKP2) expression. This observation is associated with redistribution of $Na_V1.5$ and reduction in I_{Na} amplitude. Therefore, ankyrin-G appears as a key functional component of the intercalated disc at the intersection of three complexes: $Na_V1.5$, gap junctions, and the cardiac desmosome (Sato et al. 2011). However, the fact that CV measured in neonatal cardiomyocytes monolayers and cytoskeletal architecture were not affected by ankyrin-G silencing suggests that this cytoskeleton-binding protein is not critical for junctional and cytoskeletal integrity of cardiomyocytes. Recently, using a cardiac specific ankyrin-G KO mouse, P. Mohler's group showed that ankyrin-G targets both $Na_V1.5$ and its regulatory kinase CaMKII specifically at the ID. In the absence of ankyrin-G, the CaMKII-dependent regulation of late sodium current upon β-adrenergic stimulation is lost and lethal catecholamine-induced arrhythmias occur. Interestingly, in this model, ankyrin-G deletion also leads to reorganization of the ID protein PKP2 (Makara et al. 2014).

Thus, ankyrin-G seems critical not only in the trafficking and regulation of $Na_V1.5$ at the ID but also in the organization of ID protein complexes. Hence, ankyrin-G could regulate not only the properties of the cardiac sodium current but also the integrity of mechanical and electrical coupling. As such, ankyrin-G is likely a key molecule for impulse propagation in the heart.

3.2 GAP Junctional Proteins

Connexin 43 (cx43), the most widely expressed connexin, is encoded by the GJA1 gene. Cx43 is a component of gap junctions in most cell types, which allows intercellular communication between cells to regulate proliferation, differentiation, and cell death. In the heart, cx43 is the major protein of gap junctions and plays a crucial role in the synchronized contraction of the heart. However, besides their canonical function of electrically low-resistive junctions, gap junctions and notably cx43 exert noncanonical functions by interacting with mechanical junctions and ion channel complexes. The following part will present an overview of the cross talk existing between gap junction and $Na_V1.5$ channels in the heart.

Since 2001, it has been known that changes in anisotropic conduction caused by myocyte remodeling are related to modifications in the distribution of gap junctions and $Na_V1.5$ channels (Spach et al. 2001). After birth, when myocytes mature, cx43 shifts from the sides to the ends of ventricular myocytes, thereby modifying the properties of transverse versus longitudinal propagation of the AP. As mentioned, formation of the ID in vitro follows an ordered assembly with the development of adhesive proteins first (adherens proteins and desmosomal proteins) followed by the establishment of electrochemical connections (cx43) and finally recruitment of cytoskeletal component of Na_V complexes (ankyrin-G). Once this assembly is achieved, $Na_V1.5$ is targeted to the newly formed ID structure (Geisler et al. 2010). The consistent order of the ID assembly process suggests that specific scaffoldings are prerequisite for integration of the mechanical and electrochemical elements of the disc. The dependence of functional expression and localization of $Na_V1.5$ at the ID upon cx43 localization was further supported in a conditional cx43 KO mouse showing ventricular tachycardia susceptibility. In this model, $Na_V1.5$ distribution is heterogeneous and I_{Na} is decreased suggesting a cross talk between cx43 expression and functional sodium current (Jansen et al. 2012). The complex interaction between gap junctions and sodium channels is also supported by the observation that cx43 ablation in fetal atrial myocytes reduces electrical coupling and I_{Na} (Desplantez et al. 2012).

Further insights into the relationship between $Na_V1.5$ channels and gap junctions were given by R. Gourdie's group demonstrating the organization of gap junctions in plaques versus perinexus. The former provide direct cytoplasmic connection between cells and present a strong immunopositive signal for cx43. New connexons are inserted first in regions surrounding the plaque, or the so-called perinexus, and are later delivered into the plaque. Importantly, $Na_V1.5$ channels were found to interact with perinexal cx43 but not with ZO-1 (see Sect. 3.6), suggesting that although cx43/$Na_V1.5$ and cx43/ZO-1 interactions occur in the perinexus, they are mutually exclusives (Rhett et al. 2012). Therefore, one of the potential functions of the perinexus in the heart is to facilitate conduction of AP through ephaptic coupling.

Conditional deletion of the last five amino acids in the C-terminus of cx43 in mice (the ZO-1 binding motif) leads to marked impairment of cardiac electrical activation properties and severe ventricular arrhythmias. Despite cx43 and ZO-1 co-localization at the ID and normal coupling properties of gap junction, optical mapping reveals extensive conduction delays together with decreased sodium and potassium current densities. In this mouse, $Na_V1.5$ channel expression is reduced at the ID without change in protein levels. Therefore cx43 dependent arrhythmias can develop by mechanisms other than impairment of gap junction channel function (Lübkemeier et al. 2013). M. Delmar's group investigated the ID ultrastructural organization in mutant mice and tested the hypothesis that the microtubule plus-end (EB1) is at the core of the cx43-$Na_V1.5$ relationship. Previous studies showed that $Na_V1.5$ is delivered into the membrane via the microtubule network (Casini et al. 2010) and that EB1 targets cx43 directly to adherens junctions (Shaw et al. 2007). At the super-resolution level, both EB1 and $Na_V1.5$ distribute into N-cadherin clusters at the ID. In mutant mice, the number of EB1 and $Na_V1.5$ clusters is reduced. Macropatch recordings and scanning patch clamp experiments showed reduced I_{Na} exclusively at the ID. Therefore, $Na_V1.5$ and EB1 localization at the ID appears to be dependent on cx43 (Agullo-Pascual et al. 2014a, b). The authors proposed a model in which cx43 is part of a molecular complex that captures microtubule plus-end allowing accurate delivery of $Na_V1.5$ to the ID. These observations link excitability and electrical coupling through a common molecular mechanism.

3.3 Desmosomal Proteins

3.3.1 Plakophilin 2

Plakophilin 2 (PKP2) is encoded by the PKP2 gene. Mutations in PKP2 are linked to arrhythmogenic right ventricular cardiomyopathy (ARVC), an inherited disease associated with ventricular arrhythmias and sudden death in the young. PKP2 is found primarily in cardiac myocytes and is one of several proteins that compose the desmosome.

The first evidence of a direct link between desmosome and $Na_V1.5$ channels came from M. Delmar's group. PKP2 and $Na_V1.5$ channel colocalize at the ID of freshly isolated myocytes and PKP2 interacts with $Na_V1.5$ through its head domain (Sato et al. 2009). PKP2 silencing in vitro reduces I_{Na} in cardiac myocytes and shifts the inactivation curve towards more negative potentials. Optical mapping experiments in monolayers of neonatal rat ventricular myocytes revealed that loss of PKP2 expression causes AP propagation slowing, rate-dependent activation failure, and arrhythmic behavior (Sato et al. 2009). As mentioned above, the localization of PKP2 at ID is influenced by ankyrin-G expression (Sato et al. 2011). Of note, neonatal rat ventricular myocytes silenced for PKP2 showed global reduction of ankyrin-G levels and a decrease in ankyrin-G immunoreactive signal at cell–cell contacts. Although loss of PKP2 expression does not affect $Na_V1.5$ protein levels, it is redistributed from ID-like structures to the cytoplasm (Sato et al. 2011). At the moment, whether $Na_V1.5$ re-localization is related to reduced ankyrin-G expression in PKP2-deficient cells or directly related to loss of PKP2 is not known.

The importance of PKP2 in modulating I_{Na} was demonstrated in vivo using a murine model of PKP2 haploinsufficiency. In PKP2 heterozygotes (PKP2-hz), the decrease in PKP2 protein levels is not accompanied by a reduction of other ID proteins (cx43, N-cadherin, plakoglobin) nor $Na_V1.5$. Although reduction of PKP2 expression at the ID and alteration of desmosome organization are observed, no changes in localization of $Na_V1.5$, cx43, plakoglobin, or N-cadherin were noticed. Similar to in vitro experiments, whole-cell patch clamp revealed that I_{Na} density is reduced and accompanied by a negative shift in inactivation. Super-resolution microscopy showed that I_{Na} decrease is related to a reduced number of channels at the ID and an increased separation of microtubules from cell endings (Cerrone et al. 2012). Surface ECG recordings showed that PKP2-hz mice present longer P-wave duration at baseline. In addition, flecainide challenge causes a prolongation of P-wave and QRS durations and leads to ventricular arrhythmias (Cerrone et al. 2012). Although PR and QTc intervals were also modified in the PKP2-hz mouse suggesting that other ionic current are likely affected, this study clearly identified the sodium current deficiency as the primary mechanism and PKP2 as a crucial player in I_{Na} regulation.

In adult sheep subjected to right ventricular pressure overload, cx43 undergoes extensive remodeling, with increased presence of gap junction plaques at the LM. The desmosomal/adherens junction protein plakoglobin showed similar redistribution from the ID to the LM. Desmosomal cadherins (desmocollin and Desmoglein), desmosomal components (PKP2 and desmoplakin), adherens junction protein N-cadherin, and the microtubule-associated proteins EB1 and Kif5b are also lateralized (Chkourko et al. 2012). Although $Na_V1.5$ plaques were very sporadic in pulmonary hypertension heart models, most of the lateralized cx43 do not show colocalization with $Na_V1.5$. Finally, I_{Na} is reduced together with a slight rightward shift in activation properties. This data suggests that the lateralization of cx43 induced by pressure overload is accompanied by a complex remodeling of mechanical junctions and changes in both distribution and function of $Na_V1.5$ channels (Chkourko et al. 2012).

Screening of PKP2 variants in patients with diagnosis of Brugada Syndrome (BrS) and no signs of ARVC or mutations in the BrS-related genes (SCN5A, CACNa1c, GPD1L, and MOG1) revealed five cases of single amino acid substitutions (Cerrone et al. 2014). Mutations tested in HL-1-derived cells endogenously expressing $Na_V1.5$, but not PKP2, showed that I_{Na} and $Na_V1.5$ localization were restored by transfection of WT PKP2. This did not occur when transfected with BrS-related PKP2 mutants. Similarly, hiPS-derived cardiomyocytes from a patient with a PKP2 mutation showed drastically reduced I_{Na} which could be rescued by transfection with WT PKP2, but not BrS-related PKP2 mutants (Cerrone et al. 2014). This study revealed for the first time the coexistence of sodium channelopathy and genetic PKP2 variations, suggesting that PKP2 mutations may be associated with BrS (Cerrone et al. 2014).

Whole-exome sequencing of ARVC patients without desmosomal mutations identified a rare missense variant in SCN5A (p.Arg1898His, R1898H). In hiPS-derived cardiomyocytes, marked I_{Na} reduction and reduced expression of $Na_V1.5$ and N-Cadherin at the ID were observed in the missense variant. Subsequently, SCN5A was sequenced in ARVC patients carrying desmosomal mutations and five potential pathogenic SCN5A variants were identified (p.Tyr416Cys, p.Leu729del, p.Arg1623Ter, p.Ser1787Asn, and p.Val2016Met). SCN5A variants displayed prolonged QRS duration and demonstrated major structural abnormalities on cardiac imaging. These observations suggest that $Na_V1.5$ exists in functional complexes composed of adhesion molecules, and reveal potential noncanonical mechanisms by which $Na_V1.5$ dysfunction may cause cardiomyopathy (Te Riele et al. 2017).

3.3.2 Desmoglein 2

Desmogleins are a family of cadherins consisting of proteins encoded by DSG1, DSG2, DSG3, and DSG4. Desmogleins are calcium binding transmembrane glycoprotein components of desmosomes. The encoded pre-protein is proteolytically processed to generate the mature glycoprotein. Mutations in the DSG2 gene have been associated with ARVC.

The first study demonstrating a role of desmoglein-2 (Dsg2) in the regulation of I_{Na} used an ARVC mouse model overexpressing the mutant desmoglein-2 (Dsg2-N271S) (Rizzo et al. 2012), a homologue of the DSG2-N266S mutation identified in an ARVC patient (Pilichou et al. 2006). Before 6 weeks, the mutant Dsg2 does not display cardiomyopathic changes allowing examination of the early electrophysiological phenotype prior to and in the absence of cardiac remodeling. In mutant mice aged <2 weeks, optical mapping revealed tendency towards lower CVs. At 3–4 weeks, discrete ventricular conduction slowing was observed (QRS complex fractionation). At later stages, Dsg2 mutants developed significant QRS prolongation, abnormal QRS morphology, spontaneous ventricular rhythm abnormalities, and decreased CV. The authors hypothesized that the conduction slowing observed in hearts prior to cardiomyopathic changes at 3–4 weeks of age could be related to altered localization or reduced levels of ID components. Although normal immunoreactive signals were detected for the other desmosomal, adherens, and gap junction proteins, as well as for $Na_V1.5$, cardiomyocytes from Dsg2 mutant showed lower AP upstroke velocity and reduced I_{Na} density. Morphometric analysis showed that the average intercellular space was widened in mutant mice at 3–4 weeks and continued to increase at later ages. As Dsg2 and $Na_V1.5$ interact in vivo, the authors concluded that molecular mechanisms responsible for the development of conduction slowing and arrhythmia in ARVC occur prior to gross and histological changes of the heart (Rizzo et al. 2012).

3.3.3 Plakoglobin

Plakoglobin, also known as γ-catenin, is encoded by the JUP gene. Plakoglobin is a member of the catenin protein family and homologous to β-catenin. Plakoglobin is a cytoplasmic component of desmosomes and adherens junction structures of the ID. Mutations in plakoglobin are associated with ARVC.

Using left and right ventricular free wall obtained from patients with ARVD and control patients, Noorman et al. showed reduced immunoreactive signal of plakoglobin, cx43, and $Na_V1.5$ at the ID whereas N-cadherin and desmoplakin signals and distribution were normal in ARVD patients. PKP2 signals were unaffected unless a PKP2 mutation predicting haploinsufficiency was present (Noorman et al. 2013).

3.3.4 Desmoplakin

Desmoplakin is encoded by the DSP gene. Desmoplakin is a critical component of desmosomes and mutations in desmoplakin have been shown to play a role in dilated cardiomyopathy, ARVC, and Carvajal syndrome.

Knock down of desmoplakin in HL-1 cells leads to decreased cx43 and $Na_V1.5$ expression as well as to abnormal distribution of cx43 and $Na_V1.5$. Gap junctional intercellular communication in adjacent cells is also decreased. I_{Na} is decreased in desmoplakin-silenced cells and the CV is slowed in cultured cells. This study indicates that impaired mechanical coupling potentially affects electrical function in ARVC (Zhang et al. 2013).

A new picture is emerging regarding the role of the different proteins populating the ID and their interactions with the $Na_V1.5$ channel. The current view is that the ID is the host of a protein interacting network or "connexome," where molecules classically defined as belonging to one particular structure (e.g., desmosome, gap junction, sodium channel complex) actually interact with others. Together, they control excitability, electrical coupling, and intercellular adhesion in the heart. The concept of a "connexome" could contribute to our understanding of the mechanisms leading to such inherited arrhythmias as ARVC and BrS (Agullo-Pascual et al. 2014a, b; Leo-Macias et al. 2016).

3.4 Dystrophin–Syntrophin Complex

The DMD gene, encoding the dystrophin protein, is one of the longest human genes. Dystrophin is a cytoplasmic protein, and a central partner of a protein complex (DGC detailed in Sect. 2.2.1) that connects the cytoskeleton of muscle fibers to the surrounding ECM through the cell membrane (Hoffman et al. 1987). The importance of dystrophin in the heart is illustrated by the observation that patients carrying DMD mutations exhibit severe cardiac phenotypes. Indeed, most Duchenne and Becker patients suffer from a dilated cardiomyopathy (Finsterer and Stöllberger 2003). The X-linked form of dilated cardiomyopathy is also caused by mutations in DMD (Towbin et al. 1993). Dystrophin has multiple protein–protein interaction domains, but can also interact indirectly with additional proteins via the syntrophin adaptor proteins (Albrecht and Froehner 2002). The cytosolic scaffolding product of the SNTA1 gene is the rod-shaped syntrophin α1 (SNTA1) protein. SNTA1 contains a PDZ domain, two pleckstrin homology (PH) domains, and a syntrophin unique (SU) domain. Syntrophin proteins interact with a specific domain in the C-terminus of dystrophin and, through its PDZ domain, with the C-terminal tail of various cardiac ion channels.

$Na_V1.5$ contains a PDZ-domain binding motif in its C-terminus, the SIV motif. This motif is required for indirect interaction with dystrophin via syntrophin proteins (Gavillet et al. 2006). Gavillet and coworkers have shown that in dystrophin-deficient cardiac cells from mdx mice, $Na_V1.5$ protein expression is decreased along with I_{Na} (Gavillet et al. 2006). The interaction between $Na_V1.5$ and dystrophin occurs exclusively at the LM since both dystrophin and syntrophin are virtually absent at the ID (Petitprez et al. 2011). Although the mechanism by which the DGC regulates the expression and localization of $Na_V1.5$ is still not known, it is likely by posttranslational regulation since $Na_V1.5$ mRNA levels are unaltered in dystrophin-deficient tissue (Gavillet et al. 2006). The role of the SIV motif of $Na_V1.5$ has been investigated in vivo. Knock-in mice lacking the SIV domain (ΔSIV) displays reduced $Na_V1.5$ expression and sodium current at the LM, whereas no differences were observed at the ID (Shy et al. 2014). Optical mapping of ΔSIV hearts revealed that CV was preferentially decreased in the transversal direction, leading to altered anisotropy. Shy et al. showed that internalization of ΔSIV channels is not modified in HEK293 cells. Using the proteasome inhibitor MG132, I_{Na} is rescued in ΔSIV cells, suggesting a role for this motif in the degradation process of $Na_V1.5$ (Shy et al. 2014). Interestingly, a missense mutation corresponding to the SIV motif was identified in a patient with BrS (p.V2016 M). When expressed in HEK293 cells, the mutation decreased both $Na_V1.5$ cell surface expression and I_{Na}, suggesting clinical relevance for the SIV motif in cardiac disease (Shy et al. 2014).

Recently, Delpon's group demonstrated the role of the N-terminus of $Na_V1.5$ (132 aa) in the reciprocal regulation of Kir and $Na_V1.5$ channels previously reported by J. Jalife's group (Milstein et al. 2012). The N-terminal domain of $Na_V1.5$ contains residues similar to the C-terminal consensus sequence for binding to syntrophin and is able to interact with α1-syntrophin. The $Na_V1.5$ N-terminal domain exerts a "chaperone-like" effect by increasing both sodium current and inward rectifier potassium currents by enhancing $Na_V1.5$, Kir2.1, and Kir2.2 channels' expression. Therefore, the $Na_V1.5$ N-terminal acts as an internal PDZ-like binding domain and plays a critical role in $Na_V1.5$-Kir2.x-reciprocal interactions (Matamoros et al. 2016).

Different studies identified mutations in SNTA1 in patients with congenital long QT syndrome (LQTS) (Wu et al. 2008; Ueda et al. 2008). The SNTA1 mutation p. A390V was shown to disrupt association of neuronal NOS (nNOS), the plasma membrane Ca-ATPase type 4b, syntrophin, and $Na_V1.5$. Expression of the mutated syntrophin in cardiac myocytes increases late I_{Na}, which may explain the prolongation of the QT interval in patients harboring this mutation. A proposed mechanism is that by abolishing the interaction between syntrophin, nNOS, and the Ca-ATPase, $Na_V1.5$ nitrosylation increases and subsequently stimulates the late

I_{Na} (Ueda et al. 2008). Patients harboring a double mutation in both SCN5A (R800L) and SNTA1 (A261V) display stronger clinical phenotype (Hu et al. 2013). When co-expressed with nNOS and the Ca-ATPase, the double mutant does not show changes in I_{Na} peak but rather enhanced late sodium current that can be blocked by nNOS inhibitors. Therefore, mutations in SCN5A and SNTA1 jointly exert a nNOS-dependent regulation of I_{Na} potentially leading to a prolonged APD and a LQTS phenotype (Hu et al. 2013). In the same line, drug-induced long-QT syndrome (diLQTS), often due to reduction of I_{Kr}, has been associated with a novel α1-syntrophin variant (p.E409Q). The E409Q mutation reduced the late I_{Na} without affecting I_{Na} peak, suggesting that variants within the $Na_V1.5$-interacting α1-syntrophin are potential mechanism for diLQTS (Choi et al. 2016).

3.5 Caveolins

Caveolins 1–3 are integral membrane proteins that form a hairpin loop inside the membrane, leaving both C- and N-terminals oriented towards the cytosol. Caveolins are the main constituents of caveolae, small invaginations of the plasma membrane first identified by electron microscopy as "little caves" (Palade 1953). These flask-shaped invaginations of the plasma membrane which lack an electron-dense coat are involved in endocytosis (Bastiani and Parton 2010). Caveolins also act as scaffolding proteins by concentrating signaling molecules and ion channels. Caveolin-3 (CAV3) is the main caveolin isoform expressed in the myocardium.

The first role of caveolin in the regulation of cardiac I_{Na} was demonstrated by Yarbrough and coworkers. They revealed that CAV3 co-immunoprecipitates with $Na_V1.5$ from rat cardiac tissue. β-adrenergic stimulation in cardiac myocytes increased I_{Na} through direct activation of the G-protein α-subunit (Yarbrough et al. 2002). Interestingly, the authors suggested that G-protein activation through β-adrenergic stimulation induces the opening of caveolae, and therefore, the addition of functional sodium channels to the sarcolemma. Although to date the interaction site between CAV3 and $Na_V1.5$ is still unknown, the histidine residue at position 41 of G-alpha is critical for I_{Na} increase (Palygin et al. 2008). Dystrophin has also been shown to be a component of caveolae (Doyle et al. 2000), raising the possibility that the interaction of CAV3 with $Na_V1.5$ could be indirect, likely through proteins of the DGC (Gavillet et al. 2006).

From the observation that $Na_V1.5$ channel localizes in caveolae (Shibata et al. 2006), subsequent studies investigated whether mutations in CAV3 may represent novel pathogenetic mechanisms for LQTS and Sudden Infant Death Syndrome (SIDS). Four mutations in CAV3 (F97C, S141R, T78M, and A85T) were identified in unrelated patients referred for LQTS. CAV3 mutants result in late I_{Na} increase (Vatta et al. 2006). Co-expression of SCN5A, SNTA1, nNOS, and F97C mutated-CAV3 in HEK cells resulted in enhanced *S*-nitrosylation of $Na_V1.5$. nNOS inhibitors reversed both the CAV3-F97C induced increase in late and peak I_{Na}, and decreased S-nitrosylation of SCN5A (Cheng et al. 2013). Similarly, three distinct CAV3 mutations leading to increased late I_{Na} (V14L, T78M, and L79R) were identified in SIDS (Cronk et al. 2007). These studies provide functional evidences that CAV3 mutations can induce a gain-of-function of late I_{Na} in subjects associated with LQTS or SIDS.

3.6 MAGUK Proteins

MAGUK proteins constitute a large family of multi-domain proteins. MAGUK proteins have emerged as central organizers of specialized plasma membrane domains in various cell types including cardiac myocytes. These proteins, which localize beneath the plasma membrane, regulate the surface expression of several transmembrane proteins. In 1995, Kim et al. provided the first evidence that neuronal Shaker channels are clustered by MAGUK proteins (Kim et al. 1995). Since then, MAGUK proteins have been shown to be major partners of most ion channels.

The molecular diversity of MAGUK proteins has been revealed by the study of the neuronal synapses. At the ultrastructural level, synapses are asymmetric, characterized by a dense thickening at the cytoplasmic face of the postsynaptic membrane, called postsynaptic density (PSD). In 1992, Cho et al. isolated the first MAGUK protein from an enriched PSD fraction (Cho et al. 1992). This protein, which migrates as a doublet with an apparent molecular weight of 95 kDa, was named PSD-95. Subsequently, SAP90 (Kistner et al. 1993) and SAP97 (Müller et al. 1995) were isolated from the presynaptic termini of a subset of inhibitory synapses in the rat cerebellum. Two other proteins, structurally related to SAP90 and SAP97, have been found in the adherens junction fraction of epithelial cells or in the intercalated disc of cardiac myocytes, and named ZO-1 and ZO-2 for Zona Occludens (Itoh et al. 1993, 1999). Structurally, most MAGUK proteins express one Src homology 3 (SH3) domain, one guanylate kinase-like domain (GUK), and one or several PSD-95/Dlg/ZO-1 (PDZ) domains. Besides these canonical domains, some MAGUKs express WW domains (conserved T residues), L27 domains (Lin2–Lin7), and/or a CaMKII domain. They can be classified into ten subfamilies on the basis of their structural domains (de Mendoza et al. 2010).

3.6.1 ZO1

Zonula occludens-1 (ZO-1), also known as tight junction protein-1, is a 220-kD peripheral membrane protein encoded by the TJP1 gene. ZO-1 possesses three PDZ (class I) domains, one SH3 domain, and one GUK domain.

In cardiac myocytes, ZO-1 is specifically located at the ID level where it interacts with the gap junctional protein cx43 and the adherens junction protein N-cadherin (Maass et al. 2007; Palatinus et al. 2011). Although, the precise mechanism by which cx43 is localized within cardiac myocytes is unknown, it is likely through interaction with the cytoskeleton as cx43 specifically colocalizes with the cytoskeletal proteins ZO-1 and α-spectrin (Toyofuku et al. 1998). However, co-localization between cx43 and ZO-1 is modest in ventricular myocytes as

ZO-1 is not present on the entire junctional plaque but located at the perinexus (Zhu et al. 2005; Rhett et al. 2012; Rhett and Gourdie 2012).

During development and pathophysiological remodeling, increased co-localization between cx43 and ZO-1 has been observed, suggesting a role for ZO-1 in the maintenance of gap junctions (Barker et al. 2002). Inhibition of ZO-1 in cardiac myocytes leads to a decrease in N-cadherin expression at the ID, and subsequently to destabilization of adherens junctions and lateralization of cx43 (Hunter et al. 2005; Palatinus et al. 2011; Rhett et al. 2011a, b). However, although $Na_V1.5$ was found to interact with perinexal cx43, it does not with ZO-1 (Rhett et al. 2012).

3.6.2 SAP97

SAP97, also known as synapse-associated protein 97 or disks large homolog 1 (DLG1), is encoded by the SAP97 gene. Similar to ZO-1, SAP97 is formed of three PDZ (class I) domains, one SH3 domain, one GUK domain, and one L27 domain located in N-terminus.

Immunofluorescence studies performed in myocardial sections show that SAP97 is predominantly localized at or near the ID, but is also observed laterally along the LM (El-Haou et al. 2009; Godreau et al. 2002, 2003; Peters et al. 2009; Milstein et al. 2012; Petitprez et al. 2011). Further evidence that SAP97 is enriched at the ID was obtained by using a model of cultured adult cardiac myocytes, in which formation of specialized myocyte–myocyte contacts, such as adherens and gap junctions, can be followed in vitro (Abi-Char et al. 2008). Only one study localized SAP97 to the T-tubule system in isolated ventricular myocytes (Leonoudakis et al. 2001). Coculture of cardiac myocytes and sympathetic neuronal cells has revealed another distinct localization of SAP97 at the contacts between nerve endings and myocytes, where the anchoring protein appears to be part of the β-adrenergic signaling complex (Shcherbakova et al. 2007). Taken together these studies clearly indicate a specific location for SAP97 in specialized membrane domains in cardiomyocytes. Thus, SAP97 is likely involved in the formation and regulation of specific protein complexes in specialized microdomains in cardiomyocytes.

Several cardiac ion channels have been shown to interact with SAP97, among them, $Na_V1.5$ (Petitprez et al. 2011; Milstein et al. 2012; Shy et al. 2014). The functional consequence of this interaction in vitro is an increase of I_{Na} (Petitprez et al. 2011; Milstein et al. 2012). No change in unitary conductance or open probability of $Na_V1.5$ was observed but enhancement of the functional channel density was reported (Milstein et al. 2012). At the protein level, $Na_V1.5$ surface expression is decreased in cardiac myocytes silenced for SAP97 (Petitprez et al. 2011). Thus, the stimulatory effect of SAP97 on I_{Na} is not associated with changes in biophysical properties of $Na_V1.5$ but rather enhance its surface expression. Interestingly, as for the interaction with syntrophin, the last three residues (SIV) of $Na_V1.5$ C-terminus are required for interaction with SAP97 (Shy et al. 2014). Deletion of these residues in mice results in incorrect localization of Na_V15 channels specifically at the LM. Finally, a missense mutation identified in $Na_V1.5$ C-terminal domain in a BrS patient (V2016M), not only reduces I_{Na} and channel membrane expression but also decreases the interaction between $Na_V1.5$ and

SAP97 (Shy et al. 2014). Therefore, the C-terminal SIV residues constitute a PDZ domain – binding motif that interacts with PDZ proteins at different locations within the cardiomyocyte: at the LM with syntrophin and at the ID with SAP97, thereby defining distinct $Na_V1.5$ multiprotein complexes.

How SAP97 enhances the number of functional channels in the plasma membrane is not yet completely understood. The following mechanisms have been proposed: increased forward trafficking, aggregation of channels, and formation of macromolecular complexes. Several studies indicate that SAP97 retains and stabilizes channels at the plasma membrane, such as PSD95 does with neuronal Kv1.4 channels (Jugloff et al. 2000) and as Lin-7/CASK complex does for the targeting of Kir2.3 channel at the basolateral membrane in epithelial cell lines (Olsen et al. 2002). Another mechanism underlying the effects of SAP97 on ion channels is related to its capacity to multimerize to form a large protein network, thereby facilitating interactions of channel subunits with accessory proteins. Keeping with the discovery that SAP97 is necessary for the interaction between $Na_V1.5$ and Kir2.1 (Milstein et al. 2012), MAGUK proteins could be major partners for the organization of multichannel complexes controlling cardiac excitability. The recent work of H. Abriel's group showing no SAP97-dependant regulation of I_{Na} in vivo is difficult to explain (Gillet et al. 2015). It could indicate that the function of macromolecular complexes associated with SAP97 is not conserved across tissues, species, cell types, or is modified after cell isolation.

3.6.3 CASK

The CASK (CAlcium/calmodulin-dependent Serine/threonine Kinase) protein is encoded by the CASK gene, also known as CMG2 (CAMGUK protein 2), calcium/calmodulin-dependent serine protein kinase 3, and MAGUK 2. CASK contains multidomain modules that mediate protein–protein interactions important for the establishment and maintenance of polarization in neurons and in epithelial cells (Atasoy et al. 2007; Hsueh 2009; Hsueh et al. 2000; Leonoudakis et al. 2001). As ZO-1 and SAP97, CASK possesses the SH3 and GUK domains, but differs by its possession of a CaMK domain in the N-terminal, two L27 domains, and a single PDZ (class II) domain.

CASK is expressed in many tissues and constitutive CASK-KO in mice causes cleft palate, synaptic dysfunction, and lethality (Atasoy et al. 2007). Mutations in the human gene of CASK are associated with a form of X-linked mental retardation (FG syndrome 4), mental retardation, and microcephaly with pontine and cerebellar hypoplasia (Hackett et al. 2010; Hsueh 2009; Valayannopoulos et al. 2012). Interestingly, CASK is not only located at cell-to-cell junctions like other MAGUKs, but is also able to translocate into the nucleus and to function as a regulator of transcription factors, such as Tbr-1 and Id1, to induce transcription of T-or E-element containing genes (Bredt 2000; Hsueh et al. 2000; Ojeh et al. 2008; Sun et al. 2009).

Few studies have been conducted regarding the expression and role of CASK in the myocardium despite its expression having been reported since 1998 (Cohen et al. 1998). The first work reporting CASK protein expression in the myocardium came from a proteomic study showing CASK association with the inward rectifier

Kir2 potassium channel in a multiprotein complex involving SAP97, Veli-3, and Mint1 (Leonoudakis et al. 2004). Because CASK contains a class II PDZ domain, the authors suggested that CASK could not directly interact with the channel C-terminal but is rather recruited through SAP97 or Veli. The C-terminal of Kir4.1 channels only associate with proteins of the DGC whereas Kir2.1, Kir2.2, and Kir2.3 channels can associate with both the DGC and the SAP97/CASK/Veli complexes. This data suggests that the C-terminal PDZ binding motifs of Kir channels confer specificity to their association with multicomponent protein complexes (Leonoudakis et al. 2004).

A recent study showed that CASK is restricted to the LM of myocytes, making CASK the first MAGUK excluded from the ID (Eichel et al. 2016). Interestingly, CASK localization at the LM was found to be dependent on the presence of dystrophin, as CASK was no longer expressed at the membrane in the dystrophin deficient mdx mouse myocardium. CASK and dystrophin also co-localized and co-immunoprecipitated, suggesting that CASK could be a new member of the DGC at the costamere. Contrary to all other identified $Na_V1.5$ channel partners, CASK negatively regulates the sodium current: while CASK over-expression reduces I_{Na}, CASK silencing in vitro and knock down in vivo increases I_{Na} (Eichel et al. 2016). GST pull-down experiments using the $Na_V1.5$ C-terminal (SIV motif) revealed that CASK directly interacts with $Na_V1.5$. New insights regarding the processes by which CASK regulates $Na_V1.5$ were provided. While CASK does not modify $Na_V1.5$ transcriptional or translational expressions, CASK-silencing increases the surface expression of $Na_V1.5$, and more specifically, at the LM. Treatment with brefeldin-A, an inhibitor of forward transport between the endoplasmic reticulum and the Golgi apparatus, prevents the effect, suggesting that CASK impedes early steps of $Na_V1.5$ trafficking (Eichel et al. 2016). In this regard, CASK is unique among MAGUK proteins investigated so far that usually favors anchoring/stabilization of ion channels at the sarcolemma. During atrial dilation/remodeling in both human and in a rodent model of chronic hemodynamic overload, CASK expression is reduced whereas its localization remains unchanged. This suggests that CASK down-regulation during cardiac diseases could alter the functional expression of $Na_V1.5$ at the LM. These observations not only strengthen the concept of differentially regulated populations of $Na_V1.5$ channels within the cardiomyocyte thereby playing distinct roles in cardiac physiology, but also suggest that CASK could participate in maintaining low levels of $Na_V1.5$ at the LM and consequently contribute to anisotropic conduction.

4 Conclusion

Targeting and trafficking of membrane proteins, including channels and receptors, have been extensively investigated in other excitable tissues such as the nervous system and epithelial tissues. These studies provide a body of knowledge that raises important insights into the regulation of cardiac excitability. However, due to the distinct structural and functional characteristics of cardiac myocytes, there are specificities in the trafficking and targeting of ion channels that are probably not shared with other cell types. Although they harbor specific membrane compartments and, in this sense, can be considered as polarized cells (Balse et al. 2012), they display a high level of ultrastructural organization at both the level of the ID and the LM that continues to stimulate efforts from the community of research.

In addition, important questions remain to be answered such as: how are the specialized plasma membrane domains of cardiac myocytes organized; how are channels targeted to these platforms; how are they anchored and aggregated to form large protein complexes; and via which posttranscriptional and posttranslational mechanisms are they regulated?

A central hypothesis could be that the specific targeting of the different subpopulations of $Na_V1.5$ channel to the LM or to the ID is governed by interactions with, and competition between, the various channel partners from the endoplasmic reticulum exit to the final location of $Na_V1.5$ channels at the sarcolemma. As such, one could hypothesize that $Na_V1.5$ sorting could be determined during anterograde processing steps of trafficking. Notably, early association of the channel with β auxiliary subunits and/or with other partners during biosynthesis could determine $Na_V1.5$ channel sorting and final targeting to specific subdomains. To our knowledge, the hypothesis of a sorting hub at the early stages of processing, which would orient the final targeting of the channel towards specific membrane microdomains, has never been investigated, nor the synchronous trafficking of $Na_V1.5$ channel and protein partners.

Besides the focus on the urgent need for more knowledge on the trafficking and targeting of cardiac $Na_V1.5$ channel in the heart, this chapter also showed that these processes are central to the tuning of the electrical activity of the myocardium. An important perspective of future research will be to determine the complex relationship between microdomains organization and trafficking/targeting processes of $Na_V1.5$ and protein partners, both in physiological conditions and during physiopathological remodeling associated with alterations of channel localization and changes in the ultrastructure of the myocardium.

References

Abi-Char J, El-Haou S, Balse E, Neyroud N, Vranckx R, Coulombe A, Hatem SN (2008) The anchoring protein SAP97 retains Kv1.5 channels in the plasma membrane of cardiac myocytes. Am J Physiol Heart Circ Physiol 294(4):H1851–H1861. https://doi.org/10.1152/ajpheart.01045.2007

Agullo-Pascual E, Cerrone M, Delmar M (2014a) Arrhythmogenic cardiomyopathy and brugada syndrome: diseases of the connexome. FEBS Lett 588(8):1322–1330. https://doi.org/10.1016/j.febslet.2014.02.008

Agullo-Pascual E, Lin X, Leo-Macias A, Zhang M, Liang F-X, Li Z, Pfenniger A et al (2014b) Super-resolution imaging reveals that loss of the C-terminus of connexin43 limits microtubule plus-end capture and $Na_V1.5$ localization at the intercalated disc. Cardiovasc Res 104(2):371–381. https://doi.org/10.1093/cvr/cvu195

Albrecht DE, Froehner SC (2002) Syntrophins and dystrobrevins: defining the dystrophin scaffold at synapses. Neurosignals 11(3):123–129

Allamand V, Campbell KP (2000) Animal models for muscular dystrophy: valuable tools for the development of therapies. Hum Mol Genet 9(16):2459–2467

Atasoy D, Schoch S, Ho A, Nadasy KA, Liu X, Zhang W, Mukherjee K et al (2007) Deletion of CASK in mice is lethal and impairs synaptic function. Proc Natl Acad Sci U S A 104 (7):2525–2530. https://doi.org/10.1073/pnas.0611003104

Balse E, Steele DF, Abriel H, Coulombe A, Fedida D, Hatem SN (2012) Dynamic of ion channel expression at the plasma membrane of cardiomyocytes. Physiol Rev 92(3):1317–1358. https://doi.org/10.1152/physrev.00041.2011

Barker RJ, Price RL, Gourdie RG (2002) Increased association of ZO-1 with connexin43 during remodeling of cardiac gap junctions. Circ Res 90(3):317–324

Bastiani M, Parton RG (2010) Caveolae at a glance. J Cell Sci 123(22):3831–3836. https://doi.org/10.1242/jcs.070102

Bershadsky AD, Balaban NQ, Geiger B (2003) Adhesion-dependent cell mechanosensitivity. Annu Rev Cell Dev Biol 19:677–695. https://doi.org/10.1146/annurev.cellbio.19.111301.153011

Bhargava A, Lin X, Novak P, Mehta K, Korchev Y, Delmar M, Gorelik J (2013) Super-resolution scanning patch clamp reveals clustering of functional ion channels in adult ventricular myocyte. Circ Res 112(8):1112–1120. https://doi.org/10.1161/CIRCRESAHA.111.300445

Bootman MD, Harzheim D, Smyrnias I, Conway SJ, Roderick HL (2007) Temporal changes in atrial EC-coupling during prolonged stimulation with endothelin-1. Cell Calcium 42(4-5):489–501. https://doi.org/10.1016/j.ceca.2007.05.004

Borrmann CM, Grund C, Kuhn C, Hofmann I, Pieperhoff S, Franke WW (2006) The area composita of adhering junctions connecting heart muscle cells of vertebrates. II. Colocalizations of desmosomal and fascia adhaerens molecules in the intercalated disk. Eur J Cell Biol 85(6):469–485. https://doi.org/10.1016/j.ejcb.2006.02.009

Boycott HE, Barbier CSM, Eichel CA, Costa KD, Martins RP, Louault F, Dilanian G, Coulombe A, Hatem SN, Balse E (2013) Shear stress triggers insertion of voltage-gated potassium channels from intracellular compartments in atrial myocytes. Proc Natl Acad Sci U S A 110(41):E3955–E3964. https://doi.org/10.1073/pnas.1309896110

Bredt DS (2000) Cell biology. Reeling CASK into the nucleus. Nature 404(6775):241–242. https://doi.org/10.1038/35005208.

Brenman JE, Chao DS, Xia H, Aldape K, Bredt DS (1995) Nitric oxide synthase complexed with dystrophin and absent from skeletal muscle sarcolemma in Duchenne muscular dystrophy. Cell 82(5):743–752

Casini S, Tan HL, Demirayak I, Remme CA, Amin AS, Scicluna BP, Chatyan H et al (2010) Tubulin polymerization modifies cardiac sodium channel expression and gating. Cardiovasc Res 85(4):691–700. https://doi.org/10.1093/cvr/cvp352

Cerrone M, Noorman M, Lin X, Chkourko H, Liang F-X, van der Nagel R, Hund T et al (2012) Sodium current deficit and arrhythmogenesis in a murine model of plakophilin-2-haploinsufficiency. Cardiovasc Res 95(4):460–468. https://doi.org/10.1093/cvr/cvs218

Cerrone M, Lin X, Zhang M, Agullo-Pascual E, Pfenniger A, Gusky HC, Novelli V et al (2014) Missense mutations in plakophilin-2 cause sodium current deficit and associate with a brugada syndrome phenotype. Circulation 129(10):1092–1103. https://doi.org/10.1161/CIRCULATIONAHA.113.003077

Cheng J, Valdivia CR, Vaidyanathan R, Balijepalli RC, Ackerman MJ, Makielski JC (2013) Caveolin-3 suppresses late sodium current by inhibiting nNOS-dependent S-nitrosylation of SCN5A. J Mol Cell Cardiol 61:102–110. https://doi.org/10.1016/j.yjmcc.2013.03.013

Chkourko HS, Guerrero-Serna G, Lin X, Darwish N, Pohlmann JR, Cook KE, Martens JR, Rothenberg E, Musa H, Delmar M (2012) Remodeling of mechanical junctions and of microtubule-associated proteins accompany cardiac connexin43 lateralization. Heart Rhythm 9(7):1133–1140.e6. https://doi.org/10.1016/j.hrthm.2012.03.003

Cho KO, Hunt CA, Kennedy MB (1992) The rat brain postsynaptic density fraction contains a homolog of the drosophila discs-large tumor suppressor protein. Neuron 9(5):929–942

Choi J-I, Wang C, Thomas MJ, Pitt GS (2016) α1-syntrophin variant identified in drug-induced long QT syndrome increases late sodium current. PLoS One 11(3):e0152355. https://doi.org/10.1371/journal.pone.0152355
Clark KA, McElhinny AS, Beckerle MC, Gregorio CC (2002) Striated muscle cytoarchitecture: an intricate web of form and function. Ann Rev Cell Dev Biol 18:637–706. https://doi.org/10.1146/annurev.cellbio.18.012502.105840
Cohen AR, Woods DF, Marfatia SM, Walther Z, Chishti AH, Anderson JM, Wood DF (1998) Human CASK/LIN-2 binds Syndecan-2 and protein 4.1 and localizes to the basolateral membrane of epithelial cells. J Cell Biol 142(1):129–138
Cohn RD, Campbell KP (2000) Molecular basis of muscular dystrophies. Muscle Nerve 23 (10):1456–1471
Cronk LB, Ye B, Kaku T, Tester DJ, Vatta M, Makielski JC, Ackerman MJ (2007) Novel mechanism for sudden infant death syndrome: persistent late sodium current secondary to mutations in caveolin-3. Heart Rhythm 4(2):161–166. https://doi.org/10.1016/j.hrthm.2006.11.030
Cunha SR, Mohler PJ (2006) Cardiac ankyrins: essential components for development and maintenance of excitable membrane domains in heart. Cardiovasc Res 71(1):22–29. https://doi.org/10.1016/j.cardiores.2006.03.018
Danowski BA, Imanaka-Yoshida K, Sanger JM, Sanger JW (1992) Costameres are sites of force transmission to the substratum in adult rat cardiomyocytes. J Cell Biol 118(6):1411–1420
Desplantez T, McCain ML, Beauchamp P, Rigoli G, Rothen-Rutishauser B, Parker KK, Kleber AG (2012) Connexin43 ablation in foetal atrial myocytes decreases electrical coupling, partner connexins, and sodium current. Cardiovasc Res 94:58–65
de Mendoza A, Suga H, Ruiz-Trillo I (2010) Evolution of the MAGUK protein gene family in premetazoan lineages. BMC Evol Biol 10:93. https://doi.org/10.1186/1471-2148-10-93
Doyle DD, Goings G, Upshaw-Earley J, Ambler SK, Mondul A, Palfrey HC, Page E (2000) Dystrophin associates with caveolae of rat cardiac myocytes: relationship to dystroglycan. Circ Res 87(6):480–488
Durbeej M, Campbell KP (2002) Muscular dystrophies involving the dystrophin-glycoprotein complex: an overview of current mouse models. Curr Opin Genet Dev 12(3):349–361
Eichel CA, Beuriot A, Chevalier MYE, Rougier J-S, Louault F, Dilanian G, Amour J et al (2016) Lateral membrane-specific MAGUK CASK down-regulates $Na_V1.5$ channel in cardiac myocytes. Circ Res 119(4):544–556. https://doi.org/10.1161/CIRCRESAHA.116.309254
El-Haou S, Balse E, Neyroud N, Dilanian G, Gavillet B, Abriel H, Coulombe A, Jeromin A, Hatem SN (2009) Kv4 potassium channels form a tripartite complex with the anchoring protein SAP97 and CaMKII in cardiac myocytes. Circ Res 104(6):758–769. https://doi.org/10.1161/CIRCRESAHA.108.191007
Ervasti JM (2003) Costameres: the Achilles' heel of herculean muscle. J Biol Chem 278 (16):13591–13594. https://doi.org/10.1074/jbc.R200021200
Finsterer J, Stöllberger C (2003) The heart in human dystrophinopathies. Cardiology 99(1):1–19
Franke WW, Borrmann CM, Grund C, Pieperhoff S (2006) The area composita of adhering junctions connecting heart muscle cells of vertebrates. I. Molecular definition in intercalated disks of cardiomyocytes by immunoelectron microscopy of desmosomal proteins. Eur J Cell Biol 85(2):69–82. https://doi.org/10.1016/j.ejcb.2005.11.003
Gavillet B, Rougier J-S, Domenighetti AA, Behar R, Boixel C, Ruchat P, Lehr H-A, Pedrazzini T, Abriel H (2006) Cardiac sodium channel $Na_V1.5$ is regulated by a multiprotein complex composed of syntrophins and dystrophin. Circ Res 99(4):407–414. https://doi.org/10.1161/01.RES.0000237466.13252.5e
Geisler SB, Green KJ, Isom LL, Meshinchi S, Martens JR, Delmar M, Russell MW (2010) Ordered assembly of the adhesive and electrochemical connections within newly formed intercalated disks in primary cultures of adult rat cardiomyocytes. J Biomed Biotechnol 2010:624719. https://doi.org/10.1155/2010/624719
Gillet L, Rougier J-S, Shy D, Sonntag S, Mougenot N, Essers M, Shmerling D, Balse E, Hatem SN, Abriel H (2015) Cardiac-specific ablation of synapse-associated protein SAP97 in mice

decreases potassium currents but not sodium current. Heart Rhythm 12(1):181–192. https://doi.org/10.1016/j.hrthm.2014.09.057

Godreau D, Vranckx R, Maguy A, Rücker-Martin C, Goyenvalle C, Abdelshafy S, Tessier S, Couétil JP, Hatem SN (2002) Expression, regulation and role of the MAGUK protein SAP-97 in human atrial myocardium. Cardiovasc Res 56(3):433–442

Godreau D, Vranckx R, Maguy A, Goyenvalle C, Hatem SN (2003) Different isoforms of synapse-associated protein, SAP97, are expressed in the heart and have distinct effects on the voltage-gated K+ channel Kv1.5. J Biol Chem 278(47):47046–47052. https://doi.org/10.1074/jbc.M308463200

Grady RM, Grange RW, Lau KS, Maimone MM, Nichol MC, Stull JT, Sanes JR (1999) Role for alpha-dystrobrevin in the pathogenesis of dystrophin-dependent muscular dystrophies. Nat Cell Biol 1(4):215–220. https://doi.org/10.1038/12034

Hackett A, Tarpey PS, Licata A, Cox J, Whibley A, Boyle J, Rogers C et al (2010) CASK mutations are frequent in males and cause X-linked nystagmus and variable XLMR phenotypes. Eur J Human Genet 18(5):544–552. https://doi.org/10.1038/ejhg.2009.220

Hashemi SM, Hund TJ, Mohler PJ (2009) Cardiac ankyrins in health and disease. J Mol Cell Cardiol 47(2):203–209. https://doi.org/10.1016/j.yjmcc.2009.04.010

Hoffman EP, Brown RH, Kunkel LM (1987) Dystrophin: the protein product of the duchenne muscular dystrophy locus. Cell 51(6):919–928

Hsueh Y-P (2009) Calcium/Calmodulin-dependent serine protein kinase and mental retardation. Ann Neurol 66(4):438–443. https://doi.org/10.1002/ana.21755

Hsueh YP, Wang TF, Yang FC, Sheng M (2000) Nuclear translocation and transcription regulation by the membrane-associated guanylate kinase CASK/LIN-2. Nature 404(6775):298–302. https://doi.org/10.1038/35005118.

Hu R-M, Tan B-H, Orland KM, Valdivia CR, Peterson A, Jielin P, Makielski JC (2013) Digenic inheritance novel mutations in SCN5a and SNTA1 increase late I(Na) contributing to LQT syndrome. Am J Physiol Heart Circ Physiol 304(7):H994–H1001. https://doi.org/10.1152/ajpheart.00705.2012

Hunter AW, Barker RJ, Zhu C, Gourdie RG (2005) Zonula Occludens-1 alters connexin43 gap junction size and organization by Influencing Channel accretion. Mol Biol Cell 16 (12):5686–5698. https://doi.org/10.1091/mbc.E05-08-0737

Itoh M, Nagafuchi A, Yonemura S, Kitani-Yasuda T, Tsukita S, Tsukita S (1993) The 220-kD protein colocalizing with cadherins in non-epithelial cells is identical to ZO-1, a tight junction-associated protein in epithelial cells: cDNA cloning and immunoelectron microscopy. J Cell Biol 121(3):491–502

Itoh M, Morita K, Tsukita S (1999) Characterization of ZO-2 as a MAGUK family member associated with tight as well as Adherens junctions with a binding affinity to occludin and alpha catenin. J Biol Chem 274(9):5981–5986

Jansen JA, Noorman M, Musa H, Stein M, de Jong S, van der Nagel R, Hund TJ et al (2012) Reduced heterogeneous expression of Cx43 results in decreased $Na_V1.5$ expression and reduced sodium current which accounts for arrhythmia vulnerability in conditional Cx43 knockout mice. Heart Rhythm 9(4):600–607. https://doi.org/10.1016/j.hrthm.2011.11.025

Jugloff DG, Khanna R, Schlichter LC, Jones OT (2000) Internalization of the Kv1.4 potassium channel is suppressed by clustering interactions with PSD-95. J Biol Chem 275(2):1357–1364

Kim E, Niethammer M, Rothschild A, Jan YN, Sheng M (1995) Clustering of shaker-type K+ channels by interaction with a family of membrane-associated Guanylate kinases. Nature 378 (6552):85–88. https://doi.org/10.1038/378085a0

Kistner U, Wenzel BM, Veh RW, Cases-Langhoff C, Garner AM, Appeltauer U, Voss B, Gundelfinger ED, Garner CC (1993) SAP90, a rat presynaptic protein related to the product of the drosophila tumor suppressor gene Dlg-a. J Biol Chem 268(7):4580–4583

Kumar NM, Gilula NB (1996) The gap junction communication channel. Cell 84(3):381–388

Lab MJ, Bhargava A, Wright PT, Gorelik J (2013) The scanning ion conductance microscope for cellular physiology. Am J Physiol Heart Circ Physiol 304(1):H1–H11. https://doi.org/10.1152/ajpheart.00499.2012

Lapidos KA, Kakkar R, McNally EM (2004) The dystrophin glycoprotein complex: signaling strength and integrity for the sarcolemma. Circ Res 94(8):1023–1031. https://doi.org/10.1161/01.RES.0000126574.61061.25

Leo-Macias A, Agullo-Pascual E, Delmar M (2016) The cardiac connexome: non-canonical functions of connexin43 and their role in cardiac arrhythmias. Semin Cell Dev Biol 50:13–21. https://doi.org/10.1016/j.semcdb.2015.12.002

Leonoudakis D, Mailliard W, Wingerd K, Clegg D, Vandenberg C (2001) Inward rectifier potassium channel Kir2.2 is associated with synapse-associated protein SAP97. J Cell Sci 114(Pt 5):987–998

Leonoudakis D, Conti LR, Anderson S, Radeke CM, McGuire LMM, Adams ME, Froehner SC, Yates JR, Vandenberg CA (2004) Protein trafficking and anchoring complexes revealed by proteomic analysis of inward rectifier potassium channel (Kir2.X)-associated proteins. J Biol Chem 279(21):22331–22346. https://doi.org/10.1074/jbc.M400285200

Lin X, Liu N, Lu J, Zhang J, Anumonwo JM, Isom LL, Fishman GI, Delmar M (2011) Subcellular heterogeneity of sodium current properties in adult cardiac ventricular myocytes. Heart Rhythm 8(12):1923–1930. https://doi.org/10.1016/j.hrthm.2011.07.016

Loewenstein WR (1981) Junctional intercellular communication: the cell-to-cell membrane channel. Physiol Rev 61(4):829–913

Lowe JS, Palygin O, Bhasin N, Hund TJ, Boyden PA, Shibata E, Anderson ME, Mohler PJ (2008) Voltage-gated Nav channel targeting in the heart requires an Ankyrin-G dependent cellular pathway. J Cell Biol 180(1):173–186. https://doi.org/10.1083/jcb.200710107

Lübkemeier I, Requardt RP, Lin X, Sasse P, Andrié R, Schrickel JW, Chkourko H, Bukauskas FF, Kim J-S, Frank M et al (2013) Deletion of the last five C-terminal amino acid residues of connexin43 leads to lethal ventricular arrhythmias in mice without affecting coupling via gap junction channels. Basic Res Cardiol 108:348

Maass K, Shibayama J, Chase SE, Willecke K, Delmar M (2007) C-terminal truncation of connexin43 changes number, size, and localization of cardiac gap junction plaques. Circ Res 101(12):1283–1291. https://doi.org/10.1161/CIRCRESAHA.107.162818

Mackenzie L, Llewelyn Roderick H, Berridge MJ, Conway SJ, Bootman MD (2004) The spatial pattern of atrial cardiomyocyte calcium signalling modulates contraction. J Cell Sci 117 (26):6327–6337. https://doi.org/10.1242/jcs.01559

Makara MA, Curran J, Little SC, Musa H, Polina I, Smith SA, Wright PJ et al (2014) Ankyrin-G coordinates intercalated disc signaling platform to regulate cardiac excitability in vivo. Circ Res 115(11):929–938. https://doi.org/10.1161/CIRCRESAHA.115.305154

Mansour H, de Tombe PP, Samarel AM, Russell B (2004) Restoration of resting sarcomere length after uniaxial static strain is regulated by protein kinase Cepsilon and focal adhesion kinase. Circ Res 94(5):642–649. https://doi.org/10.1161/01.RES.0000121101.32286.C8

Matamoros M, Pérez-Hernández M, Guerrero-Serna G, Amorós I, Barana A, Núñez M, Ponce-Balbuena D et al (2016) $Na_V1.5$ N-terminal domain binding to α1-syntrophin increases membrane density of human Kir2.1, Kir2.2 and $Na_V1.5$ channels. Cardiovasc Res 110 (2):279–290. https://doi.org/10.1093/cvr/cvw009

McCain ML, Parker KK (2011) Mechanotransduction: the role of mechanical stress, myocyte shape, and cytoskeletal architecture on cardiac function. Pflugers Arch Eur J Physiol 462 (1):89–104. https://doi.org/10.1007/s00424-011-0951-4

Milstein ML, Musa H, Balbuena DP, Anumonwo JMB, Auerbach DS, Furspan PB, Hou L et al (2012) Dynamic reciprocity of sodium and potassium channel expression in a macromolecular complex controls cardiac excitability and arrhythmia. Proc Natl Acad Sci U S A 109(31): E2134–E2143. https://doi.org/10.1073/pnas.1109370109

Miranti CK, Brugge JS (2002) Sensing the environment: a historical perspective on integrin signal transduction. Nat Cell Biol 4(4):E83–E90. https://doi.org/10.1038/ncb0402-e83

Mohler PJ, Schott J-J, Gramolini AO, Dilly KW, Guatimosim S, duBell WH, Song L-S et al (2003) Ankyrin-B mutation causes type 4 long-QT cardiac arrhythmia and sudden cardiac death. Nature 421(6923):634–639. https://doi.org/10.1038/nature01335

Mohler PJ, Rivolta I, Napolitano C, LeMaillet G, Lambert S, Priori SG, Bennett V (2004) $Na_V1.5$ E1053K mutation causing Brugada syndrome blocks binding to Ankyrin-G and expression of $Na_V1.5$ on the surface of cardiomyocytes. Proc Natl Acad Sci U S A 101(50):17533–17538. https://doi.org/10.1073/pnas.0403711101

Müller BM, Kistner U, Veh RW, Cases-Langhoff C, Becker B, Gundelfinger ED, Garner CC (1995) Molecular characterization and spatial distribution of SAP97, a novel presynaptic protein homologous to SAP90 and the drosophila discs-large tumor suppressor protein. J Neurosci 15(3 Pt 2):2354–2366

Noorman M, van der Heyden MAG, van Veen TAB, Cox MGPJ, Hauer RNW, de Bakker JMT, van Rijen HVM (2009) Cardiac cell-cell junctions in health and disease: electrical versus mechanical coupling. J Mol Cell Cardiol 47(1):23–31. https://doi.org/10.1016/j.yjmcc.2009.03.016

Noorman M, Hakim S, Kessler E, Groeneweg JA, Cox MGPJ, Asimaki A, van Rijen HVM et al (2013) Remodeling of the cardiac sodium channel, connexin43, and plakoglobin at the intercalated disk in patients with arrhythmogenic cardiomyopathy. Heart Rhythm 10 (3):412–419. https://doi.org/10.1016/j.hrthm.2012.11.018

Ojeh N, Pekovic V, Jahoda C, Määttä A (2008) The MAGUK-family protein CASK is targeted to nuclei of the basal epidermis and controls keratinocyte proliferation. J Cell Sci 121 (16):2705–2717. https://doi.org/10.1242/jcs.025643

Olsen O, Liu H, Wade JB, Merot J, Welling PA (2002) Basolateral membrane expression of the Kir 2.3 channel is coordinated by PDZ interaction with Lin-7/CASK complex. Am J Physiol Cell Physiol 282(1):C183–C195. https://doi.org/10.1152/ajpcell.00249.2001

Orchard C, Brette F (2008) T-tubules and sarcoplasmic reticulum function in cardiac ventricular myocytes. Cardiovasc Res 77(2):237–244. https://doi.org/10.1093/cvr/cvm002

Palade GE (1953) An electron microscope study of the mitochondrial structure. J Histochem Cytochem 1(4):188–211. https://doi.org/10.1177/1.4.188.

Palatinus JA, O'Quinn MP, Barker RJ, Harris BS, Jourdan J, Gourdie RG (2011) ZO-1 determines Adherens and gap junction localization at intercalated disks. Am J Physiol Heart Circ Physiol 300(2):H583–H594. https://doi.org/10.1152/ajpheart.00999.2010

Palygin OA, Pettus JM, Shibata EF (2008) Regulation of caveolar cardiac sodium current by a single Gsalpha histidine residue. Am J Physiol Heart Circ Physiol 294(4):H1693–H1699. https://doi.org/10.1152/ajpheart.01337.2007

Pardo JV, Siliciano JD, Craig SW (1983a) Vinculin is a component of an extensive network of myofibril-sarcolemma attachment regions in cardiac muscle fibers. J Cell Biol 97(4):1081–1088

Pardo JV, Siliciano JD, Craig SW (1983b) A Vinculin-containing cortical lattice in skeletal muscle: transverse lattice elements ('costameres') mark sites of attachment between myofibrils and sarcolemma. Proc Natl Acad Sci U S A 80(4):1008–1012

Peters CJ, Chow SS, Angoli D, Nazzari H, Cayabyab FS, Morshedian A, Accili EA (2009) In situ co-distribution and functional interactions of SAP97 with sinoatrial isoforms of HCN channels. J Mol Cell Cardiol 46(5):636–643. https://doi.org/10.1016/j.yjmcc.2009.01.010

Petitprez S, Zmoos A-F, Ogrodnik J, Balse E, Raad N, El-Haou S, Albesa M et al (2011) SAP97 and dystrophin macromolecular complexes determine two pools of cardiac sodium channels $Na_V1.5$ in cardiomyocytes. Circ Res 108(3):294–304. https://doi.org/10.1161/CIRCRESAHA.110.228312

Pilichou K, Nava A, Basso C, Beffagna G, Bauce B, Lorenzon A, Frigo G et al (2006) Mutations in desmoglein-2 gene are associated with arrhythmogenic right ventricular cardiomyopathy. Circulation 113(9):1171–1179. https://doi.org/10.1161/CIRCULATIONAHA.105.583674

Rhett JM, Gourdie RG (2012) The perinexus: a new feature of Cx43 gap junction organization. Heart Rhythm 9(4):619–623. https://doi.org/10.1016/j.hrthm.2011.10.003

Rhett JM, Jourdan J, Gourdie RG (2011) Connexin 43 connexon to gap junction transition is regulated by zonula occludens-1. Mol Biol Cell 22(9):1516–1528

Rhett JM, Ongstad EL, Jourdan J, Gourdie RG (2012) Cx43 associates with $Na_V1.5$ in the cardiomyocyte perinexus. J Membr Biol 245(7):411–422. https://doi.org/10.1007/s00232-012-9465-z

Rizzo S, Lodder EM, Verkerk AO, Wolswinkel R, Beekman L, Pilichou K, Basso C, Remme CA, Thiene G, Bezzina CR (2012) Intercalated disc abnormalities, reduced Na(+) current density, and conduction slowing in desmoglein-2 mutant mice prior to cardiomyopathic changes. Cardiovasc Res 95(4):409–418. https://doi.org/10.1093/cvr/cvs219

Rook MB, Evers MM, Vos MA, Bierhuizen MFA (2012) Biology of cardiac sodium channel $Na_V1.5$ expression. Cardiovasc Res 93(1):12–23. https://doi.org/10.1093/cvr/cvr252

Ross RS (2004) Molecular and mechanical synergy: cross-talk between Integrins and growth factor receptors. Cardiovasc Res 63(3):381–390. https://doi.org/10.1016/j.cardiores.2004.04.027

Saffitz JE (2005) Dependence of electrical coupling on mechanical coupling in cardiac myocytes: insights gained from cardiomyopathies caused by defects in cell-cell connections. Ann N Y Acad Sci 1047:336–344. https://doi.org/10.1196/annals.1341.030

Saffitz JE, Macrae CA (2010) Mutations in desmosomal protein genes and the pathogenesis of arrhythmogenic right ventricular cardiomyopathy. Heart Rhythm 7(1):30–32. https://doi.org/10.1016/j.hrthm.2009.10.028

Samarel AM (2005) Costameres, focal adhesions, and cardiomyocyte mechanotransduction. Am J Physiol Heart Circ Physiol 289(6):H2291–H2301. https://doi.org/10.1152/ajpheart.00749.2005

Sato PY, Musa H, Coombs W, Guerrero-Serna G, Patiño GA, Taffet SM, Isom LL, Delmar M (2009) Loss of plakophilin-2 expression leads to decreased sodium current and slower conduction velocity in cultured cardiac myocytes. Circ Res 105(6):523–526. https://doi.org/10.1161/CIRCRESAHA.109.201418

Sato PY, Coombs W, Lin X, Nekrasova O, Green KJ, Isom LL, Taffet SM, Delmar M (2011) Interactions between ankyrin-G, plakophilin-2, and Connexin43 at the cardiac intercalated disc. Circ Res 109(2):193–201. https://doi.org/10.1161/CIRCRESAHA.111.247023

Schotten U, Verheule S, Kirchhof P, Goette A (2011) Pathophysiological mechanisms of atrial fibrillation: a translational appraisal. Physiol Rev 91(1):265–325. https://doi.org/10.1152/physrev.00031.2009

Shai S-Y, Harpf AE, Ross RS (2002) Integrins and the myocardium. Genet Eng 24:87–105

Sharp WW, Simpson DG, Borg TK, Samarel AM, Terracio L (1997) Mechanical forces regulate focal adhesion and costamere assembly in cardiac myocytes. Am J Phys 273(2 Pt 2):H546–H556

Shaw RM, Fay AJ, Puthenveedu MA, Zastrow M v, Jan Y-N, Jan LY (2007) Microtubule plus-end-tracking proteins target gap junctions directly from the cell interior to adherens junctions. Cell 128(3):547–560. https://doi.org/10.1016/j.cell.2006.12.037

Shcherbakova OG, Hurt CM, Xiang Y, Dell'Acqua ML, Qi Z, Tsien RW, Kobilka BK (2007) Organization of beta-adrenoceptor signaling compartments by sympathetic innervation of cardiac myocytes. J Cell Biol 176(4):521–533. https://doi.org/10.1083/jcb.200604167

Shibata EF, Brown TLY, Washburn ZW, Bai J, Revak TJ, Butters CA (2006) Autonomic regulation of voltage-gated cardiac ion channels. J Cardiovasc Electrophysiol 17(s1):S34–S42. https://doi.org/10.1111/j.1540-8167.2006.00387.x

Shy D, Gillet L, Ogrodnik J, Albesa M, Verkerk AO, Wolswinkel R, Rougier J-S et al (2014) PDZ domain-binding motif regulates cardiomyocyte compartment-specific $Na_V1.5$ channel expression and function. Circulation 130(2):147–160. https://doi.org/10.1161/CIRCULATIONAHA.113.007852

Spach MS (1999) Anisotropy of cardiac tissue: a major determinant of conduction? J Cardiovasc Electrophysiol 10(6):887–890

Spach MS, Heidlage JF, Dolber PC, Barr RC (2001) Changes in anisotropic conduction caused by remodeling cell size and the cellular distribution of gap junctions and Na(+) channels. J Electrocardiol 34(Suppl):69–76

Sun R, Yongyue S, Zhao X, Qi J, Luo X, Yang Z, Yao Y, Luo X, Xia Z (2009) Human calcium/calmodulin-dependent serine protein kinase regulates the expression of p21 via the E2A transcription factor. Biochem J 419(2):457–466. https://doi.org/10.1042/BJ20080515

Te Riele ASJM, Agullo-Pascual E, James CA, Leo-Macias A, Cerrone M, Zhang M, Lin X et al (2017) Multilevel analyses of SCN5A mutations in Arrhythmogenic right ventricular dysplasia/cardiomyopathy suggest non-canonical mechanisms for disease pathogenesis. Cardiovasc Res 113(1):102–111. https://doi.org/10.1093/cvr/cvw234

Tohse N, Seki S, Kobayashi T, Tsutsuura M, Nagashima M, Yamada Y (2004) Development of excitation-contraction coupling in cardiomyocytes. Jpn J Physiol 54(1):1–6

Towbin JA, Hejtmancik JF, Brink P, Gelb B, Zhu XM, Chamberlain JS, McCabe ER, Swift M (1993) X-linked dilated cardiomyopathy. Molecular genetic evidence of linkage to the duchenne muscular dystrophy (dystrophin) gene at the Xp21 locus. Circulation 87(6):1854–1865

Toyofuku T, Yabuki M, Otsu K, Kuzuya T, Hori M, Tada M (1998) Direct association of the gap junction protein connexin-43 with ZO-1 in cardiac myocytes. J Biol Chem 273(21):12725–12731

Ueda K, Valdivia C, Medeiros-Domingo A, Tester DJ, Vatta M, Farrugia G, Ackerman MJ, Makielski JC (2008) Syntrophin mutation associated with long QT syndrome through activation of the nNOS-SCN5A macromolecular complex. Proc Natl Acad Sci U S A 105 (27):9355–9360. https://doi.org/10.1073/pnas.0801294105

Valayannopoulos V, Michot C, Rodriguez D, Hubert L, Saillour Y, Labrune P, de Laveaucoupet J et al (2012) Mutations of TSEN and CASK genes are prevalent in pontocerebellar hypoplasias type 2 and 4. Brain J Neurol 135(1):e199. https://doi.org/10.1093/brain/awr108. Author reply e200

Vatta M, Ackerman MJ, Ye B, Makielski JC, Ughanze EE, Taylor EW, Tester DJ et al (2006) Mutant caveolin-3 induces persistent late sodium current and is associated with long-QT syndrome. Circulation 114(20):2104–2112. https://doi.org/10.1161/CIRCULATIONAHA.106.635268

Verkerk AO, van Ginneken AC, van Veen TA, Tan HL (2007) Effects of heart failure on brain-type Na+ channels in rabbit ventricular myocytes. Europace 9(8):571–577. https://doi.org/10.1093/europace/eum121

Walden AP, Dibb KM, Trafford AW (2009) Differences in intracellular calcium homeostasis between atrial and ventricular myocytes. J Mol Cell Cardiol 46(4):463–473. https://doi.org/10.1016/j.yjmcc.2008.11.003

Wilde AAM, Brugada R (2011) Phenotypical manifestations of mutations in the genes encoding subunits of the cardiac sodium channel. Circ Res 108(7):884–897. https://doi.org/10.1161/CIRCRESAHA.110.238469

Wu G, Ai T, Kim JJ, Mohapatra B, Xi Y, Li Z, Abbasi S et al (2008) Alpha-1-Syntrophin mutation and the long-QT syndrome: a disease of sodium channel disruption. Circ Arrhythm Electrophysiol 1(3):193–201. https://doi.org/10.1161/CIRCEP.108.769224

Yarbrough TL, Tong L, Lee H-C, Shibata EF (2002) Localization of cardiac sodium channels in caveolin-rich membrane domains: regulation of sodium current amplitude. Circ Res 90 (4):443–449

Yeager M (1998) Structure of cardiac gap junction intercellular channels. J Struct Biol 121 (2):231–245. https://doi.org/10.1006/jsbi.1998.3972

Zemljic-Harpf AE, Ponrartana S, Avalos RT, Jordan MC, Roos KP, Dalton ND, Phan VQ, Adamson ED, Ross RS (2004) Heterozygous inactivation of the vinculin gene predisposes to stress-induced cardiomyopathy. Am J Pathol 165(3):1033–1044. https://doi.org/10.1016/S0002-9440(10)63364-0

Zhang Q, Deng C, Rao F, Modi RM, Zhu J, Liu X, Mai L et al (2013) Silencing of desmoplakin decreases connexin43/$Na_V1.5$ expression and sodium current in HL-1 cardiomyocytes. Mol Med Rep 8(3):780–786. https://doi.org/10.3892/mmr.2013.1594

Zhu C, Barker RJ, Hunter AW, Zhang Y, Jourdan J, Gourdie RG (2005) Quantitative analysis of ZO-1 colocalization with Cx43 gap junction plaques in cultures of rat neonatal cardiomyocytes. Microsc Microanal 11(03):244–248. https://doi.org/10.1017/S143192760505049X

Posttranslational Modification of Sodium Channels

Zifan Pei, Yanling Pan, and Theodore R. Cummins

Contents

Abstract

Voltage-gated sodium channels (VGSCs) are critical determinants of excitability. The properties of VGSCs are thought to be tightly controlled. However, VGSCs are also subjected to extensive modifications. Multiple posttranslational

Z. Pei
Department of Biology, Indiana University – Purdue University Indianapolis, Indianapolis, IN, USA

Department of Pharmacology and Toxicology, Indiana University – Purdue University Indianapolis, Indianapolis, IN, USA

Y. Pan
Medical Neuroscience Graduate Program, Indiana University – Purdue University Indianapolis, Indianapolis, IN, USA

T. R. Cummins (✉)
Department of Biology, Indiana University – Purdue University Indianapolis, Indianapolis, IN, USA

Department of Pharmacology and Toxicology, Indiana University – Purdue University Indianapolis, Indianapolis, IN, USA

Medical Neuroscience Graduate Program, Indiana University – Purdue University Indianapolis, Indianapolis, IN, USA
e-mail: trcummin@iupui.edu

M. Chahine (ed.), *Voltage-gated Sodium Channels: Structure, Function and Channelopathies*, Handbook of Experimental Pharmacology 246,
https://doi.org/10.1007/164_2017_69

modifications that covalently modify VGSCs in neurons and muscle have been identified. These include, but are not limited to, phosphorylation, ubiquitination, palmitoylation, nitrosylation, glycosylation, and SUMOylation. Posttranslational modifications of VGSCs can have profound impact on cellular excitability, contributing to normal and abnormal physiology. Despite four decades of research, the complexity of VGSC modulation is still being determined. While some modifications have similar effects on the various VGSC isoforms, others have isoform-specific interactions. In addition, while much has been learned about how individual modifications can impact VGSC function, there is still more to be learned about how different modifications can interact. Here we review what is known about VGSC posttranslational modifications with a focus on the breadth and complexity of the regulatory mechanisms that impact VGSC properties.

Keywords
Nav · Nitrosylation · Palmitoylation · Phosphorylation

1 Brief Overview of VGSCs

Cell membrane forms the boundary of the cell and separates cells from their exterior environment. The lipid bilayer of biological membranes has selective permeability to different molecules (ions and polar molecules), controlling the flow of substance across the membrane. There are different types of membrane transport proteins that determine the membrane permeability. Among them, active transporters facilitate an "active" transport of molecules (usually against their concentration gradient) using cellular energy. Passive transporters (e.g., ion channels), on the other hand, allow the "passive" diffusion of molecules (usually in the same direction as the concentration or electrochemical gradient) without energy consumption. Ion channels are pore-forming integral membrane proteins that gate the flow of ions across the cell membrane and therefore contribute to setting the resting membrane potential and determining cellular excitability. Landmark experiments by Hodgkin and Huxley (1952) demonstrated that the transmembrane sodium and potassium currents were responsible for the action potential generation. Moreover, they were able to identify the influence of membrane potential on sodium current kinetics and quantitatively describe the sodium conductance and gating mechanism. This work provided the foundation for understanding action potential generation and propagation and paved the way for later ion channel studies. Our knowledge of voltage-gated sodium channel (VGSC) proteins and mechanisms underlying membrane excitability has progressed significantly over the last six-plus decades. As the result of the development of patch clamp and other advanced techniques (Hamill et al. 1981; Payandeh et al. 2011; Schmidt and Catterall 1986; Messner and Catterall 1985; Costa and Catterall 1984), we now have a clearer understanding of sodium channel structure, kinetics, and function. We also know that VGSCs are subject to extensive modulation which can have significant impact on cell excitability. This review focuses on VGSC posttranslational modifications, with specific attention to more recent discoveries and studies.

VGSCs are integral membrane protein complexes that allow sodium ion flow across the membrane and conduct transmembrane sodium currents in response to changes in membrane voltage (Goldin 2001). VGSCs consist of the pore-forming α subunit and one or more β auxiliary subunits that regulate functions of channels (Catterall et al. 2005). Although β auxiliary subunits can be posttranslationally modified (Malhotra et al. 2004), here we focus on posttranslational modification (PTM) of α subunits. The VGSC α subunit consists of four transmembrane domains (DI–DIV), with each containing six transmembrane segments (S1–S6, Fig. 1). S1–S4 serve as voltage sensor and change confirmations in response to membrane potential change. S5 and S6 form the pore of the channel, which allows ion conduction through the channel (Corry and Thomas 2012). The N-terminus, C-terminus, and large cytoplasmic linkers are frequent targets of PTMs. However, smaller intracellular linkers can also be subject to modification. Extracellular regions such as the linker between specific S5 segments and the pore loops that form the selectivity filter can also be subject to important modifications. Indeed, α subunits can be covalently linked to β2 and β4 subunits via disulfide bonds between extracellular residues (Chen et al. 2012).

The VGSC family consists of nine members, and they share more than 50% common amino acid sequence in transmembrane segments and extracellular domains (Yu and Catterall 2003). The α subunit has a molecular weight of about 220–260 kDa (Payandeh et al. 2011; Catterall et al. 2005; Levinson and Ellory 1973). Some of the difference in molecular weight between isoforms and preparations can arise from extensive PTMs. The other major difference in molecular weight results from substantial differences in the length of the linker between domains I and II, with Nav1.4 and Nav1.9 having significantly shorter linkers than other mammalian isoforms. Numerous phosphorylation sites have been reported in the DI–DII linkers from the longer isoforms (Berendt et al. 2010) (Fig. 1).

In general, studies of PTM consequences have focused on modulation of VGSC gating properties and surface expression. In terms of gating, VGSCs have three basic configurations: resting (closed), activated (open), and inactivated. At resting membrane potential, VGSCs are closed. They activate (open) in response to the membrane potential depolarization. When membrane depolarization begins, several positively charged residues on the S4 segments are able to sense the membrane potential change, which leads to outward movement of S4 segments and the conformational change of VGSCs (Yarov-Yarovoy et al. 2012). The channel is then in transition to activated (open) configuration, providing a conduction pathway for the sodium currents through the channel pore (Stuhmer et al. 1989). Very soon after channel activation (usually within milliseconds), the inactivation gate that tethered to the sodium channel protein will block the channel and prevent sodium conductance (Goldin 2003). VGSCs transition to a nonconducting, inactivated conformation. This process is often referred to as "fast inactivation." Structure function studies revealed that the IFM motif (located in the intracellular linker of DIII and DIV) with three hydrophobic amino acids isoleucine, phenylalanine, and methionine is largely responsible for the channel fast inactivation (West et al. 1992). Modifications to fast inactivation can lead to altered persistent currents,

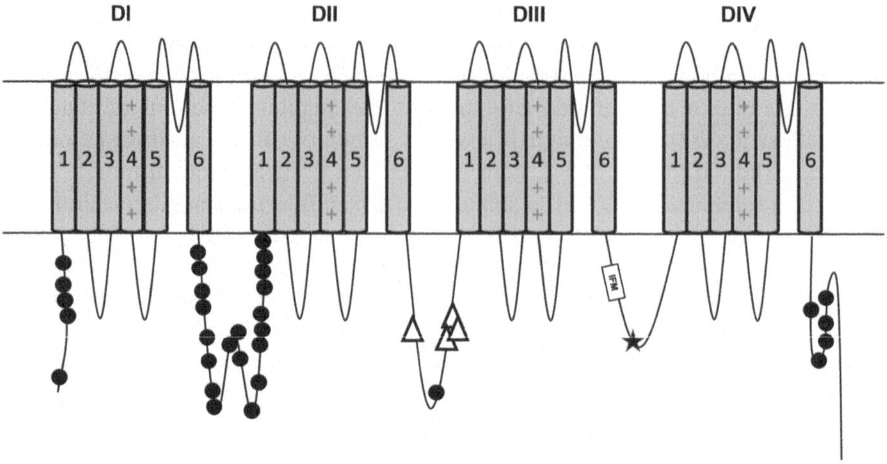

Fig. 1 Schematic representation of the linear structure of α subunit for voltage-gated sodium channels (VGSCs) with long I–II linkers. The α subunit consists of four homologous transmembrane domains (DI–DIV). Each domain contains six transmembrane segments (S1–S6). Among all, S1–S4 serve as the voltage sensor of the channel and S5–S6 form the pore. The IFM inactivation particle is located in DIII–DIV linker. The *black circles* represent likely phosphorylation sites in Nav1.2 and the *black star* is S1506. The *white triangles* represent putative palmitoylation sites in Nav1.5

also referred to as late or non-inactivating currents (Fig. 2a, b). There is also another type of inactivation that develops in response to longer depolarization time periods (usually hundreds of milliseconds) called "slow inactivation" (Silva 2014). Slow inactivation will sustain for a few seconds or even tens of seconds and is associated with altered excitability in neurons and myocytes, contributing to various disease phenotypes such as hyperkalemic periodic paralysis and long QT syndrome (Vilin and Ruben 2001). However, the detailed molecular mechanism of slow inactivation is still unknown (Ulbricht 2005). Once VGSCs enter an inactivated configuration, the channels are refractory to future stimuli and will not be able to open again until the membrane potential has repolarized to negative potentials. This process is called "recovery from inactivation," which prevents the cells from premature re-excitation and is critical in regulating action potential firing in excitable cells. Alteration in recovery rate may contribute to a disrupted action potential generation pattern. Previous study on skeletal muscle sodium channel reveals that defective recovery from fast inactivation may lead to disease phenotype producing myotonic discharges (Richmond et al. 1997). The voltage dependence of inactivation, and to a lesser extent activation, can be modulated by posttranslational modifications. This can change the degree of overlap between activation and steady-state inactivation and impact window currents that typically are active near cell resting membrane

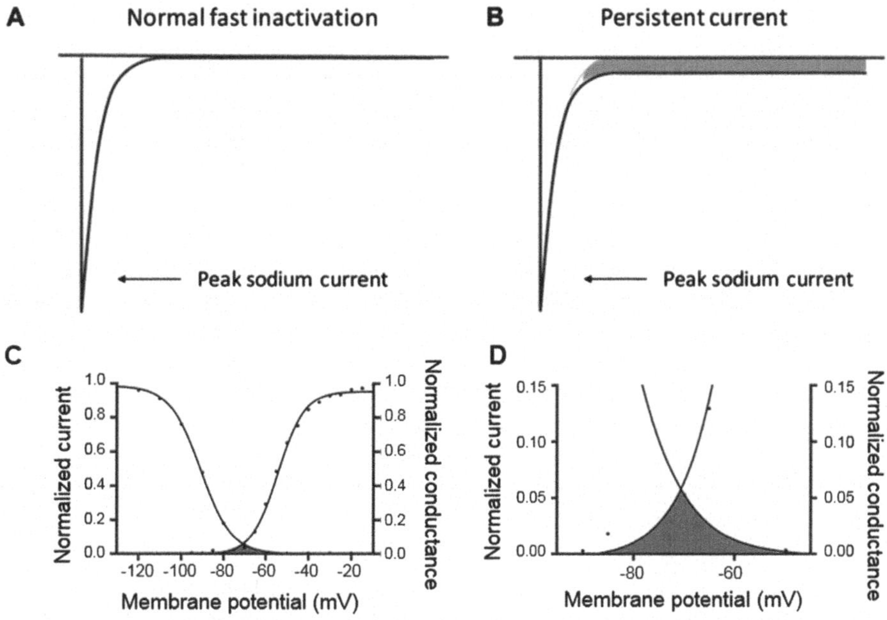

Fig. 2 Impaired fast inactivation leads to the generation of persistent (late) sodium current. (**a**) Representative current trace of normal fast inactivation. Sodium channels open and inactivate quickly after opening. (**b**) Persistent current generation when sodium channel fast inactivation is impaired, with the *gray* color indicating the non-inactivating persistent current. (**c**) Representative curves of cardiac sodium channel voltage dependence of activation and steady-state inactivation. (**d**) Representative window current (*gray* area). Window currents can be observed at voltages that are depolarized enough to activate a fraction of sodium channels but not sufficient to fully inactivate all sodium channels

potential (Fig. 2c, d). Window currents are related to classic persistent currents, but result from distinct gating alterations.

2 Posttranslational Modifications of VGSCs

Sodium channels can be modulated by a number of substances. Calcium and hydrogen ions can alter sodium channel gating. G proteins, ATP, cAMP, glycolytic metabolites, and local anesthetics have all been reported to have modulatory effects on sodium channels. VGSC function is determined by the intrinsic biophysical properties (i.e., rapid activation and inactivation), but it can also be regulated by posttranslational modifications.

Posttranslational VGSC modulation appears to be critically important in neurons and cardiac tissue. It is less clear what the physiological role of sodium channel modulation is in skeletal muscle, although it is likely to be involved in disease states such as critical illness myopathy (Teener and Rich 2006). Neuronal sodium current can be downregulated by anoxia and cyanide (Cummins et al. 1991, 1993; O'Reilly et al. 1997), and it is thought that this can be important in the anoxic response. CaMKII phosphorylation of brain VGSCs has been implicated as a major regulator of VGSC persistent currents and is likely to contribute to pathophysiologies such as epilepsy. Cardiac sodium channel modulation is believed to be important in rhythmogenesis (Schubert et al. 1990; Wagner et al. 2015). As a result, Nav1.5 modulation has been extensively studied, and Nav1.5 is the target of a multitude of posttranslational modifications including glycosylation, phosphorylation, methylation, acetylation, redox modifications, palmitoylation, and ubiquitination (Ashpole et al. 2012; Pei et al. 2016; Marionneau and Abriel 2015). Because of modern protein analysis technologies such as mass spectrometry, researchers are able to better determine the precise location of post-translational modification sites and consensus sequence among different VGSC subtypes, leading to a better understanding of the mechanistic details of these regulations.

2.1 Phosphorylation

Phosphorylation is the most extensively studied posttranslational modification of VGSCs. Various protein kinases have been identified to modulate different aspects of VGSC function through diverse pathways including Ca2+/calmodulin-dependent serine/threonine protein kinase (CaMK) (Ashpole et al. 2012; Koval et al. 2012; Aiba et al. 2010), protein kinase A and C (PKA and PKC) (Shin and Murray 2001; Murray et al. 1997; Murray et al. 1994; Hallaq et al. 2012), phosphatidylinositol 3-kinase (PI3K) (Lu et al. 2013; Lu et al. 2012), and adenosine monophosphate-activated protein kinase (APMK) (Wallace et al. 2003). These kinases can add a negatively charged phosphate group to select serine, threonine, or tyrosine residues. It is not possible to give a comprehensive review of all the findings relating to this powerful form of VGSC modulation, so we will cover only a few specific areas to provide highlights of what can be learned from the study of VGSC phosphorylation. Many of the phosphorylation sites that have been identified are located in the first intracellular linker loop (DI–DII linker, Fig. 1), and a number of these are conserved among different species and subtypes (Marionneau et al. 2012). Early estimates indicated that the α subunit could be phosphorylated at somewhere between 2 and 20 sites by protein kinase A (PKA), protein kinase C (PKC), and possibly other protein kinases. Numann et al. (1991) studied the effect of PKC on rat brain channels (Nav1.2) expressed in Chinese hamster ovary cells. They reported that activation of PKC with a membrane-permeant agent, l-oleoyl-2-acetyl-sn-glycerol (OAG), both inhibited the peak current and slowed macroscopic fast inactivation of rat neuronal sodium channels. West et al. (1991) reported that phosphorylation of serine 1506 in the III–IV linker (shown by the black star in Fig. 1) by PKC was responsible for the slowing of RIIA macroscopic inactivation. After this potential PKC site, located in the III–IV linker, was removed by mutating serine 1506 to an alanine, neither the slowing nor the inhibition by OAG was observed. Additional experiments seemingly confirmed that phosphorylation of

S1506 was necessary and sufficient for the observed slowing of inactivation. However, while phosphorylation of S1506 was also necessary for the decrease in current, it was not sufficient: phosphorylation of another site, possibly in the I–II linker, also seemed to be needed for the inhibition (Li et al. 1993). However, subsequent studies on PKC modulation of neuronal sodium channels in hippocampal neurons did not observe pronounced slowing of inactivation (Chen et al. 2005). This could be due to complex interactions between posttranslational modifications and/or accessory subunits. The putative PKC site in the III–IV linker is conserved in most VGSCs. The site is conserved in the cardiac sodium channel (Nav1.5), and PKC phosphorylation of the corresponding residue in the III–IV linker causes a major negative shift in the voltage dependence of inactivation (Qu et al. 1996). Interestingly, this site is also conserved in Nav1.4, but while PKC also induces a negative shift in the voltage dependence of inactivation for Nav1.4, this effect is not dependent on the corresponding serine in the III–IV linker (Bendahhou et al. 1995). Data from *Xenopus* oocyte experiments indicate that both Nav1.7 and Nav1.8 currents are inhibited by PKC activation and that this also shifts the voltage dependence of activation in the depolarizing direction (Vijayaragavan et al. 2004). By contrast PKA activation enhanced Nav1.8 but inhibited Nav1.7 currents in this experimental system. Recently it was shown that Nav1.7 resurgent sodium currents are enhanced by PKC, and this effect is modulated by the state of the corresponding III–IV linker residue in Nav1.7 (Tan et al. 2014). This illustrates some of the complexities of determining the functional consequences of VGSC phosphorylation and comparing the effects on various isoforms in various tissues.

Mass spectrometry is providing enhanced estimates of VGSC phosphorylation. Obtaining full coverage can be difficult with complex transmembrane proteins. In one study of Nav1.2 (Berendt et al. 2010), 66% coverage of the cytoplasmic linkers was obtained. Fifteen sites were identified, 1 in the N-terminus, 11 in the I–II linker, and 3 in the C-terminus. Unfortunately there was insufficient coverage in the III–IV linker to determine if the conserved serine residue discussed above was phosphorylated in the brain tissue. In a follow-up study on Nav1.2 (Baek et al. 2014), it was found that acute kainate-induced seizures induced a significant reduction in phosphorylation of nine sites in Nav1.2. Not surprisingly, these are primarily sites located in the I–II linker. However, this study also revealed that this downregulation was due, at least in part, to an increase in methylated arginines at three sites. Thus distinct regulation of phosphorylation and methylated arginines in Nav1.2 is likely to contribute to functional changes in Nav1.2, modulation of neuronal excitability, and perhaps seizure activity.

Calcium/calmodulin protein kinase II (CaMKII) is believed to be an important regulator of excitability in neurons and muscle. CaMKII modulation of VGSCs has also been implicated in physiological and pathophysiological control of excitability. Increased CaMKII activity has been implicated in animal models of heart failure as well as in studies of failing human hearts (Zhang et al. 2003; Hoch et al. 1999). In one study of CaMKII and Nav1.5 (Ashpole et al. 2012), it was found that a negative shift in Nav1.5 steady-state inactivation resulted from CaMKII-dependent phosphorylation of Nav1.5 at two specific phosphor sites. However a mass spectroscopy analysis of human Nav1.5 purified from HEK293 cells with >80% coverage identified 23 sites that could be phosphorylated by CaMKII in vitro (Herren et al. 2015). This suggests that CaMKII can extensively modify VGSCs by phosphorylation. A study of sodium currents in neurons from a SCN2a (Nav1.2) mutant mouse with epilepsy found that CaMKII phosphorylation of Nav1.2 increases persistent sodium currents and excitability. Maltsev et al. (2008) reported that CaMKII can also increase persistent sodium currents in cardiac myocytes. Burel et al. (2017) identified two distinct CaMKII phosphor sites in the C-terminus of Nav1.5 that contribute in part to increased Nav1.5 persistent currents; however this modulation involved altered binding of FGF13 to Nav1.5. Interestingly, these two sites were not among the 23 sites identified by Herren et al. Although several hundred studies have investigated how a multitude of kinases modulate VGSC isoforms and a substantial amount of insight has been gained, our knowledge of the interplay between different phosphor sites and how this impacts interactions with accessory proteins is still incomplete.

2.2 Arginine Methylation

As mentioned above, VGSCs can also be modified at lysine residues by arginine methylation. Much less is known about this compared to VGSC phosphorylation. Arginine methylation of Nav1.5 in stable cell lines and human ventricles has been examined using mass spectrometry (Beltran-Alvarez et al. 2015; Beltran-Alvarez et al. 2014; Beltran-Alvarez et al. 2013). Studies revealed that arginine R513, R526, and R680, located in the DI and DII intracellular linker of cardiac sodium channel, are likely subject to arginine methylation. Nav1.5 methylation also enhances cell surface expression and sodium current density. It is likely that other isoforms in addition to Nav1.2 and Nav1.5 are regulated by arginine methylation. Differential expression of protein arginine methyl transferases may play a role in how VGSCs are modulated in different tissue and cell compartments.

2.3 Glycosylation

Glycosylation, another common post-translational modification of ion channels, is an enzymatic process that attaches glycans to ion channel proteins. These sugar groups can be quite complex and extensive, making analysis complicated. Early reports (Waechter et al. 1983) indicated that glycosylation of neuronal VGSCs played a crucial role in biosynthesis, trafficking, and degradation of VGSCs. N-linked glycans (attached to a nitrogen of asparagine) and O-linked glycans (attached to the hydroxyl oxygen of serine, threonine) are often terminated by sialic acids which can, at least in theory, alter VGSC function through their negative charges. Glycosylation of Nav1.5 was determined 20 years ago by Cohen and Levitt (1993). This study indicated that Nav1.5 mass was only increased by about 5% due to glycosylation, compared to 25–30% increases observed with some other VGSC isoforms. Despite the lesser degree of modification, there is still compelling evidence that glycosylation can be an important determinant of Nav1.5 membrane trafficking (Mercier et al. 2015). Pathophysiologically, reduced cardiac sodium channel sialylation has been shown to shorten the cardiomyocyte refractory time and enhance susceptibility to ventricular arrhythmias by slowing fast inactivation and increasing the rate of recovery from inactivation (Ednie et al. 2013). Glycosylation can also impact functional properties for multiple VGSCs. It has been shown that sialylation shifts the voltage dependence of activation and inactivation toward hyperpolarized potentials, enhances the rate of fast inactivation, and reduces the rate of recovery from fast inactivation (Johnson et al. 2004; Bennett et al. 1997). The impact on Nav1.4 voltage dependence of activation and inactivation can be quite large (Ednie et al. 2015), with sialylation, N-linked glycans, and O-linked glycans likely all playing a role in this. The sensory neuronal channel Nav1.9 is also not extensively glycosylated, yet changes in glycosylation can still impact voltage dependence of steady-state inactivation, and this appears to be developmentally regulated (Tyrrell et al. 2001). In addition to α subunits, it is well established that β-subunits can also be glycosylated (Laedermann et al. 2013a). However the extent to which glycosylation is involved in control of excitability is not fully understood, and improved strategies for determining how VGSC glycosylation is controlled will undoubtedly provide invaluable insight in the near future.

2.4 Ubiquitination

Ubiquitination, another well-studied posttranslational modification of VGSCs, refers to the enzymatic process in which an ubiquitin protein is attached to the sodium channel protein. Ubiquitin is a small protein, only about 8.5 kDa, but is a fairly large addition to a protein compared to most posttranslational modifications. The addition can alter protein function in several ways. Ubiquitination can target a protein to proteasomes and induce degradation. It can also alter localization of proteins and/or their functional properties. Modifications can involve addition of a single ubiquitin subunit or a chain of ubiquitin molecules. Addition of an ubiquitin molecule involves three proteins with distinct functions. Abriel et al. (2000) investigated the functional consequences of a PY (PPXY) motif in the C-terminus region of Nav1.5. This motif can bind Nedd4, an ubiquitin-protein ligase. Interestingly Nav1.4 lacks a PY motif. Mutation of the PY motif in Nav1.5 increased current density. Overexpression of Nedd4 could decrease Nav1.5 but not Nav1.4 current. This indicated that ubiquitination can be closely associated with Nav1.5 VGSC internalization. Ubiquitination of Nav1.5 has been demonstrated from both in vitro and in vivo studies (van Bemmelen et al. 2004; Laedermann et al. 2014a). Rougier et al. (2005) noted that most VGSC isoforms contain a PY motif (the exceptions being Nav1.4 and Nav1.9). They found that Nedd4-2 could downregulate Nav1.2, Nav1.3, and Nav1.5 currents in HEK293 cells with a corresponding reduction in surface expression. Fotia et al. (2004) demonstrated that Nedd4-2 could downregulate Nav1.7 and Nav1.8 currents.

While many studies have suggested that regulation of ubiquitination of VGSCs could lead to an altered surface expression level of channels (Laedermann et al. 2014b; Rougier et al. 2013), the detailed mechanism involved in this process and the study of its clinical relevance are still lacking. Interestingly, several studies were able to demonstrate that reduced Nedd4-2 levels led to DRG hyperexcitability, and conditions that increase pain in rodents downregulate Nedd4-2 expression in DRG, indicating that reduced ubiquitination of VGSCs might be involved in the development of neuropathic pain (Laedermann et al. 2013b; Cachemaille et al. 2012). Although Nav1.6 also has a PY motif in its C-terminus, it also appears to have a potential Nedd4 binding site in the I–II linker. In an elegant study, Gasser et al. (2010) demonstrated that downregulation of Nav1.6 can be enhanced by p38 kinase phosphorylation of the region involved in the I–II linker, indicating that Nedd4-induced ubiquitination and subsequent internalization of Nav1.6 involve at least two distinct Nedd4 binding sites and may be a stress response that limits cell excitability under pathophysiological conditions.

2.5 SUMOylation

The small ubiquitin-like modifier (SUMO) protein can also be conjugated to proteins, including ion channels. SUMO proteins are roughly 12 kDa in size and, as with ubiquitin, are added to other proteins by enzymes. Plant et al. (2016) found that adding SUMO protein to the pipet solution can increase the amplitude of Nav1.2-mediated currents in HEK293 cells. They also showed that SUMO modulated Nav1.2 currents in cerebellar neurons. SUMOylation occurs on lysine residues. Plant et al. reported that mutating K38 in the N-terminus of Nav1.2 eliminated the effect of SUMO. Thus Nav1.2 seems to be directly regulated by SUMOylation. This increase in current due to SUMOylation occurred rapidly in response to hypoxic conditions and may underlie some of the initial toxicity associated with hypoxia. SUMOylation also shifted the voltage dependence of activation and steady-state inactivation in the negative direction. While a negative shift in activation can increase excitability, a negative shift in inactivation can decrease excitability. SUMOylation has also been implicated in control of excitability in sensory neurons (Dustrude et al. 2016). Reduced SUMOylation of collapsin response mediator protein 2 (CRMP2) can alter CRMP2 binding to Nav1.7. As a consequence, Nav1.7 membrane localization and current density is reduced. CRMP2 enhances ubiquitination and endocytosis of Nav1.7. This illustrates the sometimes complex web of posttranslational modifications that can influence VGSCs.

2.6 Palmitoylation

Palmitoylation has been recognized as an important post-translational mechanism for the regulation of various membrane proteins, but our knowledge of how palmitoylation regulates ion channels has been limited (Shipston 2011). Protein S-palmitoylation involves the addition of a 16-carbon palmitic acid chain to an intracellular cysteine residue through a thioester linkage (Fig. 3a). Recent studies have shown that palmitoylation can have profound impact on VGSC properties.

S-palmitoylation is a reversible process (Fig. 3b) that can dynamically regulate protein life cycle and function (Shipston 2011). Multiple enzymes that facilitate the palmitoylation process have been identified in recent years. Palmitoyltransferases (PATs), the catalytic enzymes for protein palmitoylation, form a diverse family of proteins (23 members in mammals) (Fukata et al. 2004; Korycka et al. 2012). PATs are characterized by the presence of an aspartate-histidine-histidine-cysteine (DHHC) motif within a cysteine-rich domain. The reverse process, depalmitoylation, is mediated by acyl protein thioesterases (APTs) (Zeidman et al. 2009). According to previous studies, nonenzymatic palmitoylation could be possible but is very rare and has only been reported in vitro.

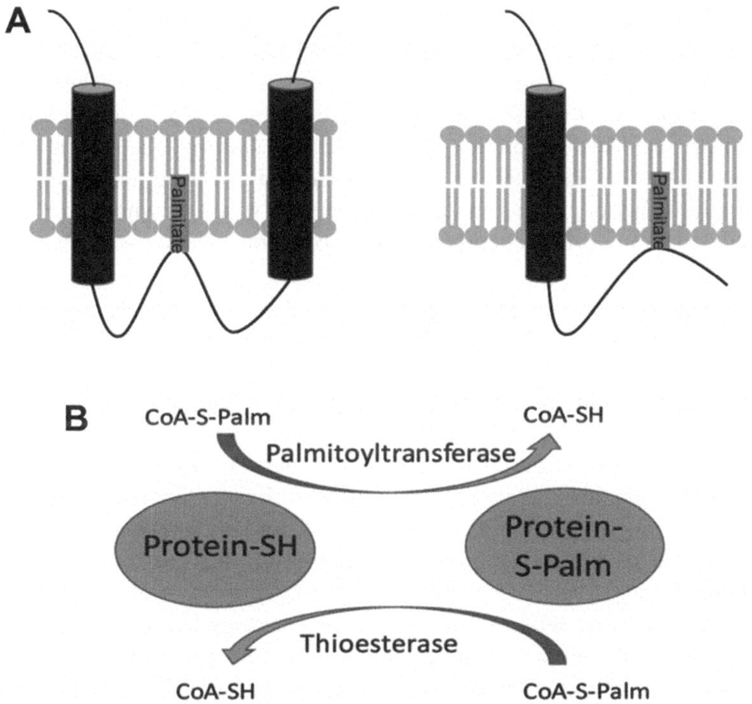

Fig. 3 (**a**) Palmitoylation can provide additional anchors for intracellular linkers or termini of channel proteins, depending on the location of the cysteine that is palmitoylated. (**b**) A cartoon illustration of the dynamic regulation of proteins by palmitoylation. Palmitoylation is a reversible process due to the labile nature of the thioester bond. Palmitoyltransferases mediate the palmitoylation process with the presence of palmitoyl-CoA. Depalmitoylation is mediated by thioesterases

Palmitoylation is involved in various phases of ion channel life cycle, including synthesis, maturation, trafficking, membrane targeting, internalization, and recycling. The formation of ligand binding sites in nicotinic acetylcholine receptors is regulated by palmitoylation (Drisdel et al. 2004; Alexander et al. 2010). Palmitoylation also regulates trafficking of AMPA receptors (Thomas et al. 2012; Hayashi et al. 2005) and the spatial organization of aquaporin channels (Suzuki et al. 2008). The addition of palmitic acid also regulates protein hydrophobicity and facilitates association to the membrane. However, few studies have focused on how palmitoylation directly alters channel biophysical activity at the membrane. Palmitoylation of the intracellular linker between S2 and S3 in Kv1.1 has been shown to increase the intrinsic voltage sensitivity of the channel (Gubitosi-Klug et al. 2005).

In 1987, brain sodium channel palmitoylation was identified to occur in the early stages of biosynthesis (Schmidt and Catterall 1987). This biochemical study indicated that brain sodium channels were palmitoylated after N-glycosylation of the channels. The addition of palmitate had a negligible impact on channel mass. It was not until over 30 years later that insight was gained into the functional consequences of palmitoylation (Bosmans et al. 2011). Interestingly, the initial functional work was on a recombinant channel that contained an early cloning artifact. An artificial cysteine in Nav1.2 that introduced a palmitoylation site had significant impact on the pharmacology of the channels. This indicated that lipid modification, at least in theory, is capable of altering the pharmacological properties of Nav1.2. Bosmans et al. also presented evidence that an endogenous cysteine might impact channel functional properties based on whether or not it was palmitoylated. However biochemical analysis of palmitoylation was not part of this short report.

The functional effect of VGSC palmitoylation is likely dependent on both the channel type and the location of the palmitoylation sites. Palmitate lipid may attach to different locations of channels, including C-terminus, N-terminus, juxta-transmembrane region, or intracellular loops (Fig. 3a), leading to the distinct alteration in various aspects of ion channel protein function (Shipston 2014). It is still unclear how palmitoylation alters the functional activity of VGSC protein. One hypothesized mechanism is that the palmitate molecules interact with the membrane lipids, changing the lipid membrane environment surrounding the targeted channel as well as impacting protein configuration, thus potentially modulating channel activity.

In recent years, the progress of ion channel palmitoylation research has been facilitated by the development of new biochemical and proteomics tools and the identification of palmitoyltransferases (Korycka et al. 2012). The traditional method of detecting protein palmitoylation involves metabolic labeling of cells with radioactive palmitate followed by immunoprecipitation and further identification of target proteins by autoradiography (Schmidt et al. 1988). This can provide direct evidence of protein palmitoylation. The development of the nonradioactive acyl-biotin exchange (ABE) assay allows more rapid detection of protein palmitoylation (Brigidi and Bamji 2013). This method provides higher sensitivity compared to traditional metabolic labeling and allows for quantitative estimates of palmitoylation. However, one caveat of this methodology is the detection of all existing S-acylated proteins, instead of more specific identification of protein palmitoylation. Another commonly used nonradioactive assay for palmitoylation detection is the click chemistry method. This method uses the alkyne fatty acid analog 17-octadecynoic acid (17-ODYA), which can be metabolically incorporated into endogenous cellular machinery at palmitoylation sites (Martin and Cravatt 2009). 17-ODYA-labeled proteins can be further linked to azide reporter tags by copper-catalyzed click chemistry and visualized by in-gel fluorescence analysis. One important advantage of this method is that it largely reduces the chance of false positive results compared with ABE assay. The most convincing demonstration of palmitoylation sites will be the mass spectrometry results. However, it is often lacking in most studies due to the complexity of this lipid modification and difficulties of resolving the palmitoylated peptides in mass spectrometry. Pei et al. (2016) demonstrated palmitoylation of Nav1.5 using three lines of biochemistry evidence. Nav1.5 palmitoylation was detected with tritiated palmitate, ABE experiments, and 17-ODYA labeling. Although most biochemical work was done in HEK293 cells with human Nav1.5, they also confirmed that rodent cardiac sodium channels were palmitoylated in myocytes (Fig. 4).

They also determined that Nav1.5 gating properties were modulated by palmitoylation (Pei et al. 2016). Surprisingly, palmitoylation did not induce a significant change in current density, indicating that palmitoylation has negligible impact on

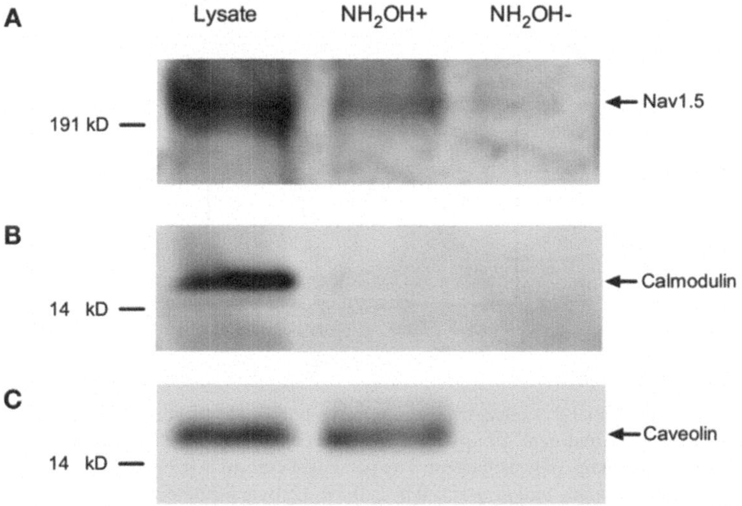

Fig. 4 Identification of Nav1.5 palmitoylation using an ABE assay. The *left lane* indicates total input lysate. The *middle lane* (NH2OH+) indicates palmitoylated protein. The *right lane* (NH2OH−) indicates the negative control group treated with tris solution. (**a**) Nav1.5 is palmitoylated in cardiac tissues. (**b**) Calmodulin is present but not palmitoylated in cardiac tissues. (**c**) Caveolin is palmitoylated in cardiac tissues. Reprinted with permission from Pei et al. (2016)

cell surface expression of Nav1.5. Instead, the data demonstrated that Nav1.5 palmitoylation has profound impact on channel availability by regulating the voltage dependence of steady-state inactivation in both HEK293 cells and cardiomyocytes. Nav1.5 inactivation was shifted in the negative direction by approximately 15–20 mV with the palmitoylation inhibitor 2-Br-palmitate (Fig. 5b). In contrast, Nav1.5 availability was significantly enhanced by palmitic acid treatment, increasing the voltage range of where window currents could be generated (Fig. 5c). Consistent with this, an elevated persistent sodium current was observed with palmitic acid treatment (Fig. 5a). As palmitic acid is the substrate for palmitoylation, this indicated that palmitoylation of Nav1.5 can enhance late sodium currents in myocytes. Together with the alteration in Nav1.5 inactivation, this could potentially lead to the reactivation of Nav1.5 and enhancement of sodium conductance during

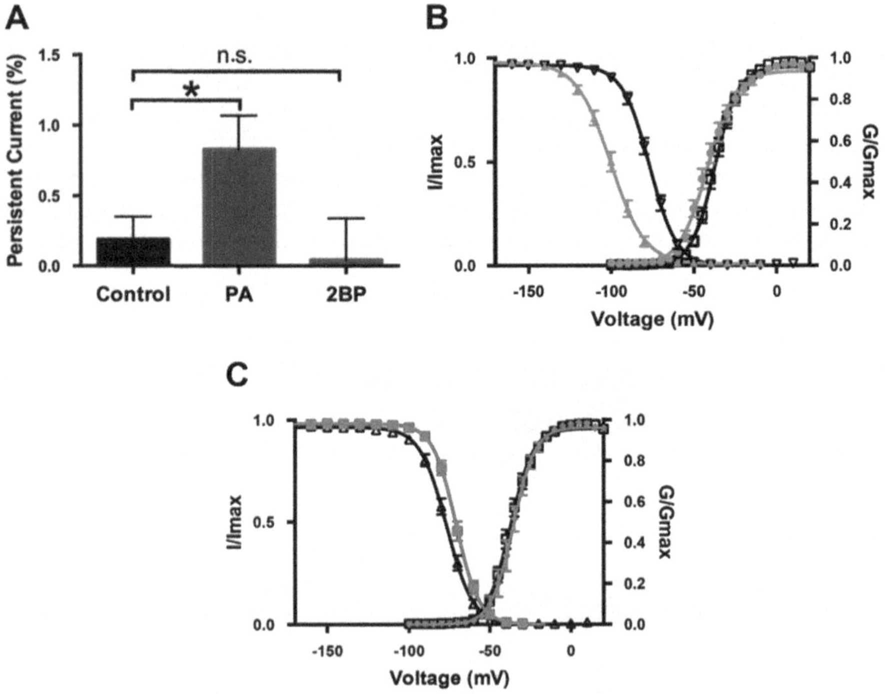

Fig. 5 (**a**) Statistical analysis comparing the relative amplitude of persistent currents from the control condition and treatment groups in neonatal cardiomyocytes. The persistent current was measured as the percentage of peak current. The persistent current is $0.2 \pm 0.2\%$ (control, *left bar*); $0.8 \pm 0.2\%$ (palmitic acid, *middle bar*); $0.0 \pm 0.3\%$ (2-Br-palmitate, *right bar*); $n = 10$. The difference between control group and palmitic acid treatment group is statistically significant ($p = 0.037$). (**b**, **c**) Comparison of sodium channel voltage dependence of activation and steady-state inactivation in neonatal cardiomyocytes. The *black* curves and symbols indicate the non-treatment group. In (**b**) the *gray* curves and symbols indicate the 2-Br-palmitate treatment group. In (**c**) the *gray* curves and symbols indicate the palmitic acid treatment group. Adapted with permission from Pei et al. (2016)

action potential phase 3 in cardiac myocytes. The opposite direction of changes in sodium channel gating properties between these two treatment groups suggests that palmitoylation can dynamically regulate Nav1.5 functions by altering channel inactivation gating. In addition, a mutant Nav1.5 with four internal cysteines mutated to alanine had a similar voltage dependence of activation to that obtained with wild-type 2-Br-palmitate treatment, indicating that specific cysteine residues on Nav1.5 were involved in the modulation of inactivation.

Pei et al. also reported that palmitoylation of Nav1.5 has a pronounced impact on cardiomyocyte excitability. Specifically, inhibiting palmitoylation greatly reduced myocyte excitability. The treatment of myocytes with 2-Br-palmitate abolished beating cells, and no action potential (spontaneously or stimulated) was observed. In contrast, enhancing palmitoylation with excess substrate increased action potential duration (Fig. 6a, b). Computer simulations indicated that changes in Nav1.5 gating due to palmitoylation are sufficient to cause the excitability changes that were observed in myocytes (Fig. 6c). Bankston and colleagues reported that the F1473C Nav1.5 mutation, a long QT mutation associated with a severe clinical phenotype, had similar biophysical consequences to those of Pei et al. observed with palmitic acid treatment (Bankston et al. 2007). The F1473C mutation shifted the midpoint of Nav1.5 inactivation in the depolarizing direction by 9 mV and increased persistent (late) currents to ~0.6% of the peak current. All together these data suggest that while depalmitoylation of Nav1.5 cysteine(s) can result in reduced cardiac excitability similar to that observed with Brugada syndrome, excessive palmitoylation of Nav1.5 can lead to enhanced cardiac activity similar to that observed with long QT syndrome mutations.

2.7 S-nitrosylation

Nitric oxide (NO) is known to be an important signaling molecule in many physiological processes, including host defense, neuronal communication, and vascular regulation. Renganathan et al. (2002) showed that an NO donor (papa-NONOate) significantly inhibited fast TTX-sensitive, slow TTX-resistant, and persistent TTX-resistant Na^+ currents in small diameter DRG neurons (Fig. 7). The NO scavenger hemoglobin blocked this inhibition, indicating that NO or a related molecule was responsible for the inhibition of multiple VGSCs (most likely Nav1.7, Nav1.8, and Nav1.9). The inhibition was independent of guanylyl cyclase and cGMP signaling pathways. Posttranslational modification of sulfhydryl groups, S-nitrosylation, can mediate the actions of NO and NO-related molecules (Stamler 1994). Renganathan et al. (2002) presented evidence indicating that S-nitrosylation was likely to mediate the inhibition of three different types of Na^+ channels, fast TTX-sensitive, slow TTX-resistant, and persistent TTX-resistant Na^+ channels in C-type DRG neurons via modification of sulfhydryl groups on the VGSCs.

Thiol sites have been considered as selective bioregulatory targets for NO and NO-related molecules. It has been proposed that protein targets of S-nitrosylation

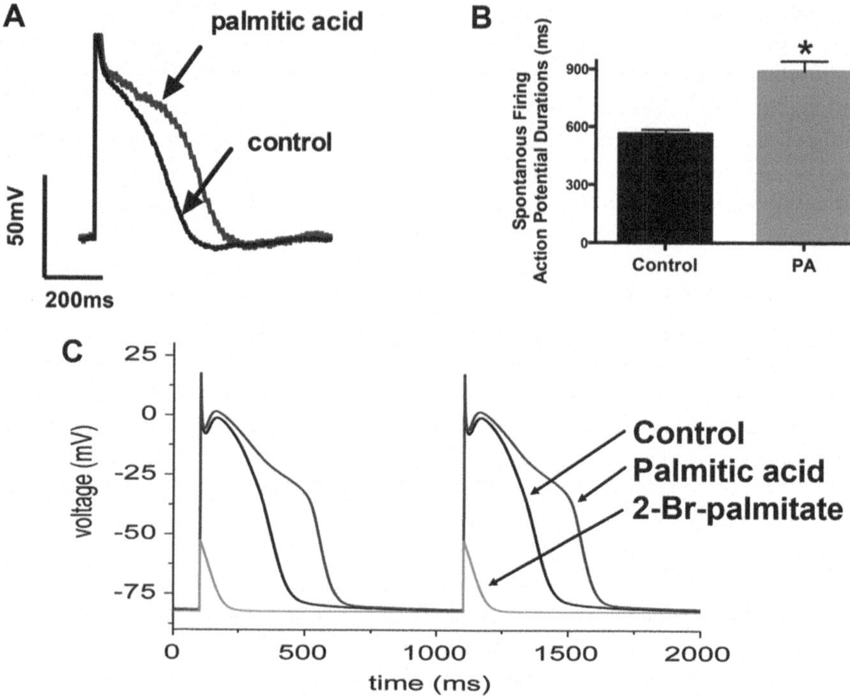

Fig. 6 Palmitoylation regulates cardiac action potential duration and firing frequency in myocytes. (**a**) Representative action potential traces of cardiomyocytes with (*blue*) and without (*black*) palmitic acid treatment. Palmitic acid (PA) treatment increases the action potential duration. (**b**) Averaged APD measurements (from the beginning of depolarization phase to the end of repolarization phase) from the spontaneous firing myocytes. Averaged APD is 563 ms in the control group and 890 ms in the palmitic acid treatment group. (**c**) Simulated action potentials from a modeled cardiac myocyte are shown. Action potentials were paced at 1 Hz in these simulations, and the second and third action potentials are shown. The action potentials from control Nav1.5 channels, Nav1.5 channels modeled to reflect the changes induced by enhanced palmitoylation (palmitic acid) and Nav1.5 channels modeled to reflect the changes induced by depalmitoylation (2-Br-palmitate). Adapted with permission from Pei et al. (2016)

often contain a consensus nitrosylation motif consisting of a cysteine (C) flanked by charged amino acids (Stamler et al. 1997). However, some proteins which do not have a consensus sequence can also be modulated by NO and NO-related products. Analysis of DRG sodium channel α subunits indicated the presence of cysteines flanked by charged amino acids at −1,2 and +1,2 positions. While both Nav1.7 and Nav1.8 are reported to have at least one consensus sequence for nitrosylation, Nav1.9 does not have a classic consensus sequence. The block of persistent Nav1.9 currents by a NO donor indicates that partial consensus sequences may be sufficient for nitrosylation of a resident cysteine and Na^+ channel block. The precise cysteine sites involved have not yet been identified, but it was speculated that

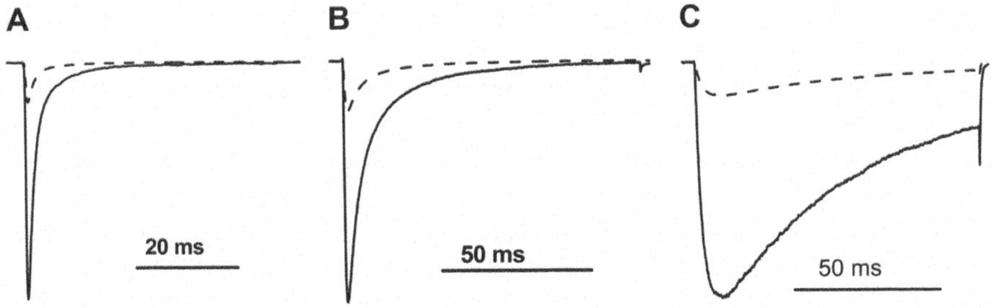

Fig. 7 Schematic illustration of relative inhibition of sensory neuron (**a**) TTX-sensitive, (**b**) slow TTX-resistant, and (**c**) persistent TTX-resistant sodium currents after putative S-nitrosylation. All three types of currents were reduced by 79–85% by nitric oxide donors in Renganathan et al. (2002)

current inhibition could play a role in regulating the excitability of the DRG neurons and might contribute to impaired impulse conduction in disease states.

As with phosphorylation, there is evidence that S-nitrosylation might differentially impact distinct VGSC isoforms. Hammarström and Gage (1999) reported that NO donors could increase persistent TTX-sensitive sodium currents in rat hippocampal neurons without impacting peak sodium current density. A two- to threefold increase in persistent current was observed, and an increase of this magnitude would undoubtedly contribute to increased excitability and thus could contribute to pathophysiological activity associated with stroke or epilepsy.

Because the increase in persistent current was relatively stable, it was proposed that a conformational change in the VGSC protein might have occurred as the result of disulfide bond formation between two closely located *S*-nitrosylated thiols in the VGSC.

S-nitrosylation has also been implicated in modulation of Nav1.5 cardiac sodium channels. Ahern et al. (2000) reported that NO donors could increase persistent currents by fivefold in both nerve terminals and myocytes. This effect was independent of cGMP, was blocked by *N*-ethylmaleimide, and was proposed to also involve S-nitrosylation. Interestingly, caveolin-3 activity seems to suppress S-nitrosylation of Nav1.5 channels and thus attenuates Nav1.5 persistent currents. However, LQT-9 mutations in caveolin-3 seem to disrupt this suppression, leading to larger Nav1.5 persistent currents and cardiac abnormalities (Cheng et al. 2013). It is not clear if S-nitrosylation and palmitoylation can both impact specific cysteine residues in VGSCs.

2.8 ROS Modifications

Reactive oxygen species (ROS) has also been implicated in regulation of VGSC activity. It has long been known that mild oxidizing agents such as chloramine-T can substantially impair sodium channel inactivation (Wang et al. 1985). Chloramine-T can impair inactivation of Nav1.2, Nav1.4, Nav1.5, and Nav1.7 channels (Kassmann et al. 2008). Kassmann et al. found that chloramine-T removed more than 50% of Nav1.4 inactivation but had negligible impact on current activation properties. The effect on inactivation was not readily reversible. ROS can oxidize the sulfur-containing amino acids cysteine and methionine. Chloramine-T is thought to preferentially target methionines. By mutating intracellular methionines, Kassmann et al. found that three specific methionines were responsible for the chloramine-T effect on inactivation. One of these was the methionine in the IFM motif. The two other methionines are located next to each other in the S4–S5 linker of DIV. Interestingly, several disease mutations that impair inactivation occur at these positions. However, it should be noted that there is indication that VGSCs may be differentially regulated by ROS. Schluter and Leffler (2016) reported that chloramine-T enhances activation in addition to impairing inactivation of sensory neuronal Nav1.7 and Nav1.8 VGSCs. Normally the cytosol is a reducing environment, but it is speculated that increased persistent sodium currents in response to ROS allow neurons and muscle to regulate firing frequency when under conditions of oxidative stress. This could be adaptive in some scenarios, but could also be problematic in other situations.

3 Conclusions

During over 40 years of research on VGSCs, we have learned that these crucial regulators of cellular excitability are themselves subject to extensive modulation. This review focused on a number of different covalent modifications of VGSCs. In addition to the modifications discussed here, it is likely that VGSCs are subject to other posttranslational modifications. Sulfation, lipoxidation, acetylation, and amidation are possibilities. Protease modification of channel proteins is also an important way to modulate their properties. The array of accessory proteins that can modulate VGSC properties continues to expand. Much has been learned about VGSCs using heterologous expression systems. However, it is clear that modulation can depend on cellular background. Differences in protein partners and enzyme activity can greatly impact the extent to which VGSCs are posttranslational modified. Furthermore, posttranslational modifications can interact, and therefore it is difficult to predict what impact a particular ensemble of modification will have on VGSC properties. Regardless, it is clear that VGSCs can be extensively modified and that this can alter trafficking, localization, gating, and pharmacology, among other properties. Active research will continue to help us understand how VGSC posttranslational modifications impact normal and abnormal excitability.

References

Abriel H, Kamynina E, Horisberger JD, Staub O (2000) Regulation of the cardiac voltage-gated Na+ channel (H1) by the ubiquitin-protein ligase Nedd4. FEBS Lett 466(2–3):377–380

Ahern GP, Hsu SF, Klyachko VA, Jackson MB (2000) Induction of persistent sodium current by exogenous and endogenous nitric oxide. J Biol Chem 275(37):28810–28815. https://doi.org/10.1074/jbc.M003090200

Aiba T, Hesketh GG, Liu T, Carlisle R, Villa-Abrille MC, O'Rourke B, Akar FG, Tomaselli GF (2010) Na+ channel regulation by Ca2+/calmodulin and Ca2+/calmodulin-dependent protein kinase II in guinea-pig ventricular myocytes. Cardiovasc Res 85(3):454–463. https://doi.org/10.1093/cvr/cvp324

Alexander JK, Govind AP, Drisdel RC, Blanton MP, Vallejo Y, Lam TT, Green WN (2010) Palmitoylation of nicotinic acetylcholine receptors. J Mol Neurosci 40(1–2):12–20. https://doi.org/10.1007/s12031-009-9246-z

Ashpole NM, Herren AW, Ginsburg KS, Brogan JD, Johnson DE, Cummins TR, Bers DM, Hudmon A (2012) Ca2+/calmodulin-dependent protein kinase II (CaMKII) regulates cardiac sodium channel NaV1.5 gating by multiple phosphorylation sites. J Biol Chem 287(24):19856–19869. https://doi.org/10.1074/jbc.M111.322537

Baek JH, Rubinstein M, Scheuer T, Trimmer JS (2014) Reciprocal changes in phosphorylation and methylation of mammalian brain sodium channels in response to seizures. J Biol Chem 289(22):15363–15373. https://doi.org/10.1074/jbc.M114.562785

Bankston JR, Yue M, Chung W, Spyres M, Pass RH, Silver E, Sampson KJ, Kass RS (2007) A novel and lethal de novo LQT-3 mutation in a newborn with distinct molecular pharmacology and therapeutic response. PLoS One 2(12):e1258. https://doi.org/10.1371/journal.pone.0001258

Beltran-Alvarez P, Espejo A, Schmauder R, Beltran C, Mrowka R, Linke T, Batlle M, Perez-Villa F, Perez GJ, Scornik FS, Benndorf K, Pagans S, Zimmer T, Brugada R (2013) Protein arginine methyl transferases-3 and -5 increase cell surface expression of cardiac sodium channel. FEBS Lett 587(19):3159–3165. https://doi.org/10.1016/j.febslet.2013.07.043

Beltran-Alvarez P, Tarradas A, Chiva C, Perez-Serra A, Batlle M, Perez-Villa F, Schulte U, Sabido E, Brugada R, Pagans S (2014) Identification of N-terminal protein acetylation and arginine methylation of the voltage-gated sodium channel in end-stage heart failure human heart. J Mol Cell Cardiol 76:126–129. https://doi.org/10.1016/j.yjmcc.2014.08.014

Beltran-Alvarez P, Feixas F, Osuna S, Diaz-Hernandez R, Brugada R, Pagans S (2015) Interplay between R513 methylation and S516 phosphorylation of the cardiac voltage-gated sodium channel. Amino Acids 47(2):429–434. https://doi.org/10.1007/s00726-014-1890-0

van Bemmelen MX, Rougier JS, Gavillet B, Apotheloz F, Daidie D, Tateyama M, Rivolta I, Thomas MA, Kass RS, Staub O, Abriel H (2004) Cardiac voltage-gated sodium channel Nav1.5 is regulated by Nedd4-2 mediated ubiquitination. Circ Res 95(3):284–291. https://doi.org/10.1161/01.RES.0000136816.05109.89

Bendahhou S, Cummins TR, Potts JF, Tong J, Agnew WS (1995) Serine-1321-independent regulation of the mu 1 adult skeletal muscle Na+ channel by protein kinase C. Proc Natl Acad Sci U S A 92(26):12003–12007

Bennett E, Urcan MS, Tinkle SS, Koszowski AG, Levinson SR (1997) Contribution of sialic acid to the voltage dependence of sodium channel gating. A possible electrostatic mechanism. J Gen Physiol 109(3):327–343

Berendt FJ, Park KS, Trimmer JS (2010) Multisite phosphorylation of voltage-gated sodium channel alpha subunits from rat brain. J Proteome Res 9(4):1976–1984. https://doi.org/10.1021/pr901171q

Bosmans F, Milescu M, Swartz KJ (2011) Palmitoylation influences the function and pharmacology of sodium channels. Proc Natl Acad Sci U S A 108(50):20213–20218. https://doi.org/10.1073/pnas.1108497108

Brigidi GS, Bamji SX (2013) Detection of protein palmitoylation in cultured hippocampal neurons by immunoprecipitation and acyl-biotin exchange (ABE). J Vis Exp (72). https://doi.org/10.3791/50031

Burel S, Coyan FC, Lorenzini M, Meyer MR, Lichti CF, Brown JH, Loussouarn G, Charpentier F, Nerbonne JM, Townsend RR, Maier LS, Marionneau C (2017) C-terminal phosphorylation of NaV1.5 impairs FGF13-dependent regulation of channel inactivation. J Biol Chem. https://doi.org/10.1074/jbc.M117.787788

Cachemaille M, Laedermann CJ, Pertin M, Abriel H, Gosselin RD, Decosterd I (2012) Neuronal expression of the ubiquitin ligase Nedd4-2 in rat dorsal root ganglia: modulation in the spared nerve injury model of neuropathic pain. Neuroscience 227:370–380. https://doi.org/10.1016/j.neuroscience.2012.09.044

Catterall WA, Goldin AL, Waxman SG (2005) International Union of Pharmacology. XLVII. Nomenclature and structure-function relationships of voltage-gated sodium channels. Pharmacol Rev 57(4):397–409

Chen Y, Cantrell AR, Messing RO, Scheuer T, Catterall WA (2005) Specific modulation of Na+ channels in hippocampal neurons by protein kinase C epsilon. J Neurosci 25(2):507–513. https://doi.org/10.1523/JNEUROSCI.4089-04.2005

Chen C, Calhoun JD, Zhang Y, Lopez-Santiago L, Zhou N, Davis TH, Salzer JL, Isom LL (2012) Identification of the cysteine residue responsible for disulfide linkage of Na+ channel alpha and beta2 subunits. J Biol Chem 287(46):39061–39069. https://doi.org/10.1074/jbc.M112.397646

Cheng J, Valdivia CR, Vaidyanathan R, Balijepalli RC, Ackerman MJ, Makielski JC (2013) Caveolin-3 suppresses late sodium current by inhibiting nNOS-dependent S-nitrosylation of SCN5A. J Mol Cell Cardiol 61:102–110. https://doi.org/10.1016/j.yjmcc.2013.03.013

Cohen SA, Levitt LK (1993) Partial characterization of the rH1 sodium channel protein from rat heart using subtype-specific antibodies. Circ Res 73(4):735–742

Corry B, Thomas M (2012) Mechanism of ion permeation and selectivity in a voltage gated sodium channel. J Am Chem Soc 134(3):1840–1846. https://doi.org/10.1021/ja210020h

Costa MR, Catterall WA (1984) Phosphorylation of the alpha subunit of the sodium channel by protein kinase C. Cell Mol Neurobiol 4(3):291–297

Cummins TR, Donnelly DF, Haddad GG (1991) Effect of metabolic inhibition on the excitability of isolated hippocampal CA1 neurons: developmental aspects. J Neurophysiol 66(5):1471–1482

Cummins TR, Jiang C, Haddad GG (1993) Human neocortical excitability is decreased during anoxia via sodium channel modulation. J Clin Invest 91(2):608–615. https://doi.org/10.1172/JCI116241

Drisdel RC, Manzana E, Green WN (2004) The role of palmitoylation in functional expression of nicotinic alpha7 receptors. J Neurosci 24(46):10502–10510. https://doi.org/10.1523/JNEUROSCI.3315-04.2004

Dustrude ET, Moutal A, Yang X, Wang Y, Khanna M, Khanna R (2016) Hierarchical CRMP2 posttranslational modifications control NaV1.7 function. Proc Natl Acad Sci U S A 113(52): E8443–E8452. doi: https://doi.org/10.1073/pnas.1610531113

Ednie AR, Horton KK, Wu J, Bennett ES (2013) Expression of the sialyltransferase, ST3Gal4, impacts cardiac voltage-gated sodium channel activity, refractory period and ventricular conduction. J Mol Cell Cardiol 59:117–127. https://doi.org/10.1016/j.yjmcc.2013.02.013

Ednie AR, Harper JM, Bennett ES (2015) Sialic acids attached to N- and O-glycans within the Nav1.4 D1S5-S6 linker contribute to channel gating. Biochim Biophys Acta 1850(2):307–317. https://doi.org/10.1016/j.bbagen.2014.10.027

Fotia AB, Ekberg J, Adams DJ, Cook DI, Poronnik P, Kumar S (2004) Regulation of neuronal voltage-gated sodium channels by the ubiquitin-protein ligases Nedd4 and Nedd4-2. J Biol Chem 279(28):28930–28935. https://doi.org/10.1074/jbc.M402820200

Fukata M, Fukata Y, Adesnik H, Nicoll RA, Bredt DS (2004) Identification of PSD-95 palmitoylating enzymes. Neuron 44(6):987–996. https://doi.org/10.1016/j.neuron.2004.12.005

Gasser A, Cheng X, Gilmore ES, Tyrrell L, Waxman SG, Dib-Hajj SD (2010) Two Nedd4-binding motifs underlie modulation of sodium channel Nav1.6 by p38 MAPK. J Biol Chem 285(34): 26149–26161. https://doi.org/10.1074/jbc.M109.098681

Goldin AL (2001) Resurgence of sodium channel research. Annu Rev Physiol 63:871–894. https://doi.org/10.1146/annurev.physiol.63.1.871
Goldin AL (2003) Mechanisms of sodium channel inactivation. Curr Opin Neurobiol 13(3): 284–290. doi: S0959438803000655 [pii]
Gubitosi-Klug RA, Mancuso DJ, Gross RW (2005) The human Kv1.1 channel is palmitoylated, modulating voltage sensing: identification of a palmitoylation consensus sequence. Proc Natl Acad Sci U S A 102(17):5964–5968. https://doi.org/10.1073/pnas.0501999102
Hallaq H, Wang DW, Kunic JD, George AL Jr, Wells KS, Murray KT (2012) Activation of protein kinase C alters the intracellular distribution and mobility of cardiac Na+ channels. Am J Physiol Heart Circ Physiol 302(3):H782–H789. https://doi.org/10.1152/ajpheart.00817.2010
Hamill OP, Marty A, Neher E, Sakmann B, Sigworth FJ (1981) Improved patch-clamp techniques for high-resolution current recording from cells and cell-free membrane patches. Pflugers Arch 391(2):85–100
Hammarstrom AK, Gage PW (1999) Nitric oxide increases persistent sodium current in rat hippocampal neurons. J Physiol 520(Pt 2):451–461
Hayashi T, Rumbaugh G, Huganir RL (2005) Differential regulation of AMPA receptor subunit trafficking by palmitoylation of two distinct sites. Neuron 47(5):709–723. https://doi.org/10.1016/j.neuron.2005.06.035
Herren AW, Weber DM, Rigor RR, Margulies KB, Phinney BS, Bers DM (2015) CaMKII phosphorylation of Na(V)1.5: novel in vitro sites identified by mass spectrometry and reduced S516 phosphorylation in human heart failure. J Proteome Res 14(5):2298–2311. https://doi.org/10.1021/acs.jproteome.5b00107
Hoch B, Meyer R, Hetzer R, Krause EG, Karczewski P (1999) Identification and expression of delta-isoforms of the multifunctional Ca2+/calmodulin-dependent protein kinase in failing and nonfailing human myocardium. Circ Res 84(6):713–721
Hodgkin AL, Huxley AF (1952) A quantitative description of membrane current and its application to conduction and excitation in nerve. J Physiol 117(4):500–544
Johnson D, Montpetit ML, Stocker PJ, Bennett ES (2004) The sialic acid component of the beta1 subunit modulates voltage-gated sodium channel function. J Biol Chem 279(43):44303–44310. https://doi.org/10.1074/jbc.M408900200
Kassmann M, Hansel A, Leipold E, Birkenbeil J, Lu SQ, Hoshi T, Heinemann SH (2008) Oxidation of multiple methionine residues impairs rapid sodium channel inactivation. Pflugers Arch 456(6):1085–1095. https://doi.org/10.1007/s00424-008-0477-6
Korycka J, Lach A, Heger E, Boguslawska DM, Wolny M, Toporkiewicz M, Augoff K, Korzeniewski J, Sikorski AF (2012) Human DHHC proteins: a spotlight on the hidden player of palmitoylation. Eur J Cell Biol 91(2):107–117. https://doi.org/10.1016/j.ejcb.2011.09.013
Koval OM, Snyder JS, Wolf RM, Pavlovicz RE, Glynn P, Curran J, Leymaster ND, Dun W, Wright PJ, Cardona N, Qian L, Mitchell CC, Boyden PA, Binkley PF, Li C, Anderson ME, Mohler PJ, Hund TJ (2012) Ca2+/calmodulin-dependent protein kinase II-based regulation of voltage-gated Na+ channel in cardiac disease. Circulation 126(17):2084–2094. https://doi.org/10.1161/CIRCULATIONAHA.112.105320
Laedermann CJ, Syam N, Pertin M, Decosterd I, Abriel H (2013a) beta1- and beta3-voltage-gated sodium channel subunits modulate cell surface expression and glycosylation of Nav1.7 in HEK293 cells. Front Cell Neurosci 7:137. https://doi.org/10.3389/fncel.2013.00137
Laedermann CJ, Cachemaille M, Kirschmann G, Pertin M, Gosselin RD, Chang I, Albesa M, Towne C, Schneider BL, Kellenberger S, Abriel H, Decosterd I (2013b) Dysregulation of voltage-gated sodium channels by ubiquitin ligase NEDD4-2 in neuropathic pain. J Clin Invest 123(7):3002–3013. https://doi.org/10.1172/JCI68996
Laedermann CJ, Decosterd I, Abriel H (2014a) Ubiquitylation of voltage-gated sodium channels. Handb Exp Pharmacol 221:231–250. https://doi.org/10.1007/978-3-642-41588-3_11
Laedermann CJ, Pertin M, Suter MR, Decosterd I (2014b) Voltage-gated sodium channel expression in mouse DRG after SNI leads to re-evaluation of projections of injured fibers. Mol Pain 10:19. https://doi.org/10.1186/1744-8069-10-19

Levinson SR, Ellory JC (1973) Molecular size of the tetrodotoxin binding site estimated by irradiation inactivation. Nat New Biol 245(143):122–123

Li M, West JW, Numann R, Murphy BJ, Scheuer T, Catterall WA (1993) Convergent regulation of sodium channels by protein kinase C and cAMP-dependent protein kinase. Science 261(5127): 1439–1442

Lu Z, Wu CY, Jiang YP, Ballou LM, Clausen C, Cohen IS, Lin RZ (2012) Suppression of phosphoinositide 3-kinase signaling and alteration of multiple ion currents in drug-induced long QT syndrome. Sci Transl Med 4(131):131ra150. https://doi.org/10.1126/scitranslmed.3003623

Lu Z, Jiang YP, Wu CY, Ballou LM, Liu S, Carpenter ES, Rosen MR, Cohen IS, Lin RZ (2013) Increased persistent sodium current due to decreased PI3K signaling contributes to QT prolongation in the diabetic heart. Diabetes 62(12):4257–4265. https://doi.org/10.2337/db13-0420

Malhotra JD, Thyagarajan V, Chen C, Isom LL (2004) Tyrosine-phosphorylated and nonphosphorylated sodium channel beta1 subunits are differentially localized in cardiac myocytes. J Biol Chem 279(39):40748–40754. https://doi.org/10.1074/jbc.M407243200

Maltsev VA, Reznikov V, Undrovinas NA, Sabbah HN, Undrovinas A (2008) Modulation of late sodium current by Ca2+, calmodulin, and CaMKII in normal and failing dog cardiomyocytes: similarities and differences. Am J Physiol Heart Circ Physiol 294(4):H1597–H1608. https://doi.org/10.1152/ajpheart.00484.2007

Marionneau C, Abriel H (2015) Regulation of the cardiac Na+ channel NaV1.5 by post-translational modifications. J Mol Cell Cardiol 82:36–47. https://doi.org/10.1016/j.yjmcc.2015.02.013

Marionneau C, Lichti CF, Lindenbaum P, Charpentier F, Nerbonne JM, Townsend RR, Merot J (2012) Mass spectrometry-based identification of native cardiac Nav1.5 channel alpha subunit phosphorylation sites. J Proteome Res 11(12):5994–6007. https://doi.org/10.1021/pr300702c

Martin BR, Cravatt BF (2009) Large-scale profiling of protein palmitoylation in mammalian cells. Nat Methods 6(2):135–138. https://doi.org/10.1038/nmeth.1293

Mercier A, Clement R, Harnois T, Bourmeyster N, Bois P, Chatelier A (2015) Nav1.5 channels can reach the plasma membrane through distinct N-glycosylation states. Biochim Biophys Acta 1850(6):1215–1223. https://doi.org/10.1016/j.bbagen.2015.02.009

Messner DJ, Catterall WA (1985) The sodium channel from rat brain. Separation and characterization of subunits. J Biol Chem 260(19):10597–10604

Murray KT, Fahrig SA, Deal KK, Po SS, Hu NN, Snyders DJ, Tamkun MM, Bennett PB (1994) Modulation of an inactivating human cardiac K+ channel by protein kinase C. Circ Res 75(6): 999–1005

Murray KT, Hu NN, Daw JR, Shin HG, Watson MT, Mashburn AB, George AL Jr (1997) Functional effects of protein kinase C activation on the human cardiac Na+ channel. Circ Res 80(3):370–376

Numann R, Catterall WA, Scheuer T (1991) Functional modulation of brain sodium channels by protein kinase C phosphorylation. Science 254(5028):115–118

O'Reilly JP, Cummins TR, Haddad GG (1997) Oxygen deprivation inhibits Na+ current in rat hippocampal neurons via protein kinase C. J Physiol 503(Pt 3):479–488

Payandeh J, Scheuer T, Zheng N, Catterall WA (2011) The crystal structure of a voltage-gated sodium channel. Nature 475(7356):353–358. https://doi.org/10.1038/nature10238

Pei Z, Xiao Y, Meng J, Hudmon A, Cummins TR (2016) Cardiac sodium channel palmitoylation regulates channel availability and myocyte excitability with implications for arrhythmia generation. Nat Commun 7:12035. https://doi.org/10.1038/ncomms12035

Plant LD, Marks JD, Goldstein SA (2016) SUMOylation of NaV1.2 channels mediates the early response to acute hypoxia in central neurons. Elife 5. https://doi.org/10.7554/eLife.20054

Qu Y, Rogers JC, Tanada TN, Catterall WA, Scheuer T (1996) Phosphorylation of S1505 in the cardiac Na+ channel inactivation gate is required for modulation by protein kinase C. J Gen Physiol 108(5):375–379

Renganathan M, Cummins TR, Waxman SG (2002) Nitric oxide blocks fast, slow, and persistent Na+ channels in C-type DRG neurons by S-nitrosylation. J Neurophysiol 87(2):761–775

Richmond JE, VanDeCarr D, Featherstone DE, George AL Jr, Ruben PC (1997) Defective fast inactivation recovery and deactivation account for sodium channel myotonia in the I1160V mutant. Biophys J 73(4):1896–1903. https://doi.org/10.1016/S0006-3495(97)78220-1

Rougier JS, van Bemmelen MX, Bruce MC, Jespersen T, Gavillet B, Apotheloz F, Cordonier S, Staub O, Rotin D, Abriel H (2005) Molecular determinants of voltage-gated sodium channel regulation by the Nedd4/Nedd4-like proteins. Am J Physiol Cell Physiol 288(3):C692–C701. https://doi.org/10.1152/ajpcell.00460.2004

Rougier JS, Gavillet B, Abriel H (2013) Proteasome inhibitor (MG132) rescues Nav1.5 protein content and the cardiac sodium current in dystrophin-deficient mdx (5cv) mice. Front Physiol 4:51. https://doi.org/10.3389/fphys.2013.00051

Schluter F, Leffler A (2016) Oxidation differentially modulates the recombinant voltage-gated Na (+) channel alpha-subunits Nav1.7 and Nav1.8. Brain Res 1648(Pt A):127–135. https://doi.org/10.1016/j.brainres.2016.07.031

Schmidt JW, Catterall WA (1986) Biosynthesis and processing of the alpha subunit of the voltage-sensitive sodium channel in rat brain neurons. Cell 46(3):437–444

Schmidt JW, Catterall WA (1987) Palmitoylation, sulfation, and glycosylation of the alpha subunit of the sodium channel. Role of post-translational modifications in channel assembly. J Biol Chem 262(28):13713–13723

Schmidt M, Schmidt MF, Rott R (1988) Chemical identification of cysteine as palmitoylation site in a transmembrane protein (Semliki Forest virus E1). J Biol Chem 263(35):18635–18639

Schubert B, Vandongen AM, Kirsch GE, Brown AM (1990) Inhibition of cardiac Na+ currents by isoproterenol. Am J Physiol 258(4 Pt 2):H977–H982

Shin HG, Murray KT (2001) Conventional protein kinase C isoforms and cross-activation of protein kinase A regulate cardiac Na+ current. FEBS Lett 495(3):154–158

Shipston MJ (2011) Ion channel regulation by protein palmitoylation. J Biol Chem 286(11): 8709–8716. https://doi.org/10.1074/jbc.R110.210005

Shipston MJ (2014) Ion channel regulation by protein S-acylation. J Gen Physiol 143(6):659–678. https://doi.org/10.1085/jgp.201411176

Silva J (2014) Slow inactivation of Na(+) channels. Handb Exp Pharmacol 221:33–49. https://doi.org/10.1007/978-3-642-41588-3_3

Stamler JS (1994) Redox signaling: nitrosylation and related target interactions of nitric oxide. Cell 78(6):931–936

Stamler JS, Toone EJ, Lipton SA, Sucher NJ (1997) (S)NO signals: translocation, regulation, and a consensus motif. Neuron 18(5):691–696

Stuhmer W, Conti F, Suzuki H, Wang XD, Noda M, Yahagi N, Kubo H, Numa S (1989) Structural parts involved in activation and inactivation of the sodium channel. Nature 339(6226): 597–603. https://doi.org/10.1038/339597a0

Suzuki H, Nishikawa K, Hiroaki Y, Fujiyoshi Y (2008) Formation of aquaporin-4 arrays is inhibited by palmitoylation of N-terminal cysteine residues. Biochim Biophys Acta 1778(4): 1181–1189. https://doi.org/10.1016/j.bbamem.2007.12.007

Tan ZY, Priest BT, Krajewski JL, Knopp KL, Nisenbaum ES, Cummins TR (2014) Protein kinase C enhances human sodium channel hNav1.7 resurgent currents via a serine residue in the domain III-IV linker. FEBS Lett 588(21):3964–3969. doi: 10.1016/j.febslet.2014.09.011 S0014-5793(14)00686-3 [pii]

Teener JW, Rich MM (2006) Dysregulation of sodium channel gating in critical illness myopathy. J Muscle Res Cell Motil 27(5–7):291–296. https://doi.org/10.1007/s10974-006-9074-5

Thomas GM, Hayashi T, Chiu SL, Chen CM, Huganir RL (2012) Palmitoylation by DHHC5/8 targets GRIP1 to dendritic endosomes to regulate AMPA-R trafficking. Neuron 73(3): 482–496. https://doi.org/10.1016/j.neuron.2011.11.021

Tyrrell L, Renganathan M, Dib-Hajj SD, Waxman SG (2001) Glycosylation alters steady-state inactivation of sodium channel Nav1.9/NaN in dorsal root ganglion neurons and is developmentally regulated. J Neurosci 21(24):9629–9637

Ulbricht W (2005) Sodium channel inactivation: molecular determinants and modulation. Physiol Rev 85(4):1271–1301. https://doi.org/10.1152/physrev.00024.2004

Vijayaragavan K, Boutjdir M, Chahine M (2004) Modulation of Nav1.7 and Nav1.8 peripheral nerve sodium channels by protein kinase A and protein kinase C. J Neurophysiol 91(4): 1556–1569. https://doi.org/10.1152/jn.00676.2003

Vilin YY, Ruben PC (2001) Slow inactivation in voltage-gated sodium channels: molecular substrates and contributions to channelopathies. Cell Biochem Biophys 35(2):171–190. https://doi.org/10.1385/CBB:35:2:171

Waechter CJ, Schmidt JW, Catterall WA (1983) Glycosylation is required for maintenance of functional sodium channels in neuroblastoma cells. J Biol Chem 258(8):5117–5123

Wagner S, Maier LS, Bers DM (2015) Role of sodium and calcium dysregulation in tachyarrhythmias in sudden cardiac death. Circ Res 116(12):1956–1970. https://doi.org/10.1161/CIRCRESAHA.116.304678

Wallace RH, Hodgson BL, Grinton BE, Gardiner RM, Robinson R, Rodriguez-Casero V, Sadleir L, Morgan J, Harkin LA, Dibbens LM, Yamamoto T, Andermann E, Mulley JC, Berkovic SF, Scheffer IE (2003) Sodium channel alpha1-subunit mutations in severe myoclonic epilepsy of infancy and infantile spasms. Neurology 61(6):765–769

Wang GK, Brodwick MS, Eaton DC (1985) Removal of sodium channel inactivation in squid axon by the oxidant chloramine-T. J Gen Physiol 86(2):289–302

West JW, Numann R, Murphy BJ, Scheuer T, Catterall WA (1991) A phosphorylation site in the Na+ channel required for modulation by protein kinase C. Science 254(5033):866–868

West JW, Patton DE, Scheuer T, Wang Y, Goldin AL, Catterall WA (1992) A cluster of hydrophobic amino acid residues required for fast Na(+)-channel inactivation. Proc Natl Acad Sci U S A 89(22):10910–10914

Yarov-Yarovoy V, DeCaen PG, Westenbroek RE, Pan CY, Scheuer T, Baker D, Catterall WA (2012) Structural basis for gating charge movement in the voltage sensor of a sodium channel. Proc Natl Acad Sci U S A 109(2):E93–102. https://doi.org/10.1073/pnas.1118434109

Yu FH, Catterall WA (2003) Overview of the voltage-gated sodium channel family. Genome Biol 4(3):207

Zeidman R, Jackson CS, Magee AI (2009) Protein acyl thioesterases (review). Mol Membr Biol 26(1):32–41. https://doi.org/10.1080/09687680802629329

Zhang T, Maier LS, Dalton ND, Miyamoto S, Ross J Jr, Bers DM, Brown JH (2003) The deltaC isoform of CaMKII is activated in cardiac hypertrophy and induces dilated cardiomyopathy and heart failure. Circ Res 92(8):912–919. https://doi.org/10.1161/01.RES.0000069686.31472.C5

Sodium Channel Trafficking

A. Mercier, P. Bois, and A. Chatelier

Contents

Abstract

Voltage-gated sodium channels (VGSC) are critical determinants of cellular electrical activity through the control of initiation and propagation of action potential. To ensure this role, these proteins are not consistently delivered to the plasma membrane but undergo drastic quality controls throughout various adaptive processes such as biosynthesis, anterograde and retrograde trafficking, and membrane targeting. In pathological conditions, this quality control could lead to the retention of functional VGSC and is therefore the target of different pharmacological approaches. The present chapter gives an overview of the current understanding of the facets of VGSC life cycle in the context of both cardiac and neuronal cell types.

A. Mercier · P. Bois · A. Chatelier (✉)
Laboratoire de Signalisation et Transports Ioniques Membranaires, Pôle Biologie Santé, Université de Poitiers, CNRS, 1 rue Georges Bonnet, TSA 51106, 86073 Poitiers Cedex 9, France
e-mail: aurelien.chatelier@univ-poitiers.fr

M. Chahine (ed.), *Voltage-gated Sodium Channels: Structure, Function and Channelopathies*, Handbook of Experimental Pharmacology 246,
https://doi.org/10.1007/164_2017_47

Keywords
Forward trafficking · Membrane targeting · Retrograde transport · Trafficking modulators · Voltage-gated sodium channel

1 Introduction

Voltage-gated sodium channels (VGSC) play a critical role in electrical signaling of excitable tissues through the control of initiation and propagation of action potentials. This function is strongly correlated to the nature of VGSC isoforms associated with their functional membrane expression and localization. To date, nine mammalian pore forming α-subunits ($Na_v1.1$–$Na_v1.9$) have been identified and divided into two categories based on their sensitivity to the puffer fish toxin tetrodotoxin (TTX). $Na_v1.1$, $Na_v1.2$, $Na_v1.3$, $Na_v1.4$, $Na_v1.6$, and $Na_v1.7$ are sensitive to TTX, while $Na_v1.5$, $Na_v1.8$, and $Na_v1.9$ are TTX-resistant. At least seven VGSC are mainly expressed in the nervous system. $Na_v1.1$, $Na_v1.2$, $Na_v1.3$, and $Na_v1.6$ are predominantly expressed in the central nervous system (CNS), while $Na_v1.7$, $Na_v1.8$, and $Na_v1.9$ are principally found in the peripheral nervous system (PNS). $Na_v1.5$ and $Na_v1.4$ are muscular sodium channels and are mainly expressed in cardiac tissue and skeletal muscle, respectively (Chahine et al. 2008; Savio-Galimberti et al. 2012). Each VGSC exists as multimeric complex at the plasma membrane presumably composed of one or several auxiliary β-subunits associated with the α-subunit. To date, four β-subunits (β_{1-4}) have been identified and are divided into two groups: β_1 and β_3 associate non-covalently with the VGSC α-subunit, whereas β_2 and β_4 bind covalently to the α-subunit through a cysteine residue on the extracellular loop (reviewed in Patino and Isom 2010).

Like most integral proteins, the membrane expression of functional Na_v channels is drastically controlled by both cell-type and common components of the trafficking machinery. These were identified as molecular chaperones, enzymes, motor, anchoring and scaffold proteins whose key functions are to ensure the effective transport of VGSC α- and β-subunits through intracellular compartments to their final functional locations. Instant membrane expression of VGSC directly reflects the cellular proteostasis environment. This balance involves various adaptive processes such as biogenesis, trafficking, and degradation of functional channels and is known to be highly cell-type- and stimulus-dependent.

In the present chapter, we provide a selective review of research focusing on the biosynthetic transport of VGSC and more specifically the general and cell-type mechanisms involved in their anterograde/retrograde trafficking, subcellular targeting and pharmacological behavior in two types of polarized cells, neurons and cardiomyocytes (see Fig. 1).

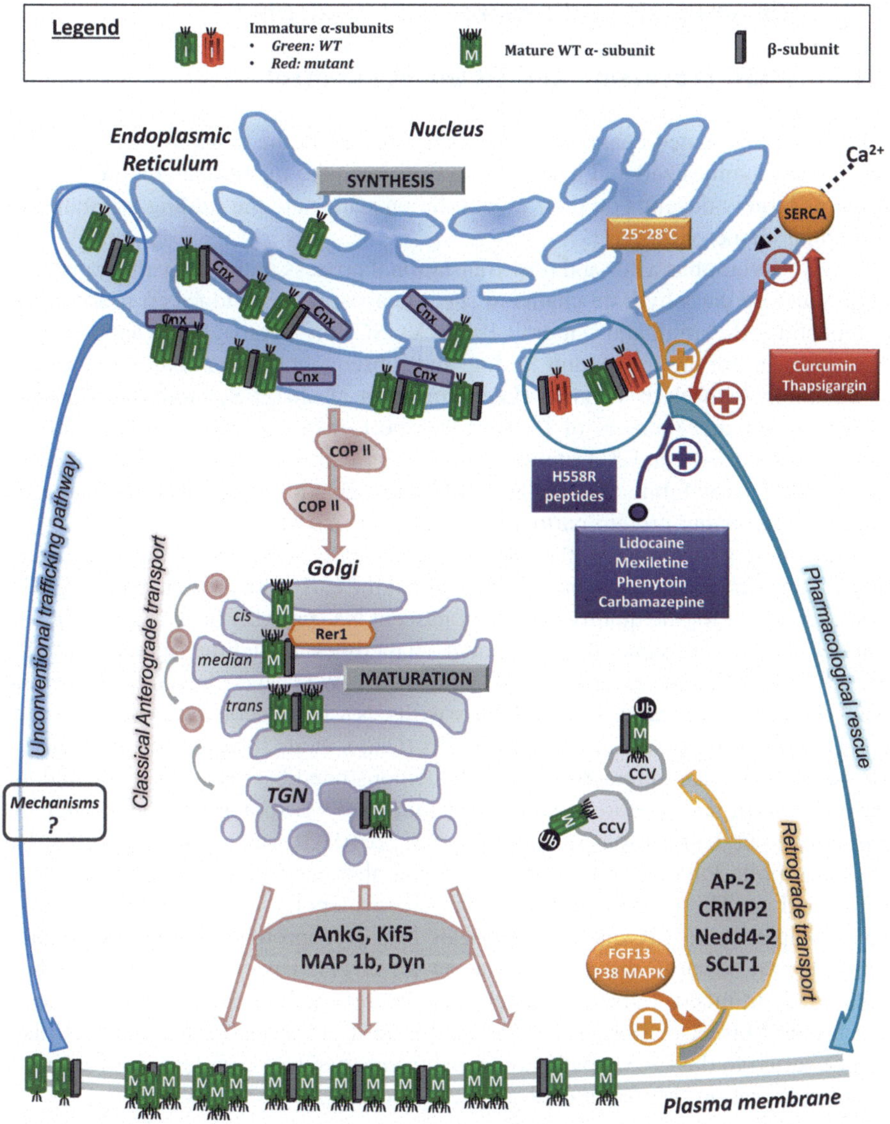

Fig. 1 Illustration of VGSC transport pathways and pharmacological trafficking modulations. This scheme summarizes the main steps and modulations reported for VGSC trafficking without any distinction between cardiac or neuronal α and β-subunit subtypes. *AnkG* Ankyrin G, *AP-2* adaptor protein complex 2, *CCV* clathrin-coated vesicle, *Cnx* calnexin, *COP II* coat protein II-coated vesicles, *CRMP2* SUMOylated collapsin response mediator protein 2, *Dyn* dynamitin, *FGF13* Fibroblast Growth Factor 13, *Kif5* kinesin 5 protein, *MAP 1b* microtubule-associated protein 1b, *Nedd4-2* precursor cell-expressed developmentally downregulated gene 4-2, *P38 MAPK* P38 mitogen-activated protein kinase, *SCLT1* sodium channel and clathrin linker 1, *TGN* trans-Golgi network, *Ub* ubiquitin, *WT* wild type

2 Biosynthesis and Anterograde Transport

2.1 VGSC Processing and ER Quality Control

Like all voltage-gated channels, VGSC's life begins in the endoplasmic reticulum (ER). Once synthesized and inserted into this organelle membrane, the fate of Na_v subunits largely depends on their glycosylation status and the folding assistance of molecular chaperones.

N-glycosylation is a co- and post-translational process initiated by the transfer of high-mannose carbohydrate chains to specific asparagine residues on nascent Na_v polypeptides (Schmidt and Catterall 1986). VGSC core-glycosylation appears to be a prerequisite for proper subunit folding and subsequent surface expression as well as assembly with $Na_v\beta$ subunits (Laedermann et al. 2013; Schmidt and Catterall 1986). As reliable markers of forward transport, different glycosylation patterns were observed for $Na_v1.5$ proteins from distinct cardiac chambers. This suggests their ability to exhibit differential trafficking behaviors, probably mediated by multiple interacting protein partners (Arakel et al. 2014).

Like many glycoproteins, they are not consistently delivered to the cell surface and undergo drastic quality controls throughout the early biosynthetic pathway. The first checkpoint for the quality control of nascent Na_v proteins occurs in the ER and implies the involvement of ER-resident chaperones including calnexin (CNX). There is substantial evidence that wild type (WT) and defective-trafficking $Na_v1.5$ proteins colocalize intracellularly with CNX in rat neonatal cardiomyocytes and Human Embryonic Kidney (HEK) cells (Clatot et al. 2012; Dulsat et al. 2017). However, a physical interaction was only demonstrated between this lectin and the $Na_v1.8$ subtype and appeared to positively influence VGSC degradation through the ER-associated degradation pathway (Li et al. 2010). It was recently reported that the Drosophila Para neuronal sodium channel also happened to be a substrate for CNX binding. These data confirm that the ER-resident lectin contributes as a molecular chaperone to ER quality control of Na_v glycoprotein folding and to their efficient exit towards the Golgi (Xiao et al. 2017). Under the action of CNX, misfolded or misassembled channels can be retained in the ER to eventually reach their correct conformation. Only after multiple unsuccessful attempts, they can be sorted to the ERAD pathway by being escorted to the proteasome for subsequent degradation (McArdle et al. 2008; Ogino et al. 2015).

2.2 VGSC ER-to-Golgi Transport

One noteworthy feature of VGSC is their predominant ER localization consistent with the existence of an intracellular reservoir of Na_v α-subunits, readily mobilizable to mediate cell adaptation to varying environmental changes (Schmidt et al. 1985; Zimmer et al. 2002). ER subpopulations of recombinant $Na_v1.5$ were shown to be redistributed to the plasma membrane upon PKA stimulation, corroborating the increase in native Na^+ current amplitude observed following β-adrenergic receptor stimulation in rat cardiomyocytes (Hallaq et al. 2006; Lu et al. 1999). As for neuronal VGSC, large somatic cytoplasmic pools of Na^+ channels have long been described such as in invertebrate Aplysia sensory neurons or in developing rat brain (Johnston et al. 1996; Schmidt et al. 1985). More recently, intracellular pools of $Na_v1.6$ proteins were identified as a ready reserve that was thought to be potentially recruited to dendritic synapses in the rat cerebral cortex (Caldwell et al. 2000).

The control of Na_v α-subunit ER exit was proposed to be partially mediated through association with $Na_v\beta$ proteins, suggesting a putative role for these regulatory subunits as chaperone-like proteins (Laedermann et al. 2013). Zhang and colleagues identified that the DI-DII linker of rat $Na_v1.8$ channels contained a functional ER-retention motif (RRR) that was shown to regulate their surface expression and be antagonized by $Na_v\beta_3$ assembly (Zhang et al. 2008). Interestingly, surface expression of human $Na_v1.8$ channels that do not harbor this rodent-conserved motif remains unaffected following PKA activation, reflecting species-dependent differences in the ER exit rate (Schirmeyer et al. 2014). Numerous putative ER retention/retrieval signals were also found in other Na_v subtypes such as $Na_v1.2$ and $Na_v1.5$ (Lee and Goldin 2009; Zhou et al. 2002). Since Na^+ current potentiation mediated by PKA activation requires the masking of $Na_v1.5$ retention motifs, the role of these sequences might contribute to the rate-limiting step in Na_v export from the ER, thus maintaining a reserve pool of channels with exposed retention signals.

Once correctly folded Na_v channels have undergone adequate posttranslational maturation, they travel from ER to the Golgi complex by being packaged in coat protein II (COPII)-coated vesicles. Although the Golgi apparatus and the trans-Golgi network exhibit pivotal roles in the maturation and polarized intracellular sorting of membrane proteins (Zhang and Wang 2016), there are currently few studies that address the VGSC fate in these compartments. To date, the vast majority of defective-trafficking Na_v mutants were found to accumulate in the ER (Baroudi et al. 2001; Clatot et al. 2012; Ruan et al. 2010). However, the functional characterization of the ataxia3 $Na_v1.6$/S21P mutants revealed evidence of their sequestration in the *cis*-Golgi compartment as a result of their defective Golgi export (Sharkey et al. 2009). Lastly, in a recent report, Valkova and colleagues have examined the impact of Rer1 deficiency on cerebellar Purkinje cells in mice (Valkova et al. 2017). Loss of this *cis*-Golgian sorting receptor was shown to disrupt $Na_v1.1$ and $Na_v1.6$ expression levels and their terminal maturation. Thus, Rer1 was proposed to act as a member of the glycoprotein quality control system in the assembly and transport of mature VGSC in a subtype-dependent manner (Valkova et al. 2017).

2.3 VGSC Microtubule-Based Delivery

From their ER exit to their plasma membrane insertion, ion channels are actively transported through the cell along microtubule tracks (see for review Steele and Fedida 2014). The post-Golgi anterograde vesicular transport of Na_v channels is driven by motor proteins including members of the kinesin superfamily proteins (KIFs). While limited data exists on the final delivery of cardiac VGSC into sarcolemma, so far most research studies have focused on neuronal Na_v channels. It has been reported that the ubiquitous KIF5B isoform was associated in a subtype-specific manner with the two neuronal $Na_v1.8$ and $Na_v1.9$ subunits in rat Dorsal Root Ganglia (DRG) (Su et al. 2013). More specifically, the functional interaction between $Na_v1.8$ and KIF5B resulted in both promoting $Na_v1.8$ forward transport and preventing its degradation. Another study provided strong evidence that the scaffolding protein Ankyrin G (AnkG) could act as a cargo adaptor protein to bind KIF5 to $Na_v1.2$ during their axonal transport (Barry et al. 2014). Interestingly, $Na_v1.6$ microtubular trafficking would be mediated by its own adaptor protein, the neuron-specific microtubule-associated protein Map 1b (O'Brien et al. 2012). As for $Na_v1.5$, Casini et al. have highlighted the importance of a functional microtubule network which is required for normal $Na_v1.5$ sarcolemmal expression (Casini et al. 2010). In fact, pre-treatment with the microtubule-stabilizing agent Taxol reduced sodium current density in primary neonatal rat cardiomyocytes. Another protein from the dynactin complex, dynamitin was found to be involved in the vectorial delivery of $Na_v1.5$-containing vesicles to the cardiomyocyte surface (Chatin et al. 2014).

2.4 VGSC Oligomerization

Until recently, it was assumed that VGSC were expressed as functional monomers associated with one or more β-subunits (Catterall 2014). However, the validity of this assumption has been challenged by experiments involving co-expressions of WT with mutant $Na_v1.5$ channels. In fact, a subset of defective-trafficking mutants was shown to exert a dominant negative effect upon Na_v channel surface expression (see for review Sottas and Abriel 2016). It now appears that $Na_v1.5$ α-subunits can interact with each other physically, as shown by immunoprecipitation experiments in HEK cells (Clatot et al. 2012; Mercier et al. 2012). Interestingly, depending on the experimental conditions, this α-α assembly was found to occur through a β_1-subunit-dependent process (Mercier et al. 2012). AFM imaging has recently provided substantial evidence that heterogeneous populations of $Na_v1.5$ oligomers could coexist at the cell surface and that relative proportions of α-subunit monomers, dimers, and trimers were largely conditioned by the presence of the β_3-subunit (Namadurai et al. 2014). As $Na_v\beta_1$ and β_3 share a high degree of amino acid sequence, it is not unreasonable to suggest that they both play a key role in facilitating $Na_v\alpha$ oligomerization, possibly through the stabilization of these macromolecular complexes. However, it remains to be clarified in which compartment(s) this interaction might occur and whether stable $Na_v\alpha$ complexes could still persist after reaching the plasma membrane.

2.5 VGSC Local Translation and Alternative Transports

Neurons, by their very nature, need to precisely control the segregation and the subsequent asymmetric distribution of plasma membrane VGSC. Over the past decade, substantial evidence has accumulated demonstrating that mature axons have the capacity for local membrane protein translation independently from the soma (see for review Cornejo et al. 2017). This biosynthetic process could be considered to overcome spatial limitations reflecting the slow axonal protein transport velocity along axons, and might allow neurons to either maintain or adjust their local proteome in distal domains (see for review Jung et al. 2012).

$Na_v1.8$ mRNA level was found to be locally up-regulated in a subtype-specific manner in sciatic nerve axons in a rat model of neuropathic pain (Ruangsri et al. 2011; Thakor et al. 2009). The authors postulated that the observed hyperexcitability of sensory neurons might be caused by the local translation of axonal accumulated $Na_v1.8$ transcripts, this hypothesis remaining speculative until recently. González and colleagues have demonstrated that mammalian peripheral axons harbor all the necessary secretory machinery required for local membrane protein synthesis (González et al. 2016). Their findings provide strong evidence that newly synthesized VGSC channels can result from both somatic and axonal biosynthetic processes. In fact, $Na_v\alpha$ subunits were shown to transit through the axonal secretory organelles before being efficiently delivered to the plasma membrane locally.

Non-conventional protein transport is not exclusive to the axonal compartment. In a recent extensive study, Hanus and colleagues revealed that a multitude of transmembrane proteins including $Na_v1.2$–1.4 and $Na_v1.7$ could reach the dendritic plasma membrane in their core-glycosylated ER forms. The authors concluded that these neuronal proteins with immature glycosylation profiles might be trafficked to the cell surface through a Golgi bypass route (Hanus et al. 2016). These findings are consistent with our previous observations showing that a pool of $Na_v1.5$ subunits that have not completed their terminal maturation could be efficiently delivered to the plasma membrane via an alternative anterograde route in HEK cells (Mercier et al. 2015). Thus, this "improper" surface expression might concern all VGSC subtypes and its functional impact should be investigated further in both cardiac and neuronal contexts. Also, in human macrophages, functional $Na_v1.5$ α-subunits were shown to be localized in late endosomal compartments, where they mediate endosomal acidification and phagocytosis (Carrithers et al. 2007). However, how this VGSC subtype is specifically targeted to these organelles and not to the plasma membrane remains to be determined.

3 Targeting and Subcellular Distribution of VGSC

The localization of VGSC at the plasma membrane is a controlled phenomenon that constitutes the final step of the trafficking process. According to their tissue function, excitable cells display various morphologies associated with specialized electrical membrane domains. VGSC subcellular targeting is therefore essential to ensure the electrophysiological function of the cell and is strongly dependent on associated proteins often involved in the intracellular trafficking of VGSC.

Neurons are polarized cells with specialized electrical microdomains such as axon initial segments (AIS), axons or somatodendritic compartment. AIS plays a key role in the action potential initiation and is therefore a site of high VGSC concentration. Axons are more specialized in action potential conduction and in myelinated axons, particular structures such as nodes of Ranvier are well known to arbor high concentrations of VGSC for saltatory conduction (for review see Nelson and Jenkins 2017). Electrophysiological function of these subcellular compartments is related to the specific functional expression of VGSC isoforms (review in Boiko et al. 2001). $Na_v1.1$ and $Na_v1.3$ have a somatodendritic localization (reviewed in Vacher et al. 2008) and could possibly control the neuronal excitability threshold through post-synaptic potential integration. On the other hand, $Na_v1.2$ and $Na_v1.6$ seem to specialize in axonal conduction. $Na_v1.2$ is present in unmyelinated axons, including the AIS and at immature nodes of Ranvier (Westenbroek et al. 1989). A recent study revealed that $Na_v1.6$ channels are preferentially inserted into the AIS membrane during hippocampal neurons development via direct vesicular trafficking and this localization is dependent on the binding with AnkG (Akin et al. 2015). In mature nodes of Ranvier of myelinated axons, $Na_v1.2$ is replaced by $Na_v1.6$, which is also present in AIS and dendrites (Boiko et al. 2001; Caldwell et al. 2000; Kaplan et al. 2001). These localizations, together with their specific biophysical properties (Rush et al. 2005), indicate that $Na_v1.2$ may be important for the conduction capabilities of unmyelinated axons, whereas $Na_v1.6$ may be more adapted to saltatory conduction. In sensory neurons, $Na_v1.7$ and $Na_v1.8$ are widely expressed from peripheral terminals in the skin to central branches and terminals in the dorsal horn of the spinal cord (Black et al. 2012).

The macromolecular complexes that underly these VGSC targeting and sequestration involve anchoring proteins such as AnkG and βIV spectrin or cell adhesion molecules and β-subunits (Jenkins and Bennett 2001; Nelson and Jenkins 2017; Zhou et al. 1998). In this macromolecular complex, contactin, a cell adhesion molecule, is an important modulator of sodium channel expression and localization. In optic nerve, $Na_v1.2$ subunits define the nodes and the axon initial segment during early postnatal development and a progressive switch from $Na_v1.2$ to $Na_v1.6$ channel subtypes occurs with maturation (Boiko et al. 2001; Kaplan et al. 2001). Experiments that knocked out the expression of contactin in optic nerve revealed a decrease in $Na_v1.6$ nodal sodium channels (Çolakoğlu et al. 2014). An interaction of contactin with $Na_v1.2$ and $Na_v1.3$ has already been described suggesting a similar mechanism (Kazarinova-Noyes et al. 2001; McEwen et al. 2004; Shah et al. 2004). AnkG is a key modulator of VGSC membrane clustering at the AIS and node of Ranvier (Kordeli et al. 1995; Leterrier et al. 2011). As described previously in this chapter, this anchoring protein could be a cargo adaptor in the trafficking pathway but it is also necessary for the stability of VGSC clusters at the plasma membranes (Barry et al. 2014; Hedstrom et al. 2008). The first interaction between AnkG and VGSC was observed in 2003 with $Na_v1.2$ and a highly conserved nine-amino acid motif in intracellular loop 2 of VGSC was identified for binding AnkG (Garrido et al. 2003; Lemaillet et al. 2003). AnkG interaction has then been characterized for other VGSC such as $Na_v1.6$ and $Na_v1.8$ (Gasser et al. 2012; Montersino et al. 2014). Proteins that interact with AnkG such as Neural cell adhesion molecule L1 (also known as L1CAM) could also contribute to functional expression and localization of Na^+ channels to the neuronal plasma membrane, ensuring correct initiation of action potential and normal firing activity (Valente et al. 2016).

Interestingly, these mechanisms of VGSC targeting seem scalable. This is the case for the interaction between VGSC and AnkG that was shown to be enhanced through the phosphorylation of the Ankyrin-binding motif of VGSC by the protein casein kinase 2 (CK2) (Bréchet et al. 2008; Hien et al. 2014). These data strongly suggest a regulatory mechanism for changes in VGSC subcellular domains such as AIS. Similarly, a recent study has shown that Fibroblast Growth Factor (FGF) 13 binds directly to VGSC in hippocampal neurons to limit their somatodendritic surface expression, although exerting little effect on VGSC within the AIS. In contrast, homologous FGF14, which is highly concentrated in the proximal axon, binds directly to VGSC to promote their axonal localization (Pablo et al. 2016). At the node of Ranvier, the role of AnkG in clustering VGSC was recently shown to be dispensable. Indeed, using a conditional knock-out model, a study revealed that in the absence of AnkG, Ankyrin R-βI spectrin protein complexes function as secondary reserve for Na^+ channel clustering machinery (Ho et al. 2014).

As for neuronal cells, cardiac myocytes display various subcellular electrophysiological needs to ensure cell–cell conduction and excitation/contraction coupling. Targeting cardiac VGSC to specific membrane areas is therefore important for the cell function. There are several lines of evidence demonstrating that the most prominent cardiac Na^+ channel is $Na_v1.5$. However, whereas their functional role is not clearly determined, numerous studies revealed that neuronal sodium channels are also present in cardiac tissues (Kaufmann et al. 2013; for review see Zimmer et al. 2014). In cardiac myocytes, VGSC membrane expression can be divided into two pools: the lateral membrane and the intercalated disc pools (for review see Abriel et al. 2015). Interestingly, the macromolecular complexes that govern these pools depend on the cardiomyocyte location. Briefly, like for neuronal AIS and node of Ranvier, targeting $Na_v1.5$ to the intercalated discs involved the AnkG anchoring protein (Makara et al. 2014) and interacting proteins such as SAP97 (Petitprez et al. 2011), N-cadherin and connexin 43 (Malhotra et al. 2004; Rhett et al. 2012) as well as plakophilin-2 and desmoglein-2 (Rizzo et al. 2012; Sato et al. 2009). At the lateral membrane of cardiomyocytes, $Na_v1.5$ targeting depends on different macromolecular complexes related to syntrophin/dystrophin expression

(Gavillet et al. 2006; Petitprez et al. 2011). This occurs through a PDZ domain binding motif on $Na_v1.5$ (Shy et al. 2014).

4 VGSC Retrograde Transport

The regulation of Na_v channel late trafficking occurs even after VGSC subunits have reached their ultimate site of function, by undergoing constitutive and dynamic turnover. Their steady-state surface expression is governed by a balance between channel anterograde trafficking, endocytic recycling, and degradative pathways. More specifically, it has been reported that sarcolemmal $Na_v1.5$ subunits have apparent long half-lives of about ~35 h whereas other "neuronal" VGSC were estimated to be turned over with half-lives ranging from 17 to 50 h, depending on the cell-type or culture conditions, but reflecting their relative stability within their macromolecular complex (Maltsev et al. 2008; Monjaraz et al. 2000; Schmidt and Catterall 1986; Sherman et al. 1985).

Whether it concerns muscle, cardiomyocyte or neuronal cells, ion channels are subject to endocytic events. These processes allow cells to tune their response to environmental stimuli by modulating their electrical excitability. To date, clathrin-mediated endocytosis (CME) is identified as the major mechanism of VGSC internalization (Dustrude et al. 2013; Garrido 2001; Liu et al. 2005). This critical process by which specific mono-ubiquitinated Na_v channels are recruited and packaged into clathrin-coated vesicles is controlled by the selective binding of different accessory proteins (see for review McMahon and Boucrot 2011).

The initial CME step related to cargo recognition requires the recruitment of adaptor protein complexes (APs) (Popova et al. 2013). Among these, the clathrin-adaptor complex AP-2 might play dual roles in vesicular VGSC transport. The first one involves a di-leucine-based motif identified in the C-terminal region of $Na_v1.2$ that can be recognized by AP-2 as an endocytic machinery component (Garrido 2001). Indeed, it appears that this internalization signal is required for the somatodendritic selective endocytosis of $Na_v1.2$ subtype, indirectly conditioning its axonal compartmentalization. It is worth adding that the $Na_v1.2$ DII-DIII loop contains an additional endocytosis signal that also promotes selective elimination of untrapped $Na_v1.2$ from the somatodendritic membrane (Fache et al. 2004).

Besides, AP-2 seems to modulate correct VGSC targeting and clustering to the AIS in early stages of neurodevelopment by preventing $Na_v1.6$ enrichment at this specialized signaling domain (Kyung et al. 2017).

Another adaptor protein, sodium channel and clathrin linker 1 (SCLT1), was identified as a subtype-specific binding partner mediating indirect $Na_v1.8$ and clathrin association (Liu et al. 2005). The formation of this tri-molecular protein complex is thought to facilitate VGSC internalization, given that SCLT1 overexpression leads to reduced current Na^+ density in murine DRG neurons.

As is the case during their forward trafficking, surface VGSC appear to interact constantly with multiple protein partners which can be involved in their endocytic internalization. αB-crystallin, a member of the small heat shock protein family, is known to be a stress-responsive protein displaying chaperone-like activities (Gangalum et al. 2012; Minami et al. 2003). Although its protective role in preventing aggregation of unstable proteins has been demonstrated in cardiac and lens cells, little is known about its potential role in the folding of VGSC channels (Horwitz 1992; Wang 2003). However, it eventually appeared that αB-crystallin can affect the stability of cell surface $Na_v1.5$ channels by reducing their Nedd4-2-dependent ubiquitylation and internalization for degradation (Huang et al. 2016).

A recent study postulates that the subcellular distribution of two members of the FGF homologous factors, FGF13 and FGF14, could be relevant for the differential modulation of VGSC membrane expression in rat hippocampal neurons (Pablo et al. 2016). While FGF14B is axonally restricted and promotes VGSC surface expression, the broadly distributed FGF13 protein has been shown to regulate VGSC selective internalization from the somatodendritic compartment via a dynamin-dependent endocytotic pathway. It is worthy to mention that FGF13 knockdown produced a cellular context-specific disruption in Na^+ current density in mouse cardiomyocytes, this being in contrast to the increase observed in hippocampal neurons (Wang et al. 2011).

The stress-activated p38 mitogen-activated protein kinase was shown to reduce $Na_v1.6$ peak currents in both native neurons and neuronal cell lines (Gasser et al. 2012; Wittmack 2005). It appeared that this effect could be attributed to the phosphorylation of a single serine residue allowing Nedd4-like (neural precursor cell-expressed developmentally downregulated gene 4) protein to bind and ubiquitylate $Na_v1.6$ channels, which in turn promotes their internalization into neurons. Subsequent diminished neuronal excitability might constitute an adaptive and protective mechanism after cellular stress or injury (Gasser et al. 2012). As for surface ubiquitylated VGSC fate, evidence suggests that the ubiquitin E3 protein ligase Nedd4-2 would induce their internalization rather than their degradation, though the occurrence of these processes seems to be subtype-and tissue-dependent (as reviewed in Laedermann et al. 2013).

Apart from classical interacting protein partners, as those described above, VGSC proteostasis was found to be directly controlled by collapsin response mediator protein 2 (CRMP2) SUMOylation (Dustrude et al. 2013). Loss of this post-translational modification results in reduced Na^+ current density in rat DRG neurons, by promoting $Na_v1.7$ endocytosis through a clathrin-mediated mechanism (Dustrude et al. 2016). The axonal CRMP2 in its SUMOylated state was demonstrated as playing a central role in recruiting endocytosis-related proteins such as Numb and Eps15, and indirectly Nedd4-2, all of these affecting $Na_v1.7$ internalization in a subtype-specific manner.

5 Trafficking Modulation in Physiopathology

The identification and characterization of VGSC mutations have considerably increased our understanding of ion channel function. Various genetic mutations lead to misfolded proteins and subsequent trafficking defects. These mutations could cover either the α-subunit or a protein involved in VGSC macromolecular complex such as β-subunits (Adsit et al. 2013; Amin et al. 2010; Vacher and Trimmer 2012). As a consequence, the reduction of the membrane functional proteins leads to electrophysiological defects and subsequent pathological manifestations. To date, most of these mutants have been associated with $Na_v1.1$ and $Na_v1.5$ sodium channels. $Na_v1.1$ is predominately expressed in the CNS and particularly in γ-aminobutyric acid (GABA)ergic neurons (Duflocq et al. 2008). Characterized mutations that affect the trafficking of $Na_v1.1$ are related to different types of epilepsia such as generalized epilepsy with febrile seizure plus (GEFS+) or severe myoclonic epilepsy of infancy (SMEI) (Rusconi et al. 2007, 2009; Thompson et al. 2012). Numerous *SCN5A* mutations leading to trafficking defective $Na_v1.5$ have been extensively studied in cardiac disease. These mutations are associated with cardiac conduction defects that could lead to sudden death as in Brugada Syndrome or Sudden Infant Death Syndrome (Amin et al. 2010).

Most of the trafficking defective proteins can be functional when they are correctly inserted into the plasma membrane. Developing strategies to restore the trafficking of these proteins is therefore of interest. Impaired trafficking of mutant ion channels is a well-established pathological mechanism of channelopathies such as Cystic Fibrosis or cardiac Long QT syndrome and many strategies used to rescue VGSC were adapted from studies on rescue of the cystic fibrosis transmembrane conductance regulator (CFTR) channel or Human ether-a-go-go-related gene (hERG) potassium channel (Farinha and Canato 2017; Ficker et al. 2002; Lukacs and Verkman 2012; Zhou et al. 1999). One method commonly used consists in the modulation of the sarcoendoplasmic reticulum calcium transport ATPase pump by treating cells with thapsigargin or curcumin (Keller et al. 2005; Moreau et al. 2012; Rusconi et al. 2007, 2009). These pharmacological treatments would act on Ca^{2+}-dependent chaperones that retain the misfolded protein in the ER (Egan et al. 2002, 2004).

Direct interaction of a molecule with VGSC α-subunit can also modulate its trafficking. These small molecules, also called pharmacological chaperones, are generally blockers of cardiac and neuronal VGSC. For example, cell incubation with class I antiarrhythmic drugs such as mexiletine or antiepileptic drugs like phenytoin and carbamazepine can functionally restore a fraction of $Na_v1.5$ trafficking defective mutants to the plasma membrane (Keller et al. 2005; Moreau et al. 2012; Rusconi et al. 2007, 2009; Thompson et al. 2012; Valdivia et al. 2004). Interestingly, whereas these strategies are almost exclusively tested on trafficking defective mutants, they could also impact wild type protein trafficking as it was observed for $Na_v1.8$ after incubation with the local anesthetic lidocaine (Zhao et al. 2007).

Cell culture at low temperature (~25–28°C) for few hours could also trigger the rescue of misfolded VGSC mutants (Keller et al. 2005; Moreau et al. 2012; Rusconi et al. 2007, 2009; Valdivia et al. 2004). While little is known concerning this effect for VGSC, it seems likely to be the consequence of a slowed folding process which acts in preventing protein misfolding and aggregation (Ulloa-Aguirre et al. 2004). Moreover, thermosensitive proteins such as Heat Shock Proteins (HSP) have been involved in protein folding processes and in the rescue of delta508 CFTR protein

(Lopes-Pacheco et al. 2015; Lukacs and Verkman 2012). These proteins could therefore be involved in VGSC correct folding processes.

Interestingly, whereas the positive influence of low temperature and calcium modulators on surface membrane channel expression could be generalized to different misfolded ion channels, the pharmacological chaperones seem to act in a substrate-specific manner. Indeed, specific molecules published to restore the trafficking of the delta F508 CTFR mutant were tested and remained without effect on rescuable folding defects of $Na_v1.1$ and $Na_v1.5$ (Norez et al. 2014; Thompson et al. 2012). Such a specificity of action is interesting since it could limit potential side effects of pharmacological rescue strategies.

The recent advances in VGSC conformational structures (Payandeh et al. 2012; Shen et al. 2017; Sula et al. 2017) will bring fundamental information to understand VGSC folding as well as misfolding consequences of mutations. This should release new pharmacological approaches for rescuing VGSC trafficking defects. An exciting strategy in the future would be to design new molecules based on VGSC structural misfolding knowledge to restore the correct structural folding of mutant VGSC and its membrane localization. Pharmacological assays could not be the only way to recover proper trafficking of VGSC. For example, a polymorphism H558R observed in $Na_v1.5$ has been shown to have a corrective effect on the trafficking defective mutation R282H associated with Brugada Syndrome (Poelzing et al. 2006). Interestingly, the expression of a 20–40 amino acid peptide that contains H558R was shown to be sufficient to restore the trafficking of R282H $Na_v1.5$ mutant (Shinlapawittayatorn et al. 2011). This peptide could be sufficient by itself to limit the misfolding of the mutant through a physical interaction. However, if trafficking rescue is interesting for therapeutic approaches, it is probably not sufficient. Indeed, in addition to their trafficking defects, VGSC mutations can display gating defects once restored to the plasma membrane (Thompson et al. 2012). Rescuing such mutants to the plasma membrane in a therapeutic perspective can therefore uncover deleterious manifestations for patients.

6 Conclusion

This chapter reviews the current understanding of the facets of VGSC trafficking (See Fig. 1). The needs of highly specialized neuronal and cardiac cells are partially fulfilled by the differential expression of key protein partners. These actors appear to determine Na_v subtypes' fate at different stages of trafficking including their anchoring to specialized targeting regions. The most interesting question is now to decipher how these protein partners dynamically and sequentially interact with VGSC. Whereas numerous studies have investigated VGSC trafficking in heterologous expression systems, this dynamic of VGSC trafficking could only be studied from native cells which represents the next challenge for research laboratories. This knowledge should result in new pharmaceutical strategies. Indeed, most of the existing approaches are focused on the modulation of biophysical properties of VGSC localized at the plasma membrane. However, understanding the different steps of VGSC life cycle and quality controls should open new avenues to modulate VGSC expression in time and space such as their plasma localization, their membrane half-life or the rescue of trafficking defective mutants.

References

Abriel H, Rougier J-S, Jalife J (2015) Ion channel macromolecular complexes in cardiomyocytes: roles in sudden cardiac death. Circ Res 116:1971–1988. https://doi.org/10.1161/CIRCRESAHA.116.305017

Adsit GS, Vaidyanathan R, Galler CM, Kyle JW, Makielski JC (2013) Channelopathies from mutations in the cardiac sodium channel protein complex. J Mol Cell Cardiol 61:34–43. https://doi.org/10.1016/j.yjmcc.2013.03.017

Akin EJ, Solé L, Dib-Hajj SD, Waxman SG, Tamkun MM (2015) Preferential targeting of $Na_v1.6$ voltage-gated Na^+ channels to the axon initial segment during development. PLoS One 10: e0124397. https://doi.org/10.1371/journal.pone.0124397

Amin AS, Asghari-Roodsari A, Tan HL (2010) Cardiac sodium channelopathies. Pflugers Arch 460:223–237. https://doi.org/10.1007/s00424-009-0761-0

Arakel EC, Brandenburg S, Uchida K, Zhang H, Lin Y-W, Kohl T, Schrul B, Sulkin MS, Efimov IR, Nichols CG, Lehnart SE, Schwappach B (2014) Tuning the electrical properties of the heart by differential trafficking of KATP ion channel complexes. J Cell Sci 127:2106–2119. https://doi.org/10.1242/jcs.141440

Baroudi G, Pouliot V, Denjoy I, Guicheney P, Shrier A, Chahine M (2001) Novel mechanism for Brugada syndrome: defective surface localization of an SCN5A mutant (R1432G). Circ Res 88:e78–e83

Barry J, Gu Y, Jukkola P, O'Neill B, Gu H, Mohler PJ, Rajamani KT, Gu C (2014) Ankyrin-G directly binds to Kinesin-1 to transport voltage-gated Na^+ channels into axons. Dev Cell 28: 117–131. https://doi.org/10.1016/j.devcel.2013.11.023

Black JA, Frézel N, Dib-Hajj SD, Waxman SG (2012) Expression of $Na_v1.7$ in DRG neurons extends from peripheral terminals in the skin to central preterminal branches and terminals in the dorsal horn. Mol Pain 8:82. https://doi.org/10.1186/1744-8069-8-82

Boiko T, Rasband MN, Levinson SR, Caldwell JH, Mandel G, Trimmer JS, Matthews G (2001) Compact myelin dictates the differential targeting of two sodium channel isoforms in the same axon. Neuron 30:91–104

Bréchet A, Fache M-P, Brachet A, Ferracci G, Baude A, Irondelle M, Pereira S, Leterrier C, Dargent B (2008) Protein kinase CK2 contributes to the organization of sodium channels in axonal membranes by regulating their interactions with ankyrin G. J Cell Biol 183:1101–1114. https://doi.org/10.1083/jcb.200805169

Caldwell JH, Schaller KL, Lasher RS, Peles E, Levinson SR (2000) Sodium channel $Na_v1.6$ is localized at nodes of Ranvier, dendrites, and synapses. Proc Natl Acad Sci 97:5616–5620. https://doi.org/10.1073/pnas.090034797

Carrithers MD, Dib-Hajj S, Carrithers LM, Tokmoulina G, Pypaert M, Jonas EA, Waxman SG (2007) Expression of the voltage-gated sodium channel $Na_V1.5$ in the macrophage late endosome regulates endosomal acidification. J Immunol 178:7822–7832

Casini S, Tan HL, Demirayak I, Remme CA, Amin AS, Scicluna BP, Chatyan H, Ruijter JM, Bezzina CR, van Ginneken ACG, Veldkamp MW (2010) Tubulin polymerization modifies cardiac sodium channel expression and gating. Cardiovasc Res 85:691–700. https://doi.org/10.1093/cvr/cvp352

Catterall WA (2014) Structure and function of voltage-gated sodium channels at atomic resolution. Exp Physiol 99:35–51. https://doi.org/10.1113/expphysiol.2013.071969

Chahine M, Chatelier A, Babich O, Krupp JJ (2008) Voltage-gated sodium channels in neurological disorders. CNS Neurol Disord Drug Targets 7:144–158

Chatin B, Colombier P, Gamblin AL, Allouis M, Le Bouffant F (2014) Dynamitin affects cell-surface expression of voltage-gated sodium channel $Na_v1.5$. Biochem J 463:339–349. https://doi.org/10.1042/BJ20140604

Clatot J, Ziyadeh-Isleem A, Maugenre S, Denjoy I, Liu H, Dilanian G, Hatem SN, Deschênes I, Coulombe A, Guicheney P, Neyroud N (2012) Dominant-negative effect of SCN5A N-terminal mutations through the interaction of $Na_v1.5$ α-subunits. Cardiovasc Res 96:53–63. https://doi.org/10.1093/cvr/cvs211

Çolakoğlu G, Bergstrom-Tyrberg U, Berglund EO, Ranscht B (2014) Contactin-1 regulates myelination and nodal/paranodal domain organization in the central nervous system. Proc Natl Acad Sci U S A 111:E394–E403. https://doi.org/10.1073/pnas.1313769110

Cornejo VH, Luarte A, Couve A (2017) Global and local mechanisms sustain axonal proteostasis of transmembrane proteins. Traffic 18:255–266. https://doi.org/10.1111/tra.12472

Duflocq A, Le Bras B, Bullier E, Couraud F, Davenne M (2008) $Na_v1.1$ is predominantly expressed in nodes of Ranvier and axon initial segments. Mol Cell Neurosci 39:180–192. https://doi.org/10.1016/j.mcn.2008.06.008

Dulsat G, Palomeras S, Cortada E, Riuró H, Brugada R, Vergés M (2017) Trafficking and localisation to the plasma membrane of $Na_v1.5$ promoted by the β2 subunit is defective due to a β2 mutation associated with Brugada syndrome: β2 in $Na_v1.5$ trafficking. Biol Cell 109: 273–291. https://doi.org/10.1111/boc.201600085

Dustrude ET, Wilson SM, Ju W, Xiao Y, Khanna R (2013) CRMP2 protein SUMOylation modulates $Na_V1.7$ channel trafficking. J Biol Chem 288:24316–24331. https://doi.org/10.1074/jbc.M113.474924

Dustrude ET, Moutal A, Yang X, Wang Y, Khanna M, Khanna R (2016) Hierarchical CRMP2 posttranslational modifications control $Na_V1.7$ function. Proc Natl Acad Sci 113: E8443–E8452. https://doi.org/10.1073/pnas.1610531113

Egan ME, Glöckner-Pagel J, Ambrose C, Cahill PA, Pappoe L, Balamuth N, Cho E, Canny S, Wagner CA, Geibel J, Caplan MJ (2002) Calcium-pump inhibitors induce functional surface expression of delta F508-CFTR protein in cystic fibrosis epithelial cells. Nat Med 8:485–492. https://doi.org/10.1038/nm0502-485

Egan ME, Pearson M, Weiner SA, Rajendran V, Rubin D, Glöckner-Pagel J, Canny S, Du K, Lukacs GL, Caplan MJ (2004) Curcumin, a major constituent of turmeric, corrects cystic fibrosis defects. Science 304:600–602. https://doi.org/10.1126/science.1093941

Fache M-P, Moussif A, Fernandes F, Giraud P, Garrido JJ, Dargent B (2004) Endocytotic elimination and domain-selective tethering constitute a potential mechanism of protein segregation at the axonal initial segment. J Cell Biol 166:571–578. https://doi.org/10.1083/jcb.200312155

Farinha CM, Canato S (2017) From the endoplasmic reticulum to the plasma membrane: mechanisms of CFTR folding and trafficking. Cell Mol Life Sci 74:39–55. https://doi.org/10.1007/s00018-016-2387-7

Ficker E, Obejero-Paz CA, Zhao S, Brown AM (2002) The binding site for channel blockers that rescue misprocessed human long QT syndrome type 2 ether-a-gogo-related gene (HERG) mutations. J Biol Chem 277:4989–4998. https://doi.org/10.1074/jbc.M107345200

Gangalum RK, Horwitz J, Kohan SA, Bhat SP (2012) αA-crystallin and αB-crystallin reside in separate subcellular compartments in the developing ocular lens. J Biol Chem 287:42407–42416. https://doi.org/10.1074/jbc.M112.414854

Garrido JJ (2001) Identification of an axonal determinant in the C-terminus of the sodium channel $Na_v1.2$. EMBO J 20:5950–5961. https://doi.org/10.1093/emboj/20.21.5950

Garrido JJ, Giraud P, Carlier E, Fernandes F, Moussif A, Fache M-P, Debanne D, Dargent B (2003) A targeting motif involved in sodium channel clustering at the axonal initial segment. Science 300:2091–2094. https://doi.org/10.1126/science.1085167

Gasser A, Ho TS-Y, Cheng X, Chang K-J, Waxman SG, Rasband MN, Dib-Hajj SD (2012) An ankyrinG-binding motif is necessary and sufficient for targeting $Na_v1.6$ sodium channels to axon initial segments and nodes of Ranvier. J Neurosci 32:7232–7243. https://doi.org/10.1523/JNEUROSCI.5434-11.2012

Gavillet B, Rougier J-S, Domenighetti AA, Behar R, Boixel C, Ruchat P, Lehr H-A, Pedrazzini T, Abriel H (2006) Cardiac sodium channel $Na_v1.5$ is regulated by a multiprotein complex composed of syntrophins and dystrophin. Circ Res 99:407–414. https://doi.org/10.1161/01.RES.0000237466.13252.5e

González C, Cánovas J, Fresno J, Couve E, Court FA, Couve A (2016) Axons provide the secretory machinery for trafficking of voltage-gated sodium channels in peripheral nerve. Proc Natl Acad Sci 113:1823–1828. https://doi.org/10.1073/pnas.1514943113

Hallaq H, Yang Z, Viswanathan P, Fukuda K, Shen W, Wang D, Wells K, Zhou J, Yi J, Murray K (2006) Quantitation of protein kinase A-mediated trafficking of cardiac sodium channels in living cells. Cardiovasc Res 72:250–261. https://doi.org/10.1016/j.cardiores.2006.08.007

Hanus C, Geptin H, Tushev G, Garg S, Alvarez-Castelao B, Sambandan S, Kochen L, Hafner A-S, Langer JD, Schuman EM (2016) Unconventional secretory processing diversifies neuronal ion channel properties. Elife 5. https://doi.org/10.7554/eLife.20609

Hedstrom KL, Ogawa Y, Rasband MN (2008) AnkyrinG is required for maintenance of the axon initial segment and neuronal polarity. J Cell Biol 183:635–640. https://doi.org/10.1083/jcb.200806112

Hien YE, Montersino A, Castets F, Leterrier C, Filhol O, Vacher H, Dargent B (2014) CK2 accumulation at the axon initial segment depends on sodium channel Na_v1. FEBS Lett 588: 3403–3408. https://doi.org/10.1016/j.febslet.2014.07.032

Ho TS-Y, Zollinger DR, Chang K-J, Xu M, Cooper EC, Stankewich MC, Bennett V, Rasband MN (2014) A hierarchy of ankyrin-spectrin complexes clusters sodium channels at nodes of Ranvier. Nat Neurosci 17:1664–1672. https://doi.org/10.1038/nn.3859

Horwitz J (1992) Alpha-crystallin can function as a molecular chaperone. Proc Natl Acad Sci U S A 89:10449–10453

Huang Y, Wang Z, Liu Y, Xiong H, Zhao Y, Wu L, Yuan C, Wang L, Hou Y, Yu G, Huang Z, Xu C, Chen Q, Wang QK (2016) αB-crystallin interacts with $Na_v1.5$ and regulates ubiquitination and internalization of cell surface $Na_v1.5$. J Biol Chem 291:11030–11041. https://doi.org/10.1074/jbc.M115.695080

Jenkins SM, Bennett V (2001) Ankyrin-G coordinates assembly of the spectrin-based membrane skeleton, voltage-gated sodium channels, and L1 CAMs at Purkinje neuron initial segments. J Cell Biol 155:739–746. https://doi.org/10.1083/jcb.200109026

Johnston WL, Dyer JR, Castellucci VF, Dunn RJ (1996) Clustered voltage-gated Na^+ channels in Aplysia axons. J Neurosci 16:1730–1739

Jung H, Yoon BC, Holt CE (2012) Axonal mRNA localization and local protein synthesis in nervous system assembly, maintenance and repair. Nat Rev Neurosci 13(5):308–324. https://doi.org/10.1038/nrn3210

Kaplan MR, Cho MH, Ullian EM, Isom LL, Levinson SR, Barres BA (2001) Differential control of clustering of the sodium channels $Na_{(v)}1.2$ and $Na_{(v)}1.6$ at developing CNS nodes of Ranvier. Neuron 30:105–119

Kaufmann SG, Westenbroek RE, Maass AH, Lange V, Renner A, Wischmeyer E, Bonz A, Muck J, Ertl G, Catterall WA, Scheuer T, Maier SKG (2013) Distribution and function of sodium channel subtypes in human atrial myocardium. J Mol Cell Cardiol 61:133–141. https://doi.org/10.1016/j.yjmcc.2013.05.006

Kazarinova-Noyes K, Malhotra JD, McEwen DP, Mattei LN, Berglund EO, Ranscht B, Levinson SR, Schachner M, Shrager P, Isom LL, Xiao ZC (2001) Contactin associates with Na^+ channels and increases their functional expression. J Neurosci 21:7517–7525

Keller DI, Rougier J-S, Kucera JP, Benammar N, Fressart V, Guicheney P, Madle A, Fromer M, Schläpfer J, Abriel H (2005) Brugada syndrome and fever: genetic and molecular characterization of patients carrying SCN5A mutations. Cardiovasc Res 67:510–519. https://doi.org/10.1016/j.cardiores.2005.03.024

Kordeli E, Lambert S, Bennett V (1995) AnkyrinG. A new ankyrin gene with neural-specific isoforms localized at the axonal initial segment and node of Ranvier. J Biol Chem 270:2352–2359
Kyung JW, Cho IH, Lee S, Song WK, Ryan TA, Hoppa MB, Kim SH (2017) Adaptor protein 2 (AP-2) complex is essential for functional axogenesis in hippocampal neurons. Sci Rep 7: 41620. https://doi.org/10.1038/srep41620
Laedermann CJ, Syam N, Pertin M, Decosterd I, Abriel H (2013) β1- and β3- voltage-gated sodium channel subunits modulate cell surface expression and glycosylation of $Na_v1.7$ in HEK293 cells. Front Cell Neurosci 7:137. https://doi.org/10.3389/fncel.2013.00137
Lee A, Goldin AL (2009) Role of the terminal domains in sodium channel localization. Channels 3:171–180. https://doi.org/10.4161/chan.3.3.8854
Lemaillet G, Walker B, Lambert S (2003) Identification of a conserved ankyrin-binding motif in the family of sodium channel alpha subunits. J Biol Chem 278:27333–27339. https://doi.org/10.1074/jbc.M303327200
Leterrier C, Brachet A, Dargent B, Vacher H (2011) Determinants of voltage-gated sodium channel clustering in neurons. Semin Cell Dev Biol 22:171–177. https://doi.org/10.1016/j.semcdb.2010.09.014
Li Q, Su Y-Y, Wang H, Li L, Wang Q, Bao L (2010) Transmembrane segments prevent surface expression of sodium channel $Na_v1.8$ and promote calnexin-dependent channel degradation. J Biol Chem 285:32977–32987. https://doi.org/10.1074/jbc.M110.143024
Liu C, Cummins TR, Tyrrell L, Black JA, Waxman SG, Dib-Hajj SD (2005) CAP-1A is a novel linker that binds clathrin and the voltage-gated sodium channel $Na_v1.8$. Mol Cell Neurosci 28:636–649. https://doi.org/10.1016/j.mcn.2004.11.007
Lopes-Pacheco M, Boinot C, Sabirzhanova I, Morales MM, Guggino WB, Cebotaru L (2015) Combination of correctors rescue ΔF508-CFTR by reducing its association with Hsp40 and Hsp27. J Biol Chem 290:25636–25645. https://doi.org/10.1074/jbc.M115.671925
Lu T, Lee H-C, Kabat JA, Shibata EF (1999) Modulation of rat cardiac sodium channel by the stimulatory G protein α subunit. J Physiol 518:371–384. https://doi.org/10.1111/j.1469-7793.1999.0371p.x
Lukacs GL, Verkman AS (2012) CFTR: folding, misfolding and correcting the ΔF508 conformational defect. Trends Mol Med 18:81–91. https://doi.org/10.1016/j.molmed.2011.10.003
Makara MA, Curran J, Little SC, Musa H, Polina I, Smith SA, Wright PJ, Unudurthi SD, Snyder J, Bennett V, Hund TJ, Mohler PJ (2014) Ankyrin-G coordinates intercalated disc signaling platform to regulate cardiac excitability in vivo. Circ Res 115:929–938. https://doi.org/10.1161/CIRCRESAHA.115.305154
Malhotra JD, Thyagarajan V, Chen C, Isom LL (2004) Tyrosine-phosphorylated and nonphosphorylated sodium channel beta1 subunits are differentially localized in cardiac myocytes. J Biol Chem 279:40748–40754. https://doi.org/10.1074/jbc.M407243200
Maltsev VA, Kyle JW, Mishra S, Undrovinas A (2008) Molecular identity of the late sodium current in adult dog cardiomyocytes identified by $Na_v1.5$ antisense inhibition. Am J Physiol Heart Circ Physiol 295:H667–H676. https://doi.org/10.1152/ajpheart.00111.2008
McArdle EJ, Kunic JD, George AL (2008) Novel SCN1A frameshift mutation with absence of truncated $Na_v1.1$ protein in severe myoclonic epilepsy of infancy. Am J Med Genet A 146A: 2421–2423. https://doi.org/10.1002/ajmg.a.32448
McEwen DP, Meadows LS, Chen C, Thyagarajan V, Isom LL (2004) Sodium channel beta1 subunit-mediated modulation of $Na_v1.2$ currents and cell surface density is dependent on interactions with contactin and ankyrin. J Biol Chem 279:16044–16049. https://doi.org/10.1074/jbc.M400856200
McMahon HT, Boucrot E (2011) Molecular mechanism and physiological functions of clathrin-mediated endocytosis. Nat Rev Mol Cell Biol 12:517–533. https://doi.org/10.1038/nrm3151
Mercier A, Clément R, Harnois T, Bourmeyster N, Faivre J-F, Findlay I, Chahine M, Bois P, Chatelier A (2012) The β1-subunit of $Na_v1.5$ cardiac sodium channel is required for a dominant negative effect through α-α interaction. PLoS One 7:e48690. https://doi.org/10.1371/journal.pone.0048690

Mercier A, Clément R, Harnois T, Bourmeyster N, Bois P, Chatelier A (2015) $Na_v1.5$ channels can reach the plasma membrane through distinct N-glycosylation states. Biochim Biophys Acta 1850:1215–1223. https://doi.org/10.1016/j.bbagen.2015.02.009

Minami M, Mizutani T, Kawanishi R, Suzuki Y, Mori H (2003) Neuronal expression of alphaB crystallin in cerebral infarction. Acta Neuropathol (Berl) 105:549–554. https://doi.org/10.1007/s00401-003-0679-0

Monjaraz E, Navarrete A, López-Santiago LF, Vega AV, Cota G (2000) L-type calcium channel activity regulates sodium channel levels in rat pituitary GH3 cells. J Physiol 523:45–55. https://doi.org/10.1111/j.1469-7793.2000.00045.x

Montersino A, Brachet A, Ferracci G, Fache M-P, Angles d'Ortoli S, Liu W, Rueda-Boroni F, Castets F, Dargent B (2014) Tetrodotoxin-resistant voltage-gated sodium channel $Na_v1.8$ constitutively interacts with ankyrin G. J Neurochem 131:33–41. https://doi.org/10.1111/jnc.12785

Moreau A, Keller DI, Huang H, Fressart V, Schmied C, Timour Q, Chahine M (2012) Mexiletine differentially restores the trafficking defects caused by two brugada syndrome mutations. Front Pharmacol 3:62. https://doi.org/10.3389/fphar.2012.00062

Namadurai S, Balasuriya D, Rajappa R, Wiemhöfer M, Stott K, Klingauf J, Edwardson JM, Chirgadze DY, Jackson AP (2014) Crystal structure and molecular imaging of the Na_v channel β3 subunit indicates a trimeric assembly. J Biol Chem 289:10797–10811. https://doi.org/10.1074/jbc.M113.527994

Nelson AD, Jenkins PM (2017) Axonal membranes and their domains: assembly and function of the axon initial segment and node of Ranvier. Front Cell Neurosci 11:136. https://doi.org/10.3389/fncel.2017.00136

Norez C, Vandebrouck C, Bertrand J, Noel S, Durieu E, Oumata N, Galons H, Antigny F, Chatelier A, Bois P, Meijer L, Becq F (2014) Roscovitine is a proteostasis regulator that corrects the trafficking defect of F508del-CFTR by a CDK-independent mechanism. Br J Pharmacol 171:4831–4849. https://doi.org/10.1111/bph.12859

O'Brien JE, Sharkey LM, Vallianatos CN, Han C, Blossom JC, Yu T, Waxman SG, Dib-Hajj SD, Meisler MH (2012) Interaction of voltage-gated sodium channel $Na_v1.6$ (*SCN8A*) with microtubule-associated protein Map 1b. J Biol Chem 287:18459–18466. https://doi.org/10.1074/jbc.M111.336024

Ogino K, Low SE, Yamada K, Saint-Amant L, Zhou W, Muto A, Asakawa K, Nakai J, Kawakami K, Kuwada JY, Hirata H (2015) RING finger protein 121 facilitates the degradation and membrane localization of voltage-gated sodium channels. Proc Natl Acad Sci U S A 112: 2859–2864. https://doi.org/10.1073/pnas.1414002112

Pablo JL, Wang C, Presby MM, Pitt GS (2016) Polarized localization of voltage-gated Na^+ channels is regulated by concerted FGF13 and FGF14 action. Proc Natl Acad Sci U S A 113:E2665–E2674. https://doi.org/10.1073/pnas.1521194113

Patino GA, Isom LL (2010) Electrophysiology and beyond: multiple roles of Na^+ channel β subunits in development and disease. Neurosci Lett 486:53–59. https://doi.org/10.1016/j.neulet.2010.06.050

Payandeh J, Gamal El-Din TM, Scheuer T, Zheng N, Catterall WA (2012) Crystal structure of a voltage-gated sodium channel in two potentially inactivated states. Nature 486:135–139. https://doi.org/10.1038/nature11077

Petitprez S, Zmoos A-F, Ogrodnik J, Balse E, Raad N, El-Haou S, Albesa M, Bittihn P, Luther S, Lehnart SE, Hatem SN, Coulombe A, Abriel H (2011) SAP97 and dystrophin macromolecular complexes determine two pools of cardiac sodium channels $Na_v1.5$ in cardiomyocytes. Circ Res 108:294–304. https://doi.org/10.1161/CIRCRESAHA.110.228312

Poelzing S, Forleo C, Samodell M, Dudash L, Sorrentino S, Anaclerio M, Troccoli R, Iacoviello M, Romito R, Guida P, Chahine M, Pitzalis M, Deschênes I (2006) SCN5A polymorphism restores trafficking of a Brugada syndrome mutation on a separate gene. Circulation 114:368–376. https://doi.org/10.1161/CIRCULATIONAHA.105.601294

Popova NV, Deyev IE, Petrenko AG (2013) Clathrin-mediated endocytosis and adaptor proteins. Acta Nat 5:62–73

Rhett JM, Ongstad EL, Jourdan J, Gourdie RG (2012) Cx43 associates with $Na_{(v)}1.5$ in the cardiomyocyte perinexus. J Membr Biol 245:411–422. https://doi.org/10.1007/s00232-012-9465-z

Rizzo S, Lodder EM, Verkerk AO, Wolswinkel R, Beekman L, Pilichou K, Basso C, Remme CA, Thiene G, Bezzina CR (2012) Intercalated disc abnormalities, reduced $Na^{(+)}$ current density, and conduction slowing in desmoglein-2 mutant mice prior to cardiomyopathic changes. Cardiovasc Res 95:409–418. https://doi.org/10.1093/cvr/cvs219

Ruan Y, Denegri M, Liu N, Bachetti T, Seregni M, Morotti S, Severi S, Napolitano C, Priori SG (2010) Trafficking defects and gating abnormalities of a novel SCN5A mutation question gene-specific therapy in long QT syndrome type 3. Circ Res 106:1374–1383. https://doi.org/10.1161/CIRCRESAHA.110.218891

Ruangsri S, Lin A, Mulpuri Y, Lee K, Spigelman I, Nishimura I (2011) Relationship of axonal voltage-gated sodium channel 1.8 ($Na_V1.8$) mRNA accumulation to sciatic nerve injury-induced painful neuropathy in rats. J Biol Chem 286:39836–39847. https://doi.org/10.1074/jbc.M111.261701

Rusconi R, Scalmani P, Cassulini RR, Giunti G, Gambardella A, Franceschetti S, Annesi G, Wanke E, Mantegazza M (2007) Modulatory proteins can rescue a trafficking defective epileptogenic $Na_v1.1$ Na^+ channel mutant. J Neurosci 27:11037–11046. https://doi.org/10.1523/JNEUROSCI.3515-07.2007

Rusconi R, Combi R, Cestèle S, Grioni D, Franceschetti S, Dalprà L, Mantegazza M (2009) A rescuable folding defective $Na_v1.1$ (SCN1A) sodium channel mutant causes GEFS+: common mechanism in $Na_v1.1$ related epilepsies? Hum Mutat 30:E747–E760. https://doi.org/10.1002/humu.21041

Rush AM, Dib-Hajj SD, Waxman SG (2005) Electrophysiological properties of two axonal sodium channels, $Na_v1.2$ and $Na_v1.6$, expressed in mouse spinal sensory neurones. J Physiol 564:803–815. https://doi.org/10.1113/jphysiol.2005.083089

Sato PY, Musa H, Coombs W, Guerrero-Serna G, Patiño GA, Taffet SM, Isom LL, Delmar M (2009) Loss of plakophilin-2 expression leads to decreased sodium current and slower conduction velocity in cultured cardiac myocytes. Circ Res 105:523–526. https://doi.org/10.1161/CIRCRESAHA.109.201418

Savio-Galimberti E, Gollob MH, Darbar D (2012) Voltage-gated sodium channels: biophysics, pharmacology, and related channelopathies. Front Pharmacol 3:124. https://doi.org/10.3389/fphar.2012.00124

Schirmeyer J, Szafranski K, Leipold E, Mawrin C, Platzer M, Heinemann SH (2014) Exon 11 skipping of *SCN10A* coding for voltage-gated sodium channels in dorsal root ganglia. Channels 8:210–215. https://doi.org/10.4161/chan.28146

Schmidt JW, Catterall WA (1986) Biosynthesis and processing of the alpha subunit of the voltage-sensitive sodium channel in rat brain neurons. Cell 46:437–445

Schmidt J, Rossie S, Catterall WA (1985) A large intracellular pool of inactive Na channel alpha subunits in developing rat brain. Proc Natl Acad Sci U S A 82:4847–4851

Shah BS, Rush AM, Liu S, Tyrrell L, Black JA, Dib-Hajj SD, Waxman SG (2004) Contactin associates with sodium channel $Na_v1.3$ in native tissues and increases channel density at the cell surface. J Neurosci 24:7387–7399. https://doi.org/10.1523/JNEUROSCI.0322-04.2004

Sharkey LM, Cheng X, Drews V, Buchner DA, Jones JM, Justice MJ, Waxman SG, Dib-Hajj SD, Meisler MH (2009) The ataxia3 mutation in the N-terminal cytoplasmic domain of sodium channel $Na_v1.6$ disrupts intracellular trafficking. J Neurosci 29:2733–2741. https://doi.org/10.1523/JNEUROSCI.6026-08.2009

Shen H, Zhou Q, Pan X, Li Z, Wu J, Yan N (2017) Structure of a eukaryotic voltage-gated sodium channel at near-atomic resolution. Science 355:eaal4326. https://doi.org/10.1126/science.aal4326

Sherman SJ, Chrivia J, Catterall WA (1985) Cyclic adenosine 3′:5′-monophosphate and cytosolic calcium exert opposing effects on biosynthesis of tetrodotoxin-sensitive sodium channels in rat muscle cells. J Neurosci 5:1570–1576

Shinlapawittayatorn K, Dudash LA, Du XX, Heller L, Poelzing S, Ficker E, Deschênes I (2011) A novel strategy using cardiac sodium channel polymorphic fragments to rescue trafficking-deficient SCN5A mutations. Circ Cardiovasc Genet 4:500–509. https://doi.org/10.1161/CIRCGENETICS.111.960633

Shy D, Gillet L, Ogrodnik J, Albesa M, Verkerk AO, Wolswinkel R, Rougier J-S, Barc J, Essers MC, Syam N, Marsman RF, van Mil AM, Rotman S, Redon R, Bezzina CR, Remme CA, Abriel H (2014) PDZ domain-binding motif regulates cardiomyocyte compartment-specific $Na_V1.5$ channel expression and function. Circulation 130:147–160. https://doi.org/10.1161/CIRCULATIONAHA.113.007852

Sottas V, Abriel H (2016) Negative-dominance phenomenon with genetic variants of the cardiac sodium channel $Na_v1.5$. Biochim Biophys Acta 1863:1791–1798. https://doi.org/10.1016/j.bbamcr.2016.02.013

Steele DF, Fedida D (2014) Cytoskeletal roles in cardiac ion channel expression. Biochim Biophys Acta 1838:665–673. https://doi.org/10.1016/j.bbamem.2013.05.001

Su Y-Y, Ye M, Li L, Liu C, Pan J, Liu W-W, Jiang Y, Jiang X-Y, Zhang X, Shu Y, Bao L (2013) KIF5B promotes the forward transport and axonal function of the voltage-gated sodium channel $Na_v1.8$. J Neurosci 33:17884–17896. https://doi.org/10.1523/JNEUROSCI.0539-13.2013

Sula A, Booker J, Ng LCT, Naylor CE, DeCaen PG, Wallace BA (2017) The complete structure of an activated open sodium channel. Nat Commun 8:14205. https://doi.org/10.1038/ncomms14205

Thakor DK, Lin A, Matsuka Y, Meyer EM, Ruangsri S, Nishimura I, Spigelman I (2009) Increased peripheral nerve excitability and local $Na_v1.8$ RNA up-regulation in painful neuropathy. Mol Pain 5:14. https://doi.org/10.1186/1744-8069-5-14

Thompson CH, Porter JC, Kahlig KM, Daniels MA, George AL (2012) Nontruncating SCN1A mutations associated with severe myoclonic epilepsy of infancy impair cell surface expression. J Biol Chem 287:42001–42008. https://doi.org/10.1074/jbc.M112.421883

Ulloa-Aguirre A, Janovick JA, Brothers SP, Conn PM (2004) Pharmacologic rescue of conformationally-defective proteins: implications for the treatment of human disease. Traffic 5:821–837. https://doi.org/10.1111/j.1600-0854.2004.00232.x

Vacher H, Trimmer JS (2012) Trafficking mechanisms underlying neuronal voltage-gated ion channel localization at the axon initial segment. Epilepsia 53(Suppl 9):21–31. https://doi.org/10.1111/epi.12032

Vacher H, Mohapatra DP, Trimmer JS (2008) Localization and targeting of voltage-dependent ion channels in mammalian central neurons. Physiol Rev 88:1407–1447. https://doi.org/10.1152/physrev.00002.2008

Valdivia CR, Tester DJ, Rok BA, Porter C-BJ, Munger TM, Jahangir A, Makielski JC, Ackerman MJ (2004) A trafficking defective, Brugada syndrome-causing SCN5A mutation rescued by drugs. Cardiovasc Res 62:53–62. https://doi.org/10.1016/j.cardiores.2004.01.022

Valente P, Lignani G, Medrihan L, Bosco F, Contestabile A, Lippiello P, Ferrea E, Schachner M, Benfenati F, Giovedì S, Baldelli P (2016) Cell adhesion molecule L1 contributes to neuronal excitability regulating the function of voltage-gated Na^+ channels. J Cell Sci 129: 1878–1891. https://doi.org/10.1242/jcs.182089

Valkova C, Liebmann L, Krämer A, Hübner CA, Kaether C (2017) The sorting receptor Rer1 controls Purkinje cell function via voltage gated sodium channels. Sci Rep 7:41248. https://doi.org/10.1038/srep41248

Wang X (2003) B-crystallin modulates protein aggregation of abnormal desmin. Circ Res 93: 998–1005. https://doi.org/10.1161/01.RES.0000102401.77712.ED

Wang C, Hennessey JA, Kirkton RD, Wang C, Graham V, Puranam RS, Rosenberg PB, Bursac N, Pitt GS (2011) Fibroblast growth factor homologous factor 13 regulates Na^+ channels and conduction velocity in murine hearts. Circ Res 109:775–782. https://doi.org/10.1161/CIRCRESAHA.111.247957

Westenbroek RE, Merrick DK, Catterall WA (1989) Differential subcellular localization of the RI and RII Na^+ channel subtypes in central neurons. Neuron 3:695–704

Wittmack EK (2005) Voltage-gated sodium channel $Na_v1.6$ is modulated by p38 mitogen-activated protein kinase. J Neurosci 25:6621–6630. https://doi.org/10.1523/JNEUROSCI.0541-05.2005

Xiao X, Chen C, Yu T-M, Ou J, Rui M, Zhai Y, He Y, Xue L, Ho MS (2017) Molecular chaperone calnexin regulates the function of drosophila sodium channel paralytic. Front Mol Neurosci 10. https://doi.org/10.3389/fnmol.2017.00057

Zhang X, Wang Y (2016) Glycosylation quality control by the Golgi structure. J Mol Biol 428: 3183–3193. https://doi.org/10.1016/j.jmb.2016.02.030

Zhang Z-N, Li Q, Liu C, Wang H-B, Wang Q, Bao L (2008) The voltage-gated Na^+ channel $Na_v1.8$ contains an ER-retention/retrieval signal antagonized by the 3 subunit. J Cell Sci 121: 3243–3252. https://doi.org/10.1242/jcs.026856

Zhao J, Ziane R, Chatelier A, O'leary ME, Chahine M (2007) Lidocaine promotes the trafficking and functional expression of $Na_{(v)}1.8$ sodium channels in mammalian cells. J Neurophysiol 98: 467–477. https://doi.org/10.1152/jn.00117.2007

Zhou D, Lambert S, Malen PL, Carpenter S, Boland LM, Bennett V (1998) AnkyrinG is required for clustering of voltage-gated Na channels at axon initial segments and for normal action potential firing. J Cell Biol 143:1295–1304

Zhou Z, Gong Q, January CT (1999) Correction of defective protein trafficking of a mutant HERG potassium channel in human long QT syndrome. Pharmacological and temperature effects. J Biol Chem 274:31123–31126

Zhou J, Shin H-G, Yi J, Shen W, Williams CP, Murray KT (2002) Phosphorylation and putative ER retention signals are required for protein kinase A-mediated potentiation of cardiac sodium current. Circ Res 91:540–546

Zimmer T, Biskup C, Dugarmaa S, Vogel F, Steinbis M, Böhle T, Wu YS, Dumaine R, Benndorf K (2002) Functional expression of GFP-linked human heart sodium channel (hH1) and subcellular localization of the a subunit in HEK293 cells and dog cardiac myocytes. J Membr Biol 186:1–12. https://doi.org/10.1007/s00232-001-0130-1

Zimmer T, Haufe V, Blechschmidt S (2014) Voltage-gated sodium channels in the mammalian heart. Glob Cardiol Sci Pract 2014:58–463. https://doi.org/10.5339/gcsp.2014.58

pH Modulation of Voltage-Gated Sodium Channels

Colin H. Peters, Mohammad-Reza Ghovanloo, Cynthia Gershome, and Peter C. Ruben

Contents

Abstract

Changes in blood and tissue pH accompany physiological and pathophysiological conditions including exercise, cardiac ischemia, ischemic stroke, and cocaine ingestion. These conditions are known to trigger the symptoms of electrical diseases in patients carrying sodium channel mutations. Protons cause a diverse set of changes to sodium channel gating, which generally lead to decreases in the amplitude of the transient sodium current and increases in the fraction of non-inactivating channels that pass persistent currents. These effects are shared with disease-causing mutants in neuronal, skeletal muscle, and cardiac tissue and may be compounded in mutants that impart greater proton sensitivity to sodium channels, suggesting a role of protons in triggering acute symptoms of electrical disease.

In this chapter, we review the mechanisms of proton block of the sodium channel pore and a suggested mode of action by which protons alter channel gating. We discuss the available data on isoform specificity of proton effects and

C. H. Peters · M.-R. Ghovanloo · C. Gershome · P. C. Ruben (✉)
Department of Biomedical Physiology and Kinesiology, Simon Fraser University, Burnaby, BC, Canada
e-mail: pruben@sfu.ca

M. Chahine (ed.), *Voltage-gated Sodium Channels: Structure, Function and Channelopathies*, Handbook of Experimental Pharmacology 246,
https://doi.org/10.1007/164_2018_99

tissue level effects. Finally, we review the role that protons play in disease and our own recent studies on proton-sensitizing mutants in cardiac and skeletal muscle sodium channels.

Keywords
Acidosis · Extracellular pH · Ischemia · Proton block · Voltage-gated sodium channel

1 Introduction

The transient sodium current through voltage-gated sodium channels initiates action potentials in neurons, skeletal muscle, and cardiac muscle. Voltage-gated sodium channels are heterotetrameric proteins formed by a single transcript that encodes four 6-transmembrane segment domains. The voltage-sensor is formed by the first four transmembrane segments of each domain and the pore is formed by the 5th and 6th segments as well as the extracellular P-loop connecting them. Any changes to the gating properties of these channels, and consequently the current passed during an action potential, can cause potentially fatal abnormalities in electrical signaling. Both gain-of-function and loss-of-function in sodium channels disrupt electrical signaling. Interestingly, several mutants display both gain- and loss-of-function, leading to multiple disease phenotypes (Webb and Cannon 2008; Makita et al. 2008).

In the primary sodium channel isoforms of the central nervous system, $Na_V1.1$, 1.2, 1.3, and 1.6, gain- and loss-of-function elicit epilepsy syndromes (Estacion et al. 2010; Catterall 2012; Veeramah et al. 2012). These include relatively mild epilepsies, like benign familial neonatal-infantile seizures, and more severe forms, such as Dravet syndrome (Heron et al. 2002; Scalmani et al. 2006; Dravet 2011). In the skeletal muscle sodium channel, $Na_V1.4$, mutants elicit myotonic and paralytic syndromes, causing an inability to relax the muscle or contract the muscle, respectively (Cannon 1996). Long QT syndrome is due to an increase in the fraction of $Na_V1.5$ cardiac sodium channels that fail to inactivate and, consequently, an increased persistent sodium current throughout the action potential plateau that delays repolarization (Wang et al. 1995). Conversely, mutants that decrease peak $Na_V1.5$ sodium current cause Brugada syndrome and other diseases of conduction (Antzelevitch et al. 2005).

Sodium channel mutants are present from birth, but interestingly these diseases can remain symptomless into adulthood. In many of these diseases, symptoms are elicited when patients perform specific activities, including exercise or use of specific drugs (Littmann et al. 2000; Miller et al. 2004; Antzelevitch et al. 2005; García-Borbolla et al. 2007; Ruan et al. 2010; Postema et al. 2011). In fact, similar symptoms and phenotypes to those found in patients with congenital sodium channel mutants can be elicited by changes to the body's internal environment in the absence of a corresponding channel mutant (McClelland et al. 2009; Anselm et al. 2014). Thus, physiological and pathophysiological changes to the body's internal environment may play an important role in triggering electrical diseases. One such change is acidaemia.

There are multiple ways in which acid-base homeostasis is maintained in the human body, including renal, ventilation, and buffering mechanisms. Under normal physiological conditions, the extracellular pH is maintained at approximately 7.4, with the intracellular pH ranging from 7.2 to 7.4. Acidaemia accompanies many physiological and pathophysiological conditions including exercise, cardiac ischemia, and hypoventilation (Hermansen and Osnes 1972; Cobbe and Poole-Wilson 1980; Epstein and Singh 2001). These conditions are known risk factors for neurological, skeletal muscle, and cardiac disorders, particularly in those with channel function compromised by the presence of a mutation (Constantinou et al. 1989; Miller et al. 2004; Di Diego et al. 2005; García-Borbolla et al. 2007; Zhan et al. 2007). During exercise, skeletal muscle tissue pH can drop to pH 6.5 (Hermansen and Osnes 1972). Postmortem measurements show SIDS cases with brain tissue pH below pH 6.5 (Constantinou et al. 1989; Butterworth and Tennant 1989). Cocaine ingestion can decrease arterial pH below pH 6.4 (Hick et al. 1999; Allam and Noble 2001; Ortega-Carnicer et al. 2001). And ischemia can cause cardiac tissue acidaemia to pH 6.0 (Cobbe and Poole-Wilson 1980; Fleet et al. 1985; Yan and Kléber 1992).

During acidaemia, protons directly block the pore of the sodium channel, alter the movement of the voltage-sensors leading to changes in channel activation, fast inactivation, and slow inactivation, and increase the fraction of non-inactivating sodium channels (Fig. 1) (Woodhull 1973; Jones et al. 2011, 2013a). The effects of protons are dependent on the channel variant, with $Na_V1.4$ being relatively resistant to changes in extracellular pH compared to $Na_V1.5$ (Vilin et al. 2012). Furthermore, mutant sodium channels may be more susceptible to changes in pH than wild-type channels, leading to an even greater exacerbation of disease symptoms (Cheng et al. 2011; Peters et al. 2016; Ghovanloo et al. 2017). To date, the effects of protons have

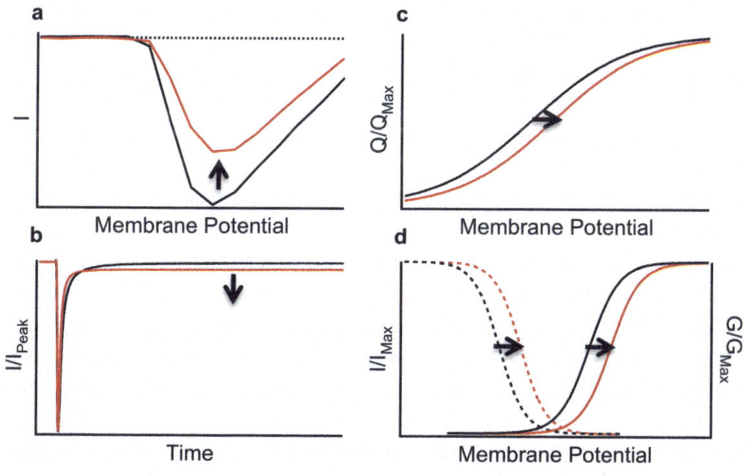

Fig. 1 Lowering extracellular pH reduces peak sodium current amplitude (**a**) and increases the fraction of channels which fail to inactivate and pass a persistent sodium current (**b**). Extracellular protons depolarize the charge-voltage relationship (**c**) as well as the conductance-voltage relationship and steady-state fast inactivation voltage-dependence of cardiac sodium channels (**d**)

not been characterized on all sodium channel variants, nor have all residues which impart proton-sensitivity been identified.

2 Molecular Mechanisms of Proton Block

Based on single channel data from guinea pig cardiomyocytes, protons decrease the conductance of individual channels (Zhang and Siegelbaum 1991). Early experiments on the effects of increasing extracellular protons indicated that proton current block is voltage-dependent based on the observation that proton block is decreased at positive potentials. Thus, Woodhull proposed that protons bind within the channel pore approximately 25% across the distance of the extracellular and intracellular membranes (Woodhull 1973). This was later disputed by Campbell, who suggested that, based on tail current analysis, proton block was independent of membrane potential and the binding site of protons lies outside the electrical field across the membrane (Campbell 1982). It is now known that a multitude of proton binding sites exist including residues in the selectivity filter, the outer charged ring, and residues C373 and H880 in cardiac sodium channels (Sun et al. 1997; Khan et al. 2002, 2006; Jones et al. 2013b).

The sodium channel selectivity filter is formed by a single residue in each of the P-loops in the four domains: D372, E898, K1419, and A1710 (Sun et al. 1997). The permeation rate of sodium is further determined by a series of carboxylate residues, E375, E901, D1423, and D1714, which form the outer charged ring (Terlau et al. 1991). These two motifs are fully conserved across all human voltage-gated sodium channels. Mutation of residues in the selectivity filter or outer charged ring shifts the pKa of proton block to more acidic pH (Sun et al. 1997; Khan et al. 2002). In particular, replacement of the selectivity filter carboxylates with alanines increases the fraction of proton independent sodium current by 25% (Sun et al. 1997). Although protonation of these carboxylates is one of the primary drivers of proton block of sodium channel conductance, proton sensitivity is not fully abolished by replacement of the entire selectivity filter with alanines or by any mutations in the outer charged ring (Sun et al. 1997; Khan et al. 2002). This suggests that other residues in the outer vestibule of the channels may play a role in proton block of sodium channels.

One such residue, which is also important in determining isoform specificity of proton block in sodium channels, is residue C373 in $Na_V1.5$ and its analogous residues Y401 in $Na_V1.4$ and F385 in $Na_V1.2$. In $Na_V1.5$ proton block nears completion at pH 4.0. By contrast, in $Na_V1.4$, 12–17% of the current is resistant to proton block (Khan et al. 2006; Jones et al. 2013b). The C373Y mutant in $Na_V1.5$ imparts a similar fraction of proton resistant current as that in $Na_V1.4$. Conversely, the Y401C mutant in $Na_V1.4$ abolishes the proton insensitive sodium current (Khan et al. 2006).

Given that the pKa of proton block is approximately 6.0, similar to the pKa of histidine, extracellular histidines may also play a role in determining proton block. Our lab previously tested two histidines in the $Na_V1.5$ domain II P-loops: H880Q and H886Q. H886Q did not produce functional channels. The H880Q mutant, however, imparts proton insensitivity to a fraction of sodium current, similar to that imparted by C373F (Jones et al. 2013b).

3 Proton Modulation of Channel Gating

The effects of acidification on sodium channel gating have been studied best in $Na_V1.5$ and to a lesser extent in $Na_V1.1$, 1.2, and 1.4. As with proton block of conductance, the effects of protons on channel gating are isoform dependent, with $Na_V1.5$ being the most sensitive (Vilin et al. 2012). Results in cardiomyocytes and heterologous expression systems show that extracellular acidification depolarizes the voltage-dependence of activation and fast inactivation in $Na_V1.5$ (Yatani et al. 1984; Vilin et al. 2012; Jones et al. 2013b). Protons also increase the fraction of non-inactivating current in $Na_V1.5$ and decrease the fraction of immobilized charge (Jones et al. 2011, 2013a; Peters et al. 2016). Protons speed recovery from, and slow onset of, fast and slow inactivation in cardiac sodium channels (Jones et al. 2011; Vilin et al. 2012). Interestingly, in heterologous expression systems intracellular acidification does not impact the gating of WT sodium channels, suggesting that protons interact with specific extracellular residues in the channel as opposed to interacting with the cell membrane to induce a charge-screening effect (Cheng et al. 2011; Hu et al. 2015).

Although the residues responsible for all proton-dependent changes in sodium channel gating have not been positively identified, structural studies and data from hERG channels suggest an important role of acidic residues in the voltage-sensing domain. The depolarization and slowed outward gating current movements in sodium channels suggest that protons directly impede the outward movement of the S4 voltage-sensors (Jones et al. 2013a). A depolarization in the movement of the S4 voltage-sensors by extracellular acidification is also seen in hERG channel voltage-sensor fluorescence recordings (Shi et al. 2014). In hERG, the effects of protons can be abolished by mutating a series of acidic residues in the voltage sensing domain: D456 and D460 in S2 and D509 in S3, which are accessible from the extracellular side (Shi et al. 2014). Crystal structures indicate that acidic residues in the voltage-sensing domains of voltage-gated sodium channels are also accessible from the extracellular fluid (Payandeh et al. 2011). Thus, like in hERG, protonation of the carboxylate residues in the voltage-sensing domains may depolarize the outward motion of the four voltage-sensors, which would depolarize the activation and fast inactivation voltage-dependence.

Currently, little is known about proton effects in $Na_V1.1$ with a single study showing only a block of current and depolarization of activation by protons (DeCaen et al. 2014). Interestingly, both $Na_V1.2$ and $Na_V1.4$ display relative insensitivity to protons compared to $Na_V1.5$. As in $Na_V1.5$, the conductance voltage-relationship in $Na_V1.2$ is depolarized by low extracellular pH (Vilin et al. 2012; Peters et al. 2013). Unlike in $Na_V1.5$, $Na_V1.2$ fast inactivation voltage-dependence is not altered by protons; however, the recovery from fast inactivation is faster and onset of fast inactivation is slower at low extracellular pH (Vilin et al. 2012; Peters et al. 2013). The literature on proton-dependent changes in slow inactivation in $Na_V1.2$ is conflicting, with one report showing an increase in slow inactivation at low pH and another showing a decrease (Vilin et al. 2012; Peters et al. 2013). $Na_V1.4$ activation and fast inactivation are not sensitive to changes in pH, nor is $Na_V1.4$ use-dependent inactivation (Vilin et al. 2012; Ghovanloo et al. 2017). Thus, of the channels studied thus far, $Na_V1.5$ is the most proton-sensitive and $Na_V1.4$ the least.

As with proton-block, residue C373 (Y401 in $Na_V1.4$ and F385 in $Na_V1.2$) plays an important role in determining isoform specificity of proton effects on channel gating. Slow inactivation in sodium channels involves both the voltage-sensing domains and the extracellular P-loops (Vilin et al. 2001; Payandeh et al. 2012; Silva and Goldstein 2013a, b). The C373F mutant abolishes the proton sensitivity of slow inactivation onset and recovery in $Na_V1.5$ (Jones et al. 2013b). C373F also removes the proton sensitivity of use-dependent inactivation in $Na_V1.5$ (Jones et al. 2013b). This suggests that the presence of aromatic acids in these positions in $Na_V1.2$ (F385) and $Na_V1.4$ (Y401) are likely responsible, at least in part, for the different proton effects in these tissues.

Although all the residues responsible for isoform-dependent differences in proton sensitivity are not known, insensitivity may be a particularly important adaptation in $Na_V1.4$. During exercise, arterial blood pH shows relatively small changes. Capillary blood pH, however, may drop by 0.2 units and skeletal muscle tissue pH may drop as low as pH 6.4 (Hermansen and Osnes 1972). Relative insensitivity to changes in pH likely plays an important role in maintaining action potential generation in working muscle (Pedersen et al. 2005).

4 Effects of Protons on Tissues

Heterologous expression and characterization of sodium channels have yielded a wealth of information on the effects of extracellular acidosis; however, in vivo, sodium channels function within multi-protein signaling complexes and associate with various proteins that modify expression and regulate gating, cytoskeletal anchoring, and signaling cascades (Meadows and Isom 2005). Thus, studies in whole tissue preparations are necessary to understand the role proton modulation of sodium plays in altering electrical signals in the body.

In ventricular cardiomyocytes, sodium channels pass a large transient current which causes the initial phase 0 depolarization. Studies in rat and canine ventricular myocytes show a similar proton-induced depolarization of conductance and decrease in peak sodium current as is seen in heterologous expression systems (Yatani et al. 1984; Watson and Gold 1995; Murphy et al. 2011). However, the effects of extracellular protons on fast inactivation differ between these studies. Furthermore, studies in rat ventricular myocytes differ from heterologous expression systems in suggesting that intracellular acidification may affect inactivation voltage-dependence and kinetics (Watson and Gold 1995). The overall effect in ventricular myocytes is a proton-induced decrease in the transient sodium current that is predicted to decrease the rate of the phase 0 depolarization, which in turn decreases conduction velocity in the heart (Yatani et al. 1984; Kléber et al. 1986; Watson and Gold 1995; Murphy et al. 2011; Jones et al. 2011).

In ventricular myocytes, the persistent sodium current is active throughout the action potential plateau and in part determines the action potential duration (Kiyosue and Arita 1989). Disease or pharmacologically induced increases in the persistent current elongate the action potential while block of the persistent sodium current shortens the action potential (Kiyosue and Arita 1989; Wang et al. 1995; Shimizu and Antzelevitch 1999). In heterologous expression systems, decreases in extracellular pH increases the fraction of sodium channels which fail to inactivate. Thus, although protons lower the overall sodium conductance, the persistent sodium current is relatively well maintained (Jones et al. 2011; Peters et al. 2016). Experiments in ventricular myocytes suggest that persistent sodium currents are either increased by ischemia, or less sensitive than peak sodium current (Murphy et al. 2011; Tang et al. 2012). In conjunction with the block of the rapid delayed rectifier potassium current by extracellular protons, persistent sodium currents likely play a large role in the elongation of the cardiac action potential during ischemia (Fry and Poole-Wilson 1981; Komukai et al. 2002; Murphy et al. 2011; Van Slyke et al. 2012).

Unlike cardiac tissue, a consensus on the effects of protons on neural tissue is complicated by the large variety of neurons. At pH 6.0 isolated rat trigeminal mesencephalic nucleus (Vmes) neurons show a depolarization of the activation and, in contrast to studies in heterologous expression systems, a depolarization of fast inactivation voltage-dependence (Vilin et al. 2012; Peters et al. 2013; Kang et al. 2016). In rat pyramidal neurons, exposure to acidic pH 6.4 resulted in the reduction of peak sodium current, depolarization of the conductance voltage relationship, and no effect on steady-state fast inactivation (Tombaugh and Somjen 1996). In both GABAergic neurons and CA1 hippocampal interneurons, extracellular acidosis reduces spike frequency (Zhan et al. 2007; Huang et al. 2015). In contrast, acidosis stimulates a subset of serotonergic neurons in the medullary raphe that may act as chemosensors (Wang et al. 2001). Overall, these studies show that while the majority of neurons studied are inhibited by extracellular protons, this is not true for all.

$Na_V1.4$ is the least proton-sensitive of the sodium channels studied in heterologous expression systems, which may be an evolutionary adaptation as during exercise the pH of skeletal muscle can decrease considerably (Hermansen and Osnes 1972; Vilin et al. 2012). Skeletal muscle tissue pH studies suggest that protons may compensate for losses of excitability that occur when extracellular potassium is elevated during exercise. This occurs by proton-dependent reductions in the inhibitory chloride currents passed by the ClC-1 channel (Pedersen et al. 2005; Bennetts et al. 2007). The reduction in chloride current increases the skeletal muscle excitability. Authors of one study suggest that the sodium current must be relatively well maintained during extracellular acidification, a theory consistent with the relative pH insensitivity of the $Na_V1.4$ channel (Pedersen et al. 2005; Vilin et al. 2012). Thus in contrast to cardiac muscle and the majority of neurons, skeletal muscle electrical excitability appears to be relatively well maintained in the face of acidaemia.

These studies show that in native tissues and intact tissues, the direct effect of protons may cause complex effects on tissue excitability. While simulations using the data from heterologous expression systems may predict some of these effects, studies in intact tissue are necessary to fully understand the role of protons in human physiology and pathophysiology.

5 Acidosis and Disease

Acidosis may trigger symptoms in sodium channel-based diseases. Increases in extracellular protons may compound with the effects of sodium channel mutants, further compromising function in tissues throughout the body (Peters et al. 2016). Protons may also act to unmask biophysical defects in mutants that function normally at physiological pH (Cheng et al. 2011). Finally, protons may alter the effects of drugs such as, for example, Ranolazine, which is itself protonatable (Peters et al. 2013; Sokolov et al. 2013).

Interestingly, acidaemia may elicit symptoms of diseases normally associated with sodium channel mutants in the absence of underlying mutants in any of the voltage-gated sodium channel genes. Patients with glutaric aciduria type 1, propionic acidaemia, and methylmalonic acidaemia may all present with epilepsy (McClelland et al. 2009; Haberlandt et al. 2009; Ma et al. 2011). One study found that over 40% of patients with methylmalonic acidaemia, a disease caused by mutations in genes involved in amino acid and fat break down, presented with epilepsy (Ma et al. 2011). This phenomenon is not unique to epilepsy; acute acidaemia is implicated in Brugada phenocopy. In Brugada phenocopy, a Brugada syndrome-like electrocardiogram (ECG) is elicited in response to environmental or clinical conditions, and is resolved when the underlying condition is resolved (Baranchuk et al. 2012). The association of acidaemia and Brugada phenocopy is particularly strong with myocardial ischemia and cocaine induced acidaemia known to elicit the characteristic Brugada ECG (Littmann et al. 2000; Ortega-Carnicer et al. 2001; Anselm et al. 2014).

In both epilepsy and Brugada phenocopy, the presence of protons may mimic the effects of mutants. Protons decrease peak sodium current through direct block of channels and depolarization of channel activation (Zhang and Siegelbaum 1991; Jones et al. 2013b). Similarly, the most common causes of Brugada syndrome are mutations in SCN5A, the gene which encodes for $Na_V1.5$, which decrease peak sodium conductance, thereby decreasing conduction velocity and altering action potential morphology (Antzelevitch et al. 2005; Wilde et al. 2010). Although many epilepsies are caused by gain-of-function mutants in sodium channels, one form, the Dravet syndrome, is caused primarily by loss-of-function in $Na_V1.1$ (Yu et al. 2006). It is believed this loss of function decreases activity of GABAergic neurons leading to disinhibition of the brain and an overall increase in excitability. Acidaemia in turn appears to decrease excitability of GABAergic neurons (Li et al. 2011; Huang et al. 2015).

Protons are also known to act in conjunction with mutants, either unmasking biophysical defects, or preferentially effecting those already present. The S1103Y polymorphism, which increases the risk of SIDS, does not produce significantly different currents at pH 7.4; however, intracellular acidosis elicits a persistent sodium current in S1103Y $Na_V1.5$, not seen in WT (Cheng et al. 2011). Similarly, the S1787N mutant in $Na_V1.5$ when expressed in conjunction with the Q1077del polymorphism produces a small persistent current that is exacerbated by decreasing intracellular protons (Hu et al. 2015). Our lab has since shown that the most common $Na_V1.5$ Brugada syndrome and long QT syndrome mutant, E1784K, is preferentially sensitive to changes in extracellular pH (Peters et al. 2016). Extracellular protons cause larger decreases in peak current and larger increases in the fraction of non-inactivating channels in the E1784K mutant compared to WT (Fig. 2), which represent exacerbation of the biophysical defects seen in Brugada syndrome and LQT3, respectively.

Preferential effects of acidosis may not be limited to cardiac sodium channels. Our lab recently studied P1158S in $Na_V1.4$, a channel mutant associated with both periodic paralysis and myotonia congenita (Webb and Cannon 2008). P1158 is a fully conserved residue located on the S4-S5 linker of DIII in $Na_V1.4$. In addition to myotonia and periodic paralysis, the mutation of this proline to leucine (P1308L) in $Na_V1.7$ also causes inherited erythromelalgia (Cheng et al. 2010).

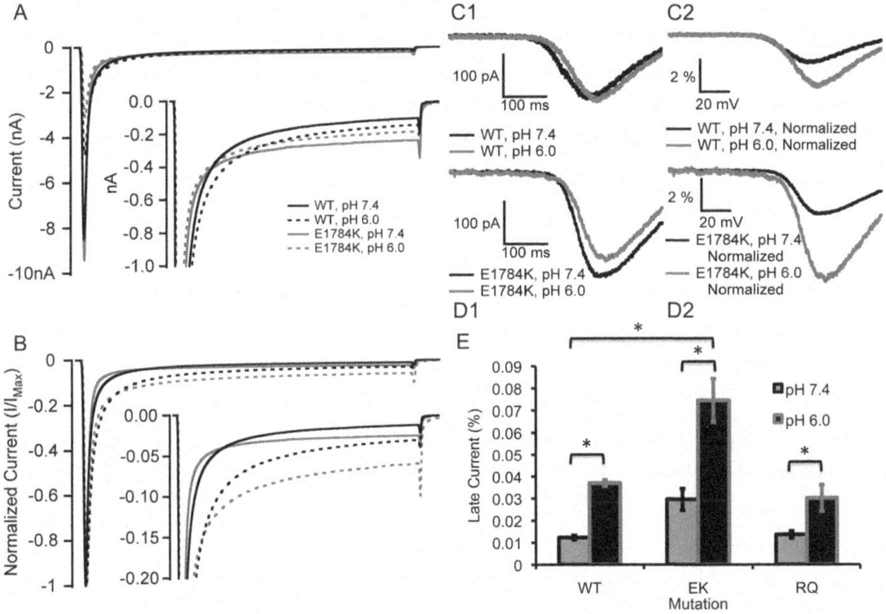

Fig. 2 Absolute (**a**) and normalized to peak (**b**) persistent current traces from wild type and E1784K $Na_V1.5$ channels at pH 7.4 and at pH 6.0. Insets show these persistent currents on a larger scale. Absolute (**c1** and **d1**) and normalized to peak (**c2** and **d2**) ramp current traces in wild type (**c**) and E1784K (**d**) $Na_V1.5$. The average percentage of persistent sodium current for wild type, R1193Q, and E1784K $Na_V1.5$ at pH 7.4 and pH 6.0 (**e**). Reprinted from Peters et al. (2016), with permission from Elsevier

We showed that the mutation of the "helix-breaker" proline to serine in the DIII S4-S5 imparts proton sensitivity to $Na_V1.4$. Action potential modeling suggests that, as proton concentrations are changed, AP morphology may shift from a periodic paralysis phenotype at physiological or alkalotic pH to a myotonia phenotype during acidosis (Ghovanloo et al. 2017).

6 Conclusion

Changes in proton concentration are associated with a wide range of physiological and pathophysiological conditions including exercise, sleep apnea, cardiac ischemia, stroke, metabolic diseases, and drug use. Protons block the pore of sodium channels and alter voltage-sensor movement to change channel gating. Protons decrease the peak sodium current while increasing the fraction of non-inactivating channels. The effects of protons can be measured at all levels of function, from single-channel recordings to electroencephalogram and electrocardiogram recordings. Protons act on both mutant and WT sodium channels and may trigger symptoms in those with congenital disease or elicit symptoms like those of the congenital disease in otherwise healthy patients. Sodium channels are not unique in this regard, as protons are known to alter the properties of both calcium and potassium channels. Thus, protons have a physiological impact whose effects are seen throughout the electrical systems of the body and whose actions alter cellular function and disease. Future studies will integrate these different effects and lead to a greater understanding of the physiological and pathophysiological roles of protons at the molecular, cellular, and tissue levels.

References

Allam S, Noble JS (2001) Cocaine-excited delirium and severe acidosis. Anaesthesia 56:385–386

Anselm DD, Evans JM, Baranchuk A (2014) Brugada phenocopy: a new electrocardiogram phenomenon. World J Cardiol 6:81–86. https://doi.org/10.4330/wjc.v6.i3.81

Antzelevitch C, Brugada P, Borggrefe M et al (2005) Brugada syndrome: report of the second consensus conference endorsed by the Heart Rhythm Society and the European Heart Rhythm Association. Circulation 111:659–670. https://doi.org/10.1161/01.CIR.0000152479.54298.51

Baranchuk A, Nguyen T, Ryu MH et al (2012) Brugada phenocopy: new terminology and proposed classification. Ann Noninvasive Electrocardiol 17:299–314. https://doi.org/10.1111/j.1542-474X.2012.00525.x

Bennetts B, Parker MW, Cromer BA (2007) Inhibition of skeletal muscle ClC-1 chloride channels by low intracellular pH and ATP. J Biol Chem 282:32780–32791. https://doi.org/10.1074/jbc.M703259200

Butterworth J, Tennant MC (1989) Postmortem human brain pH and lactate in sudden infant death syndrome. J Neurochem 53:1494–1499. https://doi.org/10.1111/j.1471-4159.1989.tb08543.x

Campbell DT (1982) Do protons block Na+ channels by binding to a site outside the pore? Nature 298:165–167. https://doi.org/10.1038/298165a0

Cannon SC (1996) Sodium channel defects in myotonia and periodic paralysis. Annu Rev Neurosci 19:141–164. https://doi.org/10.1146/annurev.ne.19.030196.001041

Catterall WA (2012) Sodium channel mutations and epilepsy. In: Noebels JL, Avoli M, Rogawski MA et al (eds) Jasper's basic mechanisms of the epilepsies, 4th edn. National Center for Biotechnology Information (US), Bethesda

Cheng X, Dib-Hajj SD, Tyrrell L et al (2010) Mutations at opposite ends of the DIII/S4-S5 linker of sodium channel Na V 1.7 produce distinct pain disorders. Mol Pain 6:24. https://doi.org/10.1186/1744-8069-6-24

Cheng J, Tester DJ, Tan B-H et al (2011) The common African American polymorphism SCN5A-S1103Y interacts with mutation SCN5A-R680H to increase late Na current. Physiol Genomics 43:461–466. https://doi.org/10.1152/physiolgenomics.00198.2010

Cobbe SM, Poole-Wilson PA (1980) The time of onset and severity of acidosis in myocardial ischaemia. J Mol Cell Cardiol 12:745–760

Constantinou JE, Gillis J, Ouvrier RA, Rahilly PM (1989) Hypoxic-ischaemic encephalopathy after near miss sudden infant death syndrome. Arch Dis Child 64:703–708. https://doi.org/10.1136/adc.64.5.703

DeCaen PG, Takahashi Y, Krulwich TA et al (2014) Ionic selectivity and thermal adaptations within the voltage-gated sodium channel family of alkaliphilic Bacillus. Elife 3:e04387. https://doi.org/10.7554/eLife.04387

Di Diego JM, Fish JM, Antzelevitch C (2005) Brugada syndrome and ischemia-induced ST-segment elevation. Similarities and differences. J Electrocardiol 38:14–17. https://doi.org/10.1016/j.jelectrocard.2005.06.003

Dravet C (2011) The core Dravet syndrome phenotype. Epilepsia 52:3–9. https://doi.org/10.1111/j.1528-1167.2011.02994.x

Epstein SK, Singh N (2001) Respiratory acidosis. Respir Care 46:366–383

Estacion M, Gasser A, Dib-Hajj SD, Waxman SG (2010) A sodium channel mutation linked to epilepsy increases ramp and persistent current of Nav1.3 and induces hyperexcitability in hippocampal neurons. Exp Neurol 224:362–368. https://doi.org/10.1016/j.expneurol.2010.04.012

Fleet WF, Johnson TA, Graebner CA, Gettes LS (1985) Effect of serial brief ischemic episodes on extracellular K+, pH, and activation in the pig. Circulation 72:922–932. https://doi.org/10.1161/01.CIR.72.4.922

Fry CH, Poole-Wilson PA (1981) Effects of acid-base changes on excitation--contraction coupling in guinea-pig and rabbit cardiac ventricular muscle. J Physiol 313:141–160

García-Borbolla M, García-Borbolla R, Valenzuela LF et al (2007) Ventricular tachycardia induced by exercise testing in a patient with Brugada syndrome. Rev Esp Cardiol 60:993–994

Ghovanloo M-R, Abdelsayed M, Peters CH, Ruben PC (2017) A mixed periodic paralysis & myotonia mutant, P1158S, imparts pH sensitivity in skeletal muscle voltage-gated sodium channels. bioRxiv 164988. https://doi.org/10.1101/164988

Haberlandt E, Canestrini C, Brunner-Krainz M et al (2009) Epilepsy in patients with propionic acidemia. Neuropediatrics 40:120–125. https://doi.org/10.1055/s-0029-1243167

Hermansen L, Osnes JB (1972) Blood and muscle pH after maximal exercise in man. J Appl Physiol 32:304–308

Heron SE, Crossland KM, Andermann E et al (2002) Sodium-channel defects in benign familial neonatal-infantile seizures. Lancet 360:851–852. https://doi.org/10.1016/S0140-6736(02)09968-3

Hick JL, Smith SW, Lynch MT (1999) Metabolic acidosis in restraint-associated cardiac arrest: a case series. Acad Emerg Med 6:239–243

Hu R-M, Tan B-H, Tester DJ et al (2015) Arrhythmogenic biophysical phenotype for SCN5A mutation S1787N depends upon splice variant background and intracellular acidosis. PLoS One 10:e0124921. https://doi.org/10.1371/journal.pone.0124921

Huang L, Zhao S, Lu W et al (2015) Acidosis-induced dysfunction of cortical GABAergic neurons through astrocyte-related excitotoxicity. PLoS One 10:e0140324. https://doi.org/10.1371/journal.pone.0140324

Jones DK, Peters CH, Tolhurst SA et al (2011) Extracellular proton modulation of the cardiac voltage-gated sodium channel, Nav1.5. Biophys J 101:2147–2156. https://doi.org/10.1016/j.bpj.2011.08.056

Jones DK, Claydon TW, Ruben PC (2013a) Extracellular protons inhibit charge immobilization in the cardiac voltage-gated sodium channel. Biophys J 105:101–107. https://doi.org/10.1016/j.bpj.2013.04.022

Jones DK, Peters CH, Allard CR et al (2013b) Proton sensors in the pore domain of the cardiac voltage-gated sodium channel. J Biol Chem 288:4782–4791. https://doi.org/10.1074/jbc.M112.434266

Kang I-S, Cho J-H, Choi I-S et al (2016) Acidic pH modulation of Na+ channels in trigeminal mesencephalic nucleus neurons. Neuroreport 27:1274–1280. https://doi.org/10.1097/WNR.0000000000000692

Khan A, Romantseva L, Lam A et al (2002) Role of outer ring carboxylates of the rat skeletal muscle sodium channel pore in proton block. J Physiol 543:71–84. https://doi.org/10.1113/jphysiol.2002.021014

Khan A, Kyle JW, Hanck DA et al (2006) Isoform-dependent interaction of voltage-gated sodium channels with protons. J Physiol 576:493–501. https://doi.org/10.1113/jphysiol.2006.115659

Kiyosue T, Arita M (1989) Late sodium current and its contribution to action potential configuration in guinea pig ventricular myocytes. Circ Res 64:389–397. https://doi.org/10.1161/01.RES.64.2.389

Kléber AG, Janse MJ, Wilms-Schopmann FJ et al (1986) Changes in conduction velocity during acute ischemia in ventricular myocardium of the isolated porcine heart. Circulation 73:189–198. https://doi.org/10.1161/01.CIR.73.1.189

Komukai K, Brette F, Pascarel C, Orchard CH (2002) Electrophysiological response of rat ventricular myocytes to acidosis. Am J Physiol Heart Circ Physiol 283:H412–H422. https://doi.org/10.1152/ajpheart.01042.2001

Li F, Liu X, Su Z, Sun R (2011) Acidosis leads to brain dysfunctions through impairing cortical GABAergic neurons. Biochem Biophys Res Commun 410:775–779. https://doi.org/10.1016/j.bbrc.2011.06.053

Littmann L, Monroe MH, Svenson RH (2000) Brugada-type electrocardiographic pattern induced by cocaine. Mayo Clin Proc 75:845–849. https://doi.org/10.4065/75.8.845

Ma X, Zhang Y, Yang Y et al (2011) Epilepsy in children with methylmalonic acidemia: electroclinical features and prognosis. Brain and Development 33:790–795. https://doi.org/10.1016/j.braindev.2011.06.001

Makita N, Behr E, Shimizu W et al (2008) The E1784K mutation in SCN5A is associated with mixed clinical phenotype of type 3 long QT syndrome. J Clin Invest 118:2219–2229. https://doi.org/10.1172/JCI34057

McClelland VM, Bakalinova DB, Hendriksz C, Singh RP (2009) Glutaric aciduria type 1 presenting with epilepsy. Dev Med Child Neurol 51:235–239. https://doi.org/10.1111/j.1469-8749.2008.03240.x

Meadows LS, Isom LL (2005) Sodium channels as macromolecular complexes: implications for inherited arrhythmia syndromes. Cardiovasc Res 67:448–458. https://doi.org/10.1016/j.cardiores.2005.04.003

Miller TM, Dias da Silva MR, Miller HA et al (2004) Correlating phenotype and genotype in the periodic paralyses. Neurology 63:1647–1655

Murphy L, Renodin D, Antzelevitch C et al (2011) Extracellular proton depression of peak and late Na^+ current in the canine left ventricle. Am J Physiol Heart Circ Physiol 301:H936–H944. https://doi.org/10.1152/ajpheart.00204.2011

Ortega-Carnicer J, Bertos-Polo J, Gutiérrez-Tirado C (2001) Aborted sudden death, transient Brugada pattern, and wide QRS dysrrhythmias after massive cocaine ingestion. J Electrocardiol 34:345–349

Payandeh J, Scheuer T, Zheng N, Catterall WA (2011) The crystal structure of a voltage-gated sodium channel. Nature 475:353–358. https://doi.org/10.1038/nature10238

Payandeh J, Gamal El-Din TM, Scheuer T et al (2012) Crystal structure of a voltage-gated sodium channel in two potentially inactivated states. Nature 486:135–139. https://doi.org/10.1038/nature11077

Pedersen TH, de Paoli F, Nielsen OB (2005) Increased excitability of acidified skeletal muscle. J Gen Physiol 125:237–246. https://doi.org/10.1085/jgp.200409173

Peters C, Sokolov S, Rajamani S, Ruben P (2013) Effects of the antianginal drug, ranolazine, on the brain sodium channel NaV1.2 and its modulation by extracellular protons. Br J Pharmacol 169:704–716. https://doi.org/10.1111/bph.12150

Peters CH, Abdelsayed M, Ruben PC (2016) Triggers for arrhythmogenesis in the Brugada and long QT 3 syndromes. Prog Biophys Mol Biol 120(1–3):77–88. https://doi.org/10.1016/j.pbiomolbio.2015.12.009

Postema PG, Vlaar APJ, DeVries JH, Tan HL (2011) Familial Brugada syndrome uncovered by hyperkalaemic diabetic ketoacidosis. Europace 13:1509–1510. https://doi.org/10.1093/europace/eur151

Ruan Y, Denegri M, Liu N et al (2010) Trafficking defects and gating abnormalities of a novel SCN5A mutation question gene-specific therapy in long QT syndrome type 3. Circ Res 106:1374–1383. https://doi.org/10.1161/CIRCRESAHA.110.218891

Scalmani P, Rusconi R, Armatura E et al (2006) Effects in neocortical neurons of mutations of the Na(v)1.2 Na+ channel causing benign familial neonatal-infantile seizures. J Neurosci 26:10100–10109. https://doi.org/10.1523/JNEUROSCI.2476-06.2006

Shi YP, Cheng YM, Van Slyke AC, Claydon TW (2014) External protons destabilize the activated voltage sensor in hERG channels. Eur Biophys J 43:59–69. https://doi.org/10.1007/s00249-013-0940-y

Shimizu W, Antzelevitch C (1999) Cellular basis for long QT, transmural dispersion of repolarization, and torsade de pointes in the long QT syndrome. J Electrocardiol 32(Suppl):177–184

Silva JR, Goldstein SAN (2013a) Voltage-sensor movements describe slow inactivation of voltage-gated sodium channels I: wild-type skeletal muscle Na(V)1.4. J Gen Physiol 141:309–321. https://doi.org/10.1085/jgp.201210909

Silva JR, Goldstein SAN (2013b) Voltage-sensor movements describe slow inactivation of voltage-gated sodium channels II: a periodic paralysis mutation in Na(V)1.4 (L689I). J Gen Physiol 141:323–334. https://doi.org/10.1085/jgp.201210910

Sokolov S, Peters CH, Rajamani S, Ruben PC (2013) Proton-dependent inhibition of the cardiac sodium channel Nav1.5 by ranolazine. Front Pharmacol 4. https://doi.org/10.3389/fphar.2013.00078

Sun YM, Favre I, Schild L, Moczydlowski E (1997) On the structural basis for size-selective permeation of organic cations through the voltage-gated sodium channel. Effect of alanine mutations at the DEKA locus on selectivity, inhibition by Ca2+ and H+, and molecular sieving. J Gen Physiol 110:693–715

Tang Q, Ma J, Zhang P et al (2012) Persistent sodium current and Na+/H+ exchange contributes to the augmentation of the reverse Na+/Ca2+ exchange during hypoxia or acute ischemia in ventricular myocytes. Pflügers Arch - Eur J Physiol 463:513–522. https://doi.org/10.1007/s00424-011-1070-y

Terlau H, Heinemann SH, Stühmer W et al (1991) Mapping the site of block by tetrodotoxin and saxitoxin of sodium channel II. FEBS Lett 293:93–96. https://doi.org/10.1016/0014-5793(91)81159-6

Tombaugh GC, Somjen GG (1996) Effects of extracellular pH on voltage-gated Na+, K+ and Ca2+ currents in isolated rat CA1 neurons. J Physiol 493:719–732

Van Slyke AC, Cheng YM, Mafi P et al (2012) Proton block of the pore underlies the inhibition of hERG cardiac K+ channels during acidosis. Am J Physiol Cell Physiol 302:C1797–C1806. https://doi.org/10.1152/ajpcell.00324.2011

Veeramah KR, O'Brien JE, Meisler MH et al (2012) De novo pathogenic SCN8A mutation identified by whole-genome sequencing of a family quartet affected by infantile epileptic encephalopathy and SUDEP. Am J Hum Genet 90:502–510. https://doi.org/10.1016/j.ajhg.2012.01.006

Vilin YY, Fujimoto E, Ruben PC (2001) A single residue differentiates between human cardiac and skeletal muscle Na+ channel slow inactivation. Biophys J 80:2221–2230. https://doi.org/10.1016/S0006-3495(01)76195-4

Vilin YY, Peters CH, Ruben PC (2012) Acidosis differentially modulates inactivation in na(v)1.2, na(v)1.4, and na(v)1.5 channels. Front Pharmacol 3:109. https://doi.org/10.3389/fphar.2012.00109

Wang Q, Shen J, Splawski I et al (1995) SCN5A mutations associated with an inherited cardiac arrhythmia, long QT syndrome. Cell 80:805–811

Wang W, Tiwari JK, Bradley SR et al (2001) Acidosis-stimulated neurons of the medullary raphe are serotonergic. J Neurophysiol 85:2224–2235

Watson CL, Gold MR (1995) Effect of intracellular and extracellular acidosis on sodium current in ventricular myocytes. Am J Phys 268:H1749–H1756

Webb J, Cannon SC (2008) Cold-induced defects of sodium channel gating in atypical periodic paralysis plus myotonia. Neurology 70:755–761. https://doi.org/10.1212/01.wnl.0000265397.70057.d8

Wilde AAM, Postema PG, Di Diego JM et al (2010) The pathophysiological mechanism underlying Brugada syndrome. J Mol Cell Cardiol 49:543–553. https://doi.org/10.1016/j.yjmcc.2010.07.012

Woodhull AM (1973) Ionic blockage of sodium channels in nerve. J Gen Physiol 61:687–708

Yan GX, Kléber AG (1992) Changes in extracellular and intracellular pH in ischemic rabbit papillary muscle. Circ Res 71:460–470

Yatani A, Brown AM, Akaike N (1984) Effect of extracellular pH on sodium current in isolated, single rat ventricular cells. J Membr Biol 78:163–168

Yu FH, Mantegazza M, Westenbroek RE et al (2006) Reduced sodium current in GABAergic interneurons in a mouse model of severe myoclonic epilepsy in infancy. Nat Neurosci 9:1142–1149. https://doi.org/10.1038/nn1754

Zhan R-Z, Nadler JV, Schwartz-Bloom RD (2007) Impaired firing and sodium channel function in CA1 hippocampal interneurons after transient cerebral ischemia. J Cereb Blood Flow Metab 27:1444–1452. https://doi.org/10.1038/sj.jcbfm.9600448

Zhang JF, Siegelbaum SA (1991) Effects of external protons on single cardiac sodium channels from guinea pig ventricular myocytes. J Gen Physiol 98:1065–1083

Regulation of Cardiac Voltage-Gated Sodium Channel by Kinases: Roles of Protein Kinases A and C

Ademuyiwa S. Aromolaran, Mohamed Chahine, and Mohamed Boutjdir

Contents

A. S. Aromolaran
Cardiovascular Research Program, VA New York Harbor Healthcare System, Brooklyn, NY, USA

Departments of Medicine, Cell Biology and Pharmacology, State University of New York Downstate Medical Center, Brooklyn, NY, USA

M. Chahine
CERVO Brain Research Center, Institut Universitaire en Santé Mentale de Québec, Quebec City, QC, Canada

Department of Medicine, Université Laval, Quebec City, QC, Canada

M. Boutjdir (✉)
Cardiovascular Research Program, VA New York Harbor Healthcare System, Brooklyn, NY, USA

Departments of Medicine, Cell Biology and Pharmacology, State University of New York Downstate Medical Center, Brooklyn, NY, USA

Department of Medicine, New York University School of Medicine, New York, NY, USA
e-mail: mboutjdir@gmail.com

M. Chahine (ed.), *Voltage-gated Sodium Channels: Structure, Function and Channelopathies*, Handbook of Experimental Pharmacology 246,
https://doi.org/10.1007/164_2017_53

Abstract

In the heart, voltage-gated sodium (Na_v) channel ($Na_v1.5$) is defined by its pore-forming α-subunit and its auxiliary β-subunits, both of which are important for its critical contribution to the initiation and maintenance of the cardiac action potential (AP) that underlie normal heart rhythm. The physiological relevance of $Na_v1.5$ is further marked by the fact that inherited or congenital mutations in $Na_v1.5$ channel gene *SCN5A* lead to altered functional expression (including expression, trafficking, and current density), and are generally manifested in the form of distinct cardiac arrhythmic events, epilepsy, neuropathic pain, migraine, and neuromuscular disorders. However, despite significant advances in defining the pathophysiology of $Na_v1.5$, the molecular mechanisms that underlie its regulation and contribution to cardiac disorders are poorly understood. It is rapidly becoming evident that the functional expression (localization, trafficking and gating) of $Na_v1.5$ may be under modulation by post-translational modifications that are associated with phosphorylation. We review here the molecular basis of cardiac Na channel regulation by kinases (PKA and PKC) and the resulting functional consequences. Specifically, we discuss: (1) recent literature on the structural, molecular, and functional properties of cardiac $Na_v1.5$ channels; (2) how these properties may be altered by phosphorylation in disease states underlain by congenital mutations in $Na_v1.5$ channel and/or subunits such as long QT and Brugada syndromes. Our expectation is that understanding the roles of these distinct and complex phosphorylation processes on the functional expression of $Na_v1.5$ is likely to provide crucial mechanistic insights into Na channel associated arrhythmogenic events and will facilitate the development of novel therapeutic strategies.

Keywords

Brugada syndrome · Long QT syndrome 3 · Protein kinase A · Protein kinase C · Voltage-gated sodium channel

1 Introduction

In human heart, inward sodium currents (I_{Na}) through voltage-activated ($Na_v1.5$) channels depolarizes the membrane potential and is responsible for phase 0 or the rapid upstroke of the cardiac action potential (AP) and therefore a critical component of the AP profile that is important for normal electrical rhythm and efficient cardiac excitability (Chen-Izu et al. 2015). $Na_v1.5$ channels also inactivate relatively quickly (within a few milliseconds) allowing the AP to rapidly propagate and engulf the entire heart. Pathological alterations that either reduce or enhance I_{Na} function underlie both cases of inherited and/or acquired arrhythmias (Ruan et al.

2009). Furthermore, during the plateau phase of an AP, a small percentage (~1–3%) of total I_{Na} remain available, and therefore contributes a small "late" Na current (or $I_{Na,L}$) (Boutjdir et al. 1994; el-Sherif et al. 1992). Under normal physiological conditions, $I_{Na,L}$ is defined as a small current that remains sustained throughout the entire duration of an AP. However there is increasing evidence that under pathological conditions involving sodium channel mutations that predispose patients to long QT syndrome type 3 or LQTS3, an increased $I_{Na,L}$ density in addition to slower I_{Na} inactivation during the plateau phase of the AP leads to a slower repolarization and prolongation of AP duration (APD) (Clancy et al. 2002).

At the molecular level, the cardiac $Na_v1.5$ channel is composed of a highly glycosylated pore-forming α-subunit (220–260 kDa) and its regulatory and/or auxiliary single transmembrane protein beta (β)-subunits (30–46 kDa) (Cusdin et al. 2008). The $Na_v1.5$ channels have been shown to be localized to distinct compartments (t-tubules, lateral sarcolemma, and the intercalated disks sites) of a cardiac cell membrane (Petitprez et al. 2011). Thus studies that distinguish among the molecular mechanisms that determine and/or control localization/density of $Na_v1.5$ channel subunits within these compartments are likely to provide novel insights into arrhythmias underlain by distinct congenital mutations in $Na_v1.5$ and/or acquired in disease settings. Furthermore, the molecular mechanisms that underlie $Na_v1.5$ channel functional regulation within these specific locations may also differ depending on the underlying pathology.

The role of $Na_v1.5$ in cardiac excitability is further underscored by unique interactions with its molecular partners including cellular accessory/regulatory proteins (Loussouarn et al. 2015), and other major cardiac ion channels (Milstein et al. 2012), suggesting that altered regulation of these various interactions may have significant clinical implications (Luo et al. 2017). Despite significant progress in understanding the structure and function of these channels, there are still significant deficits in the precise molecular mechanisms that underlie the physiological and pathophysiological processes of cardiac $Na_v1.5$ channels. To date, there is an increasing abundance of evidence showing that channel phosphorylation can influence the electrical properties of cardiomyocytes and profoundly modulate cardiac function (Dai et al. 2009). From this perspective protein kinases have been described as prime candidate mediators of pathological cardiac electrophysiological remodeling (Chahine and O'Leary 2014). The rationale is that these kinases through modulation of the phosphorylation status will lead to altered relative biophysical properties of individual or multiple combinations of ion cardiac channels. Activation of protein kinase A (PKA) and protein kinase C (PKC) has been a focal subject of great attention in the cardiovascular field since they have been implicated in pathological settings such as heart failure (Shi et al. 2017), ischemia (Khaliulin et al. 2017) and ischemic preconditioning (Khaliulin et al. 2017). The main objective of this book chapter is to review the molecular and structural outcomes of post-translational modulation (expression, gating, trafficking) of $Na_v1.5$ channel subunits through activation of PKA and PKC.

2 Ionic Basis of Cardiac AP Waveform

The cardiac AP waveform is defined by distinct cells within specialized regions of the heart. Human ventricular AP is described in five phases marked by: phase 0 due to fast I_{Na} through voltage-gated $Na_v1.5$ channels; phase 1 results from a rapid inactivation and activation of the fast transient outward potassium currents (I_{to}); phase 2 is maintained by a balance between inward calcium (Ca) current mainly through L-type ($Ca_V1.2$) channels and outward delayed rectifier potassium current (I_K) with greater contribution from the rapid component or I_{Kr}; phase 3 is largely regulated by I_{Kr} and the slowly activating (I_{Ks}) component of I_K; phase 4 is due to non-voltage-gated inwardly rectifying K (IK_{ir}) current, through I_{K1} channels a major determinant of resting membrane potential of a cardiac cell.

The normal surface density and therefore relative functional expression of Na_V, Ca_V, and K_V channels is a prerequisite for a normal heart rhythm (Abriel et al. 2015). Therefore, gain-of-function mutations in Na_V (Klaver et al. 2011) and Ca_V (Klaver et al. 2011) channels or loss-of-function mutations in K_V channels (Klaver et al. 2011) that disrupt the delicate balance lead to altered repolarization. The resultant electrical remodeling is manifested as fatal arrhythmias including long QT syndrome (LQTS) (Schwartz and Ackerman 2013) and Brugada syndrome (BrS) (Dulsat et al. 2017) serious conditions that predisposes to sudden cardiac death (Amin et al. 2013). In this context the modulating factors that control the number of functional ionic channels at the cardiac cellular membrane (t-tubules, sarcolemma, and intercalated discs) have important clinical implications.

3 Structural and Molecular Identity of Cardiac $Na_v1.5/I_{Na}$ Channel

3.1 Cardiac $Na_v1.5$ Channel Subunits

The heart $Na_v1.5$, encoded by *SCNA5*, is defined by a cytoplasmic N- and C-termini, and is marked by four homologous domains (DI–DIV), each consisting of six transmembrane segments S1–S6 linked by intracellular loops DI–II, DII–III, and DIII–IV (Kruger and Isom 2016) (Fig. 1). The first four transmembrane segments (S1–S4) comprise the voltage-sensing domain, and the last two segments (S5–S6) contain the P-loop which plays a critical role in the ionic selectivity and permeation properties of $Na_v1.5$ channel complexes (Kruger and Isom 2016). Cellular membrane depolarization is followed by concerted movements of the large number of positively charged residues localized to the S4 segment leading to voltage-dependent activation and inactivation of the channel (Chadda et al. 2017). The $Na_v1.5$ channel inactivation process which terminates the inward I_{Na} is defined in multiple stages distinguished on the basis of recovery kinetics of the channel. Therefore $Na_v1.5$ channel inactivation is marked by either a fast inactivation which occurs over tens of milliseconds (ms), an intermediate inactivation

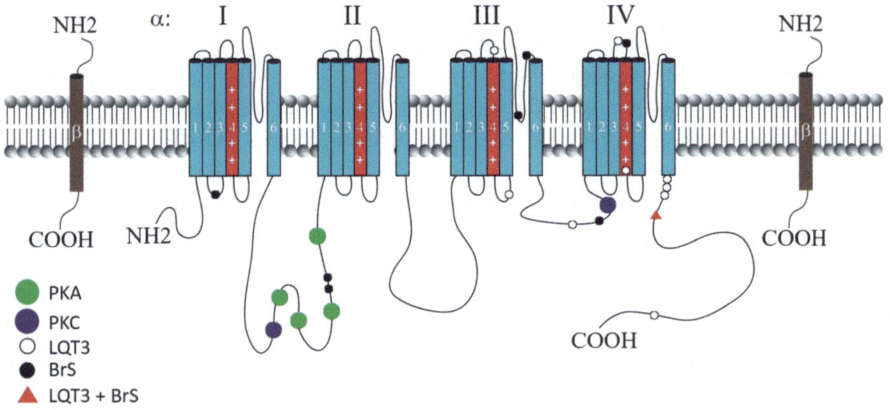

Fig. 1 Schematic illustration of the cardiac $Na_v1.5$ channel α- and β-subunits. Cardiac $Na_v1.5$ is defined by its pore-forming or α-subunit and its regulatory β-subunits. $Na_v1.5$ is defined by a cytoplasmic amino- (NH2) or N-terminus, four homologous domains (I–IV) that are connected to each other by cytoplasmic linkers, and a carboxy- (COOH) or C-terminus. The four domains are each characterized by six transmembrane domains (1–6) whereby the positively (+) charged S4 (denoted by *Red*) represents the voltage sensors. The β-subunits critically involved in the functional expression of I_{Na} consist of a single transmembrane domain, an extracellular NH2 terminus, and an intracellular carboxy (COOH) terminus. Reported sites of protein phosphorylation by PKA (*green filled circle*) and PKC (*purple filled circle*), and also the locations of congenital $Na_v1.5$ mutations that lead to LQT3 (*open circle*), BrS (*filled circle*) or both (*red filled triangle*) are also shown

within hundreds of ms (Wang et al. 2000), or a slow inactivation process requiring tens of seconds or longer timescale (Gawali and Todt 2016).

The distinct channel gating components suggest that the structural determinants or conformations that underlie $Na_v1.5$ channel kinetics may differ. In this context, the fast inactivation process has been widely reported to be mediated by the hydrophobic isoleucine-phenylalanine-methionine motif localized to the intracellular linker between domains DIII and DIV (Viswanathan et al. 2001). The carboxy-terminal of $Na_v1.5$ channel subunits has also been implicated in channel inactivation (Motoike et al. 2004). The structural configurations that underlie the slow kinetics are poorly understood although there are strong hints for a role for $Na_v1.5$ pore-forming α-subunits from disease-linked $Na_v1.5$ channel mutation studies (Ulbricht 2005).

3.2 $Na_v1.5$ and Its Associated β-Subunits

Auxiliary or regulatory β-subunits (Fig. 1) are small single transmembrane proteins that are widely expressed in the heart and are important modulators of $Na_v1.5$ functional properties (gating, localization, and expression) (Isom et al. 1992). The cardiac β-subunits exist as a family of five proteins, namely: β1/β1b (Isom et al. 1992), β2 (Isom et al. 1995), β3 (Morgan et al. 2000), and β4 (Yu et al. 2003), which are encoded by four genes, namely: *SCN1B* (encoding β1 and β1b), *SCN2B*, *SCN3B*, and *SCN4B* (Yu et al. 2003). β-subunits interact directly with and modulate $Na_v1.5$ functional expression (Wilde and Brugada 2011). The association of α-subunits of Na channels with the β-subunits can either be through non-covalent interactions (β1 and β3) (Yu et al. 2003) or disulfide bonds (β2 and β4) (Yu et al. 2003).

The β1-subunit is thought to interact within the P-loop of domain IV of the $Na_v1.5$ channel α-subunit (Qu et al. 1999). There have also been suggestions that the β3-subunit may also bind in this region since two of the three amino-acids essential for β1 interaction with $Na_v1.5$ (Qu et al. 1999) are also present in β3 (Morgan et al. 2000). In the heart, β1 and β2 are expressed throughout the atria and ventricles (Dhar Malhotra et al. 2001), whereas β3 is present in both the atria, ventricles, and Purkinje fibers (Fahmi et al. 2001). Therefore, it is likely that the isoform-specific expression profile may uniquely contribute to the electrical activity of the heart.

In this context, previous studies have shown that co-expression of β3 with cardiac Na channels in Xenopus oocytes increased I_{Na} amplitude, and resulted in a depolarizing shift in steady-state inactivation and enhanced recovery from inactivation (Fahmi et al. 2001). However, in the same study β1 co-expression increased current density and augmented the recovery from inactivation, but had no effect on the channel steady-state gating (Fahmi et al. 2001). In Chinese Hamster Ovary (CHO) cells stably expressing $Na_v1.5$ channels, co-expression of either β1 or β3 altered $Na_v1.5$ channel gating which was manifested in an accelerated current decay kinetics, and hyperpolarizing shifts in channel activation and inactivation, in addition to slowed recovery from inactivation (Ko et al. 2005). Furthermore, in these studies the effects of β3 were found to be more pronounced than those of β2 while the effects of co-expression of both β1 and β3 were more profound than the co-expression of either subunit alone. In *Scn1b* null ventricular myocytes, peak and persistent I_{Na} are also increased, an effect mediated by increased $Na_v1.5$ surface expression, with no obvious impact on channel kinetics and voltage-dependence (Lopez-Santiago et al. 2007).

Despite reported interactions of β2 with the α-subunit of $Na_v1.5$ there is still controversy about the impact of the β2-subunit on the biophysical properties of cardiac $Na_v1.5$. In tsA201 cells, β2 co-expression had negligible effects on $Na_v1.5$ current function (Dhar Malhotra et al. 2001). However more recent findings have discovered that β2 modulates $Na_v1.5$ channels through a reduction in I_{Na} density and slowed recovery from inactivation (Zimmer and Benndorf 2007). Furthermore a congenital missense mutation (β2D211G) in the β2-subunit has also been shown to reduce I_{Na} density, an effect largely due to impaired trafficking mechanisms (Riuro et al. 2013), in line with a critical role for β2-subunits in normal sinus rhythm. Therefore, studies that assess the molecular mechanisms of how distinct β-subunits modulate cardiac Na channel function are likely to have significant implications for human cardiac diseases by facilitating the development of targeted treatment options in patients.

4 Protein Kinases and Modulation of Cardiac $Na_v1.5$ Channels

4.1 Protein Phosphorylation and Cardiac $Na_v1.5$ Channel Subunits

In the heart, alterations in the normal function of $Na_v1.5$ channels that lead to changes in impaired functional expression of I_{Na} are expected to lead to the generation of life-threatening arrhythmias in humans (Grant 2001). Therefore, the identification and assessment of the molecular mechanisms of cellular factors that modulate cardiac $Na_v1.5$ channel subunits will have important implications for targeted therapeutic interventions. Ion channels are prime candidates for modulation by protein kinases (Weber et al. 2015), suggesting that pathological changes that either decrease and/or increase protein kinase levels within the cell is likely to lead to concomitant functional expression (localization, gating, and trafficking) that may underlie electrical remodeling that predisposes to impaired cardiac electrical function. $Na_v1.5$ channel subunits are regulated by protein kinases (Schreibmayer 1999), including sites for PKA- and PKC-dependent phosphorylation (Marionneau et al. 2012). Several phosphorylation sites (Fig. 1) have been identified within the first intracellular linker loop of $Na_v1.5$ (Marionneau et al. 2012) consistent with direct regulation of $Na_v1.5$ channel function by phosphorylation (Table 1).

The regulation of the $Na_v1.5$ channel by phosphorylation is complex, involving distinct individual or multiple combinations of kinase pathways that may define the pathophysiology of cardiac $Na_v1.5$ channel functional properties. Therefore, PKA/PKC-related Na channel modulation studies are likely to reveal important biophysical insights that will facilitate a clear understanding of the molecular, structural, and functional identity of the cardiac Na channel and how these mechanisms are modified in heterologous expression systems when compared to a native cardiomyocyte milieu. Therefore, an important objective of this review chapter is to discuss the roles and molecular mechanisms of $Na_v1.5$ channel regulation by PKA and PKC signaling pathways.

4.2 PKA-Dependent Phosphorylation and Cardiac $Na_v1.5$ Channel Function

Cyclic adenosine monophosphate (cAMP)-dependent PKA is a serine/threonine protein kinase that plays an important role in cardiac excitation-contraction (E–C) coupling mechanisms (Xiang and Kobilka 2003). Its relevance in heart is further underscored by its critical role as an effector molecule of the ß-adrenergic receptor (βAR) signaling pathway and further underscores its relevance in impaired E–C mechanisms associated with heart failure (Perrino et al. 2007). Increased β-adrenergic stimulation leads to enhanced levels of circulating catecholamines in the blood stream, activation of primarily G protein-coupled β-receptors on the cardiac cell membranes, and the subsequent activation of adenylate cyclase leading to the

Table 1 Summary of the common effects of PKA and PKC on I_{Na} measured in distinct cellular systems

Kinase	Effect on $Na_v1.5$	Effect on I_{Na}	Cell model	Disease	Ref.
PKA	Phosphorylation at S525, S526, and S528 sites within the intracellular linker between domains (D) I–II on $Na_v1.5$	↑ current density ↑ trafficking ↑ rate of V_{inact} Neg. shift in V_{act} and V_{inact}	HEK293 *Xenopus oocytes* Ventricular (mouse) Ventricular (rat) Ventricular (G.pig) Ventricular (rabbit) Ventricular (dog)	LQTS3 BrS	Aiba et al. (2014), Baba et al. (2004), Chandra et al. (1999), Chen et al. (2016), Frohnwieser et al. (1997), Gintant and Liu (1992), Hallaq et al. (2006), Liu et al. (2009), Matsuda et al. (1992), Murphy et al. (1996), Ono et al. (1989, 1993), Schubert et al. (1990), Schreibmayer et al. (1994), Sunami et al. (1991), Tateyama et al. (2003), and Zhou et al. (2000, 2002)
PKC	Phosphorylation at the S1503 site localized to the intracellular loop linking DIII–IV on $Na_v1.5$	↓ peak current density ↓ trafficking ↓ rate of fast V_{inact} Neg. shift in V_{inact}	CHL1610 HEK293 HL-1 *Xenopus oocytes* Ventricular (mouse) Ventricular (rat) Ventricular (G.pig) Ventricular (rabbit) hiPSC-CM	LQTS3 BrS	Tateyama et al. (2003), Hallaq et al. (2012), Liu et al. (2009, 2013, 2017), London et al. (2007), Mathieu et al. (2016), Murray et al. (1997), Nuyens et al. (2001), Qu et al. (1994), Valdivia et al. (2009), and Xiao et al. (2001)

↑ increased, ↓ decreased, *HEK* human embryonic kidney, *CHL1610* Chinese hamster lung 1610, *G.pig* guinea pig, *hiPSC-CM* induced pluripotent stem cell-derived cardiomyocytes, V_{act} voltage-dependence of activation, V_{inact} voltage-dependence of inactivation production of the second messenger cAMP, which in turn activates PKA (Fu and Xiang 2015). Therefore the stimulation of βAR represents a major mechanism to increase heart rate and contractility (Triposkiadis et al. 2009). At the structural level, the PKA holoenzyme exists as a heterotetramer characterized by a regulatory (R1 and R2) subunit dimer and two catalytic (C) subunits (Turnham and Scott 2016). The tetrameric protein complex constitutes the type I and type II PKA holoenzymes whereby the type I PKA is localized in the cytosol and the regulatory subunit of type II PKA binds to an anchor protein (AKAP), an important interaction that has been shown to underlie the preferential targeting of the type II PKA to its substrate proteins in subcellular cardiac compartments (Xiang and Kobilka 2003). Very recently, this well-established PKA activation mechanism has been revised (Smith et al. 2017). It seems that the dissociation of the holoenzyme is not necessary for activation, suggesting that PKA activity is more precisely targeted within the cell than previously thought.

Previous studies using mammalian cells expressing rat cardiac Na channel (Murphy et al. 1996), or human (Frohnwieser et al. 1997) α-subunit, used site-specific antibodies to demonstrate cAMP-dependent PKA regulation of the Na channel. It was demonstrated that two distinct sites (Fig. 1) on rat $Na_v1.5$ localized to serine positions S526 (525 in human) and S529 (528 in human) are phosphorylated by PKA both in vitro and in vivo (Murphy et al. 1996). In subsequent studies it was revealed that PKA-mediated enhancement of I_{Na} density is largely dependent on phosphorylation of S525 and S528 sites and the presence of three endoplasmic reticulum (ER) retention signals on the I–II cytoplasmic linker loop (Frohnwieser et al. 1997; Zhou et al. 2002). Further, the elimination of all three putative ER retention signals with the use of a double phosphomutant (S525A/S528A) abolished PKA effects on I_{Na} (Zhou et al. 2002). This is consistent with a molecular mechanism whereby PKA-dependent phosphorylation of S525 and S528 leads to the activation and/or recruitment of proteins leading to structural changes that allows the masking of the ER retention signals and favors trafficking of functional channels to the membrane. This conclusion is further supported by reports that mutations of five putative PKA sites (including S483/S571/S593) in I–II cytoplasmic linker loop failed to abolish PKA-dependent potentiation of rat I_{Na} (Schreibmayer et al. 1994; Zhou et al. 2002) (Table 1).

Despite these studies showing modulation of $Na_v1.5$ by PKA, there have been contrasting reports on PKA regulation of the cardiac $Na_v1.5$ channel suggesting that Na channel responses to protein kinases may be variable and that $Na_v1.5$ modulation by PKA is not well understood. In Xenopus oocytes expressing the human $Na_v1.5$ channel, the injection of cAMP and its agonists S_p-cAMPs lead to a significant increase in I_{Na} peak density (Frohnwieser et al. 1997). In the same study the authors also observed increased maximal Na conductance with no change in resting membrane potential, channel activation or steady-state inactivation. However further studies in oocytes by Zhou et al. also demonstrated that PKA activation led to increased I_{Na}, and produced hyperpolarizing shifts in the voltage-dependent activation and inactivation of the channel (Zhou et al. 2000). A similar PKA-dependent increase in I_{Na} density and hyperpolarizing shifts have also been reported in oocytes expressing rH1 (Schreibmayer et al. 1994) and in HEK293 cells expressing hH1 channel subunits (Chandra et al. 1999) respectively. In normal epicardial cells and cells isolated from remodeled infarcted canine heart, Baba et al. also reported shifts in the voltage-dependence of channel activation and inactivation in response to PKA activation using a cocktail mixture of activators including 8-chlorophenylthio cAMP, 3-isobutyl-1-methylxanthine, and forskolin (Baba et al. 2004). Furthermore, the investigators found that PKA activation slowed the time course of recovery of I_{Na} from inactivation and facilitated the closed state inactivation of Na channels (Baba et al. 2004).

These effects of PKA were also found to be prevented by agents that disrupt normal surface expression of channel subunits in both oocytes (Zhou et al. 2000) and HEK293 cells stably expressing GFP-tagged $Na_v1.5$ proteins (Hallaq et al. 2006), demonstrating that PKA also altered Na channel trafficking to increase peak I_{Na} density. Baba et al. have also shown that PKA stimulation in normal epicardial cells and a model of infarcted canine heart increased I_{Na} density, possibly by enhancing the trafficking of Na channel subunits to an active site on the cardiac cellular membrane (Baba et al. 2004). Taken together, these observations are in line with reports that protein kinases can regulate the activity of multiple receptors and transporters by modulating intracellular trafficking and the number of functional proteins at the cell surface (Bradbury and Bridges 1994).

Some of the results reported in heterologous expression systems have also been investigated in PKA studies utilizing native cardiomyocytes, albeit with contrasting outcomes. Using the inside-out configuration of the patch-clamp technique in guinea pig ventricular myocytes, Sunami et al. showed that the catalytic subunit of cAMP directly inhibited the Na channel activity in myocytes through decreased peak average I_{Na} and a slowing of the decay kinetics (Sunami et al. 1991) consistent with a direct modulation of Na channel activity through a PKA-dependent phosphorylation process. Furthermore, the authors also demonstrated a decrease in the open probability and an increase in the first opening times while the unitary current amplitude and mean open times were unaffected. However, subsequent studies in cardiomyocytes isolated from canine, rabbit, and guinea pig hearts have shown that while the maximum Na channel conductance or single-channel conductance was unchanged by PKA (Ono et al. 1989), there were shifts in the voltage-dependence of both current activation and inactivation towards hyperpolarized potentials by cAMP and isoproterenol (ISO) (Ono et al. 1989, 1993). Furthermore, in these studies the effects of PKA on Na channel kinetics were distinct and mediated by both phosphorylation dependent and independent processes.

At the macroscopic current level, β-adrenergic regulation of cardiac I_{Na} is also associated with controversial outcomes and further emphasizes the complexity of $Na_v1.5$ modulation by PKA. In an earlier demonstration of β adrenergic stimulation on Na channel function, Ono et al. demonstrated that exposure of guinea pig ventricular myocytes to the β adrenergic agonist ISO, forskolin, or dibutyryl cAMP (dBcAMP) produced a reversible depression of I_{Na} and also caused a hyperpolarizing shift in channel inactivation largely due to increased cAMP levels (Ono et al. 1989). A similar inhibition of I_{Na} with ISO in neonatal rat ventricular myocytes has also been reported (Schubert et al. 1990), and this effect was shown to be more pronounced at depolarized potentials. However, in rabbit ventricular myocytes, Matsuda et al., using whole-cell and inside-out configurations of the patch clamp technique, found that exposure to ISO, forskolin, and dBcAMP enhanced the current density and the rate of channel inactivation of both single and macroscopic I_{Na} through $Na_v1.5$ channels (Matsuda et al. 1992). Additionally, the effects of ISO were not altered with PKA inhibition consistent with both a cAMP-dependent and independent mode of Na channel regulation by ISO (Matsuda et al. 1992). Taken together, the data further underscores the importance of I_{Na} regulation by PKA. However future studies will be required to fully understand and distinguish the role of temporal contribution of PKA-dependent phosphorylation to the altered I_{Na} channel function that underlie Na channelopathies.

4.3 PKC-Dependent Phosphorylation and Cardiac $Na_v1.5$ Function

Protein kinase C (PKC) is another serine-threonine kinase that is involved in cardiac excitability and electrical remodeling in a variety of cardiovascular diseases (Ferreira et al. 2012). PKC has also been shown to exist as a group of at least ten closely related phospholipid-dependent serine-threonine protein kinases (Ferreira et al. 2011). PKC isozymes are classified based on their structure and activation requirements into three distinct subgroups, namely: conventional PKCs (α, βI, βII, and γ), which are activated by both Ca and diacylglycerol (DAG) in addition to phosphatidylserine; novel PKCs (ε, δ, η, and θ) are generally Ca-independent but are activated by DAG and phosphatidylserine; and atypical PKCs (ζ and ι/λ), which are Ca and DAG-independent but activated by phosphatidylserine. PKC isozymes, α, βI, βII, ε, θ, λ, δ, ζ, and γ have been found in the heart (Newton et al. 2016), and high levels of PKC isozymes α, βI, βII, δ, and ε have been reported in tsA201 and HL-1 cell lines (Wetsel et al. 1992). The phosphorylation of serine 1506 (S1506 brain; S1505 rodent; S1503 human) (Qu et al. 1994) in highly conserved loop connecting domains III and IV (DIII-DIV) of the brain Na channel unit has been shown to be important and necessary for PKC-related modulatory effects (Fig. 1 and Table 1).

Despite significant progress understanding $Na_v1.5$ modulation by PKC there are still gaps in understanding the molecular mechanisms that may delineate the distinct role of each PKC isoenzyme. As mentioned above, to date ten different PKC isoforms have been identified in human ventricular myocytes and in different animal species, including members of all three subfamilies (Shin et al. 2000). Therefore, a clear understanding of the functional outcome of multiple combinations of PKC isozymes in the same tissue and/or cell type may have important implications for identification of PKC pathways and for the development of isozyme specific therapeutic targets.

To fully understand the molecular mechanisms that underlie the functional interplay between PKC activation and Na channels, several studies have utilized general PKC activators such as phorbol 12-myristate 13-acetate (PMA) (Dascal and Lotan 1991) and 1-oleoyl-2-acetyl-*sn*-glycerol (OAG) (Qu et al. 1994). They showed that PKC activation by these non-isozyme specific activators leads to a reduction in I_{Na} in both brain and heart (Li et al. 1993; Murray et al. 1997; Numann et al. 1991). Similarly, heterologously expressed rat brain (Dascal and Lotan 1991; Numann et al. 1991) and human (Numann et al. 1991) $Na_v1.5$ demonstrated reduced I_{Na} in response to PKC activation. Although while it is known that both rBIIA and hH1 I_{Na} contain consensus sites for phosphorylation by PKC, most of these sites are not conserved between these two distinct Na channel isozymes. Therefore, in one study (Murray et al. 1997) whereby the conserved consensus PKC sites in the hH1 interdomain III–IV linker, containing the putative PKC site (Ser^{1503}) (Fig. 1) are deleted, it was found that the PMA-induced I_{Na} inhibition was not completely abolished suggesting a role for other potential phosphorylation sites. Nonetheless this explanation does not rule out the regulatory effects of other kinases or phosphatases (Abriel and Kass 2005) on I_{Na} channel function.

To distinguish among the functional outcomes of distinct PKC isoenzymes on Na channels, earlier studies using activators and inhibitors of distinct PKC isoforms have shown that the activation of conventional PKCs produced a depression of I_{Na} density (Shin and Murray 2001). In one of these studies it was demonstrated in Xenopus oocytes that the I_{Na} is inhibited by PMA and a novel peptide inhibitor of the εPKC εV1–2 (Xiao et al. 2001). Furthermore the data revealed that these effects on I_{Na} were mimicked by the peptide specific activator of εPKC, ψε receptor for activated C kinase (ψεRACK) (Xiao et al. 2001), while the peptide specific inhibitor of εPKC, εV1–2 (Xiao et al. 2001) prevented these effects. Similar results were also obtained in native rat ventricular myocytes and therefore taken together; the data provide strong evidence for a role for the εPKC isoform in the regulation of Na channels. Interestingly, βIIPKC and εPKC are important contributors to the molecular mechanisms that underlie cardiac hypertrophy, heart failure (Ferreira et al. 2011), and impaired cardiac excitability (Dorn et al. 1999); therefore, it is likely that studies that utilize βIIPKC and εPKC isozyme-selective inhibitors could provide crucial insights for the development of therapeutic tools useful in the treatment of chronic cardiac diseases.

Additional studies in heterologous expression systems and in native cardiac ventricular myocytes have been used to further elucidate the effects of PKC on Na channel function (Table 1). Utilizing cell attached current recordings in Chinese hamster lung 1,610 cell line stably expressing $Na_v1.5$ channel subunits (SNa-rH1), Qu et al. (1994) demonstrated that the PKC activator OAG (10 μM) caused a voltage-dependent decrease in I_{Na} and produced a 15 mV hyperpolarizing shift in the channel SSI with no obvious changes in I_{Na} channel activation (Qu et al. 1994). Similar effects on whole-cell I_{Na} inactivation have also been described in neonatal rat ventricular cardiac myocytes treated with a peptide corresponding to the catalytic subunit of PKC (Watson and Gold 1997). PKC activation with OAG also decreased the open probability of the channel while the channel conductance remained unchanged. These effects were reversed with a PKC inhibitor peptide or nonphosphorylatable mimetic, S1505A (Qu et al. 1994). Subsequent studies have also shown that the phosphorylation of S1505 (rodent; S1503) in the $Na_v1.5$ DIII-DIV inactivation loop (Hallaq et al. 2012), activation of the conventional Ca-sensitive PKC isoforms (Shin et al. 2000), in addition to trafficking defects (Hallaq et al. 2012; Shin et al. 2000), have all been proposed to contribute prominently to the depression in I_{Na} density seen with PKC activation. In another study (Murray et al. 1997) that utilized Xenopus oocytes expressing recombinant $Na_v1.5$ channel subunits (hH1), Murray et al. measured reduced I_{Na} density in response to PKC activation but there was no effect on either current decay or the voltage-dependence of channel gating. Furthermore, when the putative conserved sites in the III–IV interdomain linker were mutated, there continued to be a reduction in I_{Na} consistent with the presence of additional phosphorylation sites.

There have also been other biophysical and biological effects due to PKC activation in ventricular myocytes. In neonatal ventricular myocytes, Watson et al. demonstrated that perfusion with a peptide patterned after the catalytic subunit of PKC altered I_{Na} gating and produced a depolarizing shift in the steady-state half inactivation while the current-voltage relationship and peak I_{Na} density remained unaltered (Watson and Gold 1997). In rat ventricular myocytes, inhibition of PKC with bisindolylmaleimide prevented the effects of hydrogen peroxide on I_{Na} (Ward and Giles 1997). In rat neonatal myocytes and HEK293 cells stably expressing $Na_v1.5$, it has also been demonstrated that elevated levels of pyridine nucleotide nicotinamide adenine dinucleotide (NADH) caused a rapid decrease in I_{Na} density and that this effect is reversed by PKC inhibition (Liu et al. 2009). In another study by the same authors, it was further shown that reduction of I_{Na} by NADH is mediated through the activation of PKCδ in mouse cardiomyocytes and the effect is reversed by mutating the S1503 site (Liu et al. 2017). Prevention of PKCα isoenzyme translocation with the inhibitory peptide αV5-3 also prevented angiotensin II induced reduction in I_{Na} mouse and human-induced pluripotent stem cell-derived cardiomyocytes (Mathieu et al. 2016). Taken together these observations suggest that distinct PKC pathways may be an important therapeutic target to prevent myocardial dysfunction and/or cardiac arrhythmias associated with impaired I_{Na} functional expression. Further studies that will identify the specific targets that mediate the anti-arrhythmic PKC-mediated effects are warranted.

5 Protein Kinases and Arrhythmias

5.1 Channelopathies of the $Na_v1.5$ Channel Subunits

A number of inherited and/or congenital mutations that either lead to increased or reduced $Na_v1.5$ channel function have been mapped to the entire sequence of the gene *SCN5A*-encoded $Na_v1.5$ channel (Kapplinger et al. 2015). These include cardiac arrhythmias such as LQTS3 (Amin et al. 2013), and BrS1 (Ruan et al. 2009), serious conditions that predisposes patients to an higher risk for syncope, seizures, ventricular arrhythmias, and sudden cardiac death (Wilde et al. 2016). LQT incidence in the population worldwide is about 1:2,000 (Schwartz et al. 2009), gain-of-function mutations that lead to LQTS3 constitute 10–15% of LQTS (Crotti et al. 2008). The incidence of BrS in the general population is currently estimated at 1:2,000 (Nielsen et al. 2013) and loss-of-function mutations in *SCNA5* are associated with 20–30% of the reported cases of BrS (Crotti et al. 2012). Previous molecular and electrophysiological studies that elucidated the pathophysiology of $Na_v1.5$ have provided significant knowledge about mechanisms that underlie its functional expression (Wilde and Brugada 2011). These advances have led to improved genotype–phenotype correlations for better risk stratification approaches and treatment options for patients that harbor *SCN5A*-assocociated diseases (Wilde et al. 2016).

5.2 Kinase Regulation of Cardiac $Na_v1.5$ in Long QT Syndrome 3

To date over 80 LQTS3-causing *SCN5A* mutations have been identified in patients and ~50% of them have been studied in heterologous expression systems (Wilde and Brugada 2011). Most of these mutations have previously been identified as gain-of-function missense mutations (Bennett et al. 1995), which disrupt fast inactivation leading to the development of an abnormal sustained non-inactivating late I_{Na} during the plateau phase of the AP (Clancy and Rudy 1999) and the subsequent prolongation of the APD. Therefore, an increased I_{Na} density in LQTS3 could in principle occur either through (1) increased expression of mutant $Na_v1.5$ subunits as a result of enhanced mRNA translation or protein trafficking to the sarcolemma; (2) decreased protein degradation; (3) altered modulation by the regulatory β-subunits proteins; (4) multiple combinations of all three. In this context a slower time course of inactivation (Ruan et al. 2007), faster recovery from inactivation (Clancy et al. 2003), impaired trafficking (Moreau et al. 2012), in addition to altered gating properties (Moreau et al. 2012) have all been demonstrated as major molecular mechanisms that contribute to the pathogenesis of distinct LQT3 mutations.

5.3 PKA and Channelopathies of the $Na_v1.5$ Channel Complexes

The contribution of PKA-dependent phosphorylation of $Na_v1.5$ channel subunits has been demonstrated in the context of disease causing *SCN5A* mutations with inconsistent reports. In one of these studies Chandra et al. demonstrated that compared to normal subunits PKA stimulation had no effect on the bursting properties of I_{Na} measured in HEK293 cells expressing the LQTS-associated $Na_v1.5$ mutant ΔKPQ lacking a consensus site within DIII–DIV involved in PKA-mediated phosphorylation of the channel (Chandra et al. 1999). Also in this study, PKA activation with ISO and 8-bromo-cAMP also produced hyperpolarizing shifts in the voltage-dependence of channel activation and inactivation. In a similar study that utilized mice heterozygous ($SCN5A^{\Delta/-}$) for a knock-in KPQ-deletion of amino-acids 1,505–1,509, Nuyens et al. showed that mice displayed increased arrhythmogenetic events typical of the *Torsade de pointes* in LQTS3 patients

(Ishibashi et al. 2017), in addition to increased ventricular I_{Na} and $I_{Na,L}$ densities (Nuyens et al. 2001). However, the investigators did not investigate the role of PKA-mediated mechanisms including phosphorylation in these studies. In a subsequent study the Kass group also showed that the LQTS3 causing C-terminus D1790G mutation enhanced bursting activity of $I_{Na,L}$ through PKA-dependent phosphorylation at both the Ser^{36} in the N-terminus and Ser^{525} within the cytoplasmic linker connecting DI–DII (Tateyama et al. 2003).

More recently, Chen et al. described an LQTS3 heterozygous missense mutation V2016M with both a loss- and a gain-of-function feature (Chen et al. 2016). The authors demonstrated that heterologously expressed V2016M mutant decreased I_{Na} density, but increased $I_{Na,L}$ through a PKA-dependent activation and impaired inactivation of $Na_v1.5$ channels through both PKA- and PKC-dependent signaling pathways (Chen et al. 2016). The functional outcome of the V2016M mutation further emphasizes the complexity of the molecular mechanisms of distinct LQTS3 mutations and modulation by protein kinases. Furthermore, a BrS-associated *SCN5A* mutation (R526H) located in the S528 PKA consensus phosphorylation site has also been shown to lead to reduced basal I_{Na} density, reduced $Na_v1.5$ channel subunit surface expression, and impaired response to β-adrenergic-dependent PKA stimulation (Aiba et al. 2014).

5.4 PKC and Channelopathies of the $Na_v1.5$ Channel Complexes

Fewer studies have been aimed at elucidating the functional effects of PKC-mediated $Na_v1.5$ phosphorylation in the context of LQTS3 mutations and the gating properties of I_{Na}. LQTS3 mutations promote a burst mode of Na channel gating (Bennett et al. 1995; Numann et al. 1991) whereby a fraction of channels fail to inactivate causing a sustained $I_{Na,L}$ which can delay cardiac repolarization and predispose patients to QT prolongation on the surface ECG (Wilde and Brugada 2011). In one such study Tateyama et al. (2003) found that exposure of HEK293 cells expressing LQTS3-associated *SCN5A* mutations (Y1795C, Y1795H, and ΔKPQ) to the PKC activator OAG reduced $I_{Na,L}$ generated by mutant subunits when compared to wild-type subunits. The effect of OAG on $I_{Na,L}$ was reduced by the PKC inhibitor staurosporine but was prevented by the S1503A mutation and mimicked by the S1503D mutant (Tateyama et al. 2003) suggesting that activation of PKC may be an important contributor to a reduced risk of fatal arrhythmic events in heterozygous LQTS3 patients during elevated sympathetic nerve activity (Chinushi et al. 2003).

More evidence for PKC-mediated phosphorylation of cardiac $Na_v1.5$ channels has also been provided in the context of mutations in the GPD1L gene, which encodes the protein Glycerol 3-Phosphate Dehydrogenase 1-Like protein, that was discovered in BrS and Sudden Infant Death Syndrome (Van Norstrand et al. 2007). Initially GPD1L-associated mutant (A280V) studies from the Dudley group (London et al. 2007) demonstrated in HEK293 cells that the A280V mutant reduced I_{Na} density and the surface expression of *SCN5A* channel subunits (London et al.

2007). In a separate study it was also shown that the expression in neonatal mouse myocytes of adenoviral vectors coding for another GPD1L-associated mutant (E83K) also produced a marked decrease in I_{Na} density (Van Norstrand et al. 2007). In subsequent GPD1L studies in HEK293 cells, the effects of mutants were shown to be dependent on direct PKC phosphorylation of the S1503 site on the $Na_v1.5$ channel subunit (Valdivia et al. 2009) because the exposure of GPD1L and $Na_v1.5$ expressing cells to the PKC activator OAG or G3P (a GPD1 substrate) also caused a reduction in I_{Na} density which is reversed by the PKC inhibitor staurosporine or by mutation of S1503 site to a nonphosphorylatable alanine (Valdivia et al. 2009).

Similar PKC-dependent effects on I_{Na} functional expression independent of GPD1L mutations have also been described. Hallaq et al. (2012) monitored the movement of green fluorescent protein (GFP)-tagged $Na_v1.5$ in HEK293 cells compared with hemagglutinin-tagged $Na_v1.5$ immobilized at the membrane using confocal microscopy (Hallaq et al. 2012). They demonstrated that activation of PKC decreased surface expression and this effect was prevented by either PKC inhibition, mutation of S1503 to alanine, or reactive oxygen species inhibition. Fluorescence recovery after photobleach studies further identified that channels have decreased mobility within the membrane with PKC stimulation and conversely, an increase with PKA stimulation. Peak I_{Na} was also decreased at all test potentials after a 30-min exposure to the PKC activator phorbol 12-myristate 13-acetate (Hallaq et al. 2012).

Finally, studies by Liu et al. have further suggested that I_{Na} might be regulated directly by pyridine nucleotides, such as NADH (Liu et al. 2009). They found a twofold increase in intracellular NADH concentration in HEK293 cells stably expressing $Na_v1.5$ and infected with adenoviral vectors coding for the A280V GPD1L mutant. Moreover, there was also a dose-dependent decrease in I_{Na} with intracellular application of NADH delivered by patch pipette (Liu et al. 2009). The NADH-induced I_{Na} decrease required both PKC and ROS, and ROS was downstream of PKC because the antioxidant superoxide dismutase prevented the downregulation of current in the presence of a PKC activator, indicating that the $Na_v1.5$ channel may be directly oxidized. In contrast to Valdivia et al. (2009), Liu et al. did not see a change in $Na_v1.5$ surface expression in biotinylation and Western blot assays of cell surface $Na_v1.5$ protein or confocal experiments of $Na_v1.5$-GFP at the membrane, upon acute (2–10 min) treatment with pyruvate/lactate (to increase NADH) or phorbol 12-myristate 13-acetate (to activate PKC) (Liu et al. 2009); or investigate the role of direct PKC-dependent phosphorylation. Additionally, the findings of Liu et al. (2009) were confirmed in a mouse model of non-ischemic cardiomyopathy whereby elevated NADH levels, PKC activation, and mitochondrial ROS overproduction and a concomitant decrease in I_{Na} were observed (Liu et al. 2013). To date there are still controversies regarding the precise role of a direct phosphorylation of $Na_v1.5$ by PKC. Nonetheless, these findings suggest that the regulation of $Na_v1.5$ channels by GPD1L, NADH, and/or PKC may be an important contributor to impaired membrane excitability that underlies vulnerability to fatal arrhythmogenic episodes.

5.5 Kinase Regulation of Cardiac $Na_v1.5$ in Brugada Syndrome (BrS)

More than 100 *SCN5A* mutations are associated with BrS and there is a significant amount of evidence that BrS-associated *SCN5A* mutations cause a loss of Na channel function leading to impaired cardiac cell depolarization and excitability (Liu et al. 2017). Proposed molecular mechanisms for decreased I_{Na} in BrS include: (1) decreased expression of $Na_v1.5$ proteins (Valdivia et al. 2004) which may be due to failure of mutant channels to interact with the auxiliary β-subunits or regulatory proteins essential for normal localization of assembled channel complexes; (2) increased surface expression of non-functional $Na_v1.5$ channel subunits (Kyndt et al. 2001); (3) altered gating properties which would manifest as shifts in the voltage-dependence of activation and/or inactivation (Amin et al. 2005). Furthermore congenital mutations in the cardiac Na channel auxiliary β-subunits have also been described in BrS patients (Peeters et al. 2015). Watanabe et al. in their studies assessed the functional impact of the co-expression of mutants that lead to a truncated β1-subunit lacking the transmembrane and cytoplasmic domains on I_{Na} density (Watanabe et al. 2008). It was demonstrated that mutant β1 failed to associate with $Na_v1.5$ and therefore prevents the increase in I_{Na} normally seen with co-expression of wild-type β1-subunit. Other studies have shown that a BrS missense mutation in the *SCN3B* gene encoding β3 was found to reduce I_{Na} through both trafficking and gating defects (Hu et al. 2012).

Despite the increasing evidence for an important role for PKA- and PKC-dependent modulation of Na channel subunits in multiple tissues (Schreibmayer 1999), there is a paucity of studies that have assessed the impact of PKA and PKC on the pathophysiology of the cardiac Na channel. More specifically how and whether the activation of PKA and PKC alter the functional outcomes of LQTS3- and/or BrS-causing mutations is poorly understood.

6 Summary and Future Perspectives

The modulation of voltage-dependent cardiac Na channels by protein kinases is an emerging field that has provided novel mechanistic insights into ion channel modulation by phosphorylation-dependent and independent processes. In this review, we describe the complexity of cardiac $Na_v1.5$ channel responses emanating from distinct modulation by PKA and PKC. Current mechanistic insights (Schreibmayer 1999) demonstrate that PKA and PKC play critical roles in the altered function of $Na_v1.5$, which ultimately may have important implications for the initiation and propagation of the AP and thus cardiac excitability. For example, chronic phosphorylation by either PKA or PKC alters channel surface levels, but in antagonistic ways. In the case of PKA, the increase in myocardial conduction velocity associated with sympathetic stimulation implicates an increase in channel density at the membrane. In contrast, PKC decreases the number of functional channels at the membrane, and this may be particularly relevant for $Na_v1.5$ channel remodeling as in BrS.

The lack of progress in this area may be due to the fact that the molecular basis of regulation is controlled by a balance between distinct protein kinases and phosphatases. Furthermore, the cardiac $Na_v1.5$ channel subunits have several phosphorylation sites for different kinases, in line with the existence of a complex functional interplay that may compound the ability to distinguish the effects of distinct kinases. Therefore, to understand how Na channels are modulated by protein kinases, it is important to determine the molecular mechanisms that underlie altered levels of basal phosphorylation by PKA and PKC. It is equally important to determine how a given protein kinase activator modifies the fine balance between protein kinases, their isoforms, and distinct phosphatases. Therefore without these information, it is challenging to fully understand the functional outcome of the activation and/or inhibition of PKC and PKA on the pathophysiology of cardiac $Na_v1.5$ channels. However the feasibility of using isozyme-selective inhibitors and activators has been shown to distinguish the role of specific PKC isozymes in the regulation of Ca, Na, and K channels (Ferreira et al. 2012) under different pathological conditions such as ischemia, cardiac dysfunction, and heart failure. This knowledge is likely to inform on specific PKC isozymes as therapeutic targets for the treatment cardiovascular diseases including LQTS3 and BrS.

To date there has been enormous progress toward understanding the molecular basis of Na channel structure and function in heterologous expression systems that allow channels to be studied in isolation. There is no doubt that the data obtained have contributed significantly to our understanding of ion channel function and more specifically Na channel function. Unfortunately, heterologous systems (HEK293, CHO, Xenopus oocytes) are devoid of important cell-specific regulatory and structural interacting proteins that are critical to the modulation of cellular excitability in vivo. This is in addition to the unique cytoarchitecture of a cardiac cell and its subcellular compartments (t-tubules, intercalated discs, sarcolemma) demonstrate the critical importance of studying channels in their native cellular environments. Tissue-specific gene targeting experiments in whole animals, especially those utilizing CRISPR-9 technology (Srivastava et al. 2016), that are now widely used to investigate the role of ion channel complexes could also now be effectively applied to understanding protein kinase studies. The data is likely to reveal processes that will be better predictive of disease mechanisms. To truly understand the molecular mechanisms of $Na_v1.5$ phosphorylation by PKA and PKC in pathophysiological conditions a pre-clinical human model would be ideal. Although human-induced pluripotent stem cells derived from patients are now widely used to address fundamental questions directly related to human health, there are still significant limitations, and challenges associated with this approach (Moreau et al. 2017). However, as the iPSC technique is further refined and developed, it will provide additional, critical tools for future development of innovative and effective therapies to treat cardiac arrhythmias (Moreau et al. 2017). Nonetheless current PKA and PKC studies highlight the important role post-translational modifications play in altered trafficking and gating properties of $Na_v1.5$ and inform the scientific community that much work remains to fully unravel the complexities of post-translational regulation of $Na_v1.5$ and its role in cardiac physiology and disease.

References

Abriel H, Kass RS (2005) Regulation of the voltage-gated cardiac sodium channel $Na_v1.5$ by interacting proteins. Trends Cardiovasc Med 15:35–40

Abriel H, Rougier JS, Jalife J (2015) Ion channel macromolecular complexes in cardiomyocytes: roles in sudden cardiac death. Circ Res 116:1971–1988

Aiba T, Farinelli F, Kostecki G, Hesketh GG, Edwards D, Biswas S, Tung L, Tomaselli GF (2014) A mutation causing Brugada syndrome identifies a mechanism for altered autonomic and oxidant regulation of cardiac sodium currents. Circ Cardiovasc Genet 7:249–256

Amin AS, Verkerk AO, Bhuiyan ZA, Wilde AA, Tan HL (2005) Novel Brugada syndrome-causing mutation in ion-conducting pore of cardiac Na^+ channel does not affect ion selectivity properties. Acta Physiol Scand 185:291–301

Amin AS, Pinto YM, Wilde AA (2013) Long QT syndrome: beyond the causal mutation. J Physiol 591:4125–4139

Baba S, Dun W, Boyden PA (2004) Can PKA activators rescue Na+ channel function in epicardial border zone cells that survive in the infarcted canine heart? Cardiovasc Res 64:260–267

Bennett PB, Yazawa K, Makita N, George AL Jr (1995) Molecular mechanism for an inherited cardiac arrhythmia. Nature 376:683–685

Boutjdir M, Restivo M, Wei Y, Stergiopoulos K, el-Sherif N (1994) Early afterdepolarization formation in cardiac myocytes: analysis of phase plane patterns, action potential, and membrane currents. J Cardiovasc Electrophysiol 5:609–620

Bradbury NA, Bridges RJ (1994) Role of membrane trafficking in plasma membrane solute transport. Am J Phys 267:C1–24

Chadda KR, Jeevaratnam K, Lei M, Huang CL (2017) Sodium channel biophysics, late sodium current and genetic arrhythmic syndromes. Pflugers Arch 469:629–641

Chahine M, O'Leary ME (2014) Regulation/modulation of sensory neuron sodium channels. Handb Exp Pharmacol 221:111–135

Chandra R, Chauhan VS, Starmer CF, Grant AO (1999) Beta-adrenergic action on wild-type and KPQ mutant human cardiac Na^+ channels: shift in gating but no change in Ca^{2+}:Na^+ selectivity. Cardiovasc Res 42:490–502

Chen J, Makiyama T, Wuriyanghai Y, Ohno S, Sasaki K, Hayano M, Harita T, Nishiuchi S, Yuta Y, Ueyama T, Shimizu A, Horie M, Kimura T (2016) Cardiac sodium channel mutation associated with epinephrine-induced QT prolongation and sinus node dysfunction. Heart Rhythm 13:289–298

Chen-Izu Y, Shaw RM, Pitt GS, Yarov-Yarovoy V, Sack JT, Abriel H, Aldrich RW, Belardinelli L, Cannell MB, Catterall WA, Chazin WJ, Chiamvimonvat N, Deschenes I, Grandi E, Hund TJ, Izu LT, Maier LS, Maltsev VA, Marionneau C, Mohler PJ, Rajamani S, Rasmusson RL, Sobie EA, Clancy CE, Bers DM (2015) Na^+ channel function, regulation, structure, trafficking and sequestration. J Physiol 593:1347–1360

Chinushi M, Tagawa M, Sugiura H, Komura S, Hosaka Y, Washizuka T, Aizawa Y (2003) Ventricular tachyarrhythmias in a canine model of LQT3: arrhythmogenic effects of sympathetic activity and therapeutic effects of mexiletine. Circ J 67:263–268

Clancy CE, Rudy Y (1999) Linking a genetic defect to its cellular phenotype in a cardiac arrhythmia. Nature 400:566–569

Clancy CE, Tateyama M, Kass RS (2002) Insights into the molecular mechanisms of bradycardia-triggered arrhythmias in long QT-3 syndrome. J Clin Invest 110:1251–1262

Clancy CE, Kurokawa J, Tateyama M, Wehrens XH, Kass RS (2003) K^+ channel structure-activity relationships and mechanisms of drug-induced QT prolongation. Annu Rev Pharmacol Toxicol 43:441–461

Crotti L, Celano G, Dagradi F, Schwartz PJ (2008) Congenital long QT syndrome. Orphanet J Rare Dis 3:18

Crotti L, Marcou CA, Tester DJ, Castelletti S, Giudicessi JR, Torchio M, Medeiros-Domingo A, Simone S, Will ML, Dagradi F, Schwartz PJ, Ackerman MJ (2012) Spectrum and prevalence

of mutations involving BrS1- through BrS12-susceptibility genes in a cohort of unrelated patients referred for Brugada syndrome genetic testing: implications for genetic testing. J Am Coll Cardiol 60:1410–1418
Cusdin FS, Clare JJ, Jackson AP (2008) Trafficking and cellular distribution of voltage-gated sodium channels. Traffic 9:17–26
Dai S, Hall DD, Hell JW (2009) Supramolecular assemblies and localized regulation of voltage-gated ion channels. Physiol Rev 89:411–452
Dascal N, Lotan I (1991) Activation of protein kinase C alters voltage dependence of a Na^+ channel. Neuron 6:165–175
Dhar Malhotra J, Chen C, Rivolta I, Abriel H, Malhotra R, Mattei LN, Brosius FC, Kass RS, Isom LL (2001) Characterization of sodium channel alpha- and beta-subunits in rat and mouse cardiac myocytes. Circulation 103:1303–1310
Dorn GW 2nd, Souroujon MC, Liron T, Chen CH, Gray MO, Zhou HZ, Csukai M, Wu G, Lorenz JN, Mochly-Rosen D (1999) Sustained in vivo cardiac protection by a rationally designed peptide that causes epsilon protein kinase C translocation. Proc Natl Acad Sci U S A 96:12798–12803
Dulsat G, Palomeras S, Cortada E, Riuro H, Brugada R, Verges M (2017) Trafficking and localization to the plasma membrane of $Na_v1.5$ promoted by the beta2 subunit is defective due to a beta2 mutation associated with Brugada syndrome. Biol Cell 109:273–291
el-Sherif N, Fozzard HA, Hanck DA (1992) Dose-dependent modulation of the cardiac sodium channel by sea anemone toxin ATXII. Circ Res 70:285–301
Fahmi AI, Patel M, Stevens EB, Fowden AL, John JE 3rd, Lee K, Pinnock R, Morgan K, Jackson AP, Vandenberg JI (2001) The sodium channel beta-subunit SCN3b modulates the kinetics of SCN5a and is expressed heterogeneously in sheep heart. J Physiol 537:693–700
Ferreira JC, Brum PC, Mochly-Rosen D (2011) BetaIIPKC and epsilonPKC isozymes as potential pharmacological targets in cardiac hypertrophy and heart failure. J Mol Cell Cardiol 51:479–484
Ferreira JC, Mochly-Rosen D, Boutjdir M (2012) Regulation of cardiac excitability by protein kinase C isozymes. Front Biosci (Schol Ed) 4:532–546
Frohnwieser B, Chen LQ, Schreibmayer W, Kallen RG (1997) Modulation of the human cardiac sodium channel alpha-subunit by cAMP-dependent protein kinase and the responsible sequence domain. J Physiol 498(2):309–318
Fu Q, Xiang YK (2015) Trafficking of beta-adrenergic receptors: implications in intracellular receptor signaling. Prog Mol Biol Transl Sci 132:151–188
Gawali VS, Todt H (2016) Mechanism of inactivation in voltage-gated $Na^{(+)}$ channels. Curr Top Membr 78:409–450
Gintant GA, Liu DW (1992) Beta-adrenergic modulation of fast inward sodium current in canine myocardium. Syncytial preparations versus isolated myocytes. Circ Res 70(4):844–850
Grant AO (2001) Molecular biology of sodium channels and their role in cardiac arrhythmias. Am J Med 110:296–305
Hallaq H, Yang Z, Viswanathan PC, Fukuda K, Shen W, Wang DW, Wells KS, Zhou J, Yi J, Murray KT (2006) Quantitation of protein kinase A-mediated trafficking of cardiac sodium channels in living cells. Cardiovasc Res 72:250–261
Hallaq H, Wang DW, Kunic JD, George AL Jr, Wells KS, Murray KT (2012) Activation of protein kinase C alters the intracellular distribution and mobility of cardiac Na^+ channels. Am J Physiol Heart Circ Physiol 302:H782–H789
Hu D, Barajas-Martinez H, Medeiros-Domingo A, Crotti L, Veltmann C, Schimpf R, Urrutia J, Alday A, Casis O, Pfeiffer R, Burashnikov E, Caceres G, Tester DJ, Wolpert C, Borggrefe M, Schwartz P, Ackerman MJ, Antzelevitch C (2012) A novel rare variant in SCN1Bb linked to Brugada syndrome and SIDS by combined modulation of $Na_{(v)}1.5$ and $K_{(v)}4.3$ channel currents. Heart Rhythm 9:760–769
Ishibashi K, Aiba T, Kamiya C, Miyazaki A, Sakaguchi H, Wada M, Nakajima I, Miyamoto K, Okamura H, Noda T, Yamauchi T, Itoh H, Ohno S, Motomura H, Ogawa Y, Goto H, Minami T,

Yagihara N, Watanabe H, Hasegawa K, Terasawa A, Mikami H, Ogino K, Nakano Y, Imashiro S, Fukushima Y, Tsuzuki Y, Asakura K, Yoshimatsu J, Shiraishi I, Kamakura S, Miyamoto Y, Yasuda S, Akasaka T, Horie M, Shimizu W, Kusano K (2017) Arrhythmia risk and beta-blocker therapy in pregnant women with long QT syndrome. Heart 103(17):1374–1379

Isom LL, De Jongh KS, Patton DE, Reber BF, Offord J, Charbonneau H, Walsh K, Goldin AL, Catterall WA (1992) Primary structure and functional expression of the beta1 subunit of the rat brain sodium channel. Science 256:839–842

Isom LL, Ragsdale DS, De Jongh KS, Westenbroek RE, Reber BF, Scheuer T, Catterall WA (1995) Structure and function of the beta2 subunit of brain sodium channels, a transmembrane glycoprotein with a CAM motif. Cell 83:433–442

Kapplinger JD, Giudicessi JR, Ye D, Tester DJ, Callis TE, Valdivia CR, Makielski JC, Wilde AA, Ackerman MJ (2015) Enhanced classification of Brugada syndrome-associated and long-QT syndrome-associated genetic variants in the SCN5A-encoded $Na_{(v)}1.5$ cardiac sodium channel. Circ Cardiovasc Genet 8:582–595

Khaliulin I, Bond M, James AF, Dyar Z, Amini R, Johnson JL, Suleiman MS (2017) Functional and cardioprotective effects of simultaneous and individual activation of protein kinase A and Epac. Br J Pharmacol 174:438–453

Klaver EC, Versluijs GM, Wilders R (2011) Cardiac ion channel mutations in the sudden infant death syndrome. Int J Cardiol 152:162–170

Ko SH, Lenkowski PW, Lee HC, Mounsey JP, Patel MK (2005) Modulation of $Na_{(v)}1.5$ by beta1- and beta3-subunit co-expression in mammalian cells. Pflugers Arch 449:403–412

Kruger LC, Isom LL (2016) Voltage-gated Na^+ channels: not just for conduction. Cold Spring Harb Perspect Biol 8

Kyndt F, Probst V, Potet F, Demolombe S, Chevallier JC, Baro I, Moisan JP, Boisseau P, Schott JJ, Escande D, Le Marec H (2001) Novel SCN5A mutation leading either to isolated cardiac conduction defect or Brugada syndrome in a large French family. Circulation 104:3081–3086

Li M, West JW, Numann R, Murphy BJ, Scheuer T, Catterall WA (1993) Convergent regulation of sodium channels by protein kinase C and cAMP-dependent protein kinase. Science 261:1439–1442

Liu M, Sanyal S, Gao G, Gurung IS, Zhu X, Gaconnet G, Kerchner LJ, Shang LL, Huang CL, Grace A, London B, Dudley SC Jr (2009) Cardiac Na^+ current regulation by pyridine nucleotides. Circ Res 105:737–745

Liu M, Gu L, Sulkin MS, Liu H, Jeong EM, Greener I, Xie A, Efimov IR, Dudley SC Jr (2013) Mitochondrial dysfunction causing cardiac sodium channel downregulation in cardiomyopathy. J Mol Cell Cardiol 54:25–34

Liu M, Shi G, Yang KC, Gu L, Kanthasamy AG, Anantharam V, Dudley SC Jr (2017) Role of protein kinase C in metabolic regulation of the cardiac Na^+ channel. Heart Rhythm 14:440–447

London B, Michalec M, Mehdi H, Zhu X, Kerchner L, Sanyal S, Viswanathan PC, Pfahnl AE, Shang LL, Madhusudanan M, Baty CJ, Lagana S, Aleong R, Gutmann R, Ackerman MJ, McNamara DM, Weiss R, Dudley SC Jr (2007) Mutation in glycerol-3-phosphate dehydrogenase 1 like gene (GPD1-L) decreases cardiac Na^+ current and causes inherited arrhythmias. Circulation 116:2260–2268

Lopez-Santiago LF, Meadows LS, Ernst SJ, Chen C, Malhotra JD, McEwen DP, Speelman A, Noebels JL, Maier SK, Lopatin AN, Isom LL (2007) Sodium channel Scn1b null mice exhibit prolonged QT and RR intervals. J Mol Cell Cardiol 43:636–647

Loussouarn G, Sternberg D, Nicole S, Marionneau C, Le Bouffant F, Toumaniantz G, Barc J, Malak OA, Fressart V, Pereon Y, Baro I, Charpentier F (2015) Physiological and pathophysiological insights of $Na_v1.4$ and $Na_v1.5$ comparison. Front Pharmacol 6:314

Luo L, Ning F, Du Y, Song B, Yang D, Salvage SC, Wang Y, Fraser JA, Zhang S, Ma A, Wang T (2017) Calcium-dependent Nedd4-2 upregulation mediates degradation of the cardiac sodium channel $Na_v1.5$: implications for heart failure. Acta Physiol (Oxf) 221(1):44–58

Marionneau C, Lichti CF, Lindenbaum P, Charpentier F, Nerbonne JM, Townsend RR, Merot J (2012) Mass spectrometry-based identification of native cardiac $Na_v1.5$ channel alpha subunit phosphorylation sites. J Proteome Res 11:5994–6007
Mathieu S, El Khoury N, Rivard K, Gelinas R, Goyette P, Paradis P, Nemer M, Fiset C (2016) Reduction in $Na^{(+)}$ current by angiotensin II is mediated by PKCalpha in mouse and human-induced pluripotent stem cell-derived cardiomyocytes. Heart Rhythm 13:1346–1354
Matsuda JJ, Lee H, Shibata EF (1992) Enhancement of rabbit cardiac sodium channels by beta-adrenergic stimulation. Circ Res 70:199–207
Milstein ML, Musa H, Balbuena DP, Anumonwo JM, Auerbach DS, Furspan PB, Hou L, Hu B, Schumacher SM, Vaidyanathan R, Martens JR, Jalife J (2012) Dynamic reciprocity of sodium and potassium channel expression in a macromolecular complex controls cardiac excitability and arrhythmia. Proc Natl Acad Sci U S A 109:E2134–E2143
Moreau A, Keller DI, Huang H, Fressart V, Schmied C, Timour Q, Chahine M (2012) Mexiletine differentially restores the trafficking defects caused by two brugada syndrome mutations. Front Pharmacol 3:62
Moreau A, Boutjdir M, Chahine M (2017) Induced pluripotent stem cell-derived cardiomyocytes: cardiac applications, opportunities and challenges. Can J Physiol Pharmacol 28:1–9
Morgan K, Stevens EB, Shah B, Cox PJ, Dixon AK, Lee K, Pinnock RD, Hughes J, Richardson PJ, Mizuguchi K, Jackson AP (2000) Beta3: an additional auxiliary subunit of the voltage-sensitive sodium channel that modulates channel gating with distinct kinetics. Proc Natl Acad Sci U S A 97:2308–2313
Motoike HK, Liu H, Glaaser IW, Yang AS, Tateyama M, Kass RS (2004) The Na^+ channel inactivation gate is a molecular complex: a novel role of the COOH-terminal domain. J Gen Physiol 123:155–165
Murphy BJ, Rogers J, Perdichizzi AP, Colvin AA, Catterall WA (1996) cAMP-dependent phosphorylation of two sites in the alpha subunit of the cardiac sodium channel. J Biol Chem 271:28837–28843
Murray KT, Hu NN, Daw JR, Shin HG, Watson MT, Mashburn AB, George AL Jr (1997) Functional effects of protein kinase C activation on the human cardiac Na^+ channel. Circ Res 80:370–376
Newton AC, Antal CE, Steinberg SF (2016) Protein kinase C mechanisms that contribute to cardiac remodelling. Clin Sci (Lond) 130:1499–1510
Nielsen MW, Holst AG, Olesen SP, Olesen MS (2013) The genetic component of Brugada syndrome. Front Physiol 4:179
Numann R, Catterall WA, Scheuer T (1991) Functional modulation of brain sodium channels by protein kinase C phosphorylation. Science 254:115–118
Nuyens D, Stengl M, Dugarmaa S, Rossenbacker T, Compernolle V, Rudy Y, Smits JF, Flameng W, Clancy CE, Moons L, Vos MA, Dewerchin M, Benndorf K, Collen D, Carmeliet E, Carmeliet P (2001) Abrupt rate accelerations or premature beats cause life-threatening arrhythmias in mice with long-QT3 syndrome. Nat Med 7:1021–1027
Ono K, Kiyosue T, Arita M (1989) Isoproterenol, DBcAMP, and forskolin inhibit cardiac sodium current. Am J Phys 256:C1131–C1137
Ono K, Fozzard HA, Hanck DA (1993) Mechanism of cAMP-dependent modulation of cardiac sodium channel current kinetics. Circ Res 72:807–815
Peeters U, Scornik F, Riuro H, Perez G, Komurcu-Bayrak E, Van Malderen S, Pappaert G, Tarradas A, Pagans S, Daneels D, Breckpot K, Brugada P, Bonduelle M, Brugada R, Van Dooren S (2015) Contribution of cardiac sodium channel beta-subunit variants to Brugada syndrome. Circ J 79:2118–2129
Perrino C, Schroder JN, Lima B, Villamizar N, Nienaber JJ, Milano CA, Naga Prasad SV (2007) Dynamic regulation of phosphoinositide 3-kinase-gamma activity and beta-adrenergic receptor trafficking in end-stage human heart failure. Circulation 116:2571–2579
Petitprez S, Zmoos AF, Ogrodnik J, Balse E, Raad N, El-Haou S, Albesa M, Bittihn P, Luther S, Lehnart SE, Hatem SN, Coulombe A, Abriel H (2011) SAP97 and dystrophin macromolecular

complexes determine two pools of cardiac sodium channels $Na_V1.5$ in cardiomyocytes. Circ Res 108:294–304

Qu Y, Rogers J, Tanada T, Scheuer T, Catterall WA (1994) Modulation of cardiac Na^+ channels expressed in a mammalian cell line and in ventricular myocytes by protein kinase C. Proc Natl Acad Sci U S A 91:3289–3293

Qu Y, Rogers JC, Chen SF, McCormick KA, Scheuer T, Catterall WA (1999) Functional roles of the extracellular segments of the sodium channel alpha subunit in voltage-dependent gating and modulation by beta1 subunits. J Biol Chem 274:32647–32654

Riuro H, Beltran-Alvarez P, Tarradas A, Selga E, Campuzano O, Verges M, Pagans S, Iglesias A, Brugada J, Brugada P, Vazquez FM, Perez GJ, Scornik FS, Brugada R (2013) A missense mutation in the sodium channel beta2 subunit reveals SCN2B as a new candidate gene for Brugada syndrome. Hum Mutat 34:961–966

Ruan Y, Liu N, Bloise R, Napolitano C, Priori SG (2007) Gating properties of SCN5A mutations and the response to mexiletine in long-QT syndrome type 3 patients. Circulation 116:1137–1144

Ruan Y, Liu N, Priori SG (2009) Sodium channel mutations and arrhythmias. Nat Rev Cardiol 6:337–348

Schreibmayer W (1999) Isoform diversity and modulation of sodium channels by protein kinases. Cell Physiol Biochem 9:187–200

Schreibmayer W, Frohnwieser B, Dascal N, Platzer D, Spreitzer B, Zechner R, Kallen RG, Lester HA (1994) Beta-adrenergic modulation of currents produced by rat cardiac Na^+ channels expressed in Xenopus laevis oocytes. Receptors Channels 2:339–350

Schubert B, Vandongen AM, Kirsch GE, Brown AM (1990) Inhibition of cardiac Na^+ currents by isoproterenol. Am J Phys 258:H977–H982

Schwartz PJ, Stramba-Badiale M, Crotti L, Pedrazzini M, Besana A, Bosi G, Gabbarini F, Goulene K, Insolia R, Mannarino S, Mosca F, Nespoli L, Rimini A, Rosati E, Salice P, Spazzolini C (2009) Prevalence of the congenital long-QT syndrome. Circulation 120(18):1761–1767

Schwartz PJ, Ackerman MJ (2013) The long QT syndrome: a transatlantic clinical approach to diagnosis and therapy. Eur Heart J 34:3109–3116

Shi Q, Li M, Mika D, Fu Q, Kim S, Phan J, Shen A, Vandecasteele G, Xiang YK (2017) Heterologous desensitization of cardiac beta-adrenergic signal via hormone-induced betaAR/arrestin/PDE4 complexes. Cardiovasc Res 113:656–670

Shin HG, Murray KT (2001) Conventional protein kinase C isoforms and cross-activation of protein kinase A regulate cardiac Na^+ current. FEBS Lett 495:154–158

Shin HG, Barnett JV, Chang P, Reddy S, Drinkwater DC, Pierson RN, Wiley RG, Murray KT (2000) Molecular heterogeneity of protein kinase C expression in human ventricle. Cardiovasc Res 48:285–299

Smith FD, Esseltine JL, Nygren PJ, Veesler D, Byrne DP, Vonderach M, Strashnov I, Eyers CE, Eyers PA, Langeberg LK, Scott JD (2017) Local protein kinase A action proceeds through intact holoenzymes. Science 356:1288–1293

Srivastava U, Aromolaran AS, Fabris F, Lazaro D, Kassotis J, Qu Y, Boutjdir M (2016) Novel function of alpha1D L-type calcium channel in the atria. Biochem Biophys Res Commun 482 (4):771–776

Sunami A, Fan Z, Nakamura F, Naka M, Tanaka T, Sawanobori T, Hiraoka M (1991) The catalytic subunit of cyclic AMP-dependent protein kinase directly inhibits sodium channel activities in guinea-pig ventricular myocytes. Pflugers Arch 419:415–417

Tateyama M, Kurokawa J, Terrenoire C, Rivolta I, Kass RS (2003) Stimulation of protein kinase C inhibits bursting in disease-linked mutant human cardiac sodium channels. Circulation 107:3216–3222

Triposkiadis F, Karayannis G, Giamouzis G, Skoularigis J, Louridas G, Butler J (2009) The sympathetic nervous system in heart failure physiology, pathophysiology, and clinical implications. J Am Coll Cardiol 54:1747–1762

Turnham RE, Scott JD (2016) Protein kinase A catalytic subunit isoform PRKACA; history, function and physiology. Gene 577:101–108

Ulbricht W (2005) Sodium channel inactivation: molecular determinants and modulation. Physiol Rev 85:1271–1301
Valdivia CR, Tester DJ, Rok BA, Porter CB, Munger TM, Jahangir A, Makielski JC, Ackerman MJ (2004) A trafficking defective, Brugada syndrome-causing SCN5A mutation rescued by drugs. Cardiovasc Res 62:53–62
Valdivia CR, Ueda K, Ackerman MJ, Makielski JC (2009) GPD1L links redox state to cardiac excitability by PKC-dependent phosphorylation of the sodium channel SCN5A. Am J Physiol Heart Circ Physiol 297:H1446–H1452
Van Norstrand DW, Valdivia CR, Tester DJ, Ueda K, London B, Makielski JC, Ackerman MJ (2007) Molecular and functional characterization of novel glycerol-3-phosphate dehydrogenase 1 like gene (GPD1-L) mutations in sudden infant death syndrome. Circulation 116:2253–2259
Viswanathan PC, Bezzina CR, George AL Jr, Roden DM, Wilde AA, Balser JR (2001) Gating-dependent mechanisms for flecainide action in SCN5A-linked arrhythmia syndromes. Circulation 104:1200–1205
Wang DW, Makita N, Kitabatake A, Balser JR, George AL Jr (2000) Enhanced $Na^{(+)}$ channel intermediate inactivation in Brugada syndrome. Circ Res 87:e37–e43
Ward CA, Giles WR (1997) Ionic mechanism of the effects of hydrogen peroxide in rat ventricular myocytes. J Physiol 500(3):631–642
Watanabe H, Koopmann TT, Le Scouarnec S, Yang T, Ingram CR, Schott JJ, Demolombe S, Probst V, Anselme F, Escande D, Wiesfeld AC, Pfeufer A, Kaab S, Wichmann HE, Hasdemir C, Aizawa Y, Wilde AA, Roden DM, Bezzina CR (2008) Sodium channel beta1 subunit mutations associated with Brugada syndrome and cardiac conduction disease in humans. J Clin Invest 118:2260–2268
Watson CL, Gold MR (1997) Modulation of Na^+ current inactivation by stimulation of protein kinase C in cardiac cells. Circ Res 81:380–386
Weber S, Meyer-Roxlau S, Wagner M, Dobrev D, El-Armouche A (2015) Counteracting protein kinase activity in the heart: the multiple roles of protein phosphatases. Front Pharmacol 6:270
Wetsel WC, Khan WA, Merchenthaler I, Rivera H, Halpern AE, Phung HM, Negro-Vilar A, Hannun YA (1992) Tissue and cellular distribution of the extended family of protein kinase C isoenzymes. J Cell Biol 117:121–133
Wilde AA, Brugada R (2011) Phenotypical manifestations of mutations in the genes encoding subunits of the cardiac sodium channel. Circ Res 108:884–897
Wilde AA, Moss AJ, Kaufman ES, Shimizu W, Peterson DR, Benhorin J, Lopes C, Towbin JA, Spazzolini C, Crotti L, Zareba W, Goldenberg I, Kanters JK, Robinson JL, Qi M, Hofman N, Tester DJ, Bezzina CR, Alders M, Aiba T, Kamakura S, Miyamoto Y, Andrews ML, McNitt S, Polonsky B, Schwartz PJ, Ackerman MJ (2016) Clinical aspects of type 3 long-QT syndrome: an international multicenter study. Circulation 134:872–882
Xiang Y, Kobilka BK (2003) Myocyte adrenoceptor signaling pathways. Science 300:1530–1532
Xiao GQ, Qu Y, Sun ZQ, Mochly-Rosen D, Boutjdir M (2001) Evidence for functional role of epsilonPKC isozyme in the regulation of cardiac $Na^{(+)}$ channels. Am J Physiol Cell Physiol 281:C1477–C1486
Yu FH, Westenbroek RE, Silos-Santiago I, McCormick KA, Lawson D, Ge P, Ferriera H, Lilly J, DiStefano PS, Catterall WA, Scheuer T, Curtis R (2003) Sodium channel beta4, a new disulfide-linked auxiliary subunit with similarity to beta2. J Neurosci 23:7577–7585
Zhou J, Yi J, Hu N, George AL Jr, Murray KT (2000) Activation of protein kinase A modulates trafficking of the human cardiac sodium channel in Xenopus oocytes. Circ Res 87:33–38
Zhou J, Shin HG, Yi J, Shen W, Williams CP, Murray KT (2002) Phosphorylation and putative ER retention signals are required for protein kinase A-mediated potentiation of cardiac sodium current. Circ Res 91(6):540
Zimmer T, Benndorf K (2007) The intracellular domain of the beta2 subunit modulates the gating of cardiac $Na_v1.5$ channels. Biophys J 92:3885–3892

Part III

Drugs and Toxins Interactions with Sodium Channels

Toxins That Affect Voltage-Gated Sodium Channels

Yonghua Ji

Contents

Abstract

Voltage-gated sodium channels (VGSCs) are critical in generation and conduction of electrical signals in multiple excitable tissues. Natural toxins, produced by animal, plant, and microorganisms, target VGSCs through diverse strategies developed over millions of years of evolutions. Studying of the diverse interaction between VGSC and VGSC-targeting toxins has been contributing to the increasing understanding of molecular structure and function, pharmacology, and drug development potential of VGSCs. This chapter aims to summarize some of the current views on the VGSC-toxin interaction based on the established receptor sites of VGSC for natural toxins.

Keywords

Binding sites · Toxins · Voltage-gated sodium channel

Y. Ji (✉)
Laboratory of Neuropharmacology and Neurotoxicology, Shanghai University, Shanghai, China
e-mail: yhji@staff.shu.edu.cn

M. Chahine (ed.), *Voltage-gated Sodium Channels: Structure, Function and Channelopathies*, Handbook of Experimental Pharmacology 246,
https://doi.org/10.1007/164_2017_66

1 Introduction

Voltage-gated sodium channels (VGSCs) initiate and conduct action potentials in neurons, cardiac and skeletal muscle cells, and some endocrine cells (Hodgkin and Huxley 1952; Catterall 1992; Zimmer et al. 2014; Rogart 1981; Wada et al. 2008; Donatsch et al. 1977). In general, VGSCs are composed of a pore-forming α subunit and accessary subunits (Catterall et al. 2005; Marban et al. 1998; Scheuer 1994; Wood and Baker 2001). The α subunit of VGSCs is consisted of four functional domains (I–IV). Each domain contains six transmembrane segments (S1–S6) in which S1–4 forms a voltage sensor component and S5–6 forms a pore component. In mammals, nine subtypes of functional α subunits of VGSC (Nav1.1–1.9) are expressed differentially at tissue, cellular, and subcellular levels (Catterall et al. 2005; Noda et al. 1986; Goldin 1999; Bao 2015). VGSCs play an important role in a variety of inherited and acquired diseases and have been serving as promising drug discovery targets in recent years (Cox et al. 2006; Dib-Hajj et al. 2010; Emery et al. 2016; Faber et al. 2012; Fertleman et al. 2006; Hong et al. 2004; Kohling 2002; Lai et al. 2003; George 2005). However, despite of enormous efforts spent to develop novel therapy drugs of VGSC blockers, success has been rare (Carnevale and Klein 2017). One reason behind this problem is the lack of in-depth and complete understanding of how VGSCs work and how ligands interact with VGSC.

Natural toxins targeting VGSCs are developed by animal, plant, and micro organisms over millions of years of evolution (Moran et al. 2009; Zakon 2012). These toxins are developed to capture the prey and/or to defend the predators through diverse interaction with VGSCs (Cestele and Catterall 2000; Wang and Wang 2003). So far, six receptor sites of VGSCs have been defined based on the interaction between VGSCs and natural toxins (Li and Tomaselli 2004; Nicholson 2007; Pedraza Escalona and Possani 2013; Hille 1968). Accordingly, six types of VGSC-targeting toxins have been grouped as follows: (1) site 1 toxins block VGSCs through the outer pore; (2) site 2 toxins modulate the gating, permeability, and selectivity of VGSCs involving the inner pore; (3) site 3 toxins inhibit the fast inactivation of VGSCs with subtype selectivity; (4) site 4 toxins facilitate the steady-state activation and reduce the amplitude of the peak currents of VGSCs with species selectivity; (5) site 5 toxins enhance the activation and inhibit the fast inactivation of VGSCs; and (6) site 6 toxins inhibit the fast inactivation of VGSCs through a binding site different from site 3. These diverse VGSC-targeting toxins have been and will continue to be unique tools for studying the structure and function of VGSCs and for developing VGSC-targeting strategies in VGSC-related diseases.

2 Toxins Binding to Site 1 of VGSCs

Among all putative binding sites on sodium channels, site 1 is probably the best defined and the most straightforward site. The site 1 is composed of residues on the reentrant P-loops from all four domains. It is occupied by two different groups of neurotoxins: the water-soluble heterocyclic guanidines tetrodotoxin (TTX) and saxitoxin (STX), and the peptidic toxins μ-contoxins (Catterall et al. 2007). TTX is purified from the roe, ovaries, skin, and liver of at least 40 species of puffer fish (Fuhrman 1967). Moreover, it is also extracted from mollusks, crabs, octopus, fish, salamander, and Central American frogs (Sheumack et al. 1978; Noguchi et al. 1986; Daly et al. 1994; Mattei and Legros 2014). STX is produced by the marine dinoflagellate *Gonyaulax catenella*, and it is found in bivalves such as clams and mussels that feed on the dinoflagellates (Schantz 1986). μ-conotoxins are isolated from the venom of *Conus geographus* and related cone snails (Cruz et al. 1985; Sato et al. 1983). The binding of μ-conotoxins on receptor site 1 (Narahashi et al. 1964; Hille 1975) seems distinct from TTX. Several mutations in Nav1.4 affect the binding of TTX but not μ-conotoxins suggesting that the binding sites of TTX and μ-conotoxins are not identical (Stephan et al. 1994). Localization of TTX/STX as well as μ-conotoxins receptor sites has been instrumental in identifying the pore loop and clarifying the regions of VGSCs involved in the ion selectivity filter (Catterall et al. 2007).

TTX has been used for many decades to characterize the physiological function of sodium channels (Sato et al. 2001). According to the sensitivity to TTX, the nine VGSC subtypes can be classified into two groups, with blocking concentrations being in the nanomolar range for the TTX-S channels (Nav1.1, Nav1.2, Nav1.3, Nav1.4, Nav1.6, and Nav1.7) and in the micromolar range for the TTX-R channels (Nav1.5, Nav1.8, and Nav1.9). The structural determinant of TTX resistance has been located as a single amino acid residue found directly adjacent to the inner-ring residue in domain I. In TTX-S channels this residue is a tyrosine or phenylalanine, but in TTX-R channels it is a cysteine or serine (Goldin 2002). The neuronal TTX-R subtypes (Nav1.8 and Nav1.9) are preferentially expressed in the peripheral sensory neurons of DRG, trigeminal, and nodose ganglion. In DRG, Nav1.8 and Nav1.9 are preferentially expressed in nociceptive neurons. The activity of neuronal TTX-R channels can be studied in nociceptive neurons by application of sub-micromolar concentrations of TTX that selectively blocks TTX-S channels (Cummins et al. 2007). In addition to experimental tools, lower concentrations of TTX also show medical potentials as an anesthetic agent in treating pain, migraine, and withdrawal symptoms in heroin addicts (Narahashi 2008).

μ-Conotoxins contain several positively charged residues, which are important in interacting with the negatively charged residues of the selectivity filter. In particular, a conserved arginine at position 13 or 14 provides a guanidinium group, which appears to be the most critical for channel block (Keizer et al. 2003). On the one hand, because μ-conotoxins are much larger than TTX and STX while they bind at site 1, they are also possible to interact with other positions in VGSCs. On the other hand, residues that are important for TTX resistance in VGSCs appear not to be involved in the action of the μ-conotoxins, which suggested that the binding sites of TTX as well as μ-conotoxins are partially overlapped. Compared to TTX and STX, μ-conotoxins show subtype selectivity toward VGSCs. A group of μ-conotoxins composed by GIIIA, GIIIB, and GIIIC are highly effective blockers of the skeletal muscle sodium channel Nav1.4. However, several other μ-conotoxins, including SmIIIA, SIIIA, and KIIIA, are selective inhibitors of TTX-R sodium currents (Ekberg et al. 2008). For example, the SIIIA is able to block TTX-R sodium currents in rodent DRG neurons with little effect on TTX-S currents (Wang et al. 2006). The isoform specificity of μ-conotoxins seems to be related to specific residues in the segments 5 and 6 in domain II of the VGSC. The subtype selectivity of conotoxins might potentially enable them as valuable tools for studying the physiological roles of sodium channels. Systemic administration of KIIIA demonstrated analgesic effects in the inflammatory pain model induced by formalin, mainly through inhibition of TTX-R sodium currents (Zhang et al. 2007).

Another neurotoxin believed to affect site 1 is *Phoneutria nigriventer* toxin 1 (Tx1) from *Phoneutria nigriventer*. Tx1 blocks sodium currents in a state-dependent manner and competes for binding with μ-conotoxins GIIIB but not with TTX (Martin-Moutot et al. 2006). Therefore it was suggested that Tx1, like TTX, binds to a micro site and that the binding site of μ-conotoxins overlaps the micro sites of TTX and Tx1 (Li et al. 2003).

3 Toxins Binding to Site 2 of VGSCs

Lipid-soluble neurotoxins that target this receptor site have diverse chemical structures and are not structurally related (Catterall 1980). Their structural non-relatedness is reflected in their different sources as they can come from plants, animals, and bacteria (Wang and Wang 2003). Site 2 neurotoxins from plants are alkaloids like veratridine (VTD; from *Liliaceae*) and aconitine (from *Aconitum napellus*) and grayanotoxins (GTX; from *Ericaceae*). Batrachotoxin (BTX) is abundant in the skin of Colombian frog *Phyllobates aurotaenia*. In addition, antillatoxin (Cao et al. 2008) and hoiamide (Pereira et al. 2009) that are isolated from some cyanobacteria (*Lyngbya majuscula*) were also found to target the neurotoxin receptor site 2. These toxins are useful tools in the study of voltage-gated Na^+ channels. For example, radioactive probes using these toxins, which exhibit high affinities toward Na^+ channels with relatively low nonspecific binding characteristics, had long been used to characterize the voltage-gated sodium channels directly or the allosteric modulation by different classes of neurotoxins in various excitable membranes.

Site 2 toxins are considered to be activators as their actions on sodium channels are in such a way that sodium channels open more easily and stay open longer. When Na^+ flux assays are used, BTX is considered as a full activator, whereas veratridine and aconitine are considered as partial activators. Site 2 toxins bind preferentially to the open state of sodium channels and alter the activity of sodium channels. Upon binding, the voltage dependence of activation is shifted toward more negative potentials, causing sodium channels to open at resting potentials. The inactivation is slowed down or inhibited resulting in sustained, non-inactivating currents via an allosteric mechanism. However, the sodium conductance is reduced by site 2 binding toxins. For instance, single-channel conductance of BTX-activated Na^+ channels and veratridine-activated Na^+ channels are about 50% and 25%, respectively, of that of normal Na^+ channel. The underlying mechanism for the reduction in single-channel conductance by these two neurotoxins is not known. One possibility is a partial block of the Na^+ channel permeation pathway by these two ligands. Additionally, the ion selectivity of site 2 toxin-modified channels is altered possible due to a decreased discrimination for permeating ions (Tikhonov and Zhorov 2005; Du et al. 2011).

The mapping of the neurotoxin site 2 has shed light on the molecular determinants responsible for the alterations of channel gating upon toxin binding. By using photo-labeled BTX, it was identified that the first residues within the inner helixes of DI S6 (I433, N434, and L437) and DIV S6 (F1579 and N1584) are crucial for BTX binding in Nav1.4 (Wang and Wang 1998, 1999) and that DIVS6 (I1760 and F1764) (equivalent to F1579) are crucial for BTX binding in Nav1.2 (Linford et al. 1998). By using point mutations of the equivalent residues in Nav1.5, the key role of these residues in BTX binding was also confirmed (Wang et al. 2007). Although site 2 is generally believed to be localized mainly at the S6 of DI and DIV, it is now considered that specific amino acid residues among all four S6 segments are involved in the neurotoxin receptor site 2 by using site-directed mutagenesis (Wang et al. 2001). The exact location of all known molecular determinants contributing to the BTX binding site has been well reviewed elsewhere (Du et al. 2011). An initial allosteric model was proposed to explain the observed channel gating alterations upon site 2 toxin binding, in which the lipid-soluble toxins bind to lipid-exposed sites distinct from the pore and/or the voltage sensors (Catterall et al. 2007). Although this allosteric model provided a reasonable interpretation of modifications in channel gating, conductance, and ion selectivity, increasing evidence have emerged to promote the revision of this model. It was suggested that these site 2 toxins bind within the pore rather than at the lipid-exposed channel interface (Tikhonov and Zhorov 2005; Du et al. 2011). In a new revised model, site 2 neurotoxins, such as BTX, bind within the inner pore with residues in the S6 segments of all four domains, exposing the activator directly to the permeation pathway, which is also consistent with most studies on BTX binding (Du et al. 2011). This revised model also included the previous hypothesis that the observed inhibition of inactivation might be due to BTX interaction with the above-described residues in S6 of DIV (Du et al. 2011). The DIV S6 segment is not only involved in fast inactivation, while the altered movements of toxin-bound S6 segments also influence the movement of adjacent segments. Therefore, it is postulated that BTX binding alters the voltage-dependent movement of the DIV S4 voltage sensor and thereby modifies channel activation and its coupling to inactivation (Catterall et al. 2007). In addition, the reduced sodium conductance can be attributed to the outcome of a narrower Na^+ binding site due to the presence of a BTX in the inner pore. In contrast, the altered ion selectivity may be a direct consequence of a wider selectivity filter in the toxin-modified channels. It was established that the DEKA locus within the selectivity filter determines the sodium channel ion selectivity. Moreover, K1422 can be seen as a key residue because single point mutation of this residue into a glutamic acid conferred calcium conducting characteristics onto VGSCs (Heinemann et al. 1992). This new model suggests that BTX does not interact directly with the DEKA locus but rather causes a deficiency of water molecules in the proximity of the selectivity filter. The displacement of water molecules may lead to a shift in the PKA of the ion selectivity-determining residue K1422 and, in this way, lower the discrimination in permeating ions (Du et al. 2011).

For the lipid-soluble grayanotoxins (GTXs), several studies have contributed in locating their binding site. Similar to the BTX binding residues, it was found that S251, I433, N434, L437, I1575, and F1579 on Nav1.4 and their equivalents on Nav1.5 are involved in the binding of GTXs (Kimura et al. 2001; Ishii et al. 1999). Mutation of one specific residue at position 1586 in DIV S6 completely abolished grayanotoxin-induced effects on Nav1.4 channel. However, the same mutation did not alter BTX binding (Kimura et al. 2000). Together, it can be concluded that the GTX binding site is not completely identical to but is overlapping with the BTX binding site, as they share many molecular determinants.

In contrast with BTX, the channel modification by veratridine (VTD) is less understood. In general, it is assumed that VTD binds to the same site as BTX since VTD induces channel alterations that are similar to these of BTX (Wang and Wang 2003). However, in contrast with BTX, which does not dissociate from its receptor, the binding of VTD to receptor site is reversible, and it dissociates from its receptor upon membrane hyperpolarization (Ulbricht 1998). A bell-shaped relationship was described between the concentration of veratridine and the peak amplitude of sodium current in murine myocytes. It was also observed that increasing concentrations of VTD enhance the peak amplitude, reaching a maximum around 10 μM, while higher concentrations of VTD reduced the sodium conductance (Zhu et al. 2009a). The modulatory effects of aconitine are still poorly studied. It is found that aconitine binding to its receptor causes an incompletion of inactivation and an alteration of the ion selectivity in muscle VGSCs but not in nerve fibers (Campbell 1982).

4 Toxins Binding to Site 3 of VGSCs

Site 3 neurotoxins, including scorpion α-toxins and sea anemone toxins, slow down the inactivation and induce a prolonged opening of VGSCs. According to pharmacological binding properties, scorpion α-toxin can be divided into three groups: α-toxin, α-insect toxin, and α-like toxin. The representative toxins for these three groups include AaH II and Lqh 2, LqhαIT, and Lqh 3 and BmK I, respectively. AaH II has the most potent affinity to site 3 of mammalian neural (rNav1.1, rNav1.2) and muscular (rNav1.1–1.4) sodium channels (Legros et al. 2005). In voltage-clamp experiments in *Xenopus* oocytes, Aah II inhibited the inactivation of the activated rat brain type rNav1.2 and rat skeletal muscle type rNav1.4 and enhanced the Na^+ entry into excitable cells. EC(50) value of Aah II that induced slowing of inactivation is 2.6 ± 0.3 nM on rNav1.2 and 2.2 ± 0.2 nM on rNav1.4 (Alami et al. 2003). The receptor site of AaH II consists of the extracellular loops of DI/S5–S6, DIV/S3–S4, and DIV/S5–S6. A negatively charged amino acid residue (E1613 in rNav1.2, D1428 in rNav1.4) in the loop of DIV/S3–S4 has been shown to play a critical role in the binding of channels with toxins (Cestele and Catterall 2000; Rogers et al. 1996; Leipold et al. 2004). The crystal structure of Aah II has been elucidated, and the structure has been used as template and model for α-scorpion toxin (Gur et al. 2011; Gurevitz 2012).

Lqh αIT, a typical α-insect toxin, is 2,000-fold less active in the mouse brain than Lqh 2 which is a classical α mammal toxin (Gilles et al. 2001). Brain-type sodium channels Nav1.1–3 and Nav1.6 are weakly sensitive to Lqh αIT (Eitan et al. 1990). However, 20 nM of Lqh αIT potently impaired the inactivation of wild-type Na_v1.4 channels (Chen et al. 2000). Lqh αIT also interacts strongly with hNav1.7. Residue-swap analysis verified that two acidic residues (Asp1428 and Lys1432 in Nav1.4) within the domain IV S3–S4 extracellular loop of VGSCs were crucial for the selectivity and modulation pattern of Lqh αIT (Leipold et al. 2004). BmK I discriminates well for the three neuronal VGSCs that were independently expressed in *Xenopus* oocytes with the auxiliary β1 subunit: Nav1.6α/β1 responded with a large increase of both transient and persistent currents through inhibition of fast inactivation of sodium channels. Moreover, BmK I also accelerated the slow inactivation and delayed recovery through binding to Na_v1.6α/β1 in the open state. Nav1.2α/β1 was less affected by BmK I. At a high concentration (500 nM), BmK I increased fast time constants of inactivating current. In contrast, Nav1.3α/β1 was nearly insensitive (He et al. 2010). Residue-swap analysis verified that an acidic residue (e.g., Asp1602 in mNa_v1.6) within the domain IV S3–S4 extracellular loop of VGSCs was crucial for the selectivity and modulation pattern of BmK I, which was in agreement with other α-scorpion toxins binding to sodium channels (Zuo and Ji 2004; Bosmans and Tytgat 2007). Recently, studies showed that expression of Nav1.8 and Nav1.6 were increased in BmK I-induced pain model, indicating a potential pharmacological role of BmK I as a tool to investigate sodium channel-related pain (Ye et al. 2016).

Site 3 sea anemone toxin ATX II enhances the inward currents of Nav1.1, Nav1.2, Nav1.3, Nav1.4, Nav1.5, Nav1.6, and Nav1.7 mainly through increasing the slow component of inactivating currents (Oliveira et al. 2004; Hampl et al. 2016). In addition, ATX produces small persistent currents in Nav1.3 and Nav1.6 at high concentrations (Oliveira et al. 2004). Studies of site-directed mutagenesis found that ATX II interacts with Glu1613 on the extracellular loop of DIV/S3–S4 of the rat neuronal channel Nav1.2 (Rogers et al. 1996). Electrophysiological studies have shown that α-scorpion toxin could bind to rNav1.7 (Cestele et al. 1999), but had no effect on hNav1.7. Sequence comparison reveals that rNav1.7 and hNav1.7 are almost identical in IVS3–S4, with the exception of the acidic residue that is critical for α-toxin binding (Glu in the former vs Asp in the latter). This result also indicated that ATX II and α-scorpion toxin are bound to the site 3 of sodium channel in a different manner.

5 Toxins Binding to Site 4 of VGSCs

The receptor site 4 of voltage-gated sodium channels is mainly recognized by scorpion β-toxins, which exert their activities by altering the gating properties of VGSCs. The major effects of site 4 toxins are to reduce the amplitude of the peak Na^+ currents as well as shift the voltage-dependent activation of VGSCs to more negative potentials (Cestele et al. 2001, 2006; Stevens et al. 2011). Generally, most of these toxins initiate their modulatory effects on the VGSC activation requiring a short depolarizing prepulse (Leipold et al. 2012). Several key residues, in the extracellular loops connecting S1–S2 and S3–S4 segments of domain II, are important to the receptor site 4 (Catterall et al. 2007). Moreover, it was reported that some residues in the pore region of domain III are critical for β-toxin, such as Tz1 (*Tityus zulianus*) and BmK IT2 (*Buthus martensii* Karsch), to associate with different Na^+ channel isoforms (Leipold et al. 2006; He et al. 2011). One possible allosteric mechanism was that the voltage sensor of domain III enhances the binding of β-toxins to S4 segment in domain II (Song et al. 2011).

Site 4 toxins are always long-chain polypeptides composed of 58–76 amino acid residues with cross-linked by four disulfide bridges which belong to Csα/β motif structural superfamily (de la Vega and Possani 2007; Gurevitz et al. 2007; Possani et al. 1999). β-scorpion toxins could be divided into four groups according to their affinity and pharmacological characterizations to mammalian or insect VGSCs – classical (mammalian-selective) β-toxins, excitatory anti-insect β-toxins, depressant anti-insect β-toxins, and β-like (acting on both mammalians and insects) toxins.

1. Classical β-toxins: Classical β-toxins show high affinity to mammal Na^+ channels (Martin et al. 1987; Pintar et al. 1999). Examples of classical β-toxins include Css4 (*Centruroides suffusus suffusus*) and Cn2 (*Centruroides noxius* Hoffmann).
2. Excitatory anti-insect β-toxins: Excitatory anti-insect β-toxins fail to show any effects on mammal channels (de Dianous et al. 1987), whereas they induce a fast and repetitive activity of motor nerves by targeting insect VGSCs that results in a reversible contraction paralysis. AahIT (*Androctonus australis* Hector), BmK IT (*Buthus martensii* Karsch), LqqIT1 (*Leiurus quinquestriatus quinquestriatus*), and Bj-xtrIT (*Hottentotta judaicus*) belong to this group (Liu et al. 2011; Froy et al. 1999; Billen et al. 2008; Ji et al. 1994). Compared to other β-toxin groups, one disulfide bridge of excitatory anti-insect β-toxins is located differently, which results in different secondary structural elements. The high selectivity to insect but not mammal, combined with high potency, makes the excitatory anti-insect β-toxins promising lead compounds in the design of new insecticides (Gurevitz et al. 2007).
3. Depressant anti-insect β-toxins: These toxins induce a transient muscle contraction and followed by a slow depressant as well as flaccid paralysis (Zlotkin et al. 1991; Karbat et al. 2007). Representatives of this group are LqqIT2 (*L. q. quinquestriatus*), BjIT2 (*H. judaicus*), and BmK IT2 (*Buthus martensii* Karsch). Current-clamp experiments showed that the evoked action potentials were suppressed by toxins in this group (Strugatsky et al. 2005; Li et al. 2000). Insect-selective depressant β-toxins can also modulate mammalian VGSCs under certain conditions such as when mammalian VGSCs are excited by a long depolarizing prepulse or when there is a simultaneous binding of an α-toxin to site 3 (Cohen et al. 2007a). Furthermore, BmK IT2, a depressant anti-insect β-toxin from *Buthus martensii* Karsch, inhibited the currents of mammalian

VGSCs in DRG (dorsal root ganglion) neurons as well as hippocampal pyramidal neurons without a long depolarizing prepulse or a simultaneous binding of α-toxin to site 3 (Zhao et al. 2008; Tan et al. 2001). Combinational site 3 α-toxin (BmK I) and BmK IT2 resulted in larger peak I_{Na} and more negative half-activation voltage compared to BmK I alone. Co-applied BmK I and BmK IT2 also produced slower inactivation compared to BmK I alone (Feng et al. 2015). These results suggested that existence of depressant β-toxins in the scorpion venom may contribute to the scorpion venom-induced toxicity through enhancing effects of α-toxins.
4. β-like toxins: β-like toxins are capable of competing for binding sites on both insect and mammalian Nav channels. Tsγ (*Tityus serrulatus*), Lqhβ1 (*Leiurus quinquestriatus hebraeus*), and BmK AS (*Buthus martensii* Karsch) are well-studied examples of β-toxins (Possani et al. 1999; Ji et al. 1999; Gordon and Gurevitz 2003). Interestingly, BmK AS exhibits the pharmacological activities of both the α- and β-toxins, which could not only hyperpolarize the voltage dependence of activation but also delay the slow inactivation of VGSCs (Zhu et al. 2009b).

A bioactive surface of β-scorpion toxins is composed of two main clusters. The first cluster is associated with the α-helix, and it contains a conserved negatively charged residue, Glu24 in LqhIT2, Glu26 in Lqhβ1, Glu28 in Css4, and Glu30 in Bj-xtrIT. The conserved residue is highly important for the toxin activity and probably interacts with a positively charged counterpart of VGSC. The residues flanking the glutamate are hydrophobic and seem to form a hydrophobic environment that occludes bulk solvent from the high-energy point of interaction. The common structure of the cluster could explain the ability of β-toxins to compete for receptor 4 of different neuronal membrane preparations (Gurevitz et al. 2007). Cn2, a classical anti-mammalian β-toxin, could compete with excitatory toxin Bj-xtrIT on binding to insect VGSCs (Cohen et al. 2004). BmK IT2 shifted the activation of DmNav1, the sodium channel from *Drosophila*, to more hyperpolarized potentials, whereas it hardly affected the gating properties of rNav 1.2, rNav 1.3, and mNav 1.6, three mammalian central neuronal sodium channel subtypes. However, for DRG neurons, BmK IT2 potently blocked both tetrodotoxin-sensitive (TTX-S) and tetrodotoxin-resistant (TTX-R) components (He et al. 2011; Tan et al. 2001). The second cluster of the bioactive surface of β-scorpion toxins is composed of hydrophobic residues including Leu19, Asn22, Tyr40, Tyr42, and Phe44 in Css4, Val71, Gln72, Ile73, and Ile74 in Bj-xtrIT. These residues are associated mainly with the β2 and β3 strands and are likely to confer the specificity of β-toxins for VGSCs (Cohen et al. 2005). The commonality of the pharmacophore at the βαββ core of β-toxins leads to an attempt to evaluate the contribution of the pharmacophore to the binding affinity of β-toxins (Cohen et al. 2004, 2005). Removal of the N- and C-terminal regions in Bj-xtrIT and Css4 resulted in truncated toxin derivatives that were successfully folded in vitro and examined in binding competition assays. However, the truncated toxins did not bind at receptor site 4 nor compete with the parental toxins for this site. Unexpectedly, they were able to modify allosterically the binding and activity of site 3 scorpion toxins (anti-mammalian Lqh2 and anti-insect LqhaIT (Cohen et al. 2008).

The pharmacological properties of VGSC receptor site 4 have been well investigated and defined in more than two decades (Gurevitz et al. 2007; Cohen et al. 2006). By analyzing the bioactivity of the β-toxin Ts1 from *T. serrulatus* on chimeras made among brain, cardiac, and skeletal muscle VGSCs, domain II has been found to be important for toxin binding (Marcotte et al. 1997). This conclusion was also supported by mutation and domain substitution studies of Css4 and AahIT, respectively. Substitution of Glu779 at DII/S1–S2 or substitution of Glu837, Leu840, and Gly845 at DII/S3–S4 significantly reduced the bioactivity of Css4 on rNav1.2a (Cestele et al. 1998). Substitution of domain II in rNav1.2a with the equivalent domain from the *Drosophila* channel DmNav1 converted the insensitivity to sensitivity of AahIT (from *Androctonus australis* Hector) to rNav1.2a (Shichor et al. 2002). Based on these results, Cestèle's group constructed a model of site 4 toxins docking at the "gating module" of rNav1.2a (Cestele et al. 2006). Though the structure model described a putative face of interaction between Css4 and the rNav1.2a channel in the "gating module" of domain II, the region of VGSCs that binds the toxin surface involved in selectivity was proposed to be associated with the "pore module." Substitutions at the "pore module" of domain III (domain III SS2–S6) on rNav1.4 highlighted the involvement of Glu1251 as well as His1257 in the recognition on the β-toxin Tz1 (Leipold et al. 2006). In order to discriminate the key residues in rNav1.4, Cohen's group (Cohen et al. 2007b) employed Css4 mutants in double-mutant analysis. Close proximity between the pairs was suggested by DDG values: F14A at the toxin and E592A at domain II S1–S2 of the channel, R27Q at the toxin and E1251N at domain III SS2–S6 of the channel, and E28R at the toxin with both E650A at domain II S3–S4 and E1251N at domain III SS2–S6 of the channel. The above results suggested that receptor site 4 contains mainly the domains II and III of VGSCs. A crevice between extracellular linkers S1–S2 and S3–S4 at the gating module of domain II might be involved (Gurevitz 2012; Karbat et al. 2004). Moreover, BmK IT2 from *Buthus martensii* Karsch strongly shifted the activation of DmNav1, the sodium channel from *Drosophila*, to more hyperpolarized potentials, whereas it hardly affected the gating properties of rNav1.2a, rNav1.3, and mNav1.6, three mammalian central neuronal sodium channel subtypes. Mutations of Glu896, Leu899, and Gly904 in extracellular loop domain II S3–S4 of DmNav1 abolished the functional action of BmK IT2. In addition to domain II, Ile1529 in domain III pore loop was critical for recognition and binding of BmK IT2 (He et al. 2011). The results provided the support to the voltage sensor-trapping mechanism and offered novel insights into the molecular requirements for the development of toxin modulators.

Similar to scorpions, spiders are capable of producing neurotoxins, recognizing on the receptor site 4 of VGSCs. It is known to all that Magi 5 (*Macrothele gigas*) is the first spider toxin shown to compete with the scorpion β-toxin CssIV for receptor site 4 (Corzo et al. 2003). Other site 4 spider β-toxins include δ-palutoxins (*P. luctuosus*), curtatoxins (*Hololena curta*), μ-agatoxins (*Agelenopsis aperta*), β/δ-agatoxins (*A. orientalis*), and the recently characterized Jingzhaotoxins (*Chilobrachys jingzhao*) as well as Huwentoxins (*Ornithoctonus huwena*) (Stapleton et al. 1990; Corzo et al. 2000; Billen et al. 2010; Rong et al. 2011; Tang et al. 2014; Wei et al. 2014; Xiao et al. 2008). Spider β-toxins are composed of 34–37 residues cross-linked by four disulfide bridges, which forms an ICK (inhibitor cysteine knot) motif (Nicholson 2007). It was found that the μ-agatoxins shifted the voltage activation curve to more hyperpolarized potentials and slowed down the inactivation process of VGSCs, resulting in a non-inactivating persistent current (Adams 2004), like scorpion β-toxin BmK AS (Zhu et al. 2009b). β/δ-Agatoxins caused a bell-shaped voltage-dependent modulation on both the activation and inactivation of VGSCs (Billen et al. 2010). Interestingly, δ-palutoxins, which compete with the depressant scorpion β-toxin Bj-xtrIT for site 4, actually functioned as α-toxins by slowing down the inactivation of VGSC (Corzo et al. 2000). However, these toxins fail to displace the binding of α-toxin LqhαIT (Corzo et al. 2005). Huwentoxin-IV (HWTX-IV), a 35-residue peptide from tarantula *Ornithoctonus huwena* venom, selectively inhibits neuronal VGSC subtypes rNav1.2, rNav1.3, and hNav1.7 compared to muscle subtypes rNav1.4 and hNav1.5. The toxin docked at receptor site 4 located at the extracellular S3–S4 linker of domain II. Mutations E818Q and D816N in hNav1.7 decreased toxin affinity, whereas the reverse mutations in rNav1.4 (N655D/Q657E) and the corresponding mutations in hNav1.5 (R812D/S814E) greatly increased the sensitivity of the muscle VGSCs to HWTX-IV (Xiao et al. 2008). Jingzhaotoxin-III (JZTX-III), a well-studied 36-residue peptide from the tarantula *Chilobrachys jingzhao*, inhibited Nav1.5. It traps the DII voltage sensor of Nav1.5 by binding to the DIIS3–S4 linker (Rong et al. 2011). Two acidic residues (Asp1, Glu3) as well as four Trp residues (residues 8, 9, 28, and 30) play crucial roles in the binding of JZTX-III to Nav1.5. Mutations S799A, R800A, and L804A could additively reduce toxin sensitivity of Nav1.5. Arg800, not existent in other sodium channel subtypes, is responsible for the selective interaction of JZTX-III with Nav1.5 (Rong et al. 2011). In addition, Jingzhaotoxin-XI (JZTX-XI, *Chilobrachys jingzhao*) is also specific for Nav1.5, inhibiting the sodium conductance as well as slowing the fast inactivation (Tang et al. 2014). In conclusion, the results suggested that spider β-toxins probably have additional contact points with the extracellular loops of VGSCs, besides receptor site 4.

6 Toxins Binding to Site 5 of VGSCs

Ciguatoxins (CTX, *Gambierdiscus toxicus*) and brevetoxins (BTX, *Karenia brevis*) are highly lipophilic cyclic polyether compounds that are biosynthesized by the worldwide distributed epibenthic and planktonic dinoflagellates. The physiological and pathological, especially neurological, effects of both CTXs and BTXs are the results of their interaction with the receptor site 5 of VGSCs (Lombet et al. 1987; Rashid et al. 2013; Pearn 2001). Brevetoxins (PbTxs) consist of 10–11 transfused rings, 23 stereocenters, and an overall linear low-energy conformation (Cassell et al. 2015; Turner et al. 2015). So far, at least 14 brevetoxins have been identified. PbTx-1 and PbTx-2 are two most classical brevetoxins and the parent toxins. The parent toxins PbTx-1 and PbTx-2 are different from each other in their backbone structure (type A and type B) (Cassell et al. 2015; Turner et al. 2015). Type A includes the PbTxs 1, 7, and 10, whereas type B includes PbTxs 2, 3, 5, 8, and 9. All PbTxs possess a lactone in the A ring as well as a conservative rigid region that forms a ladder structure, which is separated from the A ring by a spacer region with limited flexibility (Gawley et al. 1995). Moreover, they all possess a side chain that allows modification at the termini of molecules (Baden et al. 2005). It is believed that the terminal, rigid four-ring system is involved in channel binding, while the functional lactone A-ring is responsible for the alterations in channel inactivation and prolongation of the mean open time (Jeglitsch et al. 1998; Purkerson-Parker et al. 2000). PbTxs interact with VGSCs by intercalating in the membrane in a head-down orientation. Previous studies have indicated that the toxins position themselves across the plasma membrane, which is parallel with the transmembrane segments, with the A ring toward the intracellular side as well as the tail terminal of the molecule facing the extracellular side (Jeglitsch et al. 1998; Trainer et al. 1994). However, the key residues involved in brevetoxin activity still remain unknown. PbTx binding at site 5 leads to distinct alterations in channel gating: (1) the activation potential is shifted toward hyperpolarized potentials; (2) channels remain longer in the open configuration which results in a longer mean open time; (3) the inactivation is slowed down or inhibited; and (4) brevetoxins have, among all known voltage-gated sodium channel modifying toxins, the unique capability to stabilize more than one conductance levels. As such brevetoxin binding induces distinct sodium ion subconductance states in addition to the normal 21 pS state (Baden et al. 2005; Jeglitsch et al. 1998; Schreibmayer and Jeglitsch 1992). Moreover, it was found that the S6 of DI and S5 of DIV participate in the formation of receptor 5 of VGSCs using a photoreactive PbTx-3 derivate as a probe (Trainer et al. 1994). $^{[3H]}$PbTx-3 is specifically bound to Nav1.4, or the Nav1.5 α subunit isoforms expressed in HEK cells. Nav1.5 appeared to be less sensitive to BTXs compared to Nav1.4. Both type A (PbTx-1) and type B (PbTx-2 as well as PbTx-3) PbTxs target both cardiac and muscle channels. Type B PbTxs exhibit a lower affinity for the heart compared to the skeletal muscle channel (Bottein Dechraoui and Ramsdell 2003). The evaluation of the relative affinity of PbTx-2 and PbTx-3 as well as CTX with VGSCs was studied by competitive binding in the presence of $^{[3H]}$PbTx-3 in the brain, heart, and skeletal muscle of rat and the marine teleost fish *Centropristis striata*. No significant differences between the rat and fish were observed in the binding of PbTxs and CTX to either the brain or skeletal muscle. However, $^{[3H]}$PbTx-3 showed a substantially lower affinity for rat heart tissue compared to fish heart tissue and other tissues.

Ciguatoxins (CTXs) are, similar to brevetoxins, lipid-soluble compounds, which are composed of 13 ether rings with a structural backbone (Yasumoto 2001). Though 29 ciguatoxin derivates have been identified, their biological activity information remains unclear, which is mainly due to difficulties in obtaining pure toxins (Perez et al. 2011). Competing with brevetoxins for site 5, CTXs induce similar modifications of channel gating. CTXs could shift the activation of VGSCs toward more negative potentials as well as inhibit the fast inactivation. Even though binding at the same receptor site, PbTxs and CTXs possibly differ from each other in their mechanism of action. CTXs were capable of producing VGSCs dependent oscillations in neuronal membrane potential (Hogg et al. 2002). Recently, it was reported that CTXs cause a concentration-dependent decrease of the amplitude of Na^+ currents in sensory neurons (Cohen et al. 2008; Perez et al. 2011; Yamaoka et al. 2009). The most powerful ciguatoxin P-CTX-1 has attracted the most attention from researchers. P-CTX-1 binds to native channels in the brain, heart, and skeletal muscle of the rat as well as the marine teleosts in the presence of $^{[3H]}$PbTx-3 (Shmukler and Nikishin 2017; Dechraoui et al. 2006). An exhausting study of the electrophysiological effects of P-CTX-1 on VGSC isoforms Nav1.1–1.9 was carried out in HEK293 cells. P-CTX-1 has shown the ability to influence all Nav isoforms. P-CTX-1 decreases the activation threshold of two key isoforms, Nav1.7 as well as Nav1.8 channels, and lengthens their active periods, contributing to the increased activity of nociceptive neurons (Inserra et al. 2017).

7 Toxins Binding to Site 6 of VGSCs

Among all sites, neurotoxin receptor site 6 is still the speculative and yet undefined site. A first proposal for this site arose when TxVIA, one kind of δ-conotoxins (δ-CTXs), was characterized from the cone snail *Conus textile*. The polypeptide toxin TxVIA has led to the identification of neurotoxin receptor site 6 on sodium channels. TxVIA causes a marked prolongation of action potentials, due to specific inhibition of sodium current inactivation (Hasson et al. 1993). So far, 19 δ-CTX sequences (such as GmVIA, PVIA, NgVIA, SVIE, and EVIA) have been identified in fish and mollusk-hunting cone snails. The δ-CTXs act at site 6 composed of a set of amino acid residues in the segment 4 of domain IV and slow down sodium channel inactivation (Fainzilber et al. 1994). The mechanism of binding and action is quite different between δ-CTXs and μ-CTXs. For example, δ-CTXs only target neuronal voltage-gated sodium channels. δ-CTXs do not reduce channel conductance associated with pore-blocking toxins. Instead δ-CTXs affect the kinetics of the channel by reducing inactivation and thus prolonging the time that the channel remains open (Fainzilber et al. 1995). In addition, δ-CTXs exhibit synergistic effect with α-scorpion toxins, probably because both of these toxins trap the IVS4 voltage sensor in a similar conformation. However, the two toxins clearly do not directly compete for the same binding site at the voltage sensor. Instead, they apparently stabilize each other at their target sites such that these toxins affect the gating of sodium channels in a highly synergistic manner, which hints that the δ-CTXs may slow inactivation by the same molecular mechanism as α-neurotoxins from scorpion, sea anemone, spider, etc. (Leipold et al. 2005).

8 Conclusion

VGSCs are important in physiology and pathology of multiple tissues. In recent years, VGSCs has been serving as promising drug development targets for multiple diseases. However, either small molecule inhibitors or monoclonal antibodies of VGSCs have not been successful in developing novel therapeutic drugs. Natural toxins targeting VGSCs provide unique tools for studying molecular structure and function, pharmacology, and drug development potential of VGSCs. With the increasing understanding of detailed mechanism of VGSC-toxin interaction, combined with development of drug-developing techniques, VGSC-targeting toxins and their functional domains could be used as the lead molecules or part of the lead molecules for the successful development of therapeutic drugs in multiple diseases in the future.

References

Adams ME (2004) Agatoxins: ion channel specific toxins from the American funnel web spider, Agelenopsis aperta. Toxicon 43:509–525

Alami M, Vacher H, Bosmans F, Devaux C, Rosso JP, Bougis PE, Tytgat J et al (2003) Characterization of Amm VIII from Androctonus mauretanicus mauretanicus: a new scorpion toxin that discriminates between neuronal and skeletal sodium channels. Biochem J 375:551–560

Baden DG, Bourdelais AJ, Jacocks H, Michelliza S, Naar J (2005) Natural and derivative brevetoxins: historical background, multiplicity, and effects. Environ Health Perspect 113:621–625

Bao L (2015) Trafficking regulates the subcellular distribution of voltage-gated sodium channels in primary sensory neurons. Mol Pain 11:61

Billen B, Bosmans F, Tytgat J (2008) Animal peptides targeting voltage-activated sodium channels. Curr Pharm Des 14:2492–2502

Billen B, Vassilevski A, Nikolsky A, Debaveye S, Tytgat J, Grishin E (2010) Unique bell-shaped voltage-dependent modulation of Na+ channel gating by novel insect-selective toxins from the spider Agelena orientalis. J Biol Chem 285:18545–18554

Bosmans F, Tytgat J (2007) Voltage-gated sodium channel modulation by scorpion alpha-toxins. Toxicon 49:142–158

Bottein Dechraoui MY, Ramsdell JS (2003) Type B brevetoxins show tissue selectivity for voltage-gated sodium channels: comparison of brain, skeletal muscle and cardiac sodium channels. Toxicon 41:919–927

Campbell DT (1982) Modified kinetics and selectivity of sodium channels in frog skeletal muscle fibers treated with aconitine. J Gen Physiol 80:713–731

Cao Z, George J, Gerwick WH, Baden DG, Rainier JD, Murray TF (2008) Influence of lipid-soluble gating modifier toxins on sodium influx in neocortical neurons. J Pharmacol Exp Ther 326:604–613

Carnevale V, Klein ML (2017) Small molecule modulation of voltage gated sodium channels. Curr Opin Struct Biol 43:156–162

Cassell RT, Chen W, Thomas S, Liu L, Rein KS (2015) Brevetoxin, the dinoflagellate neurotoxin, localizes to thylakoid membranes and interacts with the Light-Harvesting Complex II (LHCII) of photosystem II. Chembiochem 16:1060–1067

Catterall WA (1980) Neurotoxins that act on voltage-sensitive sodium channels in excitable membranes. Annu Rev Pharmacol Toxicol 20:15–43

Catterall WA (1992) Cellular and molecular biology of voltage-gated sodium channels. Physiol Rev 72:S15–S48

Catterall WA, Goldin AL, Waxman SG (2005) International Union of Pharmacology. XLVII. Nomenclature and structure-function relationships of voltage-gated sodium channels. Pharmacol Rev 57:397–409

Catterall WA, Cestele S, Yarov-Yarovoy V, Yu FH, Konoki K, Scheuer T (2007) Voltage-gated ion channels and gating modifier toxins. Toxicon 49:124–141

Cestele S, Catterall WA (2000) Molecular mechanisms of neurotoxin action on voltage-gated sodium channels. Biochimie 82:883–892

Cestele S, Qu Y, Rogers JC, Rochat H, Scheuer T, Catterall WA (1998) Voltage sensor-trapping: enhanced activation of sodium channels by beta-scorpion toxin bound to the S3-S4 loop in domain II. Neuron 21:919–931

Cestele S, Stankiewicz M, Mansuelle P, De Waard M, Dargent B, Gilles N, Pelhate M et al (1999) Scorpion alpha-like toxins, toxic to both mammals and insects, differentially interact with receptor site 3 on voltage-gated sodium channels in mammals and insects. Eur J Neurosci 11:975–985

Cestele S, Scheuer T, Mantegazza M, Rochat H, Catterall WA (2001) Neutralization of gating charges in domain II of the sodium channel alpha subunit enhances voltage-sensor trapping by a beta-scorpion toxin. J Gen Physiol 118:291–302

Cestele S, Yarov-Yarovoy V, Qu Y, Sampieri F, Scheuer T, Catterall WA (2006) Structure and function of the voltage sensor of sodium channels probed by a beta-scorpion toxin. J Biol Chem 281:21332–21344

Chen H, Gordon D, Heinemann SH (2000) Modulation of cloned skeletal muscle sodium channels by the scorpion toxins Lqh II, Lqh III, and Lqh alphaIT. Pflugers Arch 439:423–432

Cohen L, Karbat I, Gilles N, Froy O, Corzo G, Angelovici R, Gordon D et al (2004) Dissection of the functional surface of an anti-insect excitatory toxin illuminates a putative "hot spot" common to all scorpion beta-toxins affecting Na+ channels. J Biol Chem 279:8206–8211

Cohen L, Karbat I, Gilles N, Ilan N, Benveniste M, Gordon D, Gurevitz M (2005) Common features in the functional surface of scorpion beta-toxins and elements that confer specificity for insect and mammalian voltage-gated sodium channels. J Biol Chem 280:5045–5053

Cohen L, Gilles N, Karbat I, Ilan N, Gordon D, Gurevitz M (2006) Direct evidence that receptor site-4 of sodium channel gating modifiers is not dipped in the phospholipid bilayer of neuronal membranes. J Biol Chem 281:20673–20679

Cohen L, Troub Y, Turkov M, Gilles N, Ilan N, Benveniste M, Gordon D et al (2007a) Mammalian skeletal muscle voltage-gated sodium channels are affected by scorpion depressant "insect-selective" toxins when preconditioned. Mol Pharmacol 72:1220–1227

Cohen L, Ilan N, Gur M, Stuhmer W, Gordon D, Gurevitz M (2007b) Design of a specific activator for skeletal muscle sodium channels uncovers channel architecture. J Biol Chem 282:29424–29430

Cohen L, Lipstein N, Karbat I, Ilan N, Gilles N, Kahn R, Gordon D et al (2008) Miniaturization of scorpion beta-toxins uncovers a putative ancestral surface of interaction with voltage-gated sodium channels. J Biol Chem 283:15169–15176

Corzo G, Escoubas P, Stankiewicz M, Pelhate M, Kristensen CP, Nakajima T (2000) Isolation, synthesis and pharmacological characterization of delta-palutoxins IT, novel insecticidal toxins from the spider Paracoelotes luctuosus (Amaurobiidae). Eur J Biochem 267:5783–5795

Corzo G, Gilles N, Satake H, Villegas E, Dai L, Nakajima T, Haupt J (2003) Distinct primary structures of the major peptide toxins from the venom of the spider Macrothele gigas that bind to sites 3 and 4 in the sodium channel. FEBS Lett 547:43–50

Corzo G, Escoubas P, Villegas E, Karbat I, Gordon D, Gurevitz M, Nakajima T et al (2005) A spider toxin that induces a typical effect of scorpion alpha-toxins but competes with beta-toxins on binding to insect sodium channels. Biochemistry 44:1542–1549

Cox JJ, Reimann F, Nicholas AK, Thornton G, Roberts E, Springell K, Karbani G et al (2006) An SCN9A channelopathy causes congenital inability to experience pain. Nature 444:894–898

Cruz LJ, Gray WR, Olivera BM, Zeikus RD, Kerr L, Yoshikami D, Moczydlowski E (1985) Conus geographus toxins that discriminate between neuronal and muscle sodium channels. J Biol Chem 260:9280–9288
Cummins TR, Sheets PL, Waxman SG (2007) The roles of sodium channels in nociception: implications for mechanisms of pain. Pain 131:243–257
Daly JW, Gusovsky F, Myers CW, Yotsu-Yamashita M, Yasumoto T (1994) First occurrence of tetrodotoxin in a dendrobatid frog (Colostethus inguinalis), with further reports for the bufonid genus Atelopus. Toxicon 32:279–285
de Dianous S, Hoarau F, Rochat H (1987) Re-examination of the specificity of the scorpion Androctonus australis hector insect toxin towards arthropods. Toxicon 25:411–417
de la Vega RC, Possani LD (2007) Novel paradigms on scorpion toxins that affects the activating mechanism of sodium channels. Toxicon 49:171–180
Dechraoui MY, Wacksman JJ, Ramsdell JS (2006) Species selective resistance of cardiac muscle voltage gated sodium channels: characterization of brevetoxin and ciguatoxin binding sites in rats and fish. Toxicon 48:702–712
Dib-Hajj SD, Cummins TR, Black JA, Waxman SG (2010) Sodium channels in normal and pathological pain. Annu Rev Neurosci 33:325–347
Donatsch P, Lowe DA, Richardson BP, Taylor P (1977) The functional significance of sodium channels in pancreatic beta-cell membranes. J Physiol 267:357–376
Du Y, Garden DP, Wang L, Zhorov BS, Dong K (2011) Identification of new batrachotoxin-sensing residues in segment IIIS6 of the sodium channel. J Biol Chem 286:13151–13160
Eitan M, Fowler E, Herrmann R, Duval A, Pelhate M, Zlotkin E (1990) A scorpion venom neurotoxin paralytic to insects that affects sodium current inactivation: purification, primary structure, and mode of action. Biochemistry 29:5941–5947
Ekberg J, Craik DJ, Adams DJ (2008) Conotoxin modulation of voltage-gated sodium channels. Int J Biochem Cell Biol 40:2363–2368
Emery EC, Luiz AP, Wood JN (2016) Na1.7 and other voltage-gated sodium channels as drug targets for pain relief. Expert Opin Ther Targets 20:975–983
Faber CG, Hoeijmakers JG, Ahn HS, Cheng X, Han C, Choi JS, Estacion M et al (2012) Gain of function Nanu1.7 mutations in idiopathic small fiber neuropathy. Ann Neurol 71:26–39
Fainzilber M, Kofman O, Zlotkin E, Gordon D (1994) A new neurotoxin receptor site on sodium channels is identified by a conotoxin that affects sodium channel inactivation in molluscs and acts as an antagonist in rat brain. J Biol Chem 269:2574–2580
Fainzilber M, Lodder JC, Kits KS, Kofman O, Vinnitsky I, Van Rietschoten J, Zlotkin E et al (1995) A new conotoxin affecting sodium current inactivation interacts with the delta-conotoxin receptor site. J Biol Chem 270:1123–1129
Feng YJ, Feng Q, Tao J, Zhao R, Ji YH (2015) Allosteric interactions between receptor site 3 and 4 of voltage-gated sodium channels: a novel perspective for the underlying mechanism of scorpion sting-induced pain. J Venom Anim Toxins Incl Trop Dis 21:42
Fertleman CR, Baker MD, Parker KA, Moffatt S, Elmslie FV, Abrahamsen B, Ostman J et al (2006) SCN9A mutations in paroxysmal extreme pain disorder: allelic variants underlie distinct channel defects and phenotypes. Neuron 52:767–774
Froy O, Zilberberg N, Gordon D, Turkov M, Gilles N, Stankiewicz M, Pelhate M et al (1999) The putative bioactive surface of insect-selective scorpion excitatory neurotoxins. J Biol Chem 274:5769–5776
Fuhrman FA (1967) Tetrodotoxin. It is a powerful poison that is found in two almost totally unrelated kinds of animal: puffer fish and newts. It has been serving as a tool in nerve physiology and may provide a model for new local anesthetics. Sci Am 217:60–71
Gawley RE, Rein KS, Jeglitsch G, Adams DJ, Theodorakis EA, Tiebes J, Nicolaou KC et al (1995) The relationship of brevetoxin 'length' and A-ring functionality to binding and activity in neuronal sodium channels. Chem Biol 2:533–541
George AL Jr (2005) Inherited disorders of voltage-gated sodium channels. J Clin Invest 115:1990–1999

Gilles N, Leipold E, Chen H, Heinemann SH, Gordon D (2001) Effect of depolarization on binding kinetics of scorpion alpha-toxin highlights conformational changes of rat brain sodium channels. Biochemistry 40:14576–14584
Goldin AL (1999) Diversity of mammalian voltage-gated sodium channels. Ann N Y Acad Sci 868:38–50
Goldin AL (2002) Evolution of voltage-gated Na(+) channels. J Exp Biol 205:575–584
Gordon D, Gurevitz M (2003) The selectivity of scorpion alpha-toxins for sodium channel subtypes is determined by subtle variations at the interacting surface. Toxicon 41:125–128
Gur M, Kahn R, Karbat I, Regev N, Wang J, Catterall WA, Gordon D et al (2011) Elucidation of the molecular basis of selective recognition uncovers the interaction site for the core domain of scorpion alpha-toxins on sodium channels. J Biol Chem 286:35209–35217
Gurevitz M (2012) Mapping of scorpion toxin receptor sites at voltage-gated sodium channels. Toxicon 60:502–511
Gurevitz M, Karbat I, Cohen L, Ilan N, Kahn R, Turkov M, Stankiewicz M et al (2007) The insecticidal potential of scorpion beta-toxins. Toxicon 49:473–489
Hampl M, Eberhardt E, O'Reilly AO, Lampert A (2016) Sodium channel slow inactivation interferes with open channel block. Sci Rep 6:25974
Hasson A, Fainzilber M, Gordon D, Zlotkin E, Spira ME (1993) Alteration of sodium currents by new peptide toxins from the venom of a molluscivorous Conus snail. Eur J Neurosci 5:56–64
He H, Liu Z, Dong B, Zhou J, Zhu H, Ji Y (2010) Molecular determination of selectivity of the site 3 modulator (BmK I) to sodium channels in the CNS: a clue to the importance of Nav1.6 in BmK I-induced neuronal hyperexcitability. Biochem J 431:289–298
He H, Liu Z, Dong B, Zhang J, Shu X, Zhou J, Ji Y (2011) Localization of receptor site on insect sodium channel for depressant beta-toxin BmK IT2. PLoS One 6:e14510
Heinemann SH, Terlau H, Stuhmer W, Imoto K, Numa S (1992) Calcium channel characteristics conferred on the sodium channel by single mutations. Nature 356:441–443
Hille B (1968) Pharmacological modifications of the sodium channels of frog nerve. J Gen Physiol 51:199–219
Hille B (1975) The receptor for tetrodotoxin and saxitoxin. A structural hypothesis. Biophys J 15:615–619
Hodgkin AL, Huxley AF (1952) A quantitative description of membrane current and its application to conduction and excitation in nerve. J Physiol 117:500–544
Hogg RC, Lewis RJ, Adams DJ (2002) Ciguatoxin-induced oscillations in membrane potential and action potential firing in rat parasympathetic neurons. Eur J Neurosci 16:242–248
Hong S, Morrow TJ, Paulson PE, Isom LL, Wiley JW (2004) Early painful diabetic neuropathy is associated with differential changes in tetrodotoxin-sensitive and -resistant sodium channels in dorsal root ganglion neurons in the rat. J Biol Chem 279:29341–29350
Inserra MC, Israel MR, Caldwell A, Castro J, Deuis JR, Harrington AM, Keramidas A et al (2017) Multiple sodium channel isoforms mediate the pathological effects of Pacific ciguatoxin-1. Sci Rep 7:42810
Ishii H, Kinoshita E, Kimura T, Yakehiro M, Yamaoka K, Imoto K, Mori Y et al (1999) Point-mutations related to the loss of batrachotoxin binding abolish the grayanotoxin effect in Na(+) channel isoforms. Jpn J Physiol 49:457–461
Jeglitsch G, Rein K, Baden DG, Adams DJ (1998) Brevetoxin-3 (PbTx-3) and its derivatives modulate single tetrodotoxin-sensitive sodium channels in rat sensory neurons. J Pharmacol Exp Ther 284:516–525
Ji YH, Mansuelle P, Xu K, Granier C, Kopeyan C, Terakawa S, Rochat H (1994) Amino acid sequence of an excitatory insect-selective toxin (BmK IT) from venom of the scorpion Buthus martensi Karsch. Sci China B 37:42–49
Ji YH, Li YJ, Zhang JW, Song BL, Yamaki T, Mochizuki T, Hoshino M et al (1999) Covalent structures of BmK AS and BmK AS-1, two novel bioactive polypeptides purified from Chinese scorpion Buthus martensi Karsch. Toxicon 37:519–536

Karbat I, Cohen L, Gilles N, Gordon D, Gurevitz M (2004) Conversion of a scorpion toxin agonist into an antagonist highlights an acidic residue involved in voltage sensor trapping during activation of neuronal Na+ channels. FASEB J 18:683–689

Karbat I, Turkov M, Cohen L, Kahn R, Gordon D, Gurevitz M, Frolow F (2007) X-ray structure and mutagenesis of the scorpion depressant toxin LqhIT2 reveals key determinants crucial for activity and anti-insect selectivity. J Mol Biol 366:586–601

Keizer DW, West PJ, Lee EF, Yoshikami D, Olivera BM, Bulaj G, Norton RS (2003) Structural basis for tetrodotoxin-resistant sodium channel binding by mu-conotoxin SmIIIA. J Biol Chem 278:46805–46813

Kimura T, Kinoshita E, Yamaoka K, Yuki T, Yakehiro M, Seyama I (2000) On site of action of grayanotoxin in domain 4 segment 6 of rat skeletal muscle sodium channel. FEBS Lett 465:18–22

Kimura T, Yamaoka K, Kinoshita E, Maejima H, Yuki T, Yakehiro M, Seyama I (2001) Novel site on sodium channel alpha-subunit responsible for the differential sensitivity of grayanotoxin in skeletal and cardiac muscle. Mol Pharmacol 60:865–872

Kohling R (2002) Voltage-gated sodium channels in epilepsy. Epilepsia 43:1278–1295

Lai J, Hunter JC, Porreca F (2003) The role of voltage-gated sodium channels in neuropathic pain. Curr Opin Neurobiol 13:291–297

Legros C, Ceard B, Vacher H, Marchot P, Bougis PE, Martin-Eauclaire MF (2005) Expression of the standard scorpion alpha-toxin AaH II and AaH II mutants leading to the identification of some key bioactive elements. Biochim Biophys Acta 1723:91–99

Leipold E, Lu S, Gordon D, Hansel A, Heinemann SH (2004) Combinatorial interaction of scorpion toxins Lqh-2, Lqh-3, and LqhalphaIT with sodium channel receptor sites-3. Mol Pharmacol 65:685–691

Leipold E, Hansel A, Olivera BM, Terlau H, Heinemann SH (2005) Molecular interaction of delta-conotoxins with voltage-gated sodium channels. FEBS Lett 579:3881–3884

Leipold E, Hansel A, Borges A, Heinemann SH (2006) Subtype specificity of scorpion beta-toxin Tz1 interaction with voltage-gated sodium channels is determined by the pore loop of domain 3. Mol Pharmacol 70:340–347

Leipold E, Borges A, Heinemann SH (2012) Scorpion beta-toxin interference with NaV channel voltage sensor gives rise to excitatory and depressant modes. J Gen Physiol 139:305–319

Li RA, Tomaselli GF (2004) Using the deadly mu-conotoxins as probes of voltage-gated sodium channels. Toxicon 44:117–122

Li YJ, Tan ZY, Ji YH (2000) The binding of BmK IT2, a depressant insect-selective scorpion toxin on mammal and insect sodium channels. Neurosci Res 38:257–264

Li D, Xiao Y, Hu W, Xie J, Bosmans F, Tytgat J, Liang S (2003) Function and solution structure of hainantoxin-I, a novel insect sodium channel inhibitor from the Chinese bird spider Selenocosmia hainana. FEBS Lett 555:616–622

Linford NJ, Cantrell AR, Qu Y, Scheuer T, Catterall WA (1998) Interaction of batrachotoxin with the local anesthetic receptor site in transmembrane segment IVS6 of the voltage-gated sodium channel. Proc Natl Acad Sci U S A 95:13947–13952

Liu ZR, Ye P, Ji YH (2011) Exploring the obscure profiles of pharmacological binding sites on voltage-gated sodium channels by BmK neurotoxins. Protein Cell 2:437–444

Lombet A, Bidard JN, Lazdunski M (1987) Ciguatoxin and brevetoxins share a common receptor site on the neuronal voltage-dependent Na+ channel. FEBS Lett 219:355–359

Marban E, Yamagishi T, Tomaselli GF (1998) Structure and function of voltage-gated sodium channels. J Physiol 508(Pt 3):647–657

Marcotte P, Chen LQ, Kallen RG, Chahine M (1997) Effects of Tityus serrulatus scorpion toxin gamma on voltage-gated Na+ channels. Circ Res 80:363–369

Martin MF, Garcia y Perez LG, el Ayeb M, Kopeyan C, Bechis G, Jover E, Rochat H (1987) Purification and chemical and biological characterizations of seven toxins from the Mexican scorpion, Centruroides suffusus suffusus. J Biol Chem 262:4452–4459

Martin-Moutot N, Mansuelle P, Alcaraz G, Dos Santos RG, Cordeiro MN, De Lima ME, Seagar M et al (2006) Phoneutria nigriventer toxin 1: a novel, state-dependent inhibitor of neuronal sodium channels that interacts with micro conotoxin binding sites. Mol Pharmacol 69:1931–1937

Mattei C, Legros C (2014) The voltage-gated sodium channel: a major target of marine neurotoxins. Toxicon 91:84–95

Moran Y, Gordon D, Gurevitz M (2009) Sea anemone toxins affecting voltage-gated sodium channels – molecular and evolutionary features. Toxicon 54:1089–1101

Narahashi T (2008) Tetrodotoxin: a brief history. Proc Jpn Acad Ser B Phys Biol Sci 84:147–154

Narahashi T, Moore JW, Scott WR (1964) Tetrodotoxin blockage of sodium conductance increase in lobster giant axons. J Gen Physiol 47:965–974

Nicholson GM (2007) Insect-selective spider toxins targeting voltage-gated sodium channels. Toxicon 49:490–512

Noda M, Ikeda T, Kayano T, Suzuki H, Takeshima H, Kurasaki M, Takahashi H et al (1986) Existence of distinct sodium channel messenger RNAs in rat brain. Nature 320:188–192

Noguchi T, Jeon JK, Arakawa O, Sugita H, Deguchi Y, Shida Y, Hashimoto K (1986) Occurrence of tetrodotoxin and anhydrotetrodotoxin in Vibrio sp. isolated from the intestines of a xanthid crab, Atergatis floridus. J Biochem 99:311–314

Oliveira JS, Redaelli E, Zaharenko AJ, Cassulini RR, Konno K, Pimenta DC, Freitas JC et al (2004) Binding specificity of sea anemone toxins to Nav 1.1-1.6 sodium channels: unexpected contributions from differences in the IV/S3-S4 outer loop. J Biol Chem 279:33323–33335

Pearn J (2001) Neurology of ciguatera. J Neurol Neurosurg Psychiatry 70:4–8

Pedraza Escalona M, Possani LD (2013) Scorpion beta-toxins and voltage-gated sodium channels: interactions and effects. Front Biosci (Landmark Ed) 18:572–587

Pereira A, Cao Z, Murray TF, Gerwick WH (2009) Hoiamide a, a sodium channel activator of unusual architecture from a consortium of two papua new Guinea cyanobacteria. Chem Biol 16:893–906

Perez S, Vale C, Alonso E, Alfonso C, Rodriguez P, Otero P, Alfonso A et al (2011) A comparative study of the effect of ciguatoxins on voltage-dependent Na+ and K+ channels in cerebellar neurons. Chem Res Toxicol 24:587–596

Pintar A, Possani LD, Delepierre M (1999) Solution structure of toxin 2 from centruroides noxius Hoffmann, a beta-scorpion neurotoxin acting on sodium channels. J Mol Biol 287:359–367

Possani LD, Becerril B, Delepierre M, Tytgat J (1999) Scorpion toxins specific for Na+-channels. Eur J Biochem 264:287–300

Purkerson-Parker SL, Fieber LA, Rein KS, Podona T, Baden DG (2000) Brevetoxin derivatives that inhibit toxin activity. Chem Biol 7:385–393

Rashid MH, Mahdavi S, Kuyucak S (2013) Computational studies of marine toxins targeting ion channels. Mar Drugs 11:848–869

Rogart R (1981) Sodium channels in nerve and muscle membrane. Annu Rev Physiol 43:711–725

Rogers JC, Qu Y, Tanada TN, Scheuer T, Catterall WA (1996) Molecular determinants of high affinity binding of alpha-scorpion toxin and sea anemone toxin in the S3-S4 extracellular loop in domain IV of the Na+ channel alpha subunit. J Biol Chem 271:15950–15962

Rong M, Chen J, Tao H, Wu Y, Jiang P, Lu M, Su H et al (2011) Molecular basis of the tarantula toxin jingzhaotoxin-III (beta-TRTX-Cj1alpha) interacting with voltage sensors in sodium channel subtype Nav1.5. FASEB J 25:3177–3185

Sato S, Nakamura H, Ohizumi Y, Kobayashi J, Hirata Y (1983) The amino acid sequences of homologous hydroxyproline-containing myotoxins from the marine snail Conus geographus venom. FEBS Lett 155:277–280

Sato C, Ueno Y, Asai K, Takahashi K, Sato M, Engel A, Fujiyoshi Y (2001) The voltage-sensitive sodium channel is a bell-shaped molecule with several cavities. Nature 409:1047–1051

Schantz EJ (1986) Chemistry and biology of saxitoxin and related toxins. Ann N Y Acad Sci 479:15–23

Scheuer T (1994) Structure and function of voltage-gated sodium channels: regulation by phosphorylation. Biochem Soc Trans 22:479–482

Schreibmayer W, Jeglitsch G (1992) The sodium channel activator brevetoxin-3 uncovers a multiplicity of different open states of the cardiac sodium channel. Biochim Biophys Acta 1104:233–242

Sheumack DD, Howden ME, Spence I, Quinn RJ (1978) Maculotoxin: a neurotoxin from the venom glands of the octopus Hapalochlaena maculosa identified as tetrodotoxin. Science 199:188–189

Shichor I, Zlotkin E, Ilan N, Chikashvili D, Stuhmer W, Gordon D, Lotan I (2002) Domain 2 of Drosophila para voltage-gated sodium channel confers insect properties to a rat brain channel. J Neurosci 22:4364–4371

Shmukler YB, Nikishin DA (2017) Ladder-shaped ion channel ligands: current state of knowledge. Mar Drugs 15

Song W, Du Y, Liu Z, Luo N, Turkov M, Gordon D, Gurevitz M et al (2011) Substitutions in the domain III voltage-sensing module enhance the sensitivity of an insect sodium channel to a scorpion beta-toxin. J Biol Chem 286:15781–15788

Stapleton A, Blankenship DT, Ackermann BL, Chen TM, Gorder GW, Manley GD, Palfreyman MG et al (1990) Curtatoxins. Neurotoxic insecticidal polypeptides isolated from the funnel-web spider Hololena curta. J Biol Chem 265:2054–2059

Stephan MM, Potts JF, Agnew WS (1994) The microI skeletal muscle sodium channel: mutation E403Q eliminates sensitivity to tetrodotoxin but not to mu-conotoxins GIIIA and GIIIB. J Membr Biol 137:1–8

Stevens M, Peigneur S, Tytgat J (2011) Neurotoxins and their binding areas on voltage-gated sodium channels. Front Pharmacol 2:71

Strugatsky D, Zilberberg N, Stankiewicz M, Ilan N, Turkov M, Cohen L, Pelhate M et al (2005) Genetic polymorphism and expression of a highly potent scorpion depressant toxin enable refinement of the effects on insect Na channels and illuminate the key role of Asn-58. Biochemistry 44:9179–9187

Tan ZY, Xiao H, Mao X, Wang CY, Zhao ZQ, Ji YH (2001) The inhibitory effects of BmK IT2, a scorpion neurotoxin on rat nociceptive flexion reflex and a possible mechanism for modulating voltage-gated Na(+) channels. Neuropharmacology 40:352–357

Tang C, Zhou X, Huang Y, Zhang Y, Hu Z, Wang M, Chen P et al (2014) The tarantula toxin jingzhaotoxin-XI (kappa-theraphotoxin-Cj1a) regulates the activation and inactivation of the voltage-gated sodium channel Nav1.5. Toxicon 92:6–13

Tikhonov DB, Zhorov BS (2005) Sodium channel activators: model of binding inside the pore and a possible mechanism of action. FEBS Lett 579:4207–4212

Trainer VL, Baden DG, Catterall WA (1994) Identification of peptide components of the brevetoxin receptor site of rat brain sodium channels. J Biol Chem 269:19904–19909

Turner AD, Higgins C, Davidson K, Veszelovszki A, Payne D, Hungerford J, Higman W (2015) Potential threats posed by new or emerging marine biotoxins in UK waters and examination of detection methodology used in their control: brevetoxins. Mar Drugs 13:1224–1254

Ulbricht W (1998) Effects of veratridine on sodium currents and fluxes. Rev Physiol Biochem Pharmacol 133:1–54

Wada A, Wanke E, Gullo F, Schiavon E (2008) Voltage-dependent Na(v)1.7 sodium channels: multiple roles in adrenal chromaffin cells and peripheral nervous system. Acta Physiol (Oxf) 192:221–231

Wang SY, Wang GK (1998) Point mutations in segment I-S6 render voltage-gated Na+ channels resistant to batrachotoxin. Proc Natl Acad Sci U S A 95:2653–2658

Wang SY, Wang GK (1999) Batrachotoxin-resistant Na+ channels derived from point mutations in transmembrane segment D4-S6. Biophys J 76:3141–3149

Wang SY, Wang GK (2003) Voltage-gated sodium channels as primary targets of diverse lipid-soluble neurotoxins. Cell Signal 15:151–159

Wang SY, Barile M, Wang GK (2001) Disparate role of Na(+) channel D2-S6 residues in batrachotoxin and local anesthetic action. Mol Pharmacol 59:1100–1107

Wang CZ, Zhang H, Jiang H, Lu W, Zhao ZQ, Chi CW (2006) A novel conotoxin from Conus striatus, mu-SIIIA, selectively blocking rat tetrodotoxin-resistant sodium channels. Toxicon 47:122–132

Wang SY, Tikhonov DB, Mitchell J, Zhorov BS, Wang GK (2007) Irreversible block of cardiac mutant Na+ channels by batrachotoxin. Channels (Austin, TX) 1:179–188

Wei P, Xu C, Wu Q, Huang L, Liang S, Yuan C (2014) Jingzhaotoxin-35, a novel gating-modifier toxin targeting both Nav1.5 and Kv2.1 channels. Toxicon 92:90–96

Wood JN, Baker M (2001) Voltage-gated sodium channels. Curr Opin Pharmacol 1:17–21

Xiao Y, Bingham JP, Zhu W, Moczydlowski E, Liang S, Cummins TR (2008) Tarantula huwentoxin-IV inhibits neuronal sodium channels by binding to receptor site 4 and trapping the domain ii voltage sensor in the closed configuration. J Biol Chem 283:27300–27313

Yamaoka K, Inoue M, Miyazaki K, Hirama M, Kondo C, Kinoshita E, Miyoshi H et al (2009) Synthetic ciguatoxins selectively activate Nav1.8-derived chimeric sodium channels expressed in HEK293 cells. J Biol Chem 284:7597–7605

Yasumoto T (2001) The chemistry and biological function of natural marine toxins. Chem Rec 1:228–242

Ye P, Hua L, Jiao Y, Li Z, Qin S, Fu J, Jiang F et al (2016) Functional up-regulation of Nav1.8 sodium channel on dorsal root ganglia neurons contributes to the induction of scorpion sting pain. Acta Biochim Biophys Sin 48:132–144

Zakon HH (2012) Adaptive evolution of voltage-gated sodium channels: the first 800 million years. Proc Natl Acad Sci U S A 109(Suppl 1):10619–10625

Zhang MM, Green BR, Catlin P, Fiedler B, Azam L, Chadwick A, Terlau H et al (2007) Structure/function characterization of micro-conotoxin KIIIA, an analgesic, nearly irreversible blocker of mammalian neuronal sodium channels. J Biol Chem 282:30699–30706

Zhao R, Zhang XY, Yang J, Weng CC, Jiang LL, Zhang JW, Shu XQ et al (2008) Anticonvulsant effect of BmK IT2, a sodium channel-specific neurotoxin, in rat models of epilepsy. Br J Pharmacol 154:1116–1124

Zhu HL, Wassall RD, Takai M, Morinaga H, Nomura M, Cunnane TC, Teramoto N (2009a) Actions of veratridine on tetrodotoxin-sensitive voltage-gated Na currents, Na1.6, in murine vas deferens myocytes. Br J Pharmacol 157:1483–1493

Zhu MM, Tao J, Tan M, Yang HT, Ji YH (2009b) U-shaped dose-dependent effects of BmK AS, a unique scorpion polypeptide toxin, on voltage-gated sodium channels. Br J Pharmacol 158:1895–1903

Zimmer T, Haufe V, Blechschmidt S (2014) Voltage-gated sodium channels in the mammalian heart. Global Cardiol Sci Pract 2014:449–463

Zlotkin E, Eitan M, Bindokas VP, Adams ME, Moyer M, Burkhart W, Fowler E (1991) Functional duality and structural uniqueness of depressant insect-selective neurotoxins. Biochemistry 30:4814–4821

Zuo XP, Ji YH (2004) Molecular mechanism of scorpion neurotoxins acting on sodium channels: insight into their diverse selectivity. Mol Neurobiol 30:265–278

Mechanisms of Drug Binding to Voltage-Gated Sodium Channels

M. E. O'Leary and M. Chahine

Contents

Abstract

Voltage-gated sodium (Na^+) channels are expressed in virtually all electrically excitable tissues and are essential for muscle contraction and the conduction of impulses within the peripheral and central nervous systems. Genetic disorders that disrupt the function of these channels produce an array of Na^+ channelopathies resulting in neuronal impairment, chronic pain, neuromuscular pathologies, and cardiac arrhythmias. Because of their importance to the conduction of electrical

M. E. O'Leary
Cooper Medical School of Rowan University, Camden, NJ, USA
e-mail: olearym@rowan.edu

M. Chahine (✉)
CERVO Brain Research Center, Institut universitaire en santé mentale de Québec, Quebec City, QC, Canada

Department of Medicine, Université Laval, Quebec City, QC, Canada
e-mail: mohamed.chahine@phc.ulaval.ca

M. Chahine (ed.), *Voltage-gated Sodium Channels: Structure, Function and Channelopathies*, Handbook of Experimental Pharmacology 246,
https://doi.org/10.1007/164_2017_73

signals, Na^+ channels are the target of a wide variety of local anesthetic, antiarrhythmic, anticonvulsant, and antidepressant drugs. The voltage-gated family of Na^+ channels is composed of α-subunits that encode for the voltage sensor domains and the Na^+-selective permeation pore. In vivo, Na^+ channel α-subunits are associated with one or more accessory β-subunits (β_1–β_4) that regulate gating properties, trafficking, and cell-surface expression of the channels. The permeation pore of Na^+ channels is divided in two parts: the outer mouth of the pore is the site of the ion selectivity filter, while the inner cytoplasmic pore serves as the channel activation gate. The cytoplasmic lining of the permeation pore is formed by the S6 segments that include highly conserved aromatic amino acids important for drug binding. These residues are believed to undergo voltage-dependent conformational changes that alter drug binding as the channels cycle through the closed, open, and inactivated states. The purpose of this chapter is to broadly review the mechanisms of Na^+ channel gating and the models used to describe drug binding and Na^+ channel inhibition.

Keywords
Gating · Local anesthetics · Na_v · Sodium channels · Structure-function

1 Introduction

Voltage-gated Na^+ channels are expressed in electrically excitable tissues where they are essential for the initiation and propagation of action potentials. Genetic disorders that change expression or alter the gating properties of Na^+ channels have been linked to a broad spectrum of channelopathies in the neuronal, skeletal muscle, and cardiovascular systems. Because of their central role of the rapid conduction of electrical impulses, Na^+ channels have proved to be valuable targets for the therapeutic treatment of chronic pain, epilepsy, and cardiac arrhythmias. Local anesthetic, antiarrhythmic, and anticonvulsant drugs inhibit Na^+ channels and slow or completely inhibit the conduction of action potentials. Depending on the physical properties of the drug, Na^+ channel inhibition can occur by a simple pore-blocking mechanism or by preferential binding to and stabilization of the channels in nonconducting inactivated states. Both mechanisms slow the repriming of drug-modified channels under resting conditions and reduce the rate of action potential firing. Charged forms of these drugs preferentially gain access to their binding site through the internal aqueous pathway created by the state-dependent opening of the channels. Uncharged and hydrophobic drugs display less state dependence and appear to access the binding site on closed and inactivated channels through fenestrations located in the walls of the cytoplasmic pore. Mutagenesis has identified a number of conserved residues that contribute to the cytoplasmic binding site for these drugs. Drug binding is modified by membrane voltage, interaction with permeant ions, and changes in extracellular pH. Here we review Na^+ channel electrophysiology and the mechanisms of local anesthetic binding and explore the models used to describe Na^+ channel inhibition.

2 Molecular Biology of Na^+ Channels

Currently, there are ten distinct isoforms of Na^+ channels that have been identified in the human genome ($Na_v1.1$–$Na_V1.9$, Na_X). These channels have common structural motifs but differ in amino acid sequences and physiological properties (Goldin 2001). Na^+ channels are identified by the chemical symbol of the principal permeating ion (e.g., Na), the physiological regulator (voltage) indicated as a subscript (e.g., Na_V), the numerical gene subfamily (e.g., Na_V1), and a numerical value specific for each isoform (e.g., $Na_V1.1$). Splice variants of a specific isoform are designated by lowercase letters (e.g., $Na_V1.1a$) (Goldin et al. 2000). These Na^+ channels vary in their primary structures, tissue distribution, biophysical properties, and sensitivity to neurotoxins (Chahine et al. 2008). Although Na^+ channels are widely expressed, they are often classified based on the tissue in which they are most abundantly expressed. For example, $Na_V1.1$, $Na_V1.2$, $Na_V1.3$, and $Na_V1.6$ are predominantly expressed in the central nervous system, while $Na_V1.7$, $Na_v1.8$, and $Na_v1.9$ are predominantly localized within the peripheral nervous system (Catterall et al. 2005; Goldin et al. 2000). Fast inactivating TTX-sensitive ($Na_v1.7$) and a slowly inactivating TTX-insensitive ($Na_v1.8$) channels are highly expressed in dorsal root ganglion neurons where they contribute to the transmission of nociceptive signals (Chahine and O'Leary 2014; Ho and O'Leary 2011). $Na_V1.4$ is the skeletal muscle voltage-gated Na^+ channel. $Na_V1.5$ is the predominant Na^+ channel in cardiac myocytes but has also been detected in the piriform cortex and subcortical limbic nuclei (Hartmann et al. 1999). Cardiac myocytes also express several neuronal ($Na_V1.1$, $Na_V1.2$, $Na_V1.3$, $Na_V1.6$, $Na_V1.8$) (Maier et al. 2004) and muscle ($Na_V1.4$) isoforms (Qu et al. 2007). Na_x is expressed in the uterus, heart, and skeletal muscle where it appears to contribute to the regulation of salt intake (Akopian et al. 1997; Watanabe et al. 2000).

3 Sodium Channelopathies

Genetic disorders that disrupt Na^+ channel function underlie a number of inherited channelopathies of neuronal and muscle tissues. Mutations of the skeletal muscle $Na_V1.4$ channel are associated with hyperkalemic periodic paralysis (Fontaine et al. 1990; Shen et al. 2017), paramyotonia congenita (Chahine et al. 1994), potassium-aggravated myotonias (Mitrovic et al. 1994), and congenital myasthenic syndrome (Habbout et al. 2016). Mutations of $Na_v1.7$ channels resulting in their gain or loss of function have been linked to erythromelalgia, paroxysmal extreme pain disorder, and congenital insensitivity to pain (Dib-Hajj et al. 2009; Lampert et al. 2010). Mutations of $Na_V1.5$ Na^+ channels have been linked to long QT syndrome (Bennett et al. 1995b; Keller et al. 2003), Brugada syndrome (Brugada and Brugada 1992; Baroudi et al. 2004), and cardiac conduction defects (Veldkamp et al. 2000). Mutations in brain Na^+ channels cause epilepsy, mental retardation, pancerebellar atrophy, and ataxia (Meisler and Kearney 2005; Trudeau et al. 2006).

4 Structure and Function Relationships

Voltage-gated Na^+ channels generate the inward currents that underlie the rapid depolarizing phases of nerve, skeletal muscle, and cardiac action potentials (Hille 2001). Despite high sequence homology, these channels display significant differences in gating, permeation, and pharmacology (Catterall et al. 2005). The mechanisms that govern these differences are not well understood and likely reflect the different roles these Na^+ channels play in their native tissues (O'Leary 1998).

The large α-subunit (260 kDa) of Na^+ channels encodes for both ion-selective permeation and the voltage-dependent gating properties of the channels (Catterall 1986; Fozzard and Hanck 1996). Na^+ channels are organized into four homologous domains (DI–DIV) each composed of six transmembrane-spanning segments (S1–S6) (Catterall 2000) (Fig. 1). Na^+ channels have distinct pore-forming and voltage-sensing domains that combine to produce a voltage-gated permeation pathway similar to what has been described for other members of the ion channel superfamily (Yu et al. 2005; Yu and Catterall 2004). The external mouth of the pore is formed by the short extracellular linkers connecting the S5 and S6 segments (P regions) from each of the four homologous domains. The P regions form hairpin loops that dip into out of the channel core from the external side partially traversing the membrane (Doyle et al. 1998). Residues of the P regions form a narrow ring

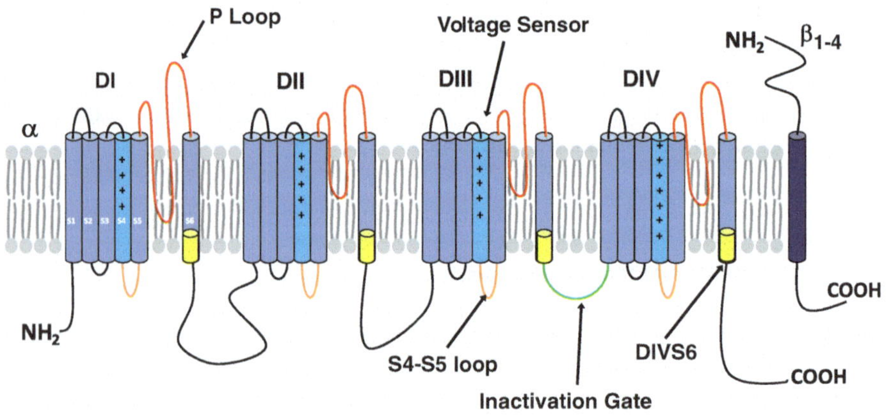

Fig. 1 Organization of voltage-gated sodium channels. The α-subunit is composed of 24 transmembrane-spanning segments organized into 4 homologous domains (DI–DIV). The S1–S4 segments include the positively charged S4 and form the voltage sensor domains. The S5–S6 segments form the inner mouth of the channel and the activation gate. The S6 segments contribute residues that line the cytoplasmic pore and contribute to drug binding (yellow). The extracellular linkers between the S5 and S6 segments form the P loops and contribute residues that determine ionic selectivity (red). The interdomain D3–D4 linker includes the IFM motif and serves as the inactivation gate of the channel (green). The S4–S5 intracellular linkers couple the voltage sensor domains to the activation gate (brown). The β-subunits (β_{1-4}) have a single transmembrane segment with an extracellular N-terminus and an intracellular C-terminus (purple). β-subunits appear to interact with the channel voltage sensors and regulate channel gating

around the external mouth of the pore (DEKA motif) that acts as the ion selectivity filter (Heinemann et al. 1992; Sun et al. 1997). The P region contributed by DI includes a phenylalanine residue that is conserved in tetrodotoxin-sensitive Na^+ channels (Satin et al. 1992; Backx et al. 1992). This residue is nonaromatic in cardiac $Na_v1.5$ and peripheral nerve $Na_v1.8/Na_v1.9$ channels accounting for the reduced tetrodotoxin sensitivity of these isoforms (Catterall et al. 2005).

The cytoplasmic aspect of the pore is formed by the S6 transmembrane segments that circumscribe a water-filled pathway that enables the rapid permeation of Na^+ ions. The S6 segments are formed an inverted teeppee with the internal ends of the S6 segments converging to form the narrow cytoplasmic entrance and the activation gate of the channel (Payandeh et al. 2011; McCusker et al. 2012; Zhang et al. 2012) (Fig. 2). Residues situated near the cytoplasmic ends of the S6 segments contribute to a binding site for the endogenous inactivation gate and have been implicated in the development of slow inactivation (McPhee et al. 1994, 1995; Vedantham and Cannon 2000; O'Reilly et al. 2001). Local anesthetic, antiarrhythmic, and anticonvulsant drugs interact with well-defined residues of the S6 segments (Ragsdale et al.

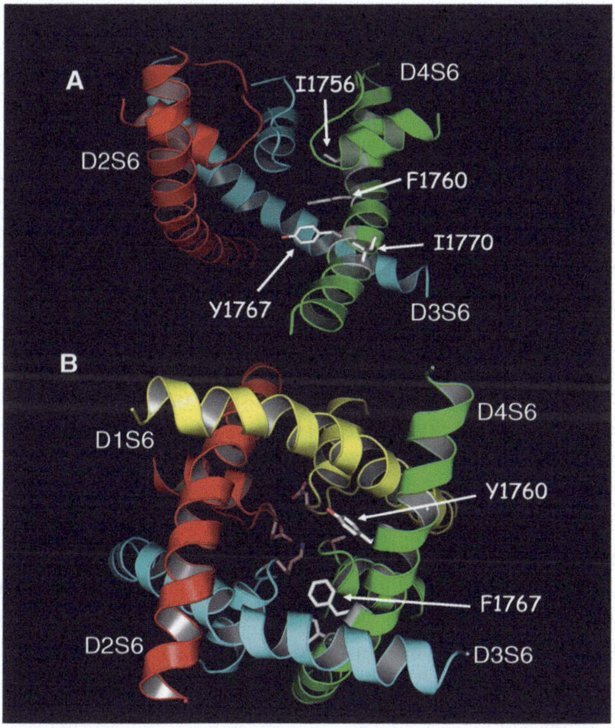

Fig. 2 Homology model of the DI–DIVS6 segments. (**A**) Side view of a homology model of the Na^+ channel DIIS6–DIVS6 segments based on the M2 segment of the MthK channel. The DIS6 was removed for more clarity. (**B**) View of the channel pore viewed from the bottom indicating that F1760 and Y1767 ($Na_V1.5$ labeling) are exposed within the central cavity

1994; Yarov-Yarovoy et al. 2001, 2002). Considering the diverse functions of the S6 segments, it is not surprising that drug binding within the inner pore of these channels both inhibits Na^+ current and modifies channel gating.

5 Voltage-Dependent Gating of Na^+ Channels

The ability of Na^+ channels to respond to changes in the membrane voltage is mediated by the S1–S4 transmembrane segments that collectively serve as the voltage sensor domains of the channel (Catterall 2014). The S4 segments of these domains have positively charged residues at every third position enabling them to sense changes in the membrane voltage (Noda et al. 1984). Depolarization causes the outward movement of 13 positive charges leading to opening of the channel (Satin et al. 1992; Schoppa et al. 1992; Hirschberg et al. 1995). This charge movement has been linked to the outward displacement of the voltage sensor domains (Yang and Horn 1995; Yang et al. 1996). The voltage sensors of domains I–III produce gating currents that have been linked to Na^+ channel activation, while domain IV appears to play a more specific role in fast inactivation (Chanda and Bezanilla 2002; Chen et al. 1996; Sheets et al. 1999; Capes et al. 2013). These findings are consistent with previous mutagenesis studies showing a close relationship between the DIV voltage sensor and fast inactivation (Chahine et al. 1994; Chen et al. 1996). The cytoplasmic S4–S5 linkers serve as structural links between the voltage-sensing and pore-forming domains (McPhee et al. 1998; Lerche et al. 1997). Conformational changes in the voltage sensors acting through the S4–S5 linkers and S5 segments translate into lateral movement of the S6 segments away from the central axis causing the channels to open (Bagneris et al. 2013, 2015). Channel opening creates a continuous aqueous pathway between the extracellular and intracellular sides of the channel that facilitates the movement of Na^+ ions across the hydrophobic core of the plasma membrane (Doyle et al. 1998).

After several milliseconds of depolarization, the activated channels enter into a nonconducting inactivated state. Early studies employing internally applied proteases determined that the region involved in fast inactivation was preferentially accessible from the cytoplasmic side of the channel (Armstrong et al. 1973; Patlak and Horn 1982). Subsequent work showed that site-directed mutations and antibodies directed against the interdomain DIII–DIV linker disrupted fast inactivation (Vassilev et al. 1989; Stuhmer et al. 1989) (Fig. 1). These studies proposed that the DIII–DIV linker acts as a hinged lid that occludes the cytoplasmic entrance of the channel during inactivation. A conserved isoleucine-phenylalanine-methionine-threonine (IFMT) motif was identified in the III–IV linker that appears to serve as a latch that keeps the inactivation gate shut until the membrane potential is returned to a hyperpolarized voltage (West et al. 1992). Cysteine modification studies demonstrated that the IFMT motif was inaccessible to intracellular hydrophilic reagent during prolonged depolarization suggesting that these residues may be buried in a nonaqueous (hydrophobic) environment during inactivation (Chahine et al. 1997; Kellenberger et al. 1996). Although the docking site for this latch has not been identified, residues near the cytoplasmic ends of the S6 segments of DI, DIII, and DIV along with the intracellular S4–S5 linkers of domains III and IV have been implicated as potential binding sites for the inactivation gate (Fig. 1) (Tang et al. 1996; McPhee et al. 1994, 1995; Smith and Goldin 1997; Lerche et al. 1997).

6 Mechanisms of Drug Binding and Channel Inhibition

Much of our current understanding of drug binding and inhibition stems from early studies of positively charged quaternary ammonium (QA) pore blockers of potassium channels (Armstrong 1966, 1971). These studies revealed that (1) QAs primarily gain access to their binding sites from the cytoplasmic side, (2) access to the binding site is regulated by the activation gate, (3) depending on the length of the QA hydrophobic tails, these compounds can either be trapped within the pore or impede the closing of the activation gate by a foot-in-the-door mechanism, and (4) raising the external concentration of K^+ ions displaces QAs from their internal sites by a trans K^+ knockoff mechanism. These findings are consistent with a QA binding site situated between the cytoplasmic activation gate and the narrow selectivity filter of the channel. Subsequent studies of voltage-gated Na^+ channels indicated a similar mechanism of QA inhibition (Cahalan 1978; Cahalan and Almers 1979a; O'Leary and Horn 1994; O'Leary et al. 1994; Strichartz 1973; Yeh and Narahashi 1977).

The observed properties of QA block of K^+ channels correlates well with structural studies of the bacterial KcsA channel (Doyle et al. 1998). Based on KcsA and subsequent work, the K^+ channel pore domain is formed by the transmembrane S5 and S6 helices that are organized as an inverted cone that converges near the cytoplasmic end to create a narrow constriction, the putative activation gate. The external end of the cytoplasmic pore is circumscribed by the S5–S6 linkers that contribute residues important for K^+ ion selectivity. A large cavity is formed between the selectivity filter and the narrow cytoplasmic ends of the S6 segments that is capable of accommodating hydrated K^+ ions and large QA compounds (Jiang et al. 2002).

In mammalian Na^+ channels, highly conserved phenylalanine and tyrosine residues of the DIVS6 are important contributors to anesthetic binding (Ragsdale et al. 1994, 1996). The DIVS6 phenylalanine is situated near the cytoplasmic entrance of the selectivity filter and stabilizes binding via cation-pi interactions with the amine moieties of these drugs (Ahern et al. 2008; Pless et al. 2011). The conserved DIVS6 tyrosine is situated near the internal end of the cytoplasmic pore just upstream from the narrow activation gate and appears to primarily associate with the hydrophobic moieties of these drugs (Ragsdale et al. 1994). Mutations of these residues generally weaken the inactivation-dependent inhibition of Na^+ channels by 10- to 100-fold. Mutagenesis has identified additional residues of the DIS6 and DIIIS6 segments that also contribute to the binding of anesthetics (Yarov-Yarovoy et al. 2001, 2002). However, mutations of the DI and DIII segments produce comparatively small reductions in anesthetic inhibition, suggesting that highly conserved phenylalanine and tyrosine residues of DIVS6 are most important for drug binding (Lipkind and Fozzard 2005).

At present, there are no high-resolution structures of mammalian Na^+ channels. However, the structures of several prokaryotic Na^+ channels have been obtained that provide accurate models of the pore-forming and voltage-sensing domains of these channels (McCusker et al. 2012; Payandeh et al. 2011; Zhang et al. 2012). Unlike eukaryotic Na^+ channels, bacterial channels are formed from four independent subunits each composed of six transmembrane-spanning segments (Payandeh et al. 2011; Ren et al. 2001). Consequently, these channels do not have an interdomain III–IV linker, which serves as the inactivation gate in eukaryotic channels, and fail to display rapid N-type inactivation (Pavlov et al. 2005). The core structure of these bacterial channels shares many common features with K^+ channels including S1–S4 voltage sensor domains, S5–S6 linker P loops that form the selectivity filter, a water-filled central cavity, and S6 segments that line the cytoplasmic pore and converge to form the activation gate near the cytoplasmic entrance of the channel (Payandeh et al. 2011). While the bacterial channels have provided important information about the structure and function relationships of Na^+ channels, they do not undergo fast inactivation, and the S6 residues important for anesthetic drug binding in eukaryotic channels are not conserved, indicating that elements previously identified to be important for state-dependent drug binding are not present in prokaryotic channels (Pavlov et al. 2005; Corry et al. 2014).

7 Modulated Receptor Hypothesis

In the presence of local anesthetics, rapid repetitive stimulation causes a progressive decrease in Na^+ current amplitude over the course of the pulse train. This inhibition preferentially occurs at depolarized voltages suggesting that anesthetic drugs may preferentially inhibit open or inactivated Na^+ channels. Several models have been proposed to account for the "use-dependent" inhibition produced by local anesthetics (Courtney 1975). The most widely accepted of these models is the modulated receptor hypothesis (MRH), the basis of which was originally formulated for local anesthetic inhibition of Na^+ channels (Courtney 1975; Hille 1977). MRH is rooted in the principle that the affinity of anesthetic binding varies with the gating state of the channel, with the open (O) and inactivated (I) states having higher affinity than the closed (C) conformation (Fig. 3). The model proposes that under resting conditions, drug binding is minimal because the binding site is in its low-affinity closed conformation (C) and drug access to the cytoplasmic binding site is blocked by the activation gate. Depolarization has two effects: (1) it opens the activation gate, allowing the drug to rapidly access the binding site through the cytoplasmic aqueous pathway (O→OB), and (2) it increases the affinity of the drug binding site (O). After several milliseconds of depolarization, Na^+ channels enter into the inactivated state (O→I), which further increases binding affinity (OB→IB) and blocks exit of the drug from the channel via the internal cytoplasmic pathway.

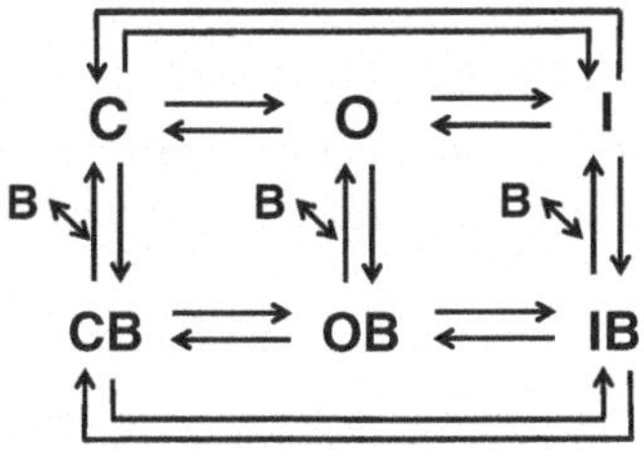

Fig. 3 State diagram of the modulated receptor hypothesis. C, O, and I are the normal closed, open, and inactivated conformations of Na^+ channels. The rate constants governing the transition between the C and O states are voltage-dependent. CB, OB, and IB are drug-associated closed, open, and inactivated conformations, respectively. Drug (B) binding to the C and I states is believed to occur via a hydrophobic pathway. Drug binding to the O state occurs through an open aqueous pathway. The O and I states bind drugs with high affinity, while C is the low-affinity conformation (I > O > C)

The time course of drug binding to inactivated channels (I→IB) is slower than the open-channel block but generally has a higher binding affinity. For example, the binding of lidocaine to $Na_V1.5$ channels displays affinities of 600 μM and 36 μM for the open and inactivated states, respectively, while closed channels are typically inhibited by higher concentrations of the drug (Bennett et al. 1995a). Enzymatic treatments and mutations that disrupt fast inactivation weaken anesthetic inhibition, supporting a role for fast inactivation in drug binding (Cahalan 1978; Bennett et al. 1995a). The inactivation gate of Na^+ channels is believed to act by plugging the cytoplasmic pore, thus inhibiting ion permeation and limiting the access of charged drugs to the cytoplasmic binding site via the aqueous pathway. To account for direct binding to closed and inactivated channels, an alternative hydrophobic pathway was initially proposed (Hille 1977). The existence of this pathway has been inferred from data showing that small uncharged and lipophilic drugs are capable of reaching the cytoplasmic binding site when the channels are closed or inactivated. Recent structural studies of bacterial Na^+ channels have identified fenestrations in the walls of the cytoplasmic pore that appears to enable drugs to gain access to the cytoplasmic binding site when the channel is closed or inactivated (Payandeh et al. 2011). Anesthetics that gain access to the binding site through the membrane lipid phase and lateral fenestrations are less dependent on state-dependent changes in accessibility and generally display weaker use-dependent inhibition than charged anesthetics (Courtney and Etter 1983; Hille 1977; Schwarz et al. 1977; Quan et al. 1996).

During rapid repetitive stimulation, Na^+ channels enter into nonconducting drug-bound states (IB, OB) from which they slowly recover when the channels return to the resting state. Anesthetic binding delays the recovery of the channels, which fail to fully recover between depolarizations within the pulse train. The progressive accumulation of the channels in these drug-modified states during successive pulses accounts for the characteristic use-dependent inhibition (Courtney 1975). The modulated receptor model proposes that high-affinity binding of anesthetics stabilizes the channels in nonconducting states from which they only slowly recover at hyperpolarized voltages. Although the mechanism underlying the slow recovery of channels at hyperpolarized voltages has not been clearly established, the mechanism likely involves interplay between recovery from inactivation and drug dissociation from the cytoplasmic binding site. These actions of clinically relevant anesthetics have been reviewed elsewhere (Butterworth and Strichartz 1990; Hille 2001; Fozzard et al. 2005).

8 Guarded Receptor Hypothesis

The guarded receptor hypothesis (GRH) proposes that access to the anesthetic binding site of Na^+ channels is regulated by the activation and inactivation gates (Starmer et al. 1984). Unlike MRH, there is no requirement that the affinity of anesthetic binding changes with the gating state of the channel. Opening creates a contiguous aqueous pathway between the internal side of the channel and the binding site located within the cytoplasmic pore. Anesthetic binding via the aqueous pathway is rapid compared to the hydrophobic membrane-delimited pathway used when the channels are closed or inactivated. Once bound within the pore, the drugs appear to inhibit the permeation of Na^+ ions via steric or electrostatic interactions (McNulty et al. 2007; Lipkind and Fozzard 2005; Courtney 1975; Tikhonov and Zhorov 2017). This is consistent with KcsA structural data showing the binding of pore-blocking tetraalkylammoniums within the central cavity just below the P loops interact with K^+ ions in the selectivity filter (Zhou et al. 2001; Faraldo-Gomez et al. 2007). For small drugs, the gating of the channel is not impaired so that the activation and inactivation gates close normally trapping the drug within the pore. Because hyperpolarization favors closing, the trapped drugs must either wait for the channels to reopen or diffuse out of the closed channels via hydrophobic pathways. Both mechanisms lead to the slow recovery of drug-modified channels, the principal driver of use-dependent inhibition. Small drugs like lidocaine and its quaternary derivatives (QX-314, QX-222) are effectively trapped within the Na^+ pore during hyperpolarization and do not impair channel gating (Yeh and Tanguy 1985). Large drugs that cannot be accommodated by the cytoplasmic binding site prevent closing by a foot-in-the-door mechanism (Armstrong 1971). Such drugs must exit the pore before the channel can fully close and produce characteristic "hooked" tail currents due to the sequential unblocking and closing of the channels at hyperpolarized voltages (Yeh and Narahashi 1977).

During repetitive pulsing, Na^+ channels transiently open before inactivating, providing a brief window (1–2 ms) for drug binding to open channels. Although very few channels bind drugs during any given pulse, over the course of a pulse train, the number of blocked channels progressively increases, resulting in use-dependent inhibition. The resulting use-dependent inhibition and the apparent hyperpolarizing shift in inactivation are similar for GRH and MRH, making it difficult to distinguish between these mechanisms. Mutations that facilitate drug escape from closed channels accelerate the recovery of channels inhibited by open-channel blockers but not those inhibited by drugs that bind to the inactivated state (O'Leary et al. 2003; Ramos and O'Leary 2004). This suggests that, unlike inactivated-state binding, trapped drugs may not tightly bind within the pore when the channels are closed. This is consistent with data showing that DIVS6 mutations have a relatively weak effect on the inhibition produced by open-channel blockers compared to drugs that bind to the inactivated state (Ragsdale et al. 1996). These findings provide additional support for studies showing that open-channel blockers do not generate strong cation-pi interactions with the highly conserved phenylalanine of the DIVS6 (Pless et al. 2011). In principal, drugs that bind to the inactivated state and open-channel blockers may interact with different but overlapping binding sites within the cytoplasmic pore. The recovery of channels inhibited by these drugs may display differences in kinetics and voltage dependence. For example, the repriming of inactivated-state binding drugs is accelerated by strong hyperpolarization that destabilizes drug binding while delaying the repriming of channels inhibited by pore blockers by favoring channel closing. It seems likely that hybrid models incorporating elements of both MRH and GRH may be necessary to fully explain drug inhibition of Na^+ channels.

9 Alternative Mechanisms of Drug Inhibition

While the open and inactivated states of Na^+ channels are well-established targets for anesthetic binding, other mechanisms have been proposed to account for the anesthetic inhibition of Na^+ channels. Anesthetics have also been proposed to either directly bind to the slow inactivated state or to promote slow inactivation by allosteric mechanisms (Balser et al. 1996; Ong et al. 2000; Chen et al. 2000). A clear determination of the role of slow inactivation in anesthetic binding has been hampered by the extensive overlap of anesthetic binding and unbinding with the onset and recovery from slow inactivation (Ulbricht 2005; Sandtner et al. 2004). Defining the role of slow inactivation is further complicated by the substantial differences in the time course and the extent of slow inactivation in different Na^+ channel isoforms (O'Reilly et al. 1999; Richmond et al. 1998). Slow binding of anesthetics to fast-inactivated channels and rapid binding to slow-inactivated channels are difficult to distinguish using standard pulsing protocols (Karoly et al. 2010; Jo and Bean 2017). Some studies have found that anesthetics impede channel entry into slow-inactivated states (Sandtner et al. 2004; Sheets et al. 2011; Jo and Bean 2017). Other studies have proposed alternative mechanisms where anesthetics preferentially bind to activated state that precede fast inactivation or alter the coupling between the channel gates and voltage sensors (Vedantham and Cannon 1999; Hanck et al. 2000; Wang et al. 2004).

10 Interaction Between Permeant Cations and Pore-Blocking Drugs

Reducing the external concentration of Na^+ is known to potentiate the inhibition produced by pore-blocking drugs (Cahalan and Almers 1979b; Gingrich et al. 1993; Ramos and O'Leary 2004). Interactions between blockers entering the channels from the cytoplasmic side and Na^+ ions entering from the external side appear to occur within the narrow confines of the permeation pore. The data are consistent with a mechanism in which charged drugs and Na^+ ions occupy adjacent but overlapping sites within the pore. Electrostatic interactions between Na^+ ions and positively charged drugs prevent both sites from being simultaneously occupied (Cahalan and Almers 1979b; Shapiro 1977). The destabilization of drug binding by raising the trans-concentration of Na^+ is consistent with a simple knockoff mechanism (Armstrong 1971). Recent structural studies of the prokaryotic Na_vMs channel revealed that these open channels have three Na^+ ions bound within the selectivity filter (Naylor et al. 2016). The innermost Na^+ binding site is adjacent to the central hydrophobic cavity and close to the putative drug binding site. Drug binding to Na_VMs channels reduces the Na^+ ion occupancy of the selectivity filter consistent with an electrostatic interaction between charged blockers and permeating Na^+ ions (Bagneris et al. 2014; Ulmschneider et al. 2013; McCusker et al. 2012; Tikhonov and Zhorov 2017). Depolarizations considered to be near the peak of neuronal action potentials favor the binding of charged compounds to the cytoplasmic pore and appear to potentiate binding by reducing the Na^+ occupancy of the selectivity filter (O'Leary et al. 1994; Baukrowitz and Yellen 1996). Drug inhibition is therefore sensitive to both the test voltage and the external concentration of Na^+ ions (O'Leary and Chahine 2002). The majority of studies investigating drug inhibition of Na^+ channels typically use test pulses to relatively hyperpolarized test pulses (<0 mV) and therefore tend to underestimate the role of voltage-dependent block of open channels (O'Leary and Chahine 2002; Wang et al. 2004).

11 Drug Inhibition Is Voltage-Dependent

Both the GRH and MRH propose that, under resting conditions, access by positively charged anesthetics to the cytoplasmic binding site of Na^+ channels is prevented by the closed activation gate. Voltage-dependent channel opening facilitates drug binding by creating a continuous aqueous pathway between the cytoplasm and the inner-pore binding site. A second component of voltage dependence is observed at more depolarized voltages where Na^+ channels are maximally activated. This component of drug binding is not dependent on channel gating but is due to the location of the cytoplasmic binding site, a portion of which is believed to be located within the membrane electric field (Strichartz 1973). Strong depolarization potentiates inhibition by driving the charged drugs onto the cytoplasmic binding site resulting in an increase in binding affinity (Strichartz 1973; Kimbrough and Gingrich 2000; Wang 1988; O'Leary and Chahine 2002; O'Leary and Horn 1994). These findings suggest that positively charged drugs enter the pore from the cytoplasmic side of the channel and traverse $\approx 50\%$ of the transmembrane electric field to reach their binding sites. Alternatively, drug binding outside the membrane electric field could lead to voltage dependence via energetic coupling of drug binding to the outward movement of permeant cations through the membrane electric field (Lu 2004). Electron density profiles of Na_VMs channels suggest that pore blockers reduce the Na^+ occupancy of the Na_VMs selectivity filter (Bagneris et al. 2014). These data raise the possibility that the voltage dependence of drug binding may arise from the outward displacement of Na^+ ions through the selectivity filter rather than a direct electrostatic interaction between the charged drug and the membrane electric field. This mechanism gains support from modeling of KcsA channels showing that the bulk of the membrane voltage is concentrated in the selectivity filter and is comparatively low within the central cavity where pore blockers appear to bind (Zhou et al. 2001; Faraldo-Gomez et al. 2007; Jogini and Roux 2005).

12 Modulation of Drug Binding by External Protons

A reduction in the external pH is known to stabilize anesthetic binding to Na^+ channels, while equivalent changes in internal pH have no effect (Grant et al. 1980, 1982; Bean et al. 1983; Courtney 1979; Schwarz et al. 1977). The data suggests that protons are able to gain access to anesthetics bound within the cytoplasmic binding site by permeating through the external mouth of the pore. H^+ binding converts the bound anesthetics to their charged forms, preventing their escape from the channel through intracellular hydrophobic pathways. This trapping effect appears to be related to the titration of the drug rather than the channel as the inhibition produced by internally applied quaternary derivatives of anesthetics is not altered by changes in external pH (Nettleton and Wang 1990). The inability of the positively charged drug to rapidly exit the channel accounts for the potent use-dependent inhibition observe in these studies. The effect of raising external H^+ concentration on drug trapping is reminiscent of the use-dependent inhibition produced by the permanently charged QX-314 analog of lidocaine (Strichartz 1973).

13 Recovery from Drug Inhibition

Voltage-dependent gating is linked to the outward translocation of the positively charged S4 segments of Na^+ channels through the membrane electric field (Armstrong and Bezanilla 1973; Keynes and Rojas 1974). The gating currents produced by the outward movement of the S4 voltage sensors produce conformational changes that result in channel opening and inactivation. For short depolarizing pulses that fail to evoke inactivation, the OFF gating charge measured after the termination of the depolarization is equal but opposite to the ON gating charge measured during the initial application of the pulse. In contrast, longer depolarizations sufficient to induce fast inactivation decrease the gating currents and immobilize $\approx$65% of the OFF gating charge (Armstrong and Bezanilla 1977). Inactivation prevents a portion of the gating charge from rapidly returning to the resting state at hyperpolarized voltages. Fast inactivation parallels the time course of gating charge immobilization, and intracellular treatments with proteases abolish both fast inactivation and charge immobilization (Armstrong and Bezanilla 1977). Fluorometric labeling studies indicate that the S4 voltage sensors of DIII and DIV are specifically linked to fast inactivation (Cha et al. 1999). Mutations of the S4–S5 linkers and the transmembrane S5–S6 segments of $Na_V1.4$ channels disrupt the electromagnetic coupling between the membrane voltage and channel opening and are consistent with an allosteric interaction between the S4 voltage sensors and the channel gates (Muroi et al. 2010). The S4–S5 linkers appear to play a central role in electromagnetic coupling. Binding of the inactivation gate both blocks the cytoplasmic entrance of the channel and prevents the DIIIS4 and DIVS4 voltage sensors from returning to their resting state. While the inactivation of Na^+ channels displays little intrinsic voltage dependence (Aldrich et al. 1983), hyperpolarization dramatically increases the rate of recovery from inactivation indicating that recovery is strongly voltage-dependent. This voltage dependence appears to result from a link between the immobilized gating charge and dissociation of the inactivation gate from its cytoplasmic binding site (Aldrich et al. 1983; Armstrong and Bezanilla 1977). In response to strong hyperpolarization, inward movement of the immobilized S4 voltage sensors appears to weaken the binding of the inactivation gate leading to rapid recovery from inactivation (Patlak 1991).

The repriming of drug-modified channels is slowed and shifted toward hyperpolarized voltages by comparison to drug-free controls. Gating current studies revealed a link between anesthetic binding, Na^+ channel inactivation, and immobilization of the gating charge (Keynes and Rojas 1974; Cahalan 1978; Cahalan and Almers 1979b). Anesthetics both slow recovery from inactivation and gating charge remobilization (Butterworth and Strichartz 1990). Like endogenous inactivation, the inhibition produced by anesthetics is reversed by strong hyperpolarization, suggesting that the repriming of drug-modified channels is similarly voltage-dependent. The effects of endogenous inactivation and anesthetic binding on gating charge immobilization are not additive, suggesting that these processes immobilize the same component of the gating charge (Cahalan and Almers 1979b; Tanguy and Yeh 1989). These findings suggest a mechanism in which anesthetics stabilize fast inactivation thereby slowing recovery at hyperpolarized voltages and delaying the remobilization of the gating charge. Synergy between the bound anesthetic and the inactivation gate appears to stabilize the channels in the fast-inactivated state thereby opposing both recovery and gating charge remobilization.

The mechanism underlying the priming of drug-modified channels has not been clearly established and may be different for drugs that tightly bind to inactivated states versus more loosely bound pore blockers. Strong hyperpolarization appears to induce a conformational change in the cytoplasmic binding site, leading to drug dissociation followed by rapid recovery from inactivation (Fig. 3, IB→I→C). We speculate that the slow dissociation of drug from inactivated channels produced by the inward movement of the immobilized domain III and IV voltage sensors is followed by rapid recovery from inactivation. This contrasts with pore-blocking drugs where strong hyperpolarization tends to trap the drugs within the cytoplasmic pore behind the closed activation gate. Weakly bound pore blockers may allow channels to rapidly recover from inactivation followed by slow untrapping from the closed state (Fig. 3, IB→CB→C). Ultimately, both mechanisms slow the recovery of drug-modified channels and produce similar use-dependent inhibition and hyperpolarizing shifts in inactivation.

14 Regulation of Drug Binding by Auxiliary β-Subunits

In vivo, these Na^+ channels are associated with one or more auxiliary β-subunits (β_1–β_4) that modulate the voltage dependence, gating properties, and expression levels of the channels (Isom 2002). Four distinct isoforms (β_1–β_4) and two splice variants (β_{1A}, β_{1B}) have been identified (Malhotra et al. 2000; Qin et al. 2003). They share a common structure consisting of a single membrane-spanning domain, a short intracellular C-terminal domain, and a large extracellular N-terminal domain incorporating an immunoglobulin-like fold similar to that found in cell adhesion molecules (Isom 2001; Yu et al. 2003). Depending on the Na^+ channel isoform and β-subunit combination, interactions between these subunits have been shown to modulate the gating kinetics, voltage dependence, and expression of the channels (Catterall 2000). β-Subunits also interact with cytoskeleton proteins, the extracellular matrix, and other molecules that regulate cell migration and aggregation (Yu et al. 2003; Brackenbury et al. 2008).

The role of auxiliary β-subunits in anesthetic inhibition is not well-established. Co-expressing the β_1-subunit with cardiac $Na_V1.5$ channels weakens anesthetic binding to the resting state and accelerates the recovery of drug-modified channels (Makielski et al. 1996). This occurs in the absence of any apparent change in the affinity of anesthetic binding to high-affinity inactivated states. Studies of truncated β_1-subunits lacking the intracellular C-terminus indicate that the cytoplasmic tail does not directly contribute to the anesthetic binding site or mediate the functional effects of the β_1-subunit on drug binding (Makielski et al. 1996). Conformational changes in the membrane-spanning and extracellular N-terminal domains appear to play a more prominent role in anesthetic binding. Recent data suggests that the β_1- and β_3-subunits differentially regulate Na^+ channels by interacting with the DIII and DIV voltage sensor domains (Zhu et al. 2017).

15 Conclusion

It has been known for decades that the anesthetic inhibition of voltage-gated Na^+ channels is dependent on both the frequency and voltage of the applied pulses. Studies of use-dependent inhibition, delayed recovery from inactivation, and hyperpolarizing shifts in the steady-state availability of Na^+ currents spawned the concept that drug binding may be dependent on the gating state of the channel. This notion was supported by work demonstrating that small hydrophobic molecules readily bound to closed and inactivated channels, while charged molecules preferentially interacted with open channels. The data indicated that the gating state of the channel, the chemical properties of drug, and the pathway used to gain access to the binding site all contributed to the observed inhibition. Furthermore, the data indicated that rather acting as a simple pore blocker, when bound these drugs appear to modify channel gating. These observations formed the foundation for the modulated receptor hypothesis in which the affinity of drug binding is determined by the gating state of the channel. Preferential binding to open and inactivated states stabilizes the channels in high-affinity nonconducting conformations that only slowly recover under resting conditions. Implicit in this formulation is that the drug binding site undergoes a voltage-dependent conformational change leading to an increase in the affinity of drug binding. We now know that the drug binding site is located within the central cavity of the channels within close proximity of the external selectivity filter. Drugs gain access to the site either through the open channel or through fenestrations in the lateral walls of the cytoplasmic pore. Aromatic and hydrophobic residues lining the cytoplasmic cavity form what can be generously described as a promiscuous binding site for drugs with varying structures and chemical properties. How this binding site accommodates the wide variety of clinically useful drugs that target these channels is the subject of ongoing work. What is currently missing is a clear understanding of the mechanisms that produce changes in drug binding affinity. Presumably, altered binding could result from state-dependent changes in the exposure of conserved binding site residues within the cytoplasmic pore. Comparing the contributions of binding site residues to drug inhibition under different gating states has provided some insights but has not yet produced a comprehensive picture of the state-dependent mechanism of drug binding. Development of novel approaches for tracking the voltage-dependent changes in the drug binding site linked to channel gating would lead to a better understanding of the state-dependent changes in drug binding.

References

Ahern CA, Eastwood AL, Dougherty DA, Horn R (2008) Electrostatic contributions of aromatic residues in the local anesthetic receptor of voltage-gated sodium channels. Circ Res 102:86–94

Akopian AN, Souslova V, Sivilotti L, Wood JN (1997) Structure and distribution of a broadly expressed atypical sodium channel. FEBS Lett 400:183–187

Aldrich RW, Corey DP, Stevens CF (1983) A reinterpretation of mammalian sodium channel gating based on single channel recording. Nature 306:436–441

Armstrong CM (1966) Time course of TEA(+)-induced anomalous rectification in squid giant axons. J Gen Physiol 50:491–503

Armstrong CM (1971) Interaction of tetraethylammonium ion derivatives with the potassium channels of giant axons. J Gen Physiol 58:413–437

Armstrong CM, Bezanilla F (1973) Currents related to movement of the gating particles of the sodium channels. Nature 242:459–461

Armstrong CM, Bezanilla F (1977) Inactivation of the sodium channel. II. Gating current experiments. J Gen Physiol 70:567–590

Armstrong CM, Bezanilla F, Rojas E (1973) Destruction of sodium conductance inactivation in squid axons perfused with pronase. J Gen Physiol 62:375–391

Backx PH, Yue DT, Lawrence JH, Marban E, Tomaselli GF (1992) Molecular localization of an ion-binding site within the pore of mammalian sodium channels. Science 257:248–251

Bagneris C, DeCaen PG, Hall BA, Naylor CE, Clapham DE, Kay CW, Wallace BA (2013) Role of the C-terminal domain in the structure and function of tetrameric sodium channels. Nat Commun 4:2465

Bagneris C, DeCaen PG, Naylor CE, Pryde DC, Nobeli I, Clapham DE, Wallace BA (2014) Prokaryotic NavMs channel as a structural and functional model for eukaryotic sodium channel antagonism. Proc Natl Acad Sci U S A 111:8428–8433

Bagneris C, Naylor CE, McCusker EC, Wallace BA (2015) Structural model of the open-closed-inactivated cycle of prokaryotic voltage-gated sodium channels. J Gen Physiol 145:5–16

Balser JR, Nuss HB, Romashko DN, Marban E, Tomaselli GF (1996) Functional consequences of lidocaine binding to slow-inactivated sodium channels. J Gen Physiol 107:643–658

Baroudi G, Napolitano C, Priori SG, Del BA, Chahine M (2004) Loss of function associated with novel mutations of the SCN5A gene in patients with Brugada syndrome. Can J Cardiol 20:425–430

Baukrowitz T, Yellen G (1996) Use-dependent blockers and exit rate of the last ion from the multi-ion pore of a K+ channel. Science 271:653–656

Bean BP, Cohen CJ, Tsien RW (1983) Lidocaine block of cardiac sodium channels. J Gen Physiol 81:613–642

Bennett PB, Valenzuela C, Chen LQ, Kallen RG (1995a) On the molecular nature of the lidocaine receptor of cardiac Na+ channels. Modification of block by alterations in the alpha-subunit III-IV interdomain. Circ Res 77:584–592

Bennett PB, Yazawa K, Makita N, George AL Jr (1995b) Molecular mechanism for an inherited cardiac arrhythmia. Nature 376:683–685

Brackenbury WJ, Djamgoz MB, Isom LL (2008) An emerging role for voltage-gated Na channels in cellular migration: regulation of central nervous system development and potentiation of invasive cancers. Neurosci 14:571–583

Brugada P, Brugada J (1992) Right bundle branch block, persistent ST segment elevation and sudden cardiac death: a distinct clinical and electrocardiographic syndrome. A multicenter report. J Am Coll Cardiol 20:1391–1396

Butterworth JF, Strichartz GR (1990) Molecular mechanisms of local anesthesia: a review. Anesthesiology 72:711–734

Cahalan MD (1978) Local anesthetic block of sodium channels in normal and pronase-treated squid giant axons. Biophys J 23:285–311

Cahalan MD, Almers W (1979a) Block of sodium conductance and gating current in squid giant axons poisoned with quaternary strychnine. Biophys J 27:57–73

Cahalan MD, Almers W (1979b) Interactions between quaternary lidocaine, the sodium channel gates, and tetrodotoxin. Biophys J 27:39–55

Capes DL, Goldschen-Ohm MP, Arcisio-Miranda M, Bezanilla F, Chanda B (2013) Domain IV voltage-sensor movement is both sufficient and rate limiting for fast inactivation in sodium channels. J Gen Physiol 142:101–112

Catterall WA (1986) Molecular properties of voltage-sensitive sodium channels. Annu Rev Biochem 55:953–985

Catterall WA (2000) From ionic currents to molecular mechanisms: the structure and function of voltage-gated sodium channels. Neuron 26:13–25

Catterall WA (2014) Structure and function of voltage-gated sodium channels at atomic resolution. Exp Physiol 99:35–51

Catterall WA, Goldin AL, Waxman SG (2005) International union of pharmacology. XLVII. Nomenclature and structure-function relationships of voltage-gated sodium channels. Pharmacol Rev 57:397–409

Cha A, Ruben PC, George AL Jr, Fujimoto E, Bezanilla F (1999) Voltage sensors in domains III and IV, but not I and II, are immobilized by Na+ channel fast inactivation. Neuron 22:73–87

Chahine M, O'Leary ME (2014) Regulation/modulation of sensory neuron sodium channels. Handb Exp Pharmacol 221:111–135

Chahine M, George AL Jr, Zhou M, Ji S, Sun W, Barchi RL, Horn R (1994) Sodium channel mutations in paramyotonia congenita uncouple inactivation from activation. Neuron 12:281–294

Chahine M, Deschenes I, Trottier E, Chen LQ, Kallen RG (1997) Restoration of fast inactivation in an inactivation-defective human heart sodium channel by the cysteine modifying reagent benzyl-MTS: analysis of IFM-ICM mutation. Biochem Biophys Res Commun 233:606–610

Chahine M, Chatelier A, Babich O, Krupp JJ (2008) Voltage-gated sodium channels in neurological disorders. CNS Neurol Disord Drug Targets 7:144–158

Chanda B, Bezanilla F (2002) Tracking voltage-dependent conformational changes in skeletal muscle sodium channel during activation. J Gen Physiol 120:629–645
Chen LQ, Santarelli V, Horn R, Kallen RG (1996) A unique role for the S4 segment of domain 4 in the inactivation of sodium channels. J Gen Physiol 108:549–556
Chen Z, Ong BH, Kambouris NG, Marban E, Tomaselli GF, Balser JR (2000) Lidocaine induces a slow inactivated state in rat skeletal muscle sodium channels. J Physiol 524:37–49
Corry B, Lee S, Ahern CA (2014) Pharmacological insights and quirks of bacterial sodium channels. Handb Exp Pharmacol 221:251–267
Courtney KR (1975) Mechanism of frequency-dependent inhibition of sodium currents in frog myelinated nerve by the lidocaine derivative GEA. J Pharmacol Exp Ther 195:225–236
Courtney KR (1979) Extracellular PH selectively modulates recovery from sodium inactivation in frog myelinated nerve. Biophys J 28:363–368
Courtney KR, Etter EF (1983) Modulated anticonvulsant block of sodium channels in nerve and muscle. Eur J Pharmacol 88:1–9
Dib-Hajj SD, Binshtok AM, Cummins TR, Jarvis MF, Samad T, Zimmermann K (2009) Voltage-gated sodium channels in pain states: role in pathophysiology and targets for treatment. Brain Res Rev 60:65–83
Doyle DA, Morais CJ, Pfuetzner RA, Kuo A, Gulbis JM, Cohen SL, Chait BT, MacKinnon R (1998) The structure of the potassium channel: molecular basis of K+ conduction and selectivity. Science 280:69–77
Faraldo-Gomez JD, Kutluay E, Jogini V, Zhao Y, Heginbotham L, Roux B (2007) Mechanism of intracellular block of the KcsA K+ channel by tetrabutylammonium: insights from X-ray crystallography, electrophysiology and replica-exchange molecular dynamics simulations. J Mol Biol 365:649–662
Fontaine B, Khurana TS, Hoffman EP, Bruns GA, Haines JL, Trofatter JA, Hanson MP, Rich J, McFarlane H, Yasek DM et al (1990) Hyperkalemic periodic paralysis and the adult muscle sodium channel alpha-subunit gene. Science 250:1000–1002
Fozzard HA, Hanck DA (1996) Structure and function of voltage-dependent sodium channels: comparison of brain II and cardiac isoforms. Physiol Rev 76:887–926
Fozzard HA, Lee PJ, Lipkind GM (2005) Mechanism of local anesthetic drug action on voltage-gated sodium channels. Curr Pharm Des 11:2671–2686
Gingrich KJ, Beardsley D, Yue DT (1993) Ultra-deep blockade of Na^+ channels by a quaternary ammonium ion: catalysis by a transition-intermediate state? J Physiol 471:319–341
Goldin AL (2001) Resurgence of sodium channel research. Annu Rev Physiol 63:871–894
Goldin AL, Barchi RL, Caldwell JH, Hofmann F, Howe JR, Hunter JC, Kallen RG, Mandel G, Meisler MH, Netter YB, Noda M, Tamkun MM, Waxman SG, Wood JN, Catterall WA (2000) Nomenclature of voltage-gated sodium channels. Neuron 28:365–368
Grant AO, Strauss LJ, Wallace AG, Strauss HC (1980) The influence of PH on Th electrophysiological effects of lidocaine in guinea pig ventricular myocardium. Circ Res 47:542–550
Grant AO, Trantham JL, Brown KK, Strauss HC (1982) PH-dependent effects of quinidine on the kinetics of DV/Dt_{max} in guinea pig ventricular myocardium. Circ Res 50:210–217
Habbout K, Poulin H, Rivier F, Giuliano S, Sternberg D, Fontaine B, Eymard B, Morales RJ, Echenne B, King L, Hanna MG, Mannikko R, Chahine M, Nicole S, Bendahhou S (2016) A recessive Nav1.4 mutation underlies congenital myasthenic syndrome with periodic paralysis. Neurology 86:161–169
Hanck DA, Makielski JC, Sheets MF (2000) Lidocaine alters activation gating of cardiac Na channels. Pflugers Arch 439:814–821
Hartmann HA, Colom LV, Sutherland ML, Noebels JL (1999) Selective localization of cardiac SCN5A sodium channels in limbic regions of rat brain. Nat Neurosci 2:593–595
Heinemann SH, Terlau H, Stuhmer W, Imoto K, Numa S (1992) Calcium channel characteristics conferred on the sodium channel by single mutations. Nature 356:441–443
Hille B (1977) Local anesthetics: hydrophilic and hydrophobic pathways for the drug-receptor reaction. J Gen Physiol 69:497–515

Hille B (2001) Ion channels in excitable membranes. Sinauer, Sunderland
Hirschberg B, Rovner A, Lieberman M, Patlak J (1995) Transfer of twelve charges is needed to open skeletal muscle Na+ channels. J Gen Physiol 106:1053–1068
Ho C, O'Leary ME (2011) Single-cell analysis of sodium channel expression in dorsal root ganglion neurons. Mol Cell Neurosci 46:159–166
Isom LL (2001) Sodium channel beta subunits: anything but auxiliary. Neurosci 7:42–54
Isom LL (2002) b-subunits: players in neuronal hyperexcitability? Novartis Found Symp 241: 124–138
Jiang Y, Lee A, Chen J, Cadene M, Chait BT, MacKinnon R (2002) The open pore conformation of potassium channels. Nature 417:523–526
Jo S, Bean BP (2017) Lacosamide inhibition of Nav1.7 voltage-gated sodium channels: slow binding to fast-inactivated states. Mol Pharmacol 91:277–286
Jogini V, Roux B (2005) Electrostatics of the intracellular vestibule of K+ channels. J Mol Biol 354:272–288
Karoly R, Lenkey N, Juhasz AO, Vizi ES, Mike A (2010) Fast- or slow-inactivated state preference of Na+ channel inhibitors: a simulation and experimental study. PLoS Comput Biol 6:e1000818
Kellenberger S, Scheuer T, Catterall WA (1996) Movement of the Na+ channel inactivation gate during inactivation. J Biol Chem 271:30971–30979
Keller DI, Acharfi S, Delacretaz E, Benammar N, Rotter M, Pfammatter JP, Fressart V, Guicheney P, Chahine M (2003) A novel mutation in SCN5A, DelQKP 1507-1509, causing long QT syndrome: role of Q1507 residue in sodium channel inactivation. J Mol Cell Cardiol 35:1513–1521
Keynes RD, Rojas E (1974) Kinetics and steady-state properties of the charged system controlling sodium conductance in the squid giant axon. J Physiol 239:393–434
Kimbrough JT, Gingrich KJ (2000) Quaternary ammonium block of mutant Na+ channels lacking inactivation: features of a transition-intermediate mechanism. J Physiol 529:93–106
Lampert A, O'Reilly AO, Reeh P, Leffler A (2010) Sodium channelopathies and pain. Pflugers Arch 460:249–263
Lerche H, Peter W, Fleischhauer R, Pika-Hartlaub U, Malina T, Mitrovic N, Lehmann-Horn F (1997) Role in fast inactivation of the IV/S4-S5 loop of the human muscle Na+ channel probed by cysteine mutagenesis. J Physiol 505:345–352
Lipkind GM, Fozzard HA (2005) Molecular modeling of local anesthetic drug binding by voltage-gated sodium channels. Mol Pharmacol 68:1611–1622
Lu Z (2004) Mechanism of rectification in inward-rectifier K+ channels. Annu Rev Physiol 66:103–129
Maier SK, Westenbroek RE, McCormick KA, Curtis R, Scheuer T, Catterall WA (2004) Distinct subcellular localization of different sodium channel alpha and beta subunits in single ventricular myocytes from mouse heart. Circulation 109:1421–1427
Makielski JC, Limberis JT, Chang SY, Fan Z, Kyle JW (1996) Coexpression of beta 1 with cardiac sodium channel alpha subunits in oocytes decreases lidocaine block. Mol Pharmacol 49:30–39
Malhotra JD, Kazen-Gillespie K, Hortsch M, Isom LL (2000) Sodium channel beta subunits mediate homophilic cell adhesion and recruit ankyrin to points of cell-cell contact. J Biol Chem 275:11383–11388
McCusker EC, Bagneris C, Naylor CE, Cole AR, D'Avanzo N, Nichols CG, Wallace BA (2012) Structure of a bacterial voltage-gated sodium channel pore reveals mechanisms of opening and closing. Nat Commun 3:1102
McNulty MM, Edgerton GB, Shah RD, Hanck DA, Fozzard HA, Lipkind GM (2007) Charge at the lidocaine binding site residue Phe-1759 affects permeation in human cardiac voltage-gated sodium channels. J Physiol 581:741–755
McPhee JC, Ragsdale DS, Scheuer T, Catterall WA (1994) A mutation in segment IVS6 disrupts fast inactivation of sodium channels. Proc Natl Acad Sci U S A 91:12346–12350

McPhee JC, Ragsdale DS, Scheuer T, Catterall WA (1995) A critical role for transmembrane segment IVS6 of the sodium channel alpha subunit in fast inactivation. J Biol Chem 270:12025–12034

McPhee JC, Ragsdale DS, Scheuer T, Catterall WA (1998) A critical role for the S4-S5 intracellular loop in domain IV of the sodium channel alpha-subunit in fast inactivation. J Biol Chem 273:1121–1129

Meisler MH, Kearney JA (2005) Sodium channel mutations in epilepsy and other neurological disorders. J Clin Invest 115:2010–2017

Mitrovic N, George AL Jr, Heine R, Wagner S, Pika U, Hartlaub U, Zhou M, Lerche H, Fahlke C, Lehmann-Horn F (1994) K(+)-aggravated myotonia: destabilization of the inactivated state of the human muscle Na+ channel by the V1589M mutation. J Physiol 478:395–402

Muroi Y, Arcisio-Miranda M, Chowdhury S, Chanda B (2010) Molecular determinants of coupling between the domain III voltage sensor and pore of a sodium channel. Nat Struct Mol Biol 17:230–237

Naylor CE, Bagneris C, DeCaen PG, Sula A, Scaglione A, Clapham DE, Wallace BA (2016) Molecular basis of ion permeability in a voltage-gated sodium channel. EMBO J 35:820–830

Nettleton J, Wang GK (1990) PH-dependent binding of local anesthetics in single batrachotoxin-activated Na+ channels. Cocaine Vs. quaternary compounds. Biophys J 58:95–106

Noda M, Shimizu S, Tanabe T, Takai T, Kayano T, Ikeda T, Takahashi H, Nakayama H, Kanaoka Y, Minamino N et al (1984) Primary structure of electrophorus electricus sodium channel deduced from CDNA sequence. Nature 312:121–127

O'Leary ME (1998) Characterization of the isoform-specific differences in the gating of neuronal and muscle sodium channels. Can J Physiol Pharmacol 76:1041–1050

O'Leary ME, Chahine M (2002) Cocaine binds to a common site on open and inactivated human heart (Na(v)1.5) sodium channels. J Physiol 541:701–716

O'Leary ME, Horn R (1994) Internal block of human heart sodium channels by symmetrical tetra-alkylammoniums. J Gen Physiol 104:507–522

O'Leary ME, Kallen RG, Horn R (1994) Evidence for a direct interaction between internal tetra-alkylammonium cations and the inactivation gate of cardiac sodium channels. J Gen Physiol 104:523–539

O'Leary ME, Digregorio M, Chahine M (2003) Closing and inactivation potentiate the cocaethylene inhibition of cardiac sodium channels by distinct mechanisms. Mol Pharmacol 64:1575–1585

O'Reilly JP, Wang SY, Kallen RG, Wang GK (1999) Comparison of slow inactivation in human heart and rat skeletal muscle Na+ channel chimaeras. J Physiol 515:61–73

O'Reilly JP, Wang SY, Wang GK (2001) Residue-specific effects on slow inactivation at V787 in D2-S6 of Na(v)1.4 sodium channels. Biophys J 81:2100–2111

Ong BH, Tomaselli GF, Balser JR (2000) A structural rearrangement in the sodium channel pore linked to slow inactivation and use dependence. J Gen Physiol 116:653–662

Patlak J (1991) Molecular kinetics of voltage-dependent Na+ channels. Physiol Rev 71:1047–1080

Patlak J, Horn R (1982) Effect of N-bromoacetamide on single sodium channel currents in excised membrane patches. J Gen Physiol 79:333–351

Pavlov E, Bladen C, Winkfein R, Diao C, Dhaliwal P, French RJ (2005) The pore, not cytoplasmic domains, underlies inactivation in a prokaryotic sodium channel. Biophys J 89:232–242

Payandeh J, Scheuer T, Zheng N, Catterall WA (2011) The crystal structure of a voltage-gated sodium channel. Nature 475:353–358

Pless SA, Galpin JD, Frankel A, Ahern CA (2011) Molecular basis for class Ib anti-arrhythmic inhibition of cardiac sodium channels. Nat Commun 2:351

Qin N, D'Andrea MR, Lubin ML, Shafaee N, Codd EE, Correa AM (2003) Molecular cloning and functional expression of the human sodium channel beta1B subunit, a novel splicing variant of the beta1 subunit. Eur J Biochem 270:4762–4770

Qu Y, Karnabi E, Chahine M, Vassalle M, Boutjdir M (2007) Expression of skeletal muscle Na(V) 1.4 Na channel isoform in canine cardiac Purkinje myocytes. Biochem Biophys Res Commun 355:28–33

Quan C, Mok WM, Wang GK (1996) Use-dependent inhibition of Na+ currents by benzocaine homologs. Biophys J 70:194–201

Ragsdale DS, McPhee JC, Scheuer T, Catterall WA (1994) Molecular determinants of state-dependent block of Na+ channels by local anesthetics. Science 265:1724–1728

Ragsdale DS, McPhee JC, Scheuer T, Catterall WA (1996) Common molecular determinants of local anesthetic, antiarrhythmic, and anticonvulsant block of voltage-gated Na+ channels. Proc Natl Acad Sci U S A 93:9270–9275

Ramos E, O'Leary ME (2004) State-dependent trapping of flecainide in the cardiac sodium channel. J Physiol 560:37–49

Ren D, Navarro B, Xu H, Yue L, Shi Q, Clapham DE (2001) A prokaryotic voltage-gated sodium channel. Science 294:2372–2375

Richmond JE, Featherstone DE, Hartmann HA, Ruben PC (1998) Slow inactivation in human cardiac sodium channels. Biophys J 74:2945–2952

Sandtner W, Szendroedi J, Zarrabi T, Zebedin E, Hilber K, Glaaser I, Fozzard HA, Dudley SC, Todt H (2004) Lidocaine: a foot in the door of the inner vestibule prevents ultra-slow inactivation of a voltage-gated sodium channel. Mol Pharmacol 66:648–657

Satin J, Kyle JW, Chen M, Bell P, Cribbs LL, Fozzard HA, Rogart RB (1992) A mutant of TTX-resistant cardiac sodium channels with TTX-sensitive properties. Science 256:1202–1205

Schoppa NE, McCormack K, Tanouye MA, Sigworth FJ (1992) The size of gating charge in wild-type and mutant Shaker potassium channels. Science 255:1712–1715

Schwarz W, Palade PT, Hille B (1977) Local anesthetics. Effect of PH on use-dependent block of sodium channels in frog muscle. Biophys J 20:343–368

Shapiro BI (1977) Effects of strychnine on the sodium conductance of the frog node of ranvier. J Gen Physiol 69:915–926

Sheets MF, Kyle JW, Kallen RG, Hanck DA (1999) The Na channel voltage sensor associated with inactivation is localized to the external charged residues of domain IV, S4. Biophys J 77:747–757

Sheets PL, Jarecki BW, Cummins TR (2011) Lidocaine reduces the transition to slow inactivation in Na(v)1.7 voltage-gated sodium channels. Br J Pharmacol 164:719–730

Shen H, Zhou Q, Pan X, Li Z, Wu J, Yan N (2017) Structure of a eukaryotic voltage-gated sodium channel at near-atomic resolution. Science 355:4326

Smith MR, Goldin AL (1997) Interaction between the sodium channel inactivation linker and domain III S4-S5. Biophys J 73:1885–1895

Starmer CF, Grant AO, Strauss HC (1984) Mechanisms of use-dependent block of sodium channels in excitable membranes by local anesthetics. Biophys J 46:15–27

Strichartz GR (1973) The inhibition of sodium currents in myelinated nerve by quaternary derivatives of lidocaine. J Gen Physiol 62:37–57

Stuhmer W, Conti F, Suzuki H, Wang XD, Noda M, Yahagi N, Kubo H, Numa S (1989) Structural parts involved in activation and inactivation of the sodium channel. Nature 339:597–603

Sun YM, Favre I, Schild L, Moczydlowski E (1997) On the structural basis for size-selective permeation of organic cations through the voltage-gated sodium channel. Effect of alanine mutations at the DEKA locus on selectivity, inhibition by Ca^{2+} and H^{+}, and molecular sieving. J Gen Physiol 110:693–715

Tang L, Kallen RG, Horn R (1996) Role of an S4-S5 linker in sodium channel inactivation probed by mutagenesis and a peptide blocker. J Gen Physiol 108:89–104

Tanguy J, Yeh JZ (1989) QX-314 restores gating charge immobilization abolished by chloramine-T treatment in squid giant axons. Biophys J 56:421–427

Tikhonov DB, Zhorov BS (2017) Mechanism of sodium channel block by local anesthetics, antiarrhythmics, and anticonvulsants. J Gen Physiol 149:465–481

Trudeau MM, Dalton JC, Day JW, Ranum LP, Meisler MH (2006) Heterozygosity for a protein truncation mutation of sodium channel SCN8A in a patient with cerebellar atrophy, ataxia, and mental retardation. J Med Genet 43:527–530

Ulbricht W (2005) Sodium channel inactivation: molecular determinants and modulation. Physiol Rev 85:1271–1301

Ulmschneider MB, Bagneris C, McCusker EC, DeCaen PG, Delling M, Clapham DE, Ulmschneider JP, Wallace BA (2013) Molecular dynamics of ion transport through the open conformation of a bacterial voltage-gated sodium channel. Proc Natl Acad Sci U S A 110:6364–6369

Vassilev P, Scheuer T, Catterall WA (1989) Inhibition of inactivation of single sodium channels by a site-directed antibody. Proc Natl Acad Sci U S A 86:8147–8151

Vedantham V, Cannon SC (1999) The position of the fast-inactivation gate during lidocaine block of voltage-gated Na^+ channels. J Gen Physiol 113:7–16

Vedantham V, Cannon SC (2000) Rapid and slow voltage-dependent conformational changes in segment IVS6 of voltage-gated Na(+) channels. Biophys J 78:2943–2958

Veldkamp MW, Viswanathan PC, Bezzina C, Baartscheer A, Wilde AA, Balser JR (2000) Two distinct congenital arrhythmias evoked by a multidysfunctional Na(+) channel. Circ Res 86: E91–E97

Wang GK (1988) Cocaine-induced closures of single batrachotoxin-activated Na^+ channels in planar lipid bilayers. J Gen Physiol 92:747–765

Wang SY, Mitchell J, Moczydlowski E, Wang GK (2004) Block of inactivation-deficient Na+ channels by local anesthetics in stably transfected mammalian cells: evidence for drug binding along the activation pathway. J Gen Physiol 124:691–701

Watanabe E, Fujikawa A, Matsunaga H, Yasoshima Y, Sako N, Yamamoto T, Saegusa C, Noda M (2000) Nav2/NaG channel is involved in control of salt-intake behavior in the CNS. J Neurosci 20:7743–7751

West JW, Patton DE, Scheuer T, Wang Y, Goldin AL, Catterall WA (1992) A cluster of hydrophobic amino acid residues required for fast Na(+)-channel inactivation. Proc Natl Acad Sci U S A 89:10910–10914

Yang N, Horn R (1995) Evidence for voltage-dependent S4 movement in sodium channels. Neuron 15:213–218

Yang N, George AL Jr, Horn R (1996) Molecular basis of charge movement in voltage-gated sodium channels. Neuron 16:113–122

Yarov-Yarovoy V, Brown J, Sharp EM, Clare JJ, Scheuer T, Catterall WA (2001) Molecular determinants of voltage-dependent gating and binding of pore-blocking drugs in transmembrane segment IIIS6 of the Na(+) channel alpha subunit. J Biol Chem 276:20–27

Yarov-Yarovoy V, McPhee JC, Idsvoog D, Pate C, Scheuer T, Catterall WA (2002) Role of amino acid residues in transmembrane segments IS6 and IIS6 of the Na+ channel alpha subunit in voltage-dependent gating and drug block. J Biol Chem 277:35393–35401

Yeh JZ, Narahashi T (1977) Kinetic analysis of pancuronium interaction with sodium channels in squid axon membranes. J Gen Physiol 69:293–323

Yeh JZ, Tanguy J (1985) Na channel activation gate modulates slow recovery from use-dependent block by local anesthetics in squid giant axons. Biophys J 47:685–694

Yu FH, Catterall WA (2004) The VGL-chanome: a protein superfamily specialized for electrical signaling and ionic homeostasis. Sci STKE 2004:15

Yu FH, Westenbroek RE, Silos-Santiago I, McCormick KA, Lawson D, Ge P, Ferriera H, Lilly J, Distefano PS, Catterall WA, Scheuer T, Curtis R (2003) Sodium channel b4, a new disulfide-linked auxiliary subunit with similarity to b2. J Neurosci 23:7577–7585

Yu FH, Yarov-Yarovoy V, Gutman GA, Catterall WA (2005) Overview of molecular relationships in the voltage-gated ion channel superfamily. Pharmacol Rev 57:387–395

Zhang X, Ren W, DeCaen P, Yan C, Tao X, Tang L, Wang J, Hasegawa K, Kumasaka T, He J, Wang J, Clapham DE, Yan N (2012) Crystal structure of an orthologue of the NaChBac voltage-gated sodium channel. Nature 486:130–134

Zhou M, Morais-Cabral JH, Mann S, MacKinnon R (2001) Potassium channel receptor site for the inactivation gate and quaternary amine inhibitors. Nature 411:657–661

Zhu W, Voelker TL, Varga Z, Schubert AR, Nerbonne JM, Silva JR (2017) Mechanisms of noncovalent beta subunit regulation of NaV channel gating. J Gen Physiol 149:813. https://doi.org/10.1085/jgp.201711802

Effects of Benzothiazolamines on Voltage-Gated Sodium Channels

Alessandro Farinato, Concetta Altamura, and Jean-François Desaphy

Contents

Abstract

Benzothiazole is a versatile fused heterocycle that aroused much interest in drug discovery as anticonvulsant, neuroprotective, analgesic, anti-inflammatory, antimicrobial, and anticancer. Two benzothiazolamines, riluzole and lubeluzole, are known blockers of voltage-gated sodium (Na_v) channels. Riluzole is clinically used as a neuroprotectant in amyotrophic lateral sclerosis. Inhibition of Na_v channels by riluzole is voltage-dependent due to preferential binding to inactivated sodium channels. Yet the drug exerts little use-dependent block, probably because it lacks protonable amine. One important property is riluzole ability to inhibit persistent Na^+ currents, which likely contributes to its neuroprotective activity. Lubeluzole showed promising neuroprotective effects in animal stroke models, but failed to show benefits in acute ischemic stroke in

A. Farinato · C. Altamura · J.-F. Desaphy (✉)
Section of Pharmacology, Department of Biomedical Sciences and Human Oncology, University of Bari Aldo Moro, Policlinico, Piazza G. Cesare 11, 70124 Bari, Italy
e-mail: jeanfrancois.desaphy@uniba.it

M. Chahine (ed.), *Voltage-gated Sodium Channels: Structure, Function and Channelopathies*, Handbook of Experimental Pharmacology 246,
https://doi.org/10.1007/164_2017_46

humans. One important concern is its propensity to prolong the cardiac QT interval, due to hERG K^+ channel block. Lubeluzole very potently inhibits Na_v channels in a voltage- and use-dependent manner, due to its great preferential affinity for inactivated channels and the presence of a protonable amine group. Patch-clamp experiments suggest that the binding sites of both drugs overlap the local anesthetic receptor within the ion-conducting pathway. Riluzole and lubeluzole displayed very potent antimyotonic activity in a rat model of myotonia, a pathological skeletal muscle condition characterized by high-frequency runs of action potentials. Such results well support the repurposing of riluzole as an antimyotonic drug, allowing the launch of a pilot study in myotonic patients. Riluzole, lubeluzole, and new Na_v channel blockers built on the benzothiazolamine scaffold will certainly continue to be investigated for possible clinical applications.

Keywords

Local anesthetic receptor · Lubeluzole · Myotonia · Riluzole

1 Overview of Voltage-Gated Sodium Channels Pharmacology

Voltage-gated sodium (Na_v) channels are primary or secondary targets of many clinically used drugs, including local anesthetics, class I antiarrhythmics, anticonvulsants, analgesics, neuroprotectants, and antimyotonic drugs (Imbrici et al. 2016). Most of these drugs share structural determinants allowing binding to a common site located within the ion-conducting pore of the channel (Catterall 2012). The common structural features of these drugs consist in a hydrophobic extremity made of one or more substituted aromatic rings and the other extremity containing a hydrophilic amine group. High affinity binding is thought to involve hydrophobic interaction with a tyrosine side chain and π-cation interaction between the charged amine of the drug and the ring of a phenylalanine (Ragsdale et al. 1994; Ahern et al. 2008; Sheets et al. 2010; Desaphy et al. 2012). Both amino acids belong to segment 6 of domain IV and line the sodium ion-conducting pathway when channel is open or inactivated, whereas the closed channel offers a less favorable conformation for drug binding. Thus, drug affinity for closed channels is by far lower than that for open/inactivated channels. Such a property can explain the voltage- and frequency-dependent inhibition of Na_v channels by these drugs. Because Na_v channel inactivation increases with depolarization, inhibition is greater in depolarized cells, thereby determining the so-called voltage-dependent inhibition. As action potentials firing increases in a cell, Na_v channels spend more time in the open and inactivated states, thereby favoring drug binding and producing the so-called frequency- or use-dependent block. The Log P and pKa of drugs are critical determinants of Na_v channel block. The higher the Log P, greater is drug hydrophobicity, and greater is the block whatever the channel state (De Luca et al. 2003a, b; Desaphy et al. 2012). The higher the pKa, greater is the ionization of drug amine group at physiological pH, and greater is the block of open/inactivated

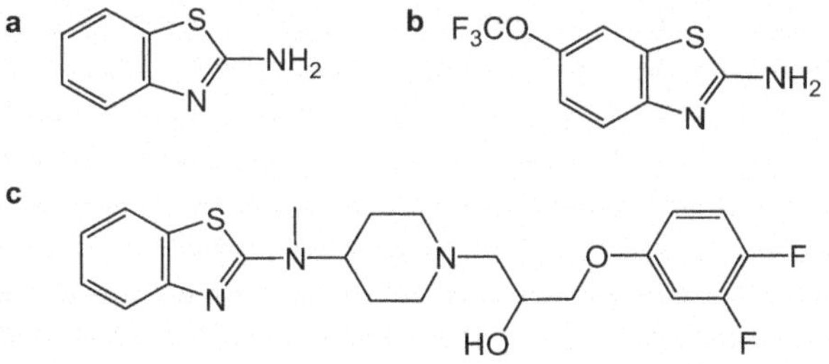

Fig. 1 Chemical structure of (**a**) 1,3-benzothiazol-2-amine, (**b**) riluzole [(6-trifluoromethoxy)1,3-benzothiazol-2-amine], and (**c**) lubeluzole [(2*S*)-1-[4-(1,3-benzothiazol-2-yl-methylamino) piperidin-1-yl]-3-(3,4-difluorophenoxy)propan-2-ol]

channels (Desaphy et al. 2010). Thus, the local anesthetic benzocaine, which amine is not protonable, blocks Na_v channels but displays little use dependence.

In mammals, nine genes are known to encode the Na_v channel α-subunits that form the ion-conducting pore (Catterall 2012). These genes are preferentially expressed in central nervous system ($Na_v1.1$, $Na_v1.2$, $Na_v1.3$, and $Na_v1.6$), skeletal muscle ($Na_v1.4$), heart ($Na_v1.5$), and peripheral nerves ($Na_v1.7$, $Na_v1.8$, and $Na_v1.9$). Both tyrosine and phenylalanine residues of the LA binding site are well conserved among Na_v channel subtypes and currently used drugs display little selectivity. Thus, voltage- and use-dependent inhibitions are fundamental properties for drug safety, as drugs can act primarily on ill depolarized or over-excited cells, while sparing normal Na_v channel function in healthy cells. A number of studies have identified new selective compounds for Na_v channel subtypes, some of which have entered clinical trials, but little information is available regarding their binding sites (Bagal et al. 2015).

Recently, two drugs called our attention because they are the only two known Na_v channel blockers containing a benzothiazolamine group; these are riluzole and lubeluzole (Fig. 1). Benzothiazoles belong to benzazole family, which encompass compounds with benzene ring fused to a five-membered ring system with two or more heteroatoms, one of which is nitrogen. Resonance contributes to the stability of the heterocyclic ring system and benzazoles show many of the characteristics of aromatic compounds (Domino et al. 1952). Benzothiazole is a versatile fused heterocyclic scaffold that aroused much interest in drug discovery as anticonvulsant, neuroprotective, analgesic, anti-inflammatory, antimicrobial, and anticancer (Kamal et al. 2015).

2 Riluzole

2.1 Pharmacology of Riluzole

Riluzole was, up to 2015, the only drug approved by the European Medicines Agency (EMA) and Food and Drug administration (FDA) for Amyotrophic Lateral Sclerosis (ALS). It was shown to prolong survival by 2–3 months in ALS patients and to delay the use of surrogate approaches, such as tracheotomy and mechanical ventilation (Bensimon et al. 1994; Lacomblez et al. 1996). ALS is characterized by a selective and progressive degeneration of lower and upper motor neurons, which lead to muscle weakness, twitching, muscle atrophy, impairment of speech and swallowing, and progressive paralysis; death eventually occurs due to respiratory failure (Tard et al. 2017). Except for a small proportion of inherited monogenic cases, ALS is a multifactorial disease, the causes of which are still unknown. One of the key pathogenic mechanisms proposed to explain ALS is excitotoxicity mediated by glutamate. It is widely acknowledged that riluzole can reduce damage of the nerve cells, because it reduces glutamate-induced excitotoxicity. The exact mechanism of action of riluzole in ALS is not well defined, but it likely involves the reduction of glutamate release, blockade of Na_v channels, and potentiation of Ca^{2+}-activated K^+ currents (Bellingham 2011).

A wealth of other pharmacological activities has been described in vitro using higher concentrations of riluzole and it is not clear whether these may influence riluzole effects in vivo. These include modulation of a plethora of potassium channels, inhibition of voltage-gated calcium (Ca_v) channels, and modulation of neurotransmitters release and receptors (Bellingham 2011). Additional targets include G protein-coupled signaling and protein kinase C (Hubert et al. 1994; Noh et al. 2000). Because of its broad range of molecular effects, riluzole has been tested in a great variety of cell and animal models of diseases for neuroprotection, anticonvulsant, anxiolytic, antidepressant, sedative, anesthetic, analgesic, anti-ischemic, and anti-cancer activities. Many of these preclinical studies have produced promising results that, however, were not easily translated to humans.

Human studies of riluzole have been reported with mitigate results in Parkinson's Disease (Jankovic and Hunter 2002; Braz et al. 2004; Bensimon et al. 2009), multiple system atrophy (Seppi et al. 2006; Bensimon et al. 2009), spinal muscular atrophy (Russman et al. 2003), Huntington's disease (Huntington Study Group 2003; Landwehrmeyer et al. 2007; Armstrong et al. 2012), mood disorders (Salardini et al. 2016; Park et al. 2017; Mathew et al. 2017), hyperalgesia and neuropathic pain (Hammer et al. 1999; Galer et al. 2000), spinal cord injury (Nagoshi et al. 2015), and irritable bowel syndrome (Mishra et al. 2014). For instance, a future clinical indication for riluzole might be cerebellar ataxia (Wood 2015). The rationale is mainly based on the known activating effect of riluzole on the small conductance Ca^{2+}-activated K^+ (SK) channels (Cao et al. 2002), which have been shown to play a critical regulatory role in the firing rate of neurons in deep cerebellar nuclei. Openers of these channels may reduce neuronal hyperexcitability and thus exert benefits in cerebellar ataxia. Yet a more pleiotropic effect of riluzole cannot be ruled out. Two RCT involving patients with cerebellar ataxia of various etiologies provided class I evidences that riluzole might prove beneficial in such conditions (Ristori et al. 2010; Romano et al. 2015). Longer confirmatory studies on larger and disease-specific populations are needed before to reach clinical practice.

2.2 Molecular Effects of Riluzole on Sodium Channels

Voltage-clamp studies of nodes of Ranvier from isolated frog nerves showed that riluzole reversibly inhibits both K^+ and Na^+ currents (Benoit and Escande 1991). Inhibition of Na^+ currents was not use dependent, but the drug shifted the voltage dependence of fast inactivation toward negative voltages likely due to a highly specific blockade of inactivated channels. Open channel blockade was negligible in physiological conditions. Thus, the mode of action of riluzole is similar to unprotonated local anesthetics, like benzocaine. Indeed, the strongest basic pKa of riluzole is about 4.57, which implies lack of protonation at physiological pH. Further studies mostly confirmed such effects in rat dorsal root ganglia (DRG) neurons (Song et al. 1997), rat cortical neurons (Stefani et al. 1997; Zona et al. 1998), cerebellar Purkinje neurons (O'Neill et al. 1997), cultured neuroendocrine cells (Beltran-Parrazal and Charles 2003), and cultured human skeletal muscle cells (Wang et al. 2008; Deflorio et al. 2014). Notably, effects of riluzole on inactivated tetrodotoxin (TTX)-sensitive and TTX-resistant Na_v channels in DRG neurons were similar, as apparent variations in current inhibition were attributed to channel differences in inactivation voltage dependence (Song et al. 1997). In addition, riluzole exerted a positive shift of activation of TTX-resistant channels in DRG neurons. Also in rat cardiac myocytes, riluzole induced a positive shift of activation of TTX-sensitive Na_v channels (Weiss et al. 2010). In heterologous systems of expression, riluzole inhibits rat $Na_v1.2$ (Hebert et al. 1994), human $Na_v1.7$ (Theile and Cummins 2011) (Fig. 2), and human $Na_v1.4$ channels (Desaphy et al. 2013a) with similar mechanism. To date, nothing is known regarding the effects of riluzole

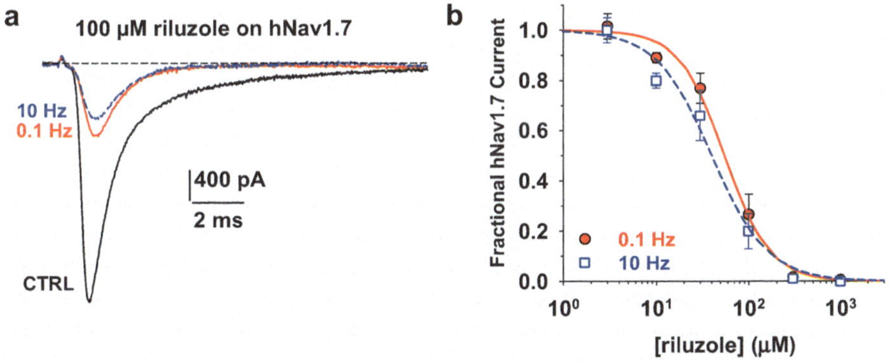

Fig. 2 Effects of riluzole on human $Na_v1.7$ sodium channel subtype. (**a**) Representative sodium current traces recorded in HEK293 cells stably transfected with $hNa_v1.7$ sodium channel subtype. Sodium currents were elicited from a holding potential of −120 mV by a 25 ms-long test pulse at −30 mV applied at 0.1 or 10 Hz stimulation frequency, in the absence of drug (ctrl) or in the presence of 100 μM riluzole (0.1 and 10 Hz). (**b**) Concentration–response relationships obtained as in (**a**) at 0.1 and 10 Hz stimulation frequencies. The relationships were fitted to a first-order binding function as described elsewhere (Desaphy et al. 2013a). The estimated half-maximum inhibitory concentration (IC_{50}) values ± the S.E. of the fit were 56 ± 4 μM at 0.1 Hz and 41 ± 5 μM at 10 Hz

on Na_v channel slow inactivation, a process that develops in tens of seconds and modulates channel availability, thereby influencing neuron and myocyte excitability.

Comparison of potencies among Na_v channel blockers is not an easy task, as IC_{50} values significantly vary depending on patch-clamp protocols, due to dependence on voltage and frequency. In Fig. 3 are reported the IC_{50} values of a number of drugs for inhibition of heterologously expressed human skeletal muscle $hNa_v1.4$ channels, all calculated in the same conditions (holding potential of −120 mV, 20 ms-long test pulse at −30 mV, 0.1 and 10 Hz frequencies) (De Bellis et al. 2017; Desaphy et al. 2003, 2009, 2010, 2012, 2014, 2016). Compared to other tested Na_v channel blockers, riluzole displays elevated potency at 0.1 Hz (IC_{50} close to 50 μM, Fig. 3a). Yet, because it lacks use dependence, riluzole effects at 10 Hz (IC_{50} close to 42 μM) are minor compared to many other examined drugs (Fig. 3b). Thus, riluzole is the drug with the weaker use-dependent behavior among the tested drugs (Fig. 3c).

An important feature of riluzole is its ability to inhibit the persistent Na^+ currents observed in many neurons, which play a critical role in modulating excitability threshold. This small persistent current is thought to be carried by the same channel as the peak current, due to sustained bursts of openings and brief recoveries from inactivated to open states (Alzheimer et al. 1993). As expected from a stabilizing effect of riluzole on Na_v channel fast inactivation, the drug was shown to inhibit persistent currents, most often at lower concentrations compared to peak currents (Urbani and Belluzzi 2000; Spadoni et al. 2002; Kononenko et al. 2004). An example of comparative inhibition of transient and persistent Na^+ currents in sympathetic neurons is shown in Fig. 4 (Lamas et al. 2009). It is however worth to note that the relative effects of riluzole on both currents are variable from a preparation to another. Interestingly, the persistent Na^+ current is increased in motor neurons from mice carrying a mutation in the superoxide dismutase causing ALS in humans, and its inhibition by riluzole likely contributes to the therapeutic efficacy of riluzole in ALS (Kuo et al. 2005; Pieri et al. 2009; Schuster et al. 2012). The drug is also able to inhibit resurgent currents carried by two $hNa_v1.7$ channel mutants responsible for inherited pain syndrome, such effect being strongly correlated with persistent current inhibition (Theile and Cummins 2011). Abnormal persistent currents have been linked to inherited diseases, such as hyperkalemic periodic paralysis (Cannon et al. 1993), long QT syndrome (Bennett et al. 1995), and epileptic encephalopathy (Lopez-Santiago et al. 2017), or acquired pathological conditions, such as cardiac ischemia (Weiss et al. 2010; Weiss and Saint 2010). Thus, the preferential inhibition of these currents by riluzole may represent an appealing mechanism for therapeutic application.

Most of the Na_v channel blockers share the same binding site within Na_v channels, which is usually named the local anesthetic (LA) receptor (Ragsdale et al. 1994, 1996). Very important for high-affinity binding is the interaction between the charged amine of drugs with the aromatic side chain of a phenylalanine in segment 6 of domain IV (Phe1586 in $hNa_v1.4$) of the channel through a π-cation interaction (Ahern et al. 2008; Desaphy et al. 2012). We demonstrated that F1579C mutation in $hNa_v1.4$ reduces riluzole effect on sodium currents by ~2.5 times,

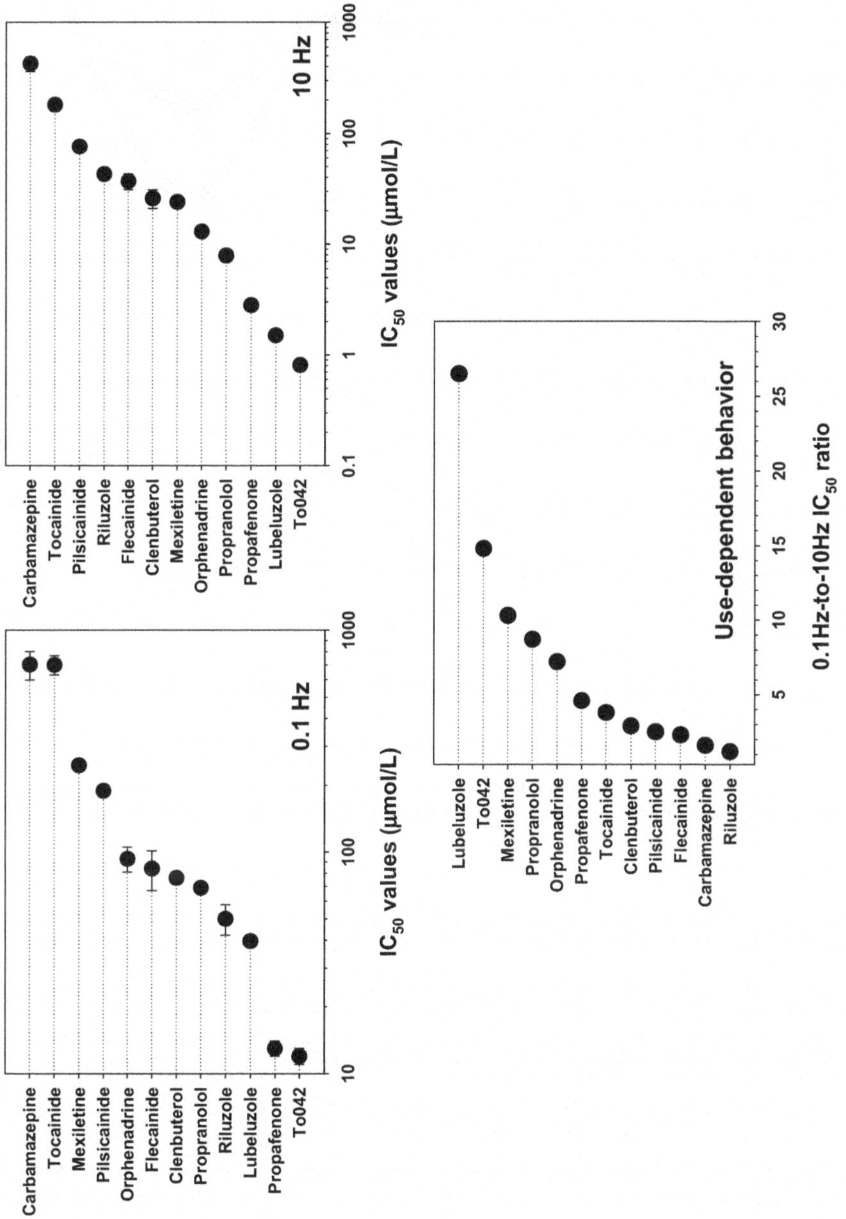

Fig. 3 Comparison of sodium channel blockers effects on $hNa_v1.4$ channels. The graphs report the IC_{50} values of sodium channel blockers for inhibition of sodium currents in HEK293 cells transfected with human skeletal muscle ($hNa_v1.4$) sodium channels. Sodium currents were elicited from a holding potential of −120 mV by a 25 ms-long test pulse at −30 mV applied at 0.1 or 10 Hz stimulation frequency

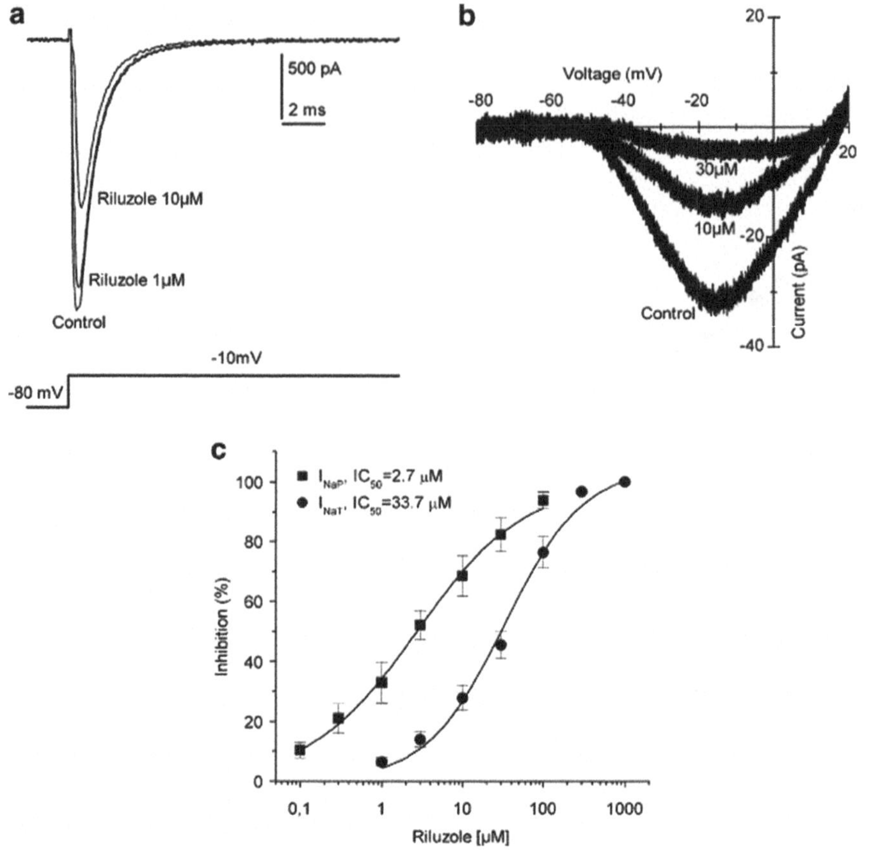

Fig. 4 Inhibition of transient (I_{NaT}) (**a**) and persistent (I_{NaP}) (**b**) sodium currents by riluzole in sympathetic neurons isolated from mouse superior cortical ganglion. (**c**) Concentration–response relationships (reproduced from Lamas et al. 2009)

thereby suggesting that riluzole also utilizes the LA receptor (Desaphy et al. 2013a). Yet, riluzole is not protonated and the nature of the interaction remains to be identified. There is a possibility that Phe1586 may strengthen hydrophobic interactions with riluzole (Sheets et al. 2010; Desaphy et al. 2012). There is too little information available today to draw definitive conclusion. Indeed, the F1579C mutation induces a positive shift in the voltage dependence of channel availability, which may contribute substantially to the variation in riluzole effect. More experiments are thus warranted to define the binding site of riluzole within Na_v channels.

Interestingly, the removal of the trifluoromethoxy (OCF_3) group from riluzole greatly reduces Na_v inhibition by 20 times (Desaphy et al. 2013a), and significantly impairs in vivo activity against glutamate-induced convulsions and maximal electroshock-induced seizures in rats (Hays et al. 1994; Jimonet et al. 1999). By testing a large series of 6-substituted-2-benzothiazolamines in vivo, two studies suggested that lipophilicity of the substituent in sixth position is a key determinant of anticonvulsant activity. Yet, correlation between lipophilicity and activity of the derivatives was not so straightforward, and activity depended on the hindrance and the electron-withdrawing capacity of the substituent, the presence of fluorine, and the OCF_3 position on the aromatic ring (Hays et al. 1994; Jimonet et al. 1999). A QSAR analysis suggested that it would not be possible to increase potency significantly by modifying aryl substituents (Hays et al. 1994). Altogether, these results suggest that this portion of the molecule is critical for riluzole activity and likely contribute to drug pharmacophores. Thus, direct testing of such derivatives on Na_v sodium channels would be helpful for the understanding of molecular interactions at the binding site.

3 Lubeluzole

3.1 Pharmacology of Lubeluzole

Lubeluzole is the name for the *S*-(+) enantiomer of 1-[4-(1,3-benzothiazo-2-yl-methylamino)piperidin-1-yl]-3-(3,4-difluorophenoxy)propan-2-ol (Fig. 1). Similar to riluzole, lubeluzole is usually considered as an anti-glutamatergic drug but exerts pharmacological effects on a multitude of targets. First experiments showed the capacity of lubeluzole to prevent the increase of glutamate concentrations in a rat model of thrombotic stroke (Scheller et al. 1997). The drug was shown to interfere with nitric oxide synthase pathway (Lesage et al. 1996; Maiese et al. 1997). Its ability to inhibit veratridine-induced toxicity suggested that lubeluzole may modulate Na_v activity (Ashton et al. 1997), which was confirmed in patch-clamp studies (Le Grand et al. 2003; Desaphy et al. 2013a). Lubeluzole and its *R*-(−) enantiomer both inhibited Ca^{2+} currents in chromaffin cells, with a preference for the non L-type Ca_v channels, as well as in isolated rat neurons (Hernández-Guijo et al. 1997; Marrannes et al. 1998). More recently, in vitro assays showed high-affinity binding to calmodulin and inhibition of Ca^{2+}/calmodulin-dependent protein kinase II (Bruno et al. 2016).

All these mechanisms may theoretically contribute to the neuroprotective effects of lubeluzole observed in various cell and animal models with a focus on ischemic stroke (Aronowski et al. 1996; De Ryck et al. 1996; Haseldonckx et al. 1997; Culmsee et al. 1998). In the thrombotic stroke rat model, lubeluzole inhibits ischemia-induced glutamate release, has a neuroprotective and anti-ischemic effect, and improved neurological function. Since these in vivo effects are stereoselective, as are the in vitro effects on glutamate release and NO pathway, it is widely acknowledged that these later mechanisms are crucial for anti-ischemic effects.

Contrary to riluzole, lubeluzole had no analgesic effect in the formalin rat model of inflammatory pain, suggesting that the two drugs may act differently on targets or affect different targets (Blackburn-Munro et al. 2002). Yet both drugs produced analgesia in acute thermal pain model, without altering motor performance.

Very low concentrations of lubeluzole were also shown to exert a synergistic effect with anticancer drugs on human tumor cell lines, an effect possibly attributable to inhibition of calmodulin, and Ca_v and Na_v channels (Cavalluzzi et al. 2013).

Unfortunately, clinical use of lubeluzole is challenged by its ability to prolong the QT interval on the electrocardiogram. This effect was first described in anesthetized dogs and further confirmed in isolated rabbit Purkinje fibers (Sugiyama et al. 1996; Le Grand et al. 2000). Lubeluzole cardiac effect was reminiscent of class III antiarrhythmic drugs, suggesting a possible inhibition of cardiac K^+ currents. Indeed, both lubeluzole and its *R*-(−) enantiomer were recently shown to be potent blockers of heterologously expressed hERG channels, with an IC_{50} close to 10 nM (Gualdani et al. 2015).

In humans, several large trials testing intravenous lubeluzole in acute ischemic stroke failed to find any improvement of survival or life quality of survivors, while the drug induced a significant increase in the odds of QT prolongation (Gandolfo et al. 2002; Muir and Lees 2003). These observations led to interruption of additional human trials.

3.2 Molecular Effects of Lubeluzole on Sodium Channels

In isolated hippocampal neurons, lubeluzole inhibited TTX-sensitive Na^+ currents with an IC_{50} of 3.1 μM but more detailed information are not available (Osikowska-Evers et al. 1995). Two studies have reported in detail the effects of lubeluzole directly on Na_v channels. In isolated guinea-pig ventricular myocytes, lubeluzole reversibly inhibited Na^+ currents elicited from a holding potential of −80 mV with an IC_{50} of 9.5 μM (Le Grand et al. 2003). The drug had no effect on activation voltage-dependence, but shifted fast inactivation toward more negative voltages. Blockade was voltage and use dependent, suggesting higher affinity for the inactivated state. Considering the plasma concentration of lubeluzole in humans after 10 mg injection a day, that is 0.14–0.27 μM (Diener et al. 1996; Herron et al. 1998), the authors suggested that Na_v channel blockade may be clinically relevant in pathological conditions in which cells are depolarized, such as cardiac ischemia (Le Grand et al. 2003). Lubeluzole exerted similar effects on heterologously expressed human skeletal muscle $Na_v1.4$ channels (Desaphy et al. 2013a). From a holding potential of −120 mV, the IC_{50} was about 35 μM at 0.1 Hz frequency stimulation but was reduced to 1.5 μM at 10 Hz, due to huge use dependency. Compared to other blockers tested in the same conditions, lubeluzole was equipotent to riluzole at 0.1 Hz, but performed among the more potent drug at 10 Hz (Fig. 3). The huge use dependence was due to the very high affinity of the drug for inactivated channels compared to closed channels, the apparent dissociation constants from the two channel states being 0.011 μM versus 840 μM, respectively (Desaphy et al. 2013a). Importantly, as for Ca_v channels (Hernández-Guijo et al. 1997), Na_v channel blockade was similar for lubeluzole and the *R*-(−) enantiomer, suggesting that Ca_v or Na_v channel inhibition is not a major contributor to the neuroprotective effects in stroke. However, use-dependent inhibition of Na_v

channels by lubeluzole might be clinically relevant in conditions of excessive action potential firing, such as myotonia (see next paragraph).

The characteristics of Na_v channel inhibition make lubeluzole resembling class I antiarrhythmic drugs, like lidocaine, suggesting binding to the LA receptor within the ion-conducting way. Indeed, the F1586C mutation in $hNa_v1.4$ reduced Na^+ current inhibition by six times at 10 Hz (Desaphy et al. 2013a). This effect was mainly due to a great reduction of affinity for inactivated channels by 60-fold. Yet the use dependence was not completely zeroed by the mutation, suggesting that interaction with other amino acids plays a significant role in high-affinity binding. In addition, we observed a 2.7-fold reduction of Na_v channel inhibition by the Y1593C mutation. Based on this information, we formulated a hypothesis for lubeluzole binding at the LA receptor considering three major interacting points (Fig. 5). With a pKa of 7.1, about half of lubeluzole molecules are neutral at physiological pH. Because lubeluzole is highly lipophilic, neutral lubeluzole may be able to reach easily its binding site within the ion-conducting pore even when the

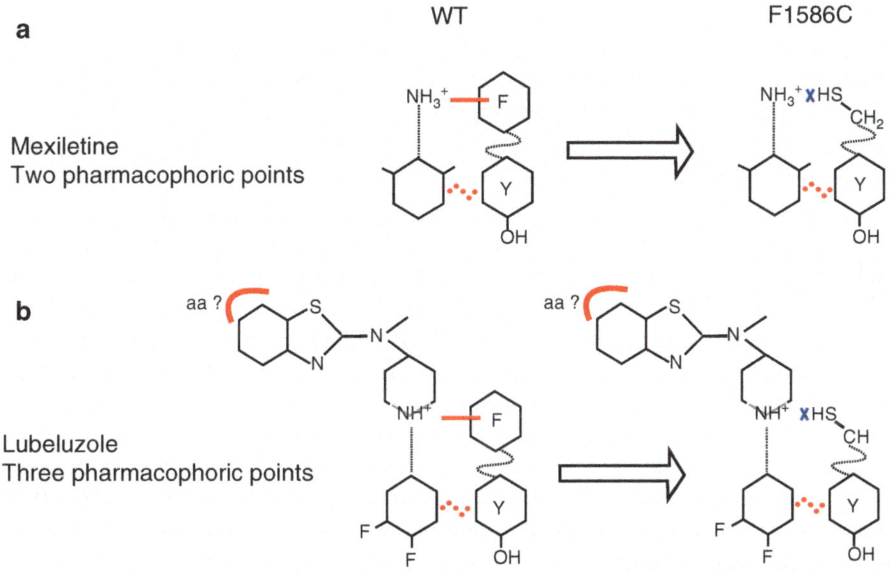

Fig. 5 Putative binding of lubeluzole to the local anesthetic (LA) receptor within the ion-conducting pathway of sodium channels. (**a**) It is widely acknowledged that the high-affinity binding site of many LA-like drugs, including mexiletine, is composed of two pharmacophoric points. The aryl moiety interacts with a tyrosine residue (Y1583 in $hNa_v1.4$) through hydrophobic interactions, while the protonated amine of drugs establishes a strong π-cation interaction with a neighboring phenylalanine (F1586 in $hNa_v1.4$), which is critical for use-dependent block. Mutation of F1586 to a nonconservative amino acid disrupts π-cation interaction and abolishes use-dependent blockade. (**b**) Lubeluzole may bind to the LA receptor through its phenoxy-propranol-amine moiety, conferring use-dependence behavior. Yet a third pharmacophoric point involving the benzothiazolamine moiety likely contributes to high-affinity binding. In this way, F1586C mutation impairs only partially use-dependent block (reproduced from Desaphy et al. 2013a)

channel is closed. This implies that use dependence is not due to a reduced access of the drug to its receptor through the closed channel, but rather mainly depends on the higher specific affinity for channels in the inactivated state, according to the modulated receptor hypothesis (Hille 1977; Hondeghem and Katzung 1977).

Lubeluzole contains two distinct chemical portions: one benzothiazolic moiety related to riluzole, and one phenoxy-propranol-amine moiety recalling propranolol. Because both riluzole and propranolol are potent blockers of skeletal muscle Na_v channels (Desaphy et al. 2003, 2013a), we wondered whether lubeluzole moieties may singly affect $hNa_v1.4$ channels. Actually, the two lubeluzole moieties were by far less efficient; the riluzole-like compound lacks the trifluoromethoxy group, while the propranolol-related compound presents two fluoride on the aryl moiety, two properties which may be detrimental (Hays et al. 1994; Jimonet et al. 1999; Desaphy et al. 2012, 2013a). Thus, the molecular integrity of lubeluzole is required to get the maximum effects on Na_v channels.

4 Riluzole and Lubeluzole as Antimyotonic Drugs?

Because riluzole and lubeluzole are potent Na_v channel inhibitors, we tested the drugs as possible antimyotonic drug (Desaphy et al. 2013a, 2014). Myotonia is a skeletal muscle disorder characterized by sarcolemma hyper-excitability and delayed relaxation after contraction (Imbrici et al. 2016). Nondystrophic myotonias are due to mutations in the skeletal muscle $hNa_v1.4$ or chloride (hClC-1) channel genes. Gain-of-function $hNa_v1.4$ mutations and loss-of-function hClC-1 mutations, both, increase sarcolemma excitability, determining myotonic runs of action potentials at elevated frequencies. Today the drug of choice is mexiletine; after two decades of off-label use, a clinical trial has confirmed its effectiveness in the treatment of myotonia (Statland et al. 2012). By blocking $hNa_v1.4$ channels in a use-dependent manner, mexiletine inhibits myotonic runs whatever the culprit gene (Statland et al. 2012; Suetterlin et al. 2015; Lo Monaco et al. 2015). Mexiletine is usually well tolerated, but between 20 and 40% of myotonic patients have only a partial or no response to the drug, likely due to pharmacogenetics mechanisms as evidenced in some patients (Desaphy et al. 2013b, 2016). It is therefore widely acknowledged that alternative drugs to mexiletine are required to meet the medical needs.

We thus developed a rat model of myotonia to evaluate antimyotonic activity of drug candidates and assayed a number of marketed Na_v channel blockers, including riluzole and lubeluzole (Desaphy et al. 2013c, 2014). In vivo, myotonia was induced in rats by intraperitoneal injection of 9AC and was evaluated by calculating the time of righting reflex, which is the time taken by the rat to return on its four limbs from the supine position. With ED_{50} values of about 0.1 mg/kg, both riluzole and lubeluzole were the more potent antimyotonic drugs among the tested drugs, being about 70 times more potent than mexiletine. Patch-clamp experiments were also performed on HEK293 cell line stably transfected with $hNa_v1.4$ channel, using voltage clamp protocol mimicking a myotonic run (Desaphy et al. 2014). The holding potential was −90 mV close to the resting sarcolemma potential, the test pulse was 5 ms long as an action potential, and the frequency stimulation was 50 Hz, as in a myotonic run. In these conditions, riluzole and lubeluzole were also the most potent drugs, with IC_{50} values close to 0.5 μM, which is almost 40 times lower than that of mexiletine.

The results suggest that both drugs might be effective in myotonic patients at lower or similar doses than that used in ALS patients for riluzole or that tested in stroke for lubeluzole. Because of its known effect on QT interval, further preclinical experiments would be required before to test lubeluzole in myotonic patients. On the other hand, the results well support the repurposing of riluzole as a potential antimyotonic drug. Based on these results, a randomized trial has been launched in Italy to verify riluzole efficacy versus mexiletine in human nondystrophic myotonia.

5 Conclusions

Riluzole and lubeluzole are among the more potent Na_v blockers in our hands. Their binding sites likely overlap the LA receptor within the ion-conducting pathway. The clinical usefulness of riluzole may rely on its ability in blocking persistent Na^+ currents. Whether lubeluzole may exert a similar effect is not known, but the huge use dependence of lubeluzole makes it a valuable candidate for membrane overexcitability disorders. Some attempt has been made to develop riluzole derivatives to improve biological activities (Coleman et al. 2015; Mancini et al. 2017). It is expected that more studies will consider the benzothiazolamine scaffold to build new Na_v channel blockers with promising pharmacological properties for the clinical setting.

Acknowledgments This study was supported by grant #19027 from A.F.M. (Association Française contre les Myopathies).

References

Ahern CA, Eastwood AL, Dougherty DA et al (2008) Electrostatic contributions of aromatic residues in the local anesthetic receptor of voltage-gated sodium channels. Circ Res 102(1): 86–94

Alzheimer C, Schwindt PC, Crill WE (1993) Modal gating of Na^+ channels as a mechanism of persistent Na^+ current in pyramidal neurons from rat and cat sensorimotor cortex. J Neurosci 13:660–673

Armstrong MJ, Miyasaki JM, American Academy of Neurology (2012) Evidence-based guideline: pharmacologic treatment of chorea in Huntington disease: report of the guideline development subcommittee of the American Academy of Neurology. Neurology 79(6):597–603

Aronowski J, Strong R, Grotta JC (1996) Treatment of experimental focal ischemia in rats with lubeluzole. Neuropharmacology 35(6):689–693

Ashton D, Willems R, Wynants J et al (1997) Altered Na(+)-channel function as an in vitro model of the ischemic penumbra: action of lubeluzole and other neuroprotective drugs. Brain Res 745(1-2):210–221

Bagal SK, Marron BE, Owen RM et al (2015) Voltage gated sodium channels as drug discovery targets. Channels (Austin) 9(6):360–366

Bellingham MC (2011) A review of the neural mechanisms of action and clinical efficiency of riluzole in treating amyotrophic lateral sclerosis: what have we learned in the last decade? CNS Neurosci Ther 17(1):4–31

Beltran-Parrazal L, Charles A (2003) Riluzole inhibits spontaneous Ca^{2+} signaling in neuroendocrine cells by activation of K^+ channels and inhibition of Na^+ channels. Br J Pharmacol 140(5): 881–888

Bennett PB, Yazawa K, Makita N et al (1995) Molecular mechanism for an inherited cardiac arrhythmia. Nature 376(6542):683–685

Benoit E, Escande D (1991) Riluzole specifically blocks inactivated Na channels in myelinated nerve fibre. Pflugers Arch 419(6):603–609

Bensimon G, Lacomblez L, Meininger V (1994) A controlled trial of riluzole in amyotrophic lateral sclerosis. ALS/Riluzole Study Group. N Eng J Med 330:585–591

Bensimon G, Ludolph A, Agid Y et al (2009) Riluzole treatment, survival and diagnostic criteria in Parkinson plus disorders: the NNIPPS study. Brain 132(Pt 1):156–171

Blackburn-Munro G, Ibsen N, Erichsen HK (2002) A comparison of the anti-nociceptive effects of voltage-activated Na^+ channel blockers in the formalin test. Eur J Pharmacol 445(3):231–238

Braz CA, Borges V, Ferraz HB (2004) Effect of riluzole on dyskinesia and duration of the on state in Parkinson disease patients: a double-blind, placebo-controlled pilot study. Clin Neuropharmacol 27(1):25–29

Bruno C, Cavalluzzi MM, Rusciano MR et al (2016) The chemosensitizing agent lubeluzole binds calmodulin and inhibits Ca(2+)/calmodulin-dependent kinase II. Eur J Med Chem 116:36–45

Cannon SC, Brown RH Jr, Corey DP (1993) Theoretical reconstruction of myotonia and paralysis caused by incomplete inactivation of sodium channels. Biophys J 65(1):270–288

Cao YJ, Dreixler JC, Couey JJ et al (2002) Modulation of recombinant and native neuronal SK channels by the neuroprotective drug riluzole. Eur J Pharmacol 449:47–54

Catterall WA (2012) Voltage-gated sodium channels at 60: structure, function and pathophysiology. J Physiol 590(11):2577–2589

Cavalluzzi MM, Viale M, Bruno C et al (2013) A convenient synthesis of lubeluzole and its enantiomer: evaluation as chemosensitizing agents on human ovarian adenocarcinoma and lung carcinoma cells. Bioorg Med Chem Lett 23(17):4820–4823

Coleman N, Nguyen HM, Cao Z et al (2015) The riluzole derivative 2-amino-6-trifluoromethylthiobenzothiazole (SKA-19), a mixed KCa2 activator and Na_V blocker, is a potent novel anticonvulsant. Neurotherapeutics 12(1):234–249

Culmsee C, Junker V, Wolz P et al (1998) Lubeluzole protects hippocampal neurons from excitotoxicity in vitro and reduces brain damage caused by ischemia. Eur J Pharmacol 342(2-3): 193–201

De Bellis M, Carbonara R, Roussel J et al (2017) Increased sodium channel use-dependent inhibition by a new potent analogue of tocainide greatly enhances in vivo antimyotonic activity. Neuropharmacology 113:206–216

De Luca A, Talon S, De Bellis M et al (2003a) Inhibition of skeletal muscle sodium currents by mexiletine analogues: specific hydrophobic interactions rather than lipophilia per se account for drug therapeutic profile. Naunyn Schmiedeberg's Arch Pharmacol 367(3):318–327

De Luca A, Talon S, De Bellis M et al (2003b) Optimal requirements for high affinity and use-dependent block of skeletal muscle sodium channel by N-benzyl analogs of tocainide-like compounds. Mol Pharmacol 64(4):932–945

De Ryck M, Keersmaekers R, Duytschaever H et al (1996) Lubeluzole protects sensorimotor function and reduces infarct size in a photochemical stroke model in rats. J Pharmacol Exp Ther 279(2): 748–758

Deflorio C, Onesti E, Lauro C et al (2014) Partial block by riluzole of muscle sodium channels in myotubes from amyotrophic lateral sclerosis patients. Neurol Res Int 2014:946073

Desaphy JF, Pierno S, De Luca A et al (2003) Different ability of clenbuterol and salbutamol to block sodium channels predicts their therapeutic use in muscle excitability disorders. Mol Pharmacol 63(3):659–670

Desaphy JF, Dipalma A, De Bellis M et al (2009) Involvement of voltage-gated sodium channels blockade in the analgesic effects of orphenadrine. Pain 142(3):225–235

Desaphy JF, Dipalma A, Costanza T et al (2010) Molecular determinants of state-dependent block of voltage-gated sodium channels by pilsicainide. Br J Pharmacol 160(6):1521–1533

Desaphy JF, Dipalma A, Costanza T et al (2012) Molecular insights into the local anesthetic receptor within voltage-gated sodium channels using hydroxylated analogs of mexiletine. Front Pharmacol 3:17. https://doi.org/10.3389/fphar.2012.00017

Desaphy JF, Carbonara R, Costanza T et al (2013a) Molecular dissection of lubeluzole use-dependent block of voltage-gated sodium channels discloses new therapeutic potentials. Mol Pharmacol 83(2):406–415

Desaphy JF, Costanza T, Carbonara R et al (2013b) In vivo evaluation of antimyotonic efficacy of β-adrenergic drugs in a rat model of myotonia. Neuropharmacology 65:21–27

Desaphy JF, Modoni A, Lomonaco M et al (2013c) Dramatic improvement of myotonia permanens with flecainide: a two-case report of a possible bench-to-bedside pharmacogenetics strategy. Eur J Clin Pharmacol 69(4):1037–1039

Desaphy JF, Carbonara R, Costanza T et al (2014) Preclinical evaluation of marketed sodium channel blockers in a rat model of myotonia discloses promising antimyotonic drugs. Exp Neurol 255: 96–102

Desaphy JF, Carbonara R, D'Amico A et al (2016) Translational approach to address therapy in myotonia permanens due to a new SCN4A mutation. Neurology 86(22):2100–2108

Diener HC, Hacke W, Hennerici M et al (1996) Lubeluzole in acute ischemic stroke. A double-blind, placebo-controlled phase II trial. Lubeluzole International Study Group. Stroke 27(1): 76–81

Domino EF, Unna KR, Kerwin J (1952) Pharmacological properties of benzazoles. I. Relationship between structure and paralyzing action. J Pharmacol Exp Ther 105(4):486–497

Galer BS, Twilling LL, Harle J et al (2000) Lack of efficacy of riluzole in the treatment of peripheral neuropathic pain conditions. Neurology 55(7):971–975

Gandolfo C, Sandercock P, Conti M (2002) Lubeluzole for acute ischaemic stroke. Cochrane Database Syst Rev 1:CD001924

Gualdani R, Cavalluzzi MM, Tadini-Buoninsegni F et al (2015) Insights on molecular determinants of hERG K^+ channel inhibition design, synthesis, and biological evaluation of lubeluzole derivatives. Biophys J 108(2):582a. (Abstract)

Hammer NA, Lillesø J, Pedersen JL et al (1999) Effect of riluzole on acute pain and hyperalgesia in humans. Br J Anaesth 82(5):718–722

Haseldonckx M, Van Reempts J, Van de Ven M et al (1997) Protection with lubeluzole against delayed ischemic brain damage in rats. A quantitative histopathologic study. Stroke 28(2): 428–432

Hays SJ, Rice MJ, Ortwine DF et al (1994) Substituted 2-benzothiazolamines as sodium flux inhibitors: quantitative structure-activity relationships and anticonvulsant activity. J Pharm Sci 83(10):1425–1432

Hebert T, Drapeau P, Pradier L et al (1994) Block of the rat brain IIA sodium channel alpha subunit by the neuroprotective drug riluzole. Mol Pharmacol 45(5):1055–1060

Hernández-Guijo JM, Gandía L, de Pascual R et al (1997) Differential effects of the neuroprotectant lubeluzole on bovine and mouse chromaffin cell calcium channel subtypes. Br J Pharmacol 122(2):275–285

Herron J, Lee P, Pesco-Koplowitz L et al (1998) Determination of the dose proportionality of single intravenous doses (5, 10, and 15 mg) of lubeluzole in healthy volunteers. Clin Ther 20(4):682–690

Hille B (1977) Local anesthetics: hydrophilic and hydrophobic pathways for the drug-receptor reaction. J Gen Physiol 69:497–515

Hondeghem LM, Katzung BG (1977) Time- and voltage-dependent interaction of antiarrhythmic drugs with cardiac sodium channels. Biochim Biophys Acta 472:373–398

Hubert JP, Delumeau JC, Glowinski J et al (1994) Antagonism by riluzole of entry of calcium evoked by NMDA and veratridine in rat cultured granule cells: evidence for a dual mechanism of action. Br J Pharmacol 113(1):261–267

Huntington Study Group (2003) Dosage effects of riluzole in Huntington's disease: a multicenter placebo-controlled study. Neurology 61(11):1551–1556

Imbrici P, Liantonio A, Camerino GM et al (2016) Therapeutic approaches to genetic ion channelopathies and perspectives in drug discovery. Front Pharmacol 7:121. https://doi.org/10.3389/fphar.2016.00121

Jankovic J, Hunter C (2002) A double-blind, placebo-controlled and longitudinal study of riluzole in early Parkinson's disease. Parkinsonism Relat Disord 8(4):271–276

Jimonet P, Audiau F, Barreau M et al (1999) Riluzole series. Synthesis and in vivo "antiglutamate" activity of 6-substituted-2-benzothiazolamines and 3-substituted-2-imino-benzothiazolines. J Med Chem 42(15):2828–2843

Kamal A, Syed MA, Mohammed SM (2015) Therapeutic potential of benzothiazoles: a patent review (2010–2014). Expert Opin Ther Pat 25(3):335–349

Kononenko NI, Shao LR, Dudek FE (2004) Riluzole-sensitive slowly inactivating sodium current in rat suprachiasmatic nucleus neurons. J Neurophysiol 91(2):710–718

Kuo JJ, Siddique T, Fu R et al (2005) Increased persistent Na(+) current and its effect on excitability in motoneurones cultured from mutant SOD1 mice. J Physiol 563:843–854

Lacomblez L, Bensimon G, Leigh PN et al (1996) Dose-ranging study of riluzole in amyotrophic lateral sclerosis. ALS/Riluzole Study Group II. Lancet 347:1425–1431

Lamas JA, Romero M, Reboreda A et al (2009) A riluzole- and valproate-sensitive persistent sodium current contributes to the resting membrane potential and increases the excitability of sympathetic neurones. Pflugers Arch 458(3):589–599

Landwehrmeyer GB, Dubois B, de Yébenes JG et al (2007) Riluzole in Huntington's disease: a 3-year, randomized controlled study. Ann Neurol 62(3):262–272

Le Grand B, Dordain-Maffre M, John GW (2000) Lubeluzole-induced prolongation of cardiac action potential in rabbit Purkinje fibres. Fundam Clin Pharmacol 14(2):159–162

Le Grand B, Talmant JM, Rieu JP et al (2003) Study of the interaction of lubeluzole with cardiac sodium channels. J Cardiovasc Pharmacol 42(5):581–587

Lesage AS, Peeters L, Leysen JE (1996) Lubeluzole, a novel long-term neuroprotectant, inhibits the glutamate-activated nitric oxide synthase pathway. J Pharmacol Exp Ther 279(2):759–766

Lo Monaco M, D'Amico A, Luigetti M et al (2015) Effect of mexiletine on transitory depression of compound motor action potential in recessive myotonia congenita. Clin Neurophysiol 126(2):399–403

Lopez-Santiago LF, Yuan Y, Wagnon JL et al (2017) Neuronal hyperexcitability in a mouse model of SCN8A epileptic encephalopathy. Proc Natl Acad Sci U S A 114(9):2383–2388

Maiese K, TenBroeke M, Kue I (1997) Neuroprotection of lubeluzole is mediated through the signal transduction pathways of nitric oxide. J Neurochem 68(2):710–714

Mancini A, Chelini A, Di Capua A et al (2017) Synthesis and biological evaluation of a new class of benzothiazines as neuroprotective agents. Eur J Med Chem 126:614–630

Marrannes R, De Prins E, Clincke G (1998) Influence of lubeluzole on voltage-sensitive Ca^{2+} channels in isolated rat neurons. J Pharmacol Exp Ther 286(1):201–214

Mathew SJ, Gueorguieva R, Brandt C et al (2017) A randomized, double-blind, placebo-controlled, sequential parallel comparison design trial of adjunctive riluzole for treatment-resistant major depressive disorder. Neuropsychopharmacology. https://doi.org/10.1038/npp.2017.106. [Epub ahead of print]

Mishra SP, Shukla SK, Pandey BL (2014) A preliminary evaluation of comparative effectiveness of riluzole in therapeutic regimen for irritable bowel syndrome. Asian Pac J Trop Biomed 4 (Suppl 1):S335–S340

Muir KW, Lees KR (2003) Excitatory amino acid antagonists for acute stroke. Cochrane Database Syst Rev 3:CD001244
Nagoshi N, Nakashima H, Fehlings MG (2015) Riluzole as a neuroprotective drug for spinal cord injury: from bench to bedside. Molecules 20:7775–7789
Noh KM, Hwang JY, Shin HC et al (2000) A novel neuroprotective mechanism of riluzole: direct inhibition of protein kinase C. Neurobiol Dis 7(4):375–383
O'Neill MJ, Bath CP, Dell CP et al (1997) Effects of Ca^{2+} and Na^{+} channel inhibitors in vitro and in global cerebral ischaemia in vivo. Eur J Pharmacol 332:121–131
Osikowska-Evers BA, Wilhelm D, Nebel U et al (1995) The effects of the novel neuroprotective compound lubeluzole on sodium current and veratridine-induced sodium load in rat brain neurons and synaptosomes. J Cereb Blood Flow Metab 15:S380. (Abstract)
Park LT, Lener MS, Hopkins M et al (2017) A double-blind, placebo-controlled, pilot study of riluzole monotherapy for acute bipolar depression. J Clin Psychopharmacol 37(3):355–358
Pieri M, Carunchio I, Curcio L et al (2009) Increased persistent sodium current determines cortical hyperexcitability in a genetic model of amyotrophic lateral sclerosis. Exp Neurol 215(2): 368–379
Ragsdale DS, McPhee JC, Scheuer T et al (1994) Molecular determinants of state-dependent block of Na^{+} channels by local anesthetics. Science 265(5179):1724–1728
Ragsdale DS, McPhee JC, Scheuer T et al (1996) Common molecular determinants of local anesthetic, antiarrhythmic, and anticonvulsant block of voltage-gated Na^{+} channels. Proc Natl Acad Sci U S A 93(17):9270–9275
Ristori G, Romano S, Visconti A et al (2010) Riluzole in cerebellar ataxia: a randomized, double-blind, placebo-controlled pilot trial. Neurology 74(10):839–845
Romano S, Coarelli G, Marcotulli C et al (2015) Riluzole in patients with hereditary cerebellar ataxia: a randomised, double-blind, placebo-controlled trial. Lancet Neurol 14(10):985–991
Russman BS, Iannaccone ST, Samaha FJ (2003) A phase 1 trial of riluzole in spinal muscular atrophy. Arch Neurol 60(11):1601–1603
Salardini E, Zeinoddini A, Mohammadinejad P et al (2016) Riluzole combination therapy for moderate-to-severe major depressive disorder: a randomized, double-blind, placebo-controlled trial. J Psychiatr Res 75:24–30
Scheller DK, De Ryck M, Kolb J et al (1997) Lubeluzole blocks increases in extracellular glutamate and taurine in the peri-infarct zone in rats. Eur J Pharmacol 338(3):243–251
Schuster JE, Fu R, Siddique T et al (2012) Effect of prolonged riluzole exposure on cultured motoneurons in a mouse model of ALS. J Neurophysiol 107(1):484–492
Seppi K, Peralta C, Diem-Zangerl A et al (2006) Placebo-controlled trial of riluzole in multiple system atrophy. Eur J Neurol 13(10):1146–1148
Sheets MF, Fozzard HA, Lipkind GM et al (2010) Sodium channel molecular conformations and antiarrhythmic drug affinity. Trends Cardiovasc Med 20(1):16–21
Song JH, Huang CS, Nagata K et al (1997) Differential action of riluzole on tetrodotoxin-sensitive and tetrodotoxin-resistant sodium channels. J Pharmacol Exp Ther 282(2):707–714
Spadoni F, Hainsworth AH, Mercuri NB et al (2002) Lamotrigine derivatives and riluzole inhibit INa,P in cortical neurons. Neuroreport 13(9):1167–1170
Statland JM, Bundy BN, Wang Y et al (2012) Mexiletine for symptoms and signs of myotonia in nondystrophic myotonia: a randomized controlled trial. JAMA 308(13):1357–1365
Stefani A, Spadoni F, Bernardi G (1997) Differential inhibition by riluzole, lamotrigine, and phenytoin of sodium and calcium currents in cortical neurons: implications for neuroprotective strategies. Exp Neurol 147(1):115–122
Suetterlin KJ, Bugiardini E, Kaski JP et al (2015) Long-term safety and efficacy of mexiletine for patients with skeletal muscle channelopathies. JAMA Neurol 72(12):1531–1533
Sugiyama A, Ni C, Arita J et al (1996) Effects of the antihypoxic and neuroprotective drug, lubeluzole, on repolarization phase of canine heart assessed by monophasic action potential recording. Toxicol Appl Pharmacol 139(1):109–114

Tard C, Defebvre L, Moreau C et al (2017) Clinical features of amyotrophic lateral sclerosis and their prognostic value. Rev Neurol (Paris) 173(5):263–272
Theile JW, Cummins TR (2011) Inhibition of $Na_v\beta4$ peptide-mediated resurgent sodium currents in $Na_v1.7$ channels by carbamazepine, riluzole, and anandamide. Mol Pharmacol 80(4):724–734
Urbani A, Belluzzi O (2000) Riluzole inhibits the persistent sodium current in mammalian CNS neurons. Eur J Neurosci 12(10):3567–3574
Wang YJ, Lin MW, Lin AA et al (2008) Riluzole-induced block of voltage-gated Na^+ current and activation of BKCa channels in cultured differentiated human skeletal muscle cells. Life Sci 82(1-2):11–20
Weiss SM, Saint DA (2010) The persistent sodium current blocker riluzole is antiarrhythmic and anti-ischaemic in a pig model of acute myocardial infarction. PLoS One 5(11):e14103
Weiss S, Benoist D, White E et al (2010) Riluzole protects against cardiac ischaemia and reperfusion damage via block of the persistent sodium current. Br J Pharmacol 160(5):1072–1082
Wood H (2015) Movement disorders: repurposing riluzole to treat hereditary cerebellar ataxia. Nat Rev Neurol 11(10):547
Zona C, Siniscalchi A, Mercuri NB et al (1998) Riluzole interacts with voltage-activated sodium and potassium currents in cultured rat cortical neurons. Neuroscience 85(3):931–938

Structural Models of Ligand-Bound Sodium Channels

Boris S. Zhorov

Contents

Abstract

X-ray and cryo-EM structures of tetrameric and pseudo-tetrameric P-loop channels are used to elaborate homology models of mammalian voltage-gated sodium channels with drugs and neurotoxins. Such models integrate experimental data, assist in planning new experiments, and may facilitate drug design. This chapter outlines sodium channel models with local anesthetics, anticonvulsants, and antiarrhythmics, which are used to manage pain and treat sodium channelopathies. Further summarized are sodium channel models with tetrodotoxin, mu-conotoxins, batrachotoxin, scorpion toxins, and insecticides. Possible involvement of sodium ions in the action of some ligands is discussed.

B. S. Zhorov (✉)
Department of Biochemistry and Biomedical Sciences, McMaster University, Hamilton, ON, Canada

I. M. Sechenov Institute of Evolutionary Physiology and Biochemistry, RAS, St. Petersburg, Russia

Institute of Molecular Biology and Genetics, Almazov Federal Heart, Blood and Endocrinology Centre, St. Petersburg, Russia
e-mail: zhorov@mcmaster.ca

M. Chahine (ed.), *Voltage-gated Sodium Channels: Structure, Function and Channelopathies*, Handbook of Experimental Pharmacology 246,
https://doi.org/10.1007/164_2017_44

Keywords
Antiarrhythmics · Anticonvulsants · Batrachotoxin · Conotoxins · Homology modeling · Insecticides · Ligand docking · Local anesthetics · Pyrethroids · Tetrodotoxin

Abbreviations

DDT	Dichlorodiphenyltrichloroethane
EM	Electron microscopy
LAs	Local anesthetics
MC	Monte Carlo
MCM	MC minimization
MD	Molecular dynamics
PyR	Pyrethroid receptor
SCBIs	Sodium channel blocker insecticides
STX	Saxitoxin
TTX	Tetrodotoxin
VSD	Voltage-sensing domain

1 Structure of Sodium Channels

Voltage-gated sodium channels play key roles in physiology. They are targets for naturally occurring toxins (Stevens et al. 2011), various drugs used to treat pain and health disorders (Catterall 2014; Catterall and Swanson 2015), and insecticides (Silver et al. 2014). The α_1 subunit of these channels folds from a polypeptide chain of four homologous repeats that circumvent the ion-permeating pore (Fig. 1). Each repeat contains six transmembrane helical segments (S1–S6), which are connected by extracellular and intracellular loops, and contributes to the pore domain an outer helix (S5), a pore-lining inner helix (S6), and an extracellular membrane re-entering P-loop (P) with descending (P1) and ascending (P2) helices. Residues at the C-ends of S6s contribute to the activation gate. Segments S1–S4 form voltage-sensing domains. Upon membrane depolarization S4s move in the extracellular direction. S4–S5 linker helices follow this motion initiating the activation gate opening. D, E, K, and A residues in homologous positions of the P1–P2 linkers contribute to the selectivity filter. The DEKA ring separates the outer pore, which is exposed to the extracellular space, and the inner pore, which opens to the cytoplasm upon the channel activation.

Universal labels help appreciate homologous positions of residues in P-loop channels (Zhorov and Tikhonov 2004). A label shows repeat number, segment type (*p*, P-loop; *i*, Inner helix), and relative residue number in the segment. For example, F^{4i15} designates phenylalanine in repeat four inner helix, position 15 (Fig. 2).

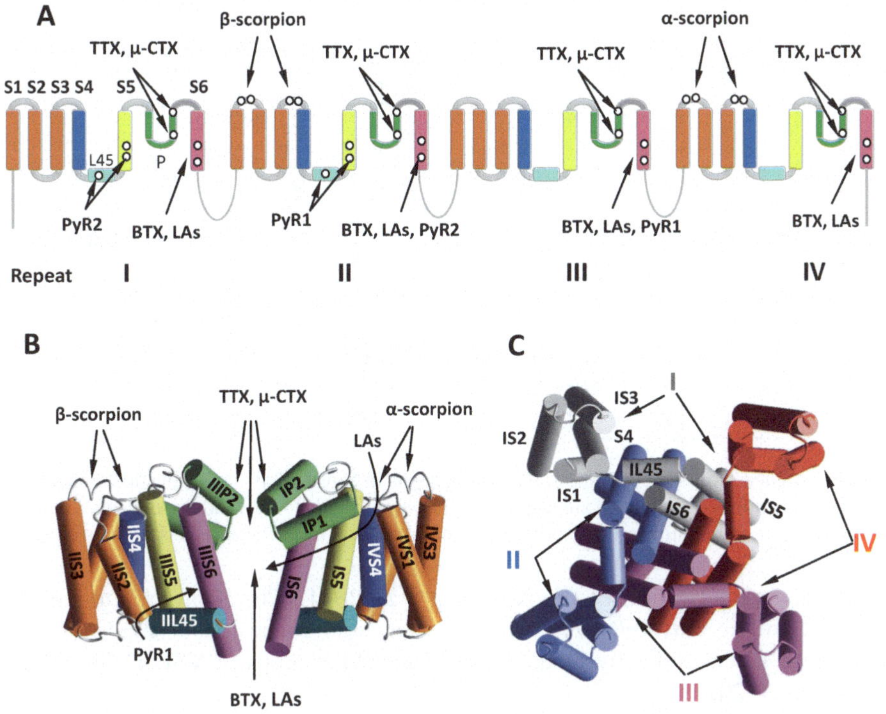

Fig. 1 Ligand binding sites in sodium channels. (**a**) Transmembrane topology of eukaryotic channels. (**b**) Side view of the NavAb X-ray structure (Payandeh et al. 2011) with two subunits removed for clarity and helices colored as in (**a**). (**c**) Intracellular view of NavAb. Individual subunits have different colors. Reprinted from Zhorov and Tikhonov (2016) with permission from Elsevier

X-ray structures of closed prokaryotic sodium channels NavAb (Payandeh et al. 2011) and NavRh (Zhang et al. 2012b) confirmed common folding with potassium channels and revealed specific features such as P2 helices, wide selectivity filters, and large subunit interfaces (fenestrations). The latter may provide hydrophobic access pathway for drugs. Recently, X-ray structures of the open prokaryotic channels NavAb and NavMs (Lenaeus et al. 2017; Sula et al. 2017), and cryo-EM structure of the closed eukaryotic channel NavPaS (Shen et al. 2017) become available.

In the X-ray structure of the NavMs channel with a LA-like molecule PL1, a single bromine atom of PL1 is seen in a fenestration and modeling suggests a horizontal PL1 orientation and electrostatic mechanism of the channel block (Bagneris et al. 2014). In the X-ray structure of a bacterial calcium channel CavAb with a verapamil derivative (Tang et al. 2016) the pore-facing carbonyls at the C-ends of P1 helices are 5.6–7 Å from the center of the ligand ammonium group. The latter apparently displaces a calcium ion, which is seen between these

Channel	P-loop	p40 · p50 · p60
NavMs		ISL YTLFQVM**T**LE SWSMGIVRPV MN
Nav1.4	I	WAF LALFRLM**TQD** YW-**E**NLFQLT LR
	II	HSF LIVFRIL**CGE** -WI**E**TMWDCM EV
	III	LGY LSLLQVA**TFK** GW-M**D**IMYAA VD
	IV	NSI ICLFEIT**TSA** GW-**D**GLLNPI LN

	S6	i10 · i20 · i30
NavMs		NAWVFFIPF IMLTTFTVLN LFIGIIVDAM
Nav1.4	I	TYMIFFVVI IFLG**S**FYLIN LILAVVAMAY
	II	MCLTVFLMV MVIG**N**LVVLN LFLALLLSSF
	III	YMYLYFVIF IIFGS**F**FTLN LFIGVIIDNF
	IV	IGICFFCSY IIIS**F**LI**V**VN M**Y**IAIILENF

Fig. 2 Aligned sequences of P-loops and S6 segments in NavMs and Nav1.4 channels. Residues, which are mentioned in the text, are underlined. Rows above the P-loop and S6 sequences indicate segment types and relative residue positions, which are used in universal labels of the residues

carbonyls in the ligand-free CavAb (Tang et al. 2014). The X-ray structure of the homotetrameric NavAb/Nav1.7 chimera was used to design a small-molecule antagonist that specifically binds to the extracellular loops of VSD-IV in Nav1.7 (Ahuja et al. 2015).

2 Homology Modeling and Ligand Docking

In the absence of high-resolution structures of eukaryotic sodium channels with drugs and toxins computer-based homology modeling is used to predict such structures. The approach is based on the assumption that homologous proteins have similar folding. Computational methods use semi-empirical energy functions (force fields) and energy optimizations that eliminate unrealistic atomic clashes and favor atom–atom attractions. The search for optimal geometry involves random sampling of many 3D structures by using molecular dynamics (MD) or/and Monte Carlo energy minimizations (MCM). MD provides numerical solutions of the Newton equations of motion of atoms. All atoms move simultaneously, an important advantage to simulate large systems, e.g., ion channels in the lipid and water environment. However, a very small (femtosecond) time step requires huge computational resources to simulate millisecond trajectories (Jensen et al. 2012). During MCM, the molecular system "jumps" over energy barriers and this facilitates docking of ligands (Garden and Zhorov 2010). A promising approach is combining MCM to find energetically optimal structures and MD to simulate motions of the structures (Marzian et al. 2013).

Data from mutational, electrophysiological, and other experimental studies greatly facilitate ligand docking. However, results should be treated with caution due to several problems. Firstly, a residue substitution can affect ligand action either directly or indirectly (allosterically). Secondly, the sequence alignment of the query and template channels may be ambiguous. Thirdly, experimental data may be consistent with various low-energy structures. Furthermore, a computational model may inadequately represent the environment, precision of energy calculations is limited, and energy sampling protocols may overlook important structures. Reliable models are expected to integrate diverse experimental data and provide testable predictions. Some models, which seem consistent with various experimental data, are outlined below.

3 Inner Pore Blockers

The inner pore of sodium channels is blocked by various drugs including local anesthetics (LAs), anticonvulsants, and antiarrhythmics (Fig. 3). Typical LAs, e.g., lidocaine, are flexible molecules with an aromatic ring and protonatable amino group. Bulky semi-rigid molecules such as cocaine and quinidine also block the channel apparently at the site that overlaps with the LA receptor (Ragsdale et al. 1996; O'Leary and Chahine 2002). Mutational analyses suggest that these drugs interact with phenylalanine F^{4i15} (Ragsdale et al. 1994; Yarov-Yarovoy et al. 2002; Ahern et al. 2008) and residues in the pore-lining helices IS6, IIIS6, and IVS6

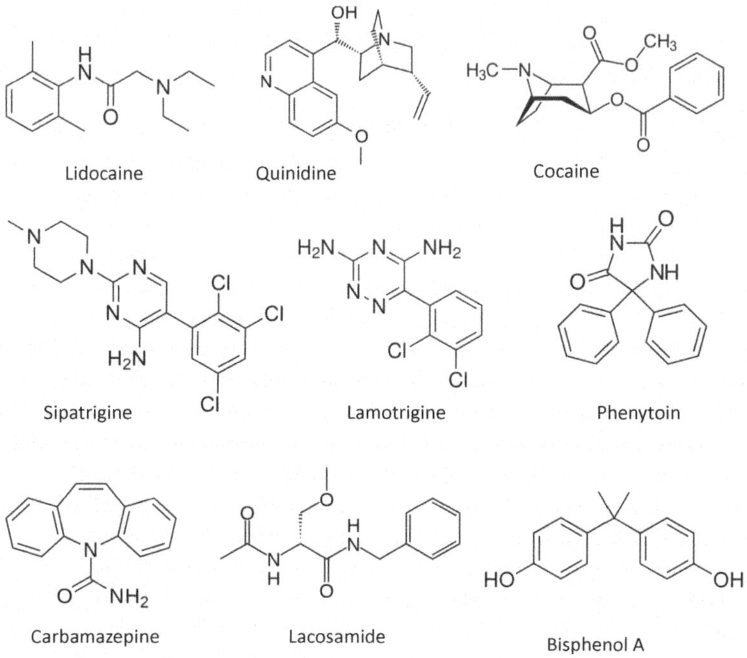

Fig. 3 Inner pore-blocking drugs

(Mike and Lukacs 2010), P1 helices (Yamagishi et al. 2009), and the selectivity-filter DEKA locus (Sunami et al. 1997). In contrast to protonatable LA-like molecules, typical anticonvulsants, e.g., phenytoin and carbamazepine, are electroneutral molecules with an aromatic moiety and non-ionizable polar groups (Liu et al. 2003). Intriguingly, despite highly diverse chemical structures, various drugs have a common binding region within the inner pore (Ragsdale et al. 1996; Kuo 1998) and similar mechanisms of action (Catterall 1987, 2012).

Early Models of Ligand–Channel Complexes Computational docking of cationic ligands in homology models allowed to rationalize various experimental data (Lipkind and Fozzard 2005; Scheib et al. 2006; Tikhonov and Zhorov 2007; Bruhova et al. 2008; Browne et al. 2009). The drugs are proposed to block the ion permeation by the electrostatic mechanism (Lipkind and Fozzard 2005; Tikhonov et al. 2006). Two principal binding modes were proposed. In the "vertical" mode, the ligand positive charge approaches the outer pore and the hydrophobic moiety extends along the inner pore (Tikhonov and Zhorov 2007). In the "horizontal" mode the charged moiety also approaches the outer pore, while the opposite end protrudes into the III/IV repeat interface, which is proposed to serve a hydrophobic access pathway for ligands to the closed channel (Tikhonov et al. 2006; Bruhova et al. 2008). Existence of the hydrophobic pathway has long been predicted (Hille 1977) and such pathways are now seen as wide fenestrations in the X-ray structure of NavAb (Payandeh et al. 2011) and other prokaryotic sodium channels. In contrast to cationic drugs, electroneutral anticonvulsants were proposed to block the ion permeation by sterically occluding the inner pore (Lipkind and Fozzard 2010).

Electrostatics of Drug-Channel Interactions In the low-dielectric environment of membrane proteins, electrostatic interactions are expected to provide large contributions to the ligand binding energy. Therefore, the fact that cationic ligands target the cation-attractive permeation pathway is not surprising. But how electroneutral ligands could block the cation-attractive pore? Permeant ions are hypothesized to interact with electronegative groups of such ligands (Zhorov and Ananthanarayanan 1996; Tikhonov et al. 2006; Zhorov and Tikhonov 2013). These interactions are weak in the bulk solvent, but they may be strengthened in the low-dielectric environment of the confined permeation pathway where the ligand and a permeant ion can be favorably positioned against each other due to interactions with the channel residues. Experimental structures of ion channels with ion-bound ligands are lacking. In the X-ray structure of the NavMs channel, a completely hydrated sodium ion (Na_{III}) is seen between four backbone carbonyls *p48* at the C-end turns of P1 helices (Naylor et al. 2016). This structure was used to build a homology model of the Nav1.4 channel and dock cationic and electroneutral ligands (Tikhonov and Zhorov 2017). Intensive docking using Monte Carlo-energy minimizations predicted ensembles of low-energy binding modes rather than unique ligand–channel complexes. Representative ligand–channel complexes are briefly described below.

Cationic Ligands In the channel model with lidocaine and a sodium ion, the organic and inorganic cations repelled each other and were scattered over the pore domain. However, in the sodium-free model, the lidocaine ammonium group often occurred at site Na_{III} due to electrostatic attraction to backbone carbonyls *p48* and side chains *p49* (Fig. 4a, b). In alternative binding modes, the ammonium group formed cation-π contacts with F^{4i15} or approached a putative position Na_{IV} at the focus of P1 helices, which is analogous to site 5 for potassium ions in the KcsA potassium channel (Zhou et al. 2001). The aromatic moiety usually formed face-to-face or edge-to-face contacts with F^{4i15} and approached Y^{4i22}. In addition to the vertical orientations, horizontal orientations were found with the ammonium group

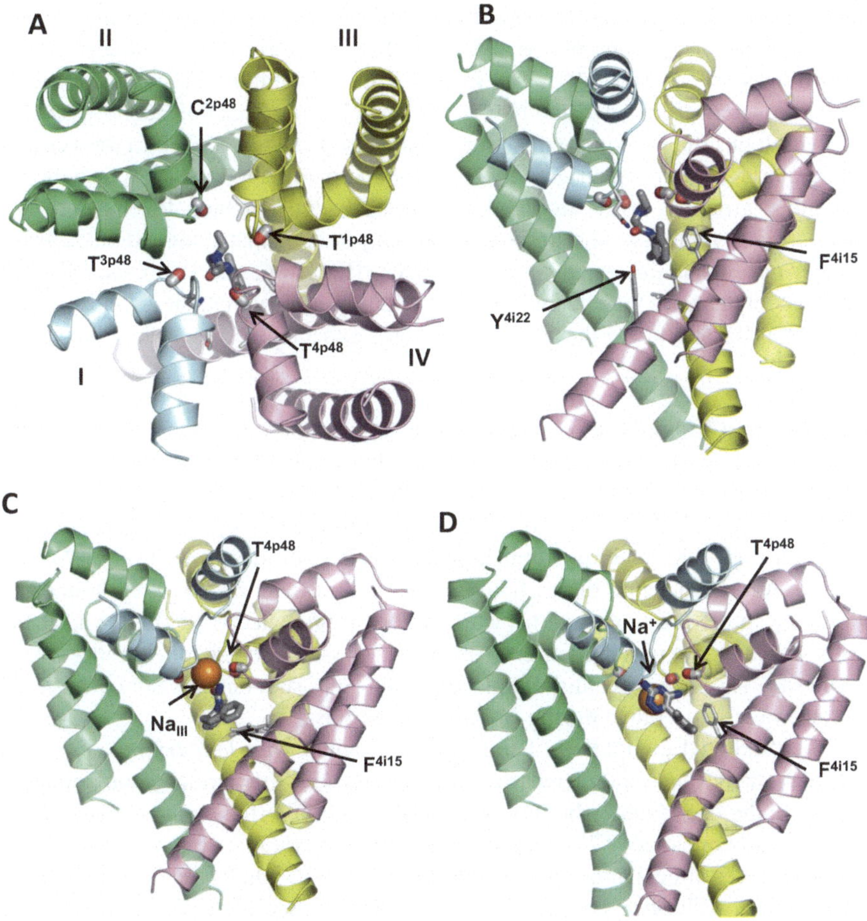

Fig. 4 NavMs-based model of the Nav1.4 channel with ligands. Repeats I, II, III, and IV are *cyan*, *green*, *yellow*, and *magenta*, respectively. Backbone carbonyls in position *p48* are *thick sticks*, a sodium ion is an *orange sphere* and side chains F^{4i15} and Y^{4i22} are *thin sticks*. (**a**) Extracellular view of lidocaine. (**b–d**) Side views of lidocaine, carbamazepine, and lamotrigine, respectively

below site Na_{III} and aromatic moiety extending in the III/IV fenestration as predicted before (Bruhova et al. 2008). In most of the binding modes, lidocaine would electrostatically block the ion permeation. Rather low lidocaine affinity (Bean et al. 1983) is consistent with the diversity of predicted binding modes. These results justified docking of other cationic ligands in the sodium-free channel model.

QX-314 is a quaternary lidocaine analog that blocks sodium channels (Qu et al. 1995). In the low-energy binding modes, the triethylammonium group of QX-314 fits between four backbone carbonyls *p48* or approached side chains in position *p49* in agreement with the data that mutations $Q^{1p49}C$ and $F^{3p49}C$ decrease the ligand potency (Yamagishi et al. 2009). Mutational data suggest that bulky *cocaine and quinidine* bind in the inner pore (Ragsdale et al. 1996; O'Leary and Chahine 2002). In the channel models with these ligands, the protonated amino group approached site Na_{III}, but it was also found at level *i15*. The aromatic group bound between F^{4i15} and Y^{4i22}. The hydroxyl group of quinidine interacted with polar groups in positions *p48* and *p49*. *Sipatrigine* is a protonatable neuroprotective drug that blocks the inner pore of sodium channels (Liu et al. 2003). In the vertical binding mode, the long sipatrigine molecule bound parallel to the pore axis with the ammonium group at level Na_{III} and the opposite end reaching level *i23* in the activation-gate region. In the horizontal binding mode, sipatrigine extended from the inner pore to the interface between helices IIIP, IIIS6, and IVS6, establishing multiple contacts with the channel residues. Numerous protonatable ligands, which are structurally similar to the above drugs, are expected to block the channel in similar ways.

Electroneutral Ligands Experimental data suggest the inner pore as the binding region for various electroneutral ligands (Ragsdale et al. 1996; Kuo 1998; Liu et al. 2003; O'Reilly et al. 2012). In the sodium-free channel, carbamazepine binding modes were scattered over the pore domain as observed in MD simulations of the NavAb channel with electroneutral ligands (Boiteux et al. 2014). In the channel model with a sodium ion, the carbonyl oxygen of carbamazepine often bound the sodium ion at the pore axis and tricyclic moiety approached F^{4i15} (Fig. 4c). These results justified docking of other electroneutral ligands in the channel model with a sodium ion. The latter was initially placed at the Na_{III} site, but was not constrained to it and was free to move during MC-minimizations.

In the predicted complexes of the sodium channel with *lamotrigine* (Fig. 4d), the triazine-ring plane interacted with the sodium ion, the amino groups donated H-bonds to the backbone carbonyls *p48*, and the aromatic ring bound between F^{4i15} and Y^{4i22} and contacted V^{4i18} in agreement with the data that alanine substitutions of these residues affect lamotrigine action (Liu et al. 2003). In low-energy complexes of the channel with *phenytoin*, the sodium ion approached the ligand aromatic ring and bound to the carbonyl oxygen between two NH groups, which donated H-bonds to residues *p48* and *p49*. For example, NH groups donated H-bonds to the T^{3p48} backbone and Q^{1p49} side chain. Ligands like phenytoin, carbamazepine, and lamotrigine have conserved mutual disposition of the O=C–NH or =N–C–NH moieties (Fig. 3). The models predict that the electronegative atoms would bind Na_{III}, while polar hydrogens would donate H-bonds to residues in positions *p48* and *p49*. In such binding modes, a ligand would displace water molecules from the Na_{III} hydration shell and bridge Na_{III} to the channel oxygen atoms, thus strengthening the channel block.

Lacosamide does not block batrachotoxin-activated Nav1.5 channels (Wang and Wang 2014) indicating that its binding site overlaps with that of batrachotoxin, which is located in the inner pore (Tikhonov and Zhorov 2005c; Du et al. 2011). In some binding modes, (R)-lacosamide chelated a sodium ion at the Na_{III} site by two carbonyl oxygens, the NH groups approached oxygen atoms in positions *p48* and *p49*, while the aromatic ring bound between F^{4i15} and Y^{4i22}.

Bisphenol A is a pollutant whose action is sensitive to mutation $F^{4i15}A$ (O'Reilly et al. 2012). In the lowest-energy structure, two aromatic rings chelated Na_{III} by π-cations, whereas two hydroxyl groups donated H-bonds to diagonally opposed backbone carbonyls in position *p48*.

Common Binding Modes of Cationic and Electroneutral Ligands Superposition of the channel-bound ligands (Fig. 5a) shows the ammonium nitrogen (blue) or a ligand-bound sodium ion (yellow) approaching the Na_{III} site and the most remote from this site atom (gray) below side chain of F^{4i15}. Schematically these binding modes are shown in Fig. 5b where the ligand ammonium groups (top row) of ligand-bound sodium ion (bottom row) are at the level of cation-attractive site Na_{III} (red rectangle), whereas hydrophobic groups are at the levels of predominantly hydrophobic residues in the inner pore (gray rectangle). These binding modes suggest the electrostatic mechanism of the channel block. The ammonium group of a charged ligand would displace a permeant ion, whereas a neutral ligand would clamp an ion at its binding site. In any case, the ligand-associated charge would repel permeant cations and thus block the ion permeation. The binding modes are consistent with the pharmacophore model for LAs (Khodorov 1981), where the ammonium group and an aromatic ring are separated by four bonds (Fig. 5c). Many inner-pore blockers, e.g., ranolazine, do not fit this pharmacophore. However, compounds, which are structurally similar to the above-considered, are expected to have analogous binding modes and mechanisms of the channel block. It should be noted that the proposed mechanism is a hypothesis that needs testing by further experimental and computational studies.

Inner Pore Blockers in Therapy of Sodium Channelopathies More than a thousand disease-causing mutations are identified in Nav1.X channels (Huang et al. 2017). The mutations are mapped in the pore domain, voltage-sensing domains, and extracellular and intracellular loops (Catterall 2014; Shen et al. 2017). For most of the channelopathy mutations, atomic mechanisms of the channel malfunctioning remain to be elucidated. Regardless, many sodium channelopathies are treated with small-molecule drugs (El-Sherif and Boutjdir 2015; Imbrici et al. 2016) whose channel-bound models are outlined above. For example, antiepileptic

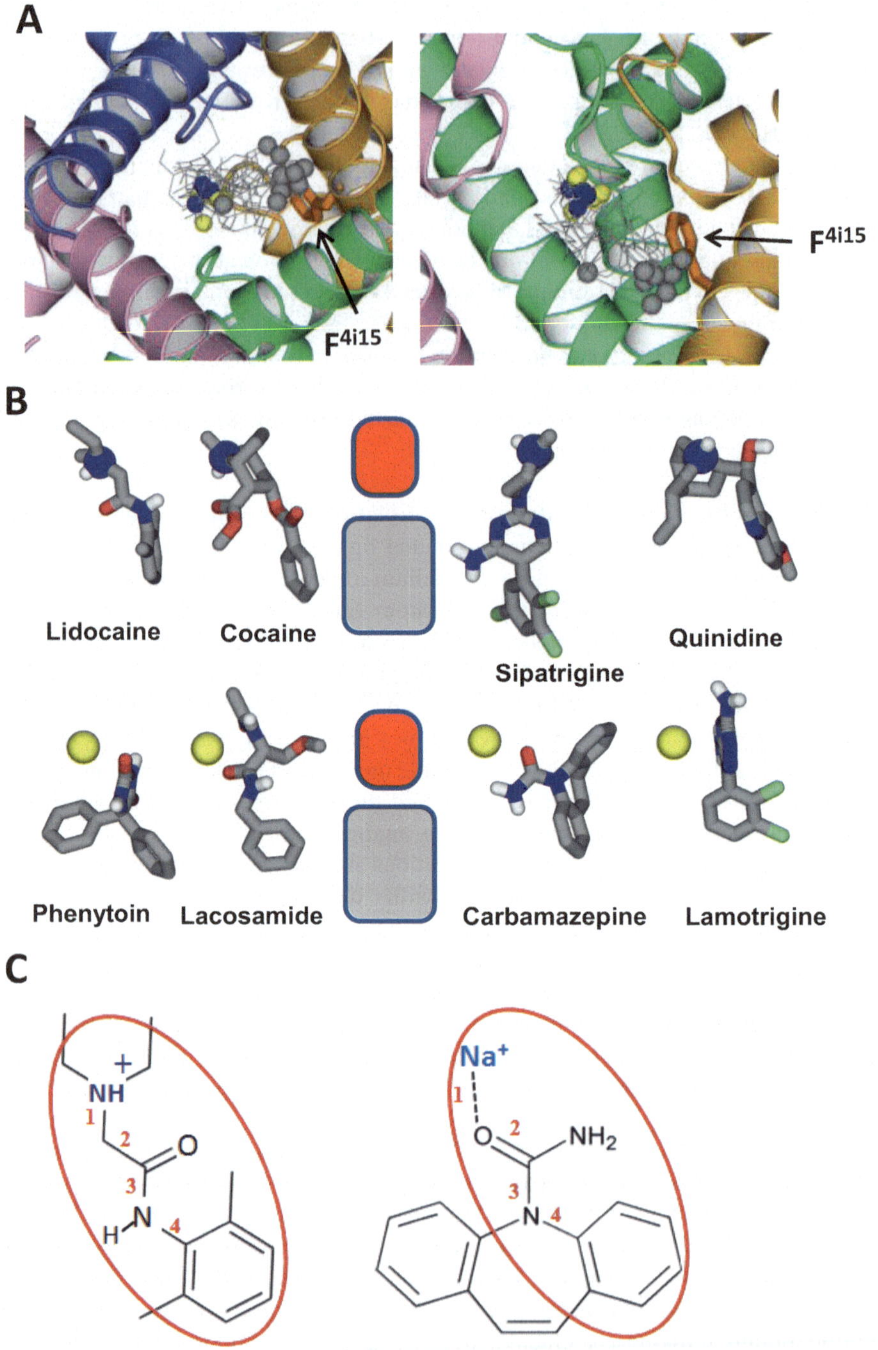

Fig. 5 (**a**) Cytoplasmic and side views at superposed channel-bound ligands. *Blue* and *yellow spheres* show, respectively, the ammonium nitrogens and ligand-bound sodium ions, which are

drugs like carbamazepine, phenytoin, and lamotrigine, which reduce neuronal hyperexcitability, are used not only for the therapy of various epileptic syndromes caused by mutations in the Nav1.1, Nav1.2, Nav1.3, or Nav1.6 channels, but also to treat migraine (Nav1.1 mutations), skeletal-muscle channelopathies (Nav1.4 mutations), and painful syndromes (mutations in Nav1.7, Nav1.8, or Nav1.9). Cardiac channelopathies are caused by mutations in various genes, including Nav1.5-encoding SCN5A. The first-choice treatment for the long-QT syndrome is β-adrenoceptor antagonists, but sodium channel blockers including mexiletine or ranolazine can be used as add-on therapy.

New gene- and mutation-specific drugs are desired. The 3D models of drug-channel complexes can assist in the drug development at the stages of lead discovery and lead optimization. Indeed, high-throughput *in silico* screening of millions purchasable molecules is used to select promising lead candidates for experimental high-throughput screening. The proposed sodium ion involvement in the action of electroneutral ligands suggests that different models should be used for *in silico* high-throughput ligand screenings. Models with sodium ions at the Na_{III} or Na_{IV} sites should be used to find electroneutral lead hits, whereas models without ions at these sites should be used to find lead hits among cationic ligands. Repurposing of marketed drugs is another promising approach to treat sodium channelopathies. Such drugs do not need expensive safety tests, and their 3D complementarity to the sodium channel models can be tested *in silico*.

4 Neurotoxins

Tetrodotoxin binds in the outer pore of TTX-sensitive sodium channels. A sodium channel model with TTX and STX was proposed before any crystal structures of P-loop channels become available (Lipkind and Fozzard 1994). The model is based on mutational data that residues D^{1p50}, E^{2p50}, E^{1p54}, E^{2p54}, and D^{4p54} are important for the action of TTX and STX (Terlau et al. 1991). These data were further used to build TTX- and STX-bound sodium channel models based on X-ray structures of potassium channels (Lipkind and Fozzard 2000; Tikhonov and Zhorov 2005a). More recently, the same experimental data (Terlau et al. 1991) necessitated introduction of insertions–deletions in the NavAb/Nav1.4 sequence alignment (Fig. 2), which underlines a NavAb-based model of Nav1.4 with TTX (Tikhonov and

Fig. 5 (continued) clustered at the Na_{III} site. *Gray spheres* show carbon atoms, which are most remote from the Na_{III} site. These atoms are clustered below F^{4i15} (except for bisphenol A). (**b**) Scheme of ligand–channel interactions. The ligand-bound sodium ion or ammonium nitrogen are attracted to the Na_{III} site (*red rectangle*) and aromatic groups to the inner pore hydrophobic residues (*gray rectangle*). (**c**) A common pharmacophore of lidocaine and sodium-bound carbamazepine. Reproduced from Tikhonov and Zhorov (2017) (https://doi.org/10.1085/jgp.201611668) where coordinates of the models are available

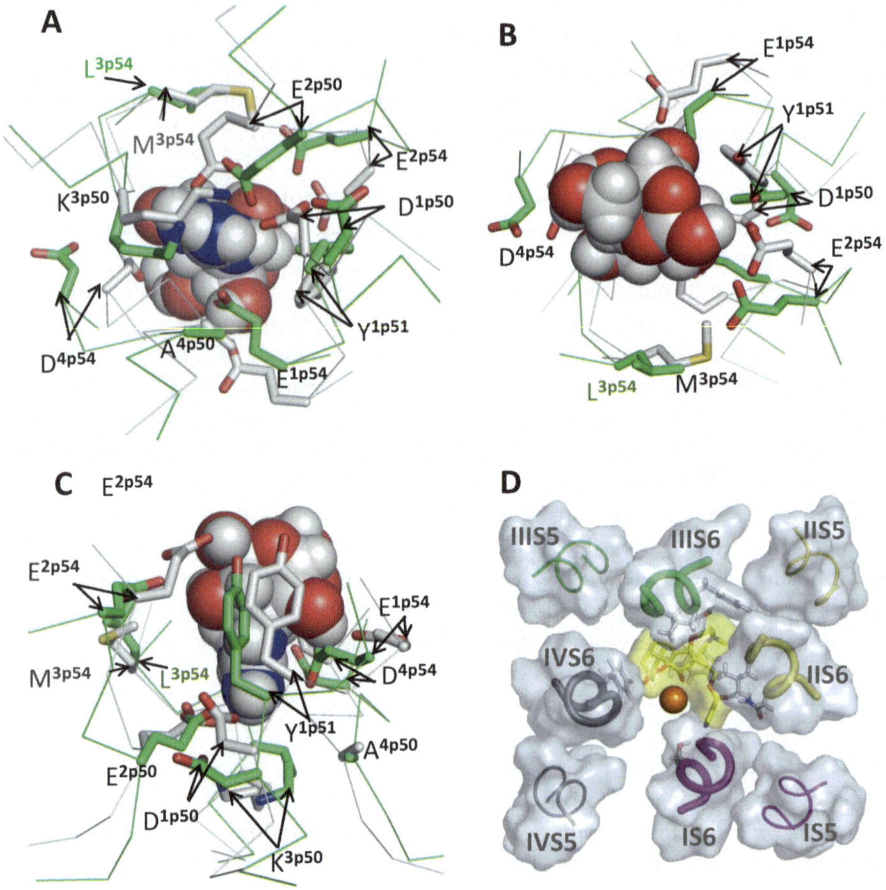

Fig. 6 Intracellular (**a**), extracellular (**b**), and side (**c**) views at the TTX-bound model of Nav1.4 (*gray bonds*) and the NavPaS cryo-EM structure (*green bonds*) (Shen et al. 2017). TTX is space-filled and backbones are indicated as alpha carbon tracings. Coordinates of the TTX-bound model are available (Tikhonov and Zhorov 2012). (**d**) Cytoplasmic view at BTX in the inner pore. For clarity, only middle parts of S5s and S6s helices are shown. Semi-transparent van der Waals surfaces of the helices and BTX are *gray* and *yellow*, respectively. A sodium ion (*orange sphere*) binds between BTX oxygens and S^{1i15}. BTX fits into the inner pore, but does not block it [Figure **d** was originally published in the *Journal of Biological Chemistry* (Du et al. 2011). © The American Society for Biochemistry and Molecular Biology]

Zhorov 2012). Superposition with the NavPaS structure (Shen et al. 2017) supports this model (Fig. 6a–c).

μ-Conotoxins The TTX receptor model (Tikhonov and Zhorov 2012) was further employed to dock peptide toxins GIIIA, PIIIA, and KIIIA (Korkosh et al. 2014). The mutant cycle analysis, which provides energies of pairwise interactions between individual residues in the channel and toxin (Chang et al. 1998; Choudhary et al.

2007), facilitated docking of GIIIA and resulted in the model where computed and experimental pairwise energies correlate. The four outer carboxylates (E^{1p53}, E^{2p53}, D^{3p54}, and D^{4p53}), which reportedly contribute to binding of permeant ions (Khan et al. 2002), are involved in the GIIIA binding. Computations further predicted how deep the charged residues of PIIIA penetrate from the extracellular space into the outer pore (Korkosh et al. 2014) and these estimates are in good agreement with the experimental data (McArthur et al. 2011b). The intriguing data that TTX and KIIIA can simultaneously bind to the sodium channel (Zhang et al. 2009) are rationalized in the model where TTX can bypass the channel-bound KIIIA and reach its receptor in the outer pore (Korkosh et al. 2014). The model further explained why some native and mutant conotoxins incompletely block the channel (Hui et al. 2002; McArthur et al. 2011a; Wilson et al. 2011). A toxin completely blocks the current if its basic residues salt-bridge with the four outer carboxylates of the channel. If a toxin lacks some of the basic residues, at least one of the outer carboxylates does not salt-bridge with the toxin and sodium ions can permeate, but slower than in the toxin-free channel.

Batrachotoxin (BTX) is a sodium channel agonist that has long been used in electrophysiological studies. The bulky steroidal molecule initially was thought to bind at the channel-lipid interface and allosterically activate the channel. However, BTX-sensing residues were found in all the four pore-lining inner helices, partially overlapping with LA-sensing residues (Wang and Wang 2003). Subsequent modeling study proposed that BTX and other sodium channel agonists (veratridine and aconitine) would bind in the inner pore, prevent the activation-gate closure, and allow sodium ions to move between the agonist hydrophilic face and the channel hydrophilic residues (Tikhonov and Zhorov 2005b). The model predicted BTX-binding residues within the pore and experiments confirmed these predictions (Wang et al. 2006, 2007a, 2007b). In particular, when the pore-facing N^{2i15} was replaced with lysine, BTX irreversibly blocked the channel (Wang et al. 2007a) implying involvement of N^{2i15} in the BTX action. Later, additional BTX-sensing residues were found and a model was elaborated to integrate available experimental data (Du et al. 2011). In this model, the horseshoe-shaped BTX is engaged in cation-π interactions with F^{3i16} and contacts BTX-sensing residues in all four repeats (Fig. 6d). Oxygen atoms at the horseshoe inner surface constitute a transient binding site for permeant ions, while the bulky BTX would resist the activation gate closure. To some extent, the model of BTX in the sodium channel resembles a surgical stent in a blood vessel.

Peptide Toxins Targeting VSDs Peptide toxins produced by scorpions, spiders, and sea anemones bind to the extracellular loops in VSDs thus modifying the channel gating. α-Toxins produced by scorpions (Wang et al. 2011; Zhang et al. 2011, 2012a), spiders (Bosmans and Swartz 2010; Minassian et al. 2014), and sea anemones (Xiao et al. 2014) slow down inactivation, whereas β-toxins trap VSD-II in the activated state, causing a negative shift of activation. The extracellular loops of VSD-II and VSD-IV are involved in binding of α- and β-scorpion toxins, respectively (Thomsen and Catterall 1989; Rogers et al. 1996; Cestele et al.

2006) (Fig. 1). β-Toxin sensing residues are also found in loop IIIP2–S6 (Zhang et al. 2012a). Mutational studies helped elaborate models of scorpion toxin action (Zhang et al. 2012a). β-Toxin CssIV is proposed to bind deeply in a slot between loops IIS1–S2 and IIS3–S4, whereas α-toxins appear to fit a tapered slot between extracellular loops IVS1–S2 and IVS3–S4 (Wang et al. 2011). The different effects of α-toxins (inhibition of fast-inactivation) and β-toxins (enhancement of activation) may be due to functional asymmetry of the channel repeats. Thus, the VSD-I and VSD-II rapidly move in response to depolarization, whereas VSD-IV moves slowly (Chanda and Bezanilla 2002). A model with Huwentoxin-IV in the cleft between extracellular loops IIS1–S2 and IIS3–S4 (Minassian et al. 2014) is similar to that with the α-scorpion toxin.

Insecticides Pyrethroid insecticides are synthetic analogs of natural pyrethrins from the *pyrethrum daisy* flowers. They are widely used to control deleterious arthropod pests and mosquito-borne diseases, including malaria and dengue (Dong et al. 2014). Intensive use of DDT and pyrethroids has led to the development of knockdown resistance (*kdr)* in many arthropod pests (Silver et al. 2014), motivating development of new insecticides. Structural models integrate mutational data, including *kdr* mutations, and predict two pyrethroid receptors, PyR1 and PyR2, in the lipid-exposed interfaces between repeats II/III and I/II, respectively (O'Reilly et al. 2006; Du et al. 2013, 2015). Ligand binding to both PyR1 and PyR2 seems necessary to activate the channel. DDT is further predicted to bind at two sites that overlap with PyR1 and PyR2 (Du et al. 2016). The model-driven mutagenesis allowed to discover new pyrethroid- and DDT-sensing residues (Du et al. 2015, 2016).

Indoxacarb, its active metabolite DCJW, and metaflumizone are electroneutral non-ionizable ligands that belong to a relatively new class of sodium channel blocker insecticides (SCBIs). In the open insect sodium channel model, SCBIs bind in the inner pore, interact with a sodium ion at the focus of P1 helices, and extend an aromatic moiety into the III/IV fenestration. Model-driven mutagenesis revealed new SCBI-sensing residues, including insect-specific ones (Zhang et al. 2016).

5 Conclusion

In the absence of experimental 3D structures of eukaryotic sodium channels with ligands, computational structural modeling can be used to visualize ligand–channel complexes, provide mechanistic rationale for mutational, electrophysiological, ligand-binding, and other experimental data, help design new experiments, and may assist in development of new drugs. However, precision of ligand-bound sodium channel models is limited and some models may be incorrect. A key criterion of a model validity is its consistency with a large body of experimental data and, especially, the ability to rationalize experimental data, which were not used at the stage of the model building.

Acknowledgments This work was supported by grant 17-15-01292 from the Russian Science Foundation.

References

Ahern CA, Eastwood AL, Dougherty DA, Horn R (2008) Electrostatic contributions of aromatic residues in the local anesthetic receptor of voltage-gated sodium channels. Circ Res 102 (1):86–94

Ahuja S, Mukund S, Deng L, Khakh K, Chang E, Ho H et al (2015) Structural basis of Nav1.7 inhibition by an isoform-selective small-molecule antagonist. Science 350(6267):aac5464

Bagneris C, DeCaen PG, Naylor CE, Pryde DC, Nobeli I, Clapham DE et al (2014) Prokaryotic NavMs channel as a structural and functional model for eukaryotic sodium channel antagonism. Proc Natl Acad Sci U S A 111(23):8428–8433

Bean BP, Cohen CJ, Tsien RW (1983) Lidocaine block of cardiac sodium channels. J Gen Physiol 81(5):613–642

Boiteux C, Vorobyov I, French RJ, French C, Yarov-Yarovoy V, Allen TW (2014) Local anesthetic and antiepileptic drug access and binding to a bacterial voltage-gated sodium channel. Proc Natl Acad Sci U S A 111(36):13057–13062

Bosmans F, Swartz KJ (2010) Targeting voltage sensors in sodium channels with spider toxins. Trends Pharmacol Sci 31(4):175–182

Browne LE, Blaney FE, Yusaf SP, Clare JJ, Wray D (2009) Structural determinants of drugs acting on the Nav1.8 channel. J Biol Chem 284(16):10523–10536

Bruhova I, Tikhonov DB, Zhorov BS (2008) Access and binding of local anesthetics in the closed sodium channel. Mol Pharmacol 74(4):1033–1045

Catterall WA (1987) Common modes of drug action on Na+ channels: local anesthetics, antiarrhythmics and anticonvulsants. Trends Pharmacol Sci 8(2):57–65

Catterall WA (2012) Voltage-gated sodium channels at 60: structure, function and pathophysiology. J Physiol 590(11):2577–2589

Catterall WA (2014) Sodium channels, inherited epilepsy, and antiepileptic drugs. Annu Rev Pharmacol Toxicol 54:317–338

Catterall WA, Swanson TM (2015) Structural basis for pharmacology of voltage-gated sodium and calcium channels. Mol Pharmacol 88(1):141–150

Cestele S, Yarov-Yarovoy V, Qu Y, Sampieri F, Scheuer T, Catterall WA (2006) Structure and function of the voltage sensor of sodium channels probed by a beta-scorpion toxin. J Biol Chem 281(30):21332–21344

Chanda B, Bezanilla F (2002) Tracking voltage-dependent conformational changes in skeletal muscle sodium channel during activation. J Gen Physiol 120(5):629–645

Chang NS, French RJ, Lipkind GM, Fozzard HA, Dudley S Jr (1998) Predominant interactions between mu-conotoxin Arg-13 and the skeletal muscle Na+ channel localized by mutant cycle analysis. Biochemistry 37(13):4407–4419

Choudhary G, Aliste MP, Tieleman DP, French RJ, Dudley SC Jr (2007) Docking of mu-conotoxin GIIIA in the sodium channel outer vestibule. Channels (Austin) 1(5):344–352

Dong K, Du Y, Rinkevich F, Nomura Y, Xu P, Wang L et al (2014) Molecular biology of insect sodium channels and pyrethroid resistance. Insect Biochem Mol Biol 50:1–17

Du Y, Garden D, Wang L, Zhorov BS, Dong K (2011) Identification of new batrachotoxin-sensing residues in segment IIIS6 of sodium channel. J Biol Chem 286(15):13151–13160

Du Y, Nomura Y, Satar G, Hu Z, Nauen R, He SY et al (2013) Molecular evidence for dual pyrethroid-receptor sites on a mosquito sodium channel. Proc Natl Acad Sci U S A 110 (29):11785–11790

Du Y, Nomura Y, Zhorov BS, Dong K (2015) Rotational symmetry of two pyrethroid receptor sites in the mosquito sodium channel. Mol Pharmacol 88(2):273–280

Du Y, Nomura Y, Zhorov BS, Dong K (2016) Evidence for dual binding sites for 1,1,1-Trichloro-2,2-bis(p-chlorophenyl)ethane (DDT) in insect sodium channels. J Biol Chem 291(9):4638–4648

El-Sherif N, Boutjdir M (2015) Role of pharmacotherapy in cardiac ion channelopathies. Pharmacol Ther 155:132–142

Garden DP, Zhorov BS (2010) Docking flexible ligands in proteins with a solvent exposure- and distance-dependent dielectric function. J Comput Aided Mol Des 24(2):91–105

Hille B (1977) Local anesthetics: hydrophilic and hydrophobic pathways for the drug-receptor reaction. J Gen Physiol 69(4):497–515

Huang W, Liu M, Yan SF, Yan N (2017) Structure-based assessment of disease-related mutations in human voltage-gated sodium channels. Protein Cell 8(6):401–438

Hui K, Lipkind G, Fozzard HA, French RJ (2002) Electrostatic and steric contributions to block of the skeletal muscle sodium channel by mu-conotoxin. J Gen Physiol 119(1):45–54

Imbrici P, Liantonio A, Camerino GM, De Bellis M, Camerino C, Mele A et al (2016) Therapeutic approaches to genetic ion channelopathies and perspectives in drug discovery. Front Pharmacol 7:121

Jensen MO, Jogini V, Borhani DW, Leffler AE, Dror RO, Shaw DE (2012) Mechanism of voltage gating in potassium channels. Science 336(6078):229–233

Khan A, Romantseva L, Lam A, Lipkind G, Fozzard HA (2002) Role of outer ring carboxylates of the rat skeletal muscle sodium channel pore in proton block. J Physiol 543(Pt 1):71–84

Khodorov BI (1981) Sodium inactivation and drug-induced immobilization of the gating charge in nerve membrane. Prog Biophys Mol Biol 37(2):49–89

Korkosh VS, Zhorov BS, Tikhonov DB (2014) Folding similarity of the outer pore region in prokaryotic and eukaryotic sodium channels revealed by docking of conotoxins GIIIA, PIIIA, and KIIIA in a NavAb-based model of Nav1.4. J Gen Physiol 144(3):231–244

Kuo CC (1998) A common anticonvulsant binding site for phenytoin, carbamazepine, and lamotrigine in neuronal Na+ channels. Mol Pharmacol 54(4):712–721

Lenaeus MJ, Gamal El-Din TM, Ing C, Ramanadane K, Pomes R, Zheng N et al (2017) Structures of closed and open states of a voltage-gated sodium channel. Proc Natl Acad Sci U S A 114 (15):E3051–E3060

Lipkind GM, Fozzard HA (1994) A structural model of the tetrodotoxin and saxitoxin binding site of the Na+ channel. Biophys J 66(1):1–13

Lipkind GM, Fozzard HA (2000) KcsA crystal structure as framework for a molecular model of the Na(+) channel pore. Biochemistry 39(28):8161–8170

Lipkind GM, Fozzard HA (2005) Molecular modeling of local anesthetic drug binding by voltage-gated sodium channels. Mol Pharmacol 68(6):1611–1622

Lipkind GM, Fozzard HA (2010) Molecular model of anticonvulsant drug binding to the voltage-gated sodium channel inner pore. Mol Pharmacol 78(4):631–638

Liu G, Yarov-Yarovoy V, Nobbs M, Clare JJ, Scheuer T, Catterall WA (2003) Differential interactions of lamotrigine and related drugs with transmembrane segment IVS6 of voltage-gated sodium channels. Neuropharmacology 44(3):413–422

Marzian S, Stansfeld PJ, Rapedius M, Rinne S, Nematian-Ardestani E, Abbruzzese JL et al (2013) Side pockets provide the basis for a new mechanism of Kv channel-specific inhibition. Nat Chem Biol 9(8):507–513

McArthur JR, Ostroumov V, Al-Sabi A, McMaster D, French RJ (2011a) Multiple, distributed interactions of mu-conotoxin PIIIA associated with broad targeting among voltage-gated sodium channels. Biochemistry 50(1):116–124

McArthur JR, Singh G, O'Mara ML, McMaster D, Ostroumov V, Tieleman DP et al (2011b) Orientation of mu-conotoxin PIIIA in a sodium channel vestibule, based on voltage dependence of its binding. Mol Pharmacol 80(2):219–227

Mike A, Lukacs P (2010) The enigmatic drug binding site for sodium channel inhibitors. Curr Mol Pharmacol 3(3):129–144

Minassian NA, Gibbs A, Shih AY, Liu Y, Neff RA, Sutton SW et al (2014) Analysis of the structural and molecular basis of voltage-sensitive sodium channel inhibition by the spider toxin huwentoxin-IV (mu-TRTX-Hh2a). J Biol Chem 288(31):22707–22720

Naylor CE, Bagneris C, DeCaen PG, Sula A, Scaglione A, Clapham DE et al (2016) Molecular basis of ion permeability in a voltage-gated sodium channel. EMBO J 35:820–830

O'Leary ME, Chahine M (2002) Cocaine binds to a common site on open and inactivated human heart (Na(v)1.5) sodium channels. J Physiol 541(Pt 3):701–716

O'Reilly AO, Khambay BP, Williamson MS, Field LM, Wallace BA, Davies TG (2006) Modelling insecticide-binding sites in the voltage-gated sodium channel. Biochem J 396 (2):255–263

O'Reilly AO, Eberhardt E, Weidner C, Alzheimer C, Wallace BA, Lampert A (2012) Bisphenol A binds to the local anesthetic receptor site to block the human cardiac sodium channel. PLoS One 7(7):e41667

Payandeh J, Scheuer T, Zheng N, Catterall WA (2011) The crystal structure of a voltage-gated sodium channel. Nature 475(7356):353–358

Qu Y, Rogers J, Tanada T, Scheuer T, Catterall WA (1995) Molecular determinants of drug access to the receptor site for antiarrhythmic drugs in the cardiac Na+ channel. Proc Natl Acad Sci U S A 92(25):11839–11843

Ragsdale DS, McPhee JC, Scheuer T, Catterall WA (1994) Molecular determinants of state-dependent block of Na+ channels by local anesthetics. Science 265(5179):1724–1728

Ragsdale DS, McPhee JC, Scheuer T, Catterall WA (1996) Common molecular determinants of local anesthetic, antiarrhythmic, and anticonvulsant block of voltage-gated Na+ channels. Proc Natl Acad Sci U S A 93(17):9270–9275

Rogers JC, Qu Y, Tanada TN, Scheuer T, Catterall WA (1996) Molecular determinants of high affinity binding of alpha-scorpion toxin and sea anemone toxin in the S3-S4 extracellular loop in domain IV of the Na+ channel alpha subunit. J Biol Chem 271(27):15950–15962

Scheib H, McLay I, Guex N, Clare JJ, Blaney FE, Dale TJ et al (2006) Modeling the pore structure of voltage-gated sodium channels in closed, open, and fast-inactivated conformation reveals details of site 1 toxin and local anesthetic binding. J Mol Model 12(6):813–822

Shen H, Zhou Q, Pan X, Li Z, Wu J, Yan N (2017) Structure of a eukaryotic voltage-gated sodium channel at near-atomic resolution. Science 355(6328):eaal4326

Silver KS, Du Y, Nomura Y, Oliveira EE, Salgado VL, Zhorov BS et al (2014) Voltage-gated sodium channels as insecticide targets. In: Cohen E (ed) Advances in insect physiology, vol 46. Academic Press, Oxford

Stevens M, Peigneur S, Tytgat J (2011) Neurotoxins and their binding areas on voltage-gated sodium channels. Front Pharmacol 2:71

Sula A, Booker J, Ng LC, Naylor CE, DeCaen PG, Wallace BA (2017) The complete structure of an activated open sodium channel. Nat Commun 8:14205

Sunami A, Dudley SC Jr, Fozzard HA (1997) Sodium channel selectivity filter regulates antiarrhythmic drug binding. Proc Natl Acad Sci U S A 94(25):14126–14131

Tang L, Gamal El-Din TM, Payandeh J, Martinez GQ, Heard TM, Scheuer T et al (2014) Structural basis for Ca2+ selectivity of a voltage-gated calcium channel. Nature 505(7481):56–61

Tang L, Gamal El-Din TM, Swanson TM, Pryde DC, Scheuer T, Zheng N et al (2016) Structural basis for inhibition of a voltage-gated Ca2+ channel by Ca2+ antagonist drugs. Nature 537 (7618):117–121

Terlau H, Heinemann SH, Stuhmer W, Pusch M, Conti F, Imoto K et al (1991) Mapping the site of block by tetrodotoxin and saxitoxin of sodium channel II. FEBS Lett 293(1–2):93–96

Thomsen WJ, Catterall WA (1989) Localization of the receptor site for alpha-scorpion toxins by antibody mapping: implications for sodium channel topology. Proc Natl Acad Sci U S A 86 (24):10161–10165

Tikhonov DB, Zhorov BS (2005a) Modeling P-loops domain of sodium channel: homology with potassium channels and interaction with ligands. Biophys J 88(1):184–197

Tikhonov DB, Zhorov BS (2005b) Sodium channel activators: model of binding inside the pore and a possible mechanism of action. FEBS Lett 579(20):4207–4212

Tikhonov DB, Zhorov BS (2007) Sodium channels: ionic model of slow inactivation and state-dependent drug binding. Biophys J 93(5):1557–1570

Tikhonov DB, Zhorov BS (2012) Architecture and pore block of eukaryotic voltage-gated sodium channels in view of NavAb bacterial sodium channel structure. Mol Pharmacol 82(1):97–104

Tikhonov DB, Zhorov BS (2017) Mechanism of sodium channel block by local anesthetics, antiarrhythmics, and anticonvulsants. J Gen Physiol 149(4):465–481

Tikhonov DB, Bruhova I, Zhorov BS (2006) Atomic determinants of state-dependent block of sodium channels by charged local anesthetics and benzocaine. FEBS Lett 580(26):6027–6032

Wang SY, Wang GK (2003) Voltage-gated sodium channels as primary targets of diverse lipid-soluble neurotoxins. Cell Signal 15(2):151–159

Wang GK, Wang SY (2014) Block of human cardiac sodium channels by lacosamide: evidence for slow drug binding along the activation pathway. Mol Pharmacol 85(5):692–702

Wang SY, Mitchell J, Tikhonov DB, Zhorov BS, Wang GK (2006) How batrachotoxin modifies the sodium channel permeation pathway: computer modeling and site-directed mutagenesis. Mol Pharmacol 69(3):788–795

Wang SY, Tikhonov DB, Mitchell J, Zhorov BS, Wang GK (2007a) Irreversible block of cardiac mutant Na+ channels by batrachotoxin. Channels (Austin) 1(3):179–188

Wang SY, Tikhonov DB, Zhorov BS, Mitchell J, Wang GK (2007b) Serine-401 as a batrachotoxin- and local anesthetic-sensing residue in the human cardiac Na+ channel. Pflugers Arch 454(2):277–287

Wang J, Yarov-Yarovoy V, Kahn R, Gordon D, Gurevitz M, Scheuer T et al (2011) Mapping the receptor site for alpha-scorpion toxins on a Na+ channel voltage sensor. Proc Natl Acad Sci U S A 108(37):15426–15431

Wilson MJ, Yoshikami D, Azam L, Gajewiak J, Olivera BM, Bulaj G et al (2011) Mu-conotoxins that differentially block sodium channels NaV1.1 through 1.8 identify those responsible for action potentials in sciatic nerve. Proc Natl Acad Sci U S A 108(25):10302–10307

Xiao Y, Blumenthal K, Cummins TR (2014) Gating-pore currents demonstrate selective and specific modulation of individual sodium channel voltage-sensors by biological toxins. Mol Pharmacol 86(2):159–167

Yamagishi T, Xiong W, Kondratiev A, Velez P, Mendez-Fitzwilliam A, Balser JR et al (2009) Novel molecular determinants in the pore region of sodium channels regulate local anesthetic binding. Mol Pharmacol 76(4):861–871

Yarov-Yarovoy V, McPhee JC, Idsvoog D, Pate C, Scheuer T, Catterall WA (2002) Role of amino acid residues in transmembrane segments IS6 and IIS6 of the Na+ channel alpha subunit in voltage-dependent gating and drug block. J Biol Chem 277(38):35393–35401

Zhang MM, McArthur JR, Azam L, Bulaj G, Olivera BM, French RJ et al (2009) Synergistic and antagonistic interactions between tetrodotoxin and mu-conotoxin in blocking voltage-gated sodium channels. Channels (Austin) 3(1):32–38

Zhang JZ, Yarov-Yarovoy V, Scheuer T, Karbat I, Cohen L, Gordon D et al (2011) Structure-function map of the receptor site for beta-scorpion toxins in domain II of voltage-gated sodium channels. J Biol Chem 286(38):33641–33651

Zhang X, Ren W, DeCaen P, Yan C, Tao X, Tang L et al (2012a) Crystal structure of an orthologue of the NaChBac voltage-gated sodium channel. Nature 486(7401):130–134

Zhang JZ, Yarov-Yarovoy V, Scheuer T, Karbat I, Cohen L, Gordon D et al (2012b) Mapping the interaction site for a beta-scorpion toxin in the pore module of domain III of voltage-gated Na(+) channels. J Biol Chem 287(36):30719–30728

Zhang Y, Du Y, Jiang D, Behnke C, Nomura Y, Zhorov BS et al (2016) The receptor site and mechanism of action of sodium channel blocker insecticides. J Biol Chem 291(38):20113–20124

Zhorov BS, Ananthanarayanan VS (1996) Structural model of a synthetic Ca2+ channel with bound Ca2+ ions and dihydropyridine ligand. Biophys J 70(1):22–37

Zhorov BS, Tikhonov DB (2004) Potassium, sodium, calcium and glutamate-gated channels: pore architecture and ligand action. J Neurochem 88(4):782–799

Zhorov BS, Tikhonov DB (2013) Ligand action on sodium, potassium, and calcium channels: role of permeant ions. Trends Pharmacol Sci 34(3):154–161

Zhorov BS, Tikhonov DB (2016) Computational structural pharmacology and toxicology of voltage-gated sodium channels. Curr Top Membr 78:117–144

Zhou Y, Morais-Cabral JH, Kaufman A, MacKinnon R (2001) Chemistry of ion coordination and hydration revealed by a K+ channel-Fab complex at 2.0 a resolution. Nature 414(6859):43–48

Selective Ligands and Drug Discovery Targeting the Voltage-Gated Sodium Channel Nav1.7

Jian Payandeh and David H. Hackos

Contents

Abstract

The voltage-gated sodium (Nav) channel Nav1.7 has been the focus of intense investigation in recent years. Human genetics studies of individuals with gain-of-function and loss-of-function mutations in the Nav1.7 channel have implicated

J. Payandeh (✉)
Department of Structural Biology, Genentech Inc., South San Francisco, CA, USA
e-mail: payandeh.jian@gene.com

D. H. Hackos (✉)
Department of Neuroscience, Genentech Inc., South San Francisco, CA, USA
e-mail: hackos.david@gene.com

M. Chahine (ed.), *Voltage-gated Sodium Channels: Structure, Function and Channelopathies*, Handbook of Experimental Pharmacology 246,
https://doi.org/10.1007/164_2018_97

Nav1.7 as playing a critical role in pain. Therefore, selective inhibition of Nav1.7 represents a potentially new analgesic strategy that is expected to be devoid of the significant liabilities associated with available treatment options. Although the identification and development of selective Nav channel modulators have historically been challenging, a number of recent publications has demonstrated progression of increasingly subtype-selective small molecules and peptides toward potential use in preclinical or clinical studies. In this respect, we focus on three binding sites that appear to offer the highest potential for the discovery and optimization of Nav1.7-selective inhibitors: the extracellular vestibule of the pore, the extracellular loops of voltage-sensor domain II (VSD2), and the extracellular loops of voltage-sensor domain IV (VSD4). Notably, these three receptor sites on Nav1.7 can all be defined as extracellular druggable sites, suggesting that non-small molecule formats are potential therapeutic options. In this chapter, we will review specific considerations and challenges underlying the identification and optimization of selective, potential therapeutics targeting Nav1.7 for chronic pain indications.

Keywords

Drug discovery · Nav1.7 · Pain · Subtype-selectivity · Voltage-gated sodium channel

1 Considerations for Selective, Therapeutic Targeting of Nav1.7

The voltage-gated sodium (Nav) channel Nav1.7 has been the focus of intense study in recent years. Numerous gain-of-function and loss-of-function mutations in the Nav1.7 channel have been implicated in inherited or spontaneous human pain syndromes (Cox et al. 2006; Goldberg et al. 2007; Vetter et al. 2017). Accordingly, our current understanding of the physiology of Nav1.7 has been the subject of many quality reviews (Vetter et al. 2017; Habib et al. 2015; Dib-Hajj et al. 2013; Fischer and Waxman 2010). Since Nav1.7 is primarily expressed in the peripheral nervous system, selective inhibition of Nav1.7 represents a potentially new analgesic strategy that is expected to be devoid of the significant liabilities associated with available treatment options, along with the potential to maintain high levels of efficacy and safety (Vetter et al. 2017; Skerratt and West 2015; Sun et al. 2014). Although the identification and development of selective Nav channel modulators have historically been challenging, a number of recent publications have summarized the extensive work that has progressed increasingly subtype-selective small molecule and peptide scaffolds forward for potential use in preclinical or clinical studies (Sun et al. 2014; Focken et al. 2016; Price et al. 2017; Marx et al. 2016; Flinspach et al. 2017; Deuis et al. 2017; Wu et al. 2017a; Kornecook et al. 2017; Swain et al. 2017; Storer et al. 2017; Murray et al. 2015a, b; Biswas et al. 2017; Schenkel et al. 2017; La et al. 2017; Graceffa et al. 2017; Shcherbatko et al. 2016; Alexandrou et al. 2016; Roecker et al. 2017; Yang et al. 2014; Pero et al. 2017; Pineda et al. 2014; Bagal et al. 2015). It will not be the focus of this chapter, therefore, to exhaustively summarize work already presented by our colleagues in the field. Rather, we aim to focus on specific considerations and challenges underlying the identification of selective, potential therapeutics targeting Nav1.7 for chronic pain indications.

A diverse array of Nav channel modulators that target a vast number of receptor sites on Nav channels has been described and characterized in the literature (Cestele and Catterall 2000; de Lera Ruiz and Kraus 2015). Here, we focus primarily on the three binding sites that appear to offer the highest potential for the discovery and optimization of Nav1.7-selective inhibitors. The primary motivation in prioritizing these receptor sites for consideration includes available data that demonstrates that inhibition of Nav1.7 is possible at these sites, and that molecular selectivity can be achieved over closely related Nav channel subtypes. Accordingly, we focus on available chemical matter which is known to target: (1) the extracellular vestibule of the pore, the infamous tetrodotoxin (TTX) and saxitoxin (STX) binding site (Hille 1975); (2) the extracellular loops of voltage-sensor domain II (or VSD2); and (3) the extracellular loops of voltage-sensor domain IV (or VSD4). It is noteworthy that all three prioritized receptor sites on Nav1.7 can be defined as extracellular druggable sites, suggesting that non-small molecule formats are potential therapeutic options. However, we are immediately cautious about the challenges associated with subcutaneous or intravenous injection that would likely be necessary for the delivery of macrocycles, peptides, or antibody formats as required for long-term dosing and delivery in any potential chronic or neuropathic pain indications. Nevertheless, all molecular formats targeting these three prioritized Nav1.7 receptor sites are worthy of consideration and review, if for no other reason, as proof-of-concept modulators that may require further exploration, optimization, or development.

Focusing on these three receptor sites that appear to offer the most direct path to the discovery of a highly subtype-selective Nav1.7 inhibitor, it is important to point out the considerable physiological and functional distinctions that exist between these sites. First, inhibitors that target the extracellular vestibule of the pore, such as TTX and STX (Hille 1975), block in a largely voltage-independent manner (Cohen et al. 1981); therefore state-dependence of channel block would not be expected to contribute therapeutically or provide a potential increase in the safety window. Second, in the case of modulators targeting the extracellular loops of voltage-sensor domain II (or VSD2) (Bosmans et al. 2008), specific classes of peptide toxins are known that can inhibit channel activation, such as protoxin-II (ProTx2) (Middleton et al. 2002), which affords the distinct opportunity to inhibit a Nav1.7 channel that has not previously opened (Flinspach et al. 2017). Finally, while peptides such as α-scorpion toxins bind to the extracellular loops of voltage-sensor domain IV (or VSD4) to suppress fast-inactivation (Bosmans et al. 2008; Campos et al. 2008; Thomsen and Catterall 1989), a recently characterized class of highly subtype-selective small molecule inhibitor has been found to bind to this region of VSD4 (McCormack et al. 2013), but only after the channel has entered the inactivated state either following channel activation or via direct entry from a closed state (i.e., closed-state inactivation) (McCormack et al. 2013; Ahuja et al. 2015). How these differentiating properties may be favorable or unfavorable for efficacy and safety in preclinical and clinical models for different pain indications remains unknown, but these are important considerations for the future as the field aims to develop and clinically validate useful and improved analgesics.

2 Introduction to Nav Channels

In a series of landmark studies (Hille 2001), Hodgkin and Huxley first posited that the membrane of an excitable cell must undergo rapid changes in the selective permeability of Na^+ and K^+ ions during an action potential, and that the movement of voltage-dependent gating particles controlled these changes (Hodgkin and Huxley 1952). The subsequent isolation, characterization, and molecular cloning of founding members of the voltage-gated ion channel (VGIC) superfamily confirmed the existence of membrane embedded pores that open and close ion selective passageways in response to small changes in the membrane potential (Papazian et al. 1987; Noda et al. 1986). Different classes of VGICs are now known to have distinct and diverse functional roles in neuronal and cellular excitability (Catterall et al. 2005a, b; Gutman et al. 2005). In humans, 10 voltage-gated sodium (Nav) channel, 40 voltage-gated potassium (Kv) channel, and 10 voltage-gated calcium (Cav) channel pore-forming α-subunit genes have been identified (Yu and Catterall 2004).

Nav channels initiate the upstroke of the action potential and are optimized for this specialized function by virtue of their rapid activation and inactivation properties (Hille 2001; Hodgkin and Huxley 1952). Nine Nav channel subtypes are functionally expressed in humans, Nav1.1–Nav1.9, whereas Nav2.1 or the Nax channel (encoded by the *SCN7A* gene) represents an atypical subtype whose biophysical and physiological properties remain poorly defined (Catterall et al. 2005a). The distinctive expression pattern of Nav channel subtypes along axons and within cell bodies helps to shape the firing properties of each excitable cell (Black et al. 1996; Fry et al. 2007; Ho and O'Leary 2011), but Nav channels are still often classified by their primary tissue and cellular expression profiles (Catterall et al. 2005a). The Nav1.1, Nav1.2, and Nav1.6 channel subtypes are best known for their predominant expression within the central nervous system (CNS), whereas Nav1.4 and Nav1.5 are most often referred to as the muscle and cardiac Nav channels, respectively (Catterall et al. 2005a). Not surprisingly, inherited or spontaneous mutations in Nav channel genes and their accessory subunits have been linked to various disorders in the brain, muscle, and heart (Escayg and Goldin 2010; O'Brien and Meisler 2013; Nicole and Fontaine 2015; Savio-Galimberti et al. 2017; O'Malley and Isom 2015).

3 Nav Channel Structure, Biophysics, and Receptor Sites

For decades, our structural-level understanding of Nav channels was rendered only through the analysis of rigorous biophysical studies that often utilized pharmacological and molecular biology tools as further probes of channel architecture (Hille

2001). At the turn of the century, the first high-resolution crystallographic structures of model K^+ channels began to serve as structural templates for homology modeling and further hypothesis-driven Nav channel research (Doyle et al. 1998; Long et al. 2005). Within the last decade, however, our molecular level view of Nav channels has been dramatically sharpened, first through the availability of ~3 Å resolution crystallographic models of simple homotetrameric bacterial Nav (BacNav) channel homologues (Payandeh et al. 2011, 2012; Zhang et al. 2012a; McCusker et al. 2012; Shaya et al. 2014), and more recently, further extended through advances in cryo-electron microscopy (cryoEM) techniques which have produced ~4 Å resolution structural models of intact eukaryotic Nav channels from insect and eel (Shen et al. 2017; Yan et al. 2017), respectively. These breakthrough structural data now set the stage for Nav channel pharmacology and drug discovery to finally enter the modern age of structure-guided drug design (Ahuja et al. 2015; Clairfeuille et al. 2017).

The pore-forming α-subunit of VGICs share a conserved architecture where four subunits or homologous domains create a central ion-conducting pore surrounded by four voltage-sensor domains (VSDs) (Yu and Catterall 2004; Long et al. 2005; Payandeh et al. 2011; Shen et al. 2017; Yan et al. 2017; Wu et al. 2015). The α-subunit of eukaryotic Nav channels contains 24-transmembrane (TM) segments linked in four homologous repeat domains (DI-DIV), where each domain contains six TM segments (S1–S6) (Fig. 1a, b). The S5 and S6 segments from each domain associate to form the central pore module (PM) which scaffolds the ion selectivity filter containing the signature sequence Asp-Glu-Lys-Ala (DEKA) locus that forms the Na^+ selectivity filter near the extracellular side (Shen et al. 2017; Yan et al. 2017). As anticipated from early receptor site mapping studies (Thomsen and Catterall 1989; Cohen et al. 2007; Leipold et al. 2007), the S1–S4 segments form the peripheral VSDs that exist in a domain-swapped arrangement around the pore in Nav channels (Fig. 1a, b). Mammalian Nav channels also associate with a family of immunoglobulin-like single-TM anchored auxiliary β-subunits that are known to modulate the trafficking, gating, and pharmacological profiles of Nav channels, often in a subtype or context-dependent manner (O'Malley and Isom 2015; Gilchrist et al. 2013; Chahine and O'Leary 2011; Ulbricht 2005; Das et al. 2016; Wilson et al. 2011a; Zhang et al. 2013). Recently, structural elucidation of the eel Nav1.4 channel-β1 subunit complex provided the first high-resolution insights into how β-subunits may exert these biochemical and biophysical properties onto the α-subunit (Yan et al. 2017) (Fig. 1b, c). Mammalian Nav channels are further known to interact with an array of intracellular proteins to form a signaling hub (Leterrier et al. 2010; Abriel and Kass 2005), and some of these interactions have also been characterized in structural detail (Pitt and Lee 2016). Although additional targetable sites may exist within these Nav channel-auxiliary subunit interfaces, we focus here on the potentially druggable receptor sites that have been established within the pore-forming α-subunit itself.

Central to the voltage-dependent gating of Nav channels, VSDs are capable of sensing changes in the membrane voltage by virtue of positively charged arginine (or lysine) residues known as gating charges that are found in a conserved RxxR motif along the S4 helix (Noda et al. 1986; Stuhmer et al. 1989). The architecture of Nav channels places the S4 gating charges within the membrane electric field, where

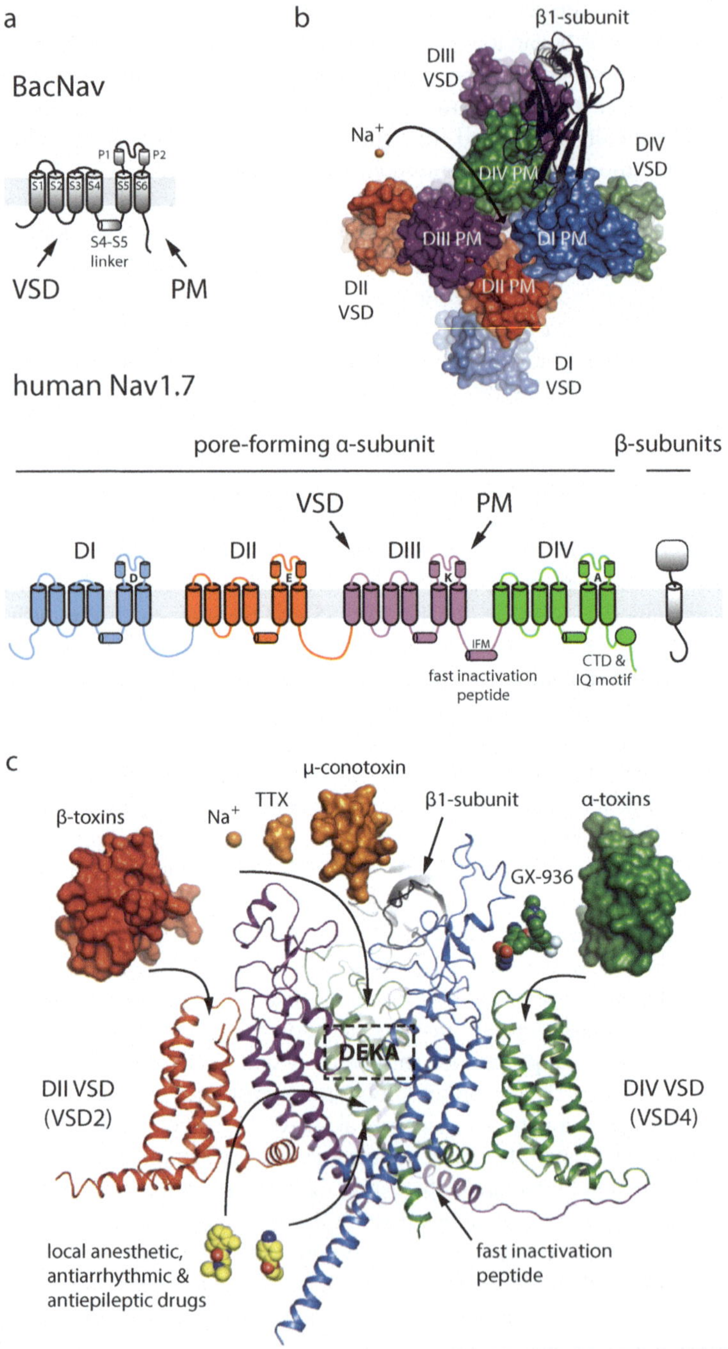

Fig. 1 Eukaryotic Nav channel architecture and major inhibitory receptor sites. (**a**) In gray, a schematic of the bacterial Nav (BacNav) channel subunit highlights the voltage-sensor domain (VSD, S1–S4) and the pore module (PM, S5–S6) connected through the S4–S5 linker. In color,

their outward displacement in response to membrane depolarization initiates opening of the central ion pore via an electromechanical coupling mechanism involving the intervening S4–S5 linker and associated VSD-PM contacts (Payandeh et al. 2011, 2012; Zhang et al. 2012a; Shen et al. 2017; Yan et al. 2017). High-resolution electrophysiological measurements of gating currents generated by the movement of gating charges have helped to confirm that the S4 segments are responsible for initiating processes of voltage-dependent activation and inactivation in Nav channels (Stuhmer et al. 1989; Armstrong and Bezanilla 1973; Bezanilla 2000). Using site-specific probes incorporated along the S4 and other regions within the VSD and channel, fluorometry measurements have further corroborated movement of the S4 during voltage-dependent activation and fast inactivation (Chanda and Bezanilla 2002; Chanda et al. 2004; Cha et al. 1999). In fact, a consensus in the field has begun to emerge suggesting displacements of the S4 on the order of 5–10 Å during gating (Vargas et al. 2012; Guo et al. 2016), although other extremes have been proposed (Chanda et al. 2005). However, the precise vector or mechanism of S4 displacement is still somewhat unsettled, since high-resolution structures of different VGICs so far hint that important differences likely exist between major channel subtypes (Guo et al. 2016; Whicher and MacKinnon 2016; Lee and MacKinnon 2017; Sun and MacKinnon 2017).

Physiological recordings and spectroscopic studies have confirmed that there is a functional specialization of the VSDs within Nav channels (Chanda and Bezanilla 2002; Chanda et al. 2004; Cha et al. 1999). Specifically, voltage-dependent activation of VSD1–3 is sufficient to open the central pore in mammalian Nav channels, whereas the activation of VSD4 is necessary to promote inactivation (Chanda and Bezanilla 2002; Capes et al. 2013). This is a key distinguishing functional feature between eukaryotic Nav channels and their homotetramic Kv channel counterparts (Ahern et al. 2016), where functional specialization of the VSDs is clearly permitted by the amino acid differences found within the eukaryotic Nav channel VSDs (Bosmans et al. 2008; Lacroix et al. 2013; Pless et al. 2014). Moreover, the sequence differences and resulting functional specialization of the four VSDs likely allowed for eukaryotic Nav channels to gain their contemporary role in initiating the action potential. Accordingly, pathogenic disease mutations and pharmacological modulators are known to target

Fig. 1 (continued) below, a schematic of the human Nav1.7 channel with each homologous domain (DI–DIV) shown in a different color and residues forming the "DEKA" motif of the selectivity filter indicated between the P1 and P2 helices. The fast inactivation peptide, which contains an IFM motif, is located between DIII and DIV; and the structured C-terminal domain (CTD) and presence of a calmodulin binding IQ motif are indicated. (**b**) Top view (extracellular) of the eel Nav1.4-β1 subunit complex structure (PDB 5XSY) with the pore forming α-subunit shown in surface representation colored according to the schematic of Nav1.7. The β1 subunit is shown in cartoon rendering (black). (**c**) Side view of the eel Nav1.4-β1 subunit complex structure colored as in part (**b**) with the VSD of DI and PM of DII removed for clarity. Representative peptide toxin and small molecule inhibitors are indicated with arrows pointing to the general location of their respective primary receptor sites. Note, the location of the selectivity filter is indicated by a dashed box (DEKA). PDB codes for peptide toxins are β-scorpion toxin (1BCG), μ-conotoxin (1TCG), and α-scorpion toxin (2ASC) the VSDs of Nav channels with diverse effects on channel gating and biophysics (McCormack et al. 2013; Ahuja et al. 2015; Catterall 2010; Cannon 2010; Bosmans and Swartz 2010; Catterall et al. 2007). Here, sequence differences within the extracellular loops of the VSDs of Nav channels offer the distinct potential for subtype-selective channel targeting and pharmacological modulation.

Nav channels are targeted by a variety of clinically relevant drugs and naturally occurring toxins. These modulators represent important tools to probe the underlying structural and biophysical characteristics of Nav channels (Hille 2001); but in general terms, known Nav channel antagonists alter channel function through one of two distinct mechanisms: they either block ion conduction directly or modify channel gating (Bagal et al. 2015; de Lera Ruiz and Kraus 2015; Ahern et al. 2016; Israel et al. 2017). Historically, a number of distinct and non-overlapping binding sites for various drugs and toxins have been identified on Nav channels (Cestele and Catterall 2000); but these modulators are most easily considered in three general classes: (1) inner vestibule binders, (2) extracellular selectivity filter blockers, and (3) peripheral VSD binders (Fig. 1c). Here, in the following sections, we highlight recent literature describing modulators with the potential to selectively target Nav1.7 (Sun et al. 2014; Focken et al. 2016; Price et al. 2017; Marx et al. 2016; Flinspach et al. 2017; Deuis et al. 2017; Wu et al. 2017a; Kornecook et al. 2017; Swain et al. 2017; Storer et al. 2017; Murray et al. 2015a, b; Biswas et al. 2017; Schenkel et al. 2017; La et al. 2017; Graceffa et al. 2017; Shcherbatko et al. 2016; Alexandrou et al. 2016; Roecker et al. 2017; Yang et al. 2014; Pero et al. 2017; Pineda et al. 2014; Bagal et al. 2015) and the progress made towards ligand and drug discovery efforts on this emerging target for the potential treatment of chronic pain.

4 Introduction to Nav1.7 Physiology and Channelopathies

The critical role that Nav1.7 plays in the sensation of pain was initially discovered through genetics studies when it was found that individuals lacking functional Nav1.7 channels experienced congenital insensitivity to pain (CIP) (Cox et al. 2006; Goldberg et al. 2007). These individuals have a striking inability to feel both mechanical and thermal pain while apparently lacking defects in non-noxious somatosensation such as touch, proprioception, and the ability to distinguish (e.g.) hot, cold, vibration, and pinprick. Such individuals with Nav1.7-dependent CIP continuously injure themselves as a result of this lack of pain sensation. Among the most common types of injuries observed are biting of the tongue, self-mutilation of digits, frequent bone fractures, and frequent burn-related injuries (Cox et al. 2006, 2010; Goldberg et al. 2007; Kurban et al. 2010; Shorer et al. 2014; Sawal et al. 2016; Mansouri et al. 2014).

The expression pattern of Nav1.7 is consistent with its role in pain as this channel is highly expressed in the majority of sensory neurons present in the dorsal root ganglion (DRG) and the trigeminal ganglion (TG) (Black et al. 1996, 2012; Dib-Hajj et al. 2010). Anosmia or hyposmia is also a noted feature of the CIP phenotype, which is consistent with the observation that Nav1.7 is also highly expressed in olfactory epithelial neurons and olfactory neuron-specific knockout of Nav1.7

results in anosmia in mice (Weiss et al. 2011). Other locations where Nav1.7 is expressed are the sympathetic ganglia neurons as well as in restricted areas in the brain such as the hypothalamic/preoptic area, the subfornical organ, and several brainstem nuclei (Ahmad et al. 2007; Morinville et al. 2007), though autonomic and CNS-related defects have not been observed in CIP individuals, perhaps due to compensatory effects from other Nav channels.

It is interesting to consider the reason why Nav1.7 plays such an essential role in pain. This would be a simple question to answer if Nav1.7 were the only voltage-gated sodium channel expressed in sensory neurons. However, this is not the case as Nav1.1, Nav1.6, Nav1.8, and Nav1.9 are also expressed in sensory neurons in addition to Nav1.7 (Black et al. 1996; Dib-Hajj et al. 1998). This is particularly true for the TTX-resistant sodium channels Nav1.8 and Nav1.9, which are expressed in the majority of small-diameter DRG neurons involved in nociception. The leading hypothesis is that Nav1.7 is required for the initial generation of action potentials at the peripheral terminals of sensory neurons, where the detection of noxious stimuli takes place. The biophysical properties of Nav1.7 make this channel particularly suited for this role, as Nav1.7 is characterized by a comparatively slow closed state-inactivation (i.e., closed state-to-inactivated state transitions), which means that Nav1.7 is available to boost subthreshold stimuli arising from small, slow depolarizations of the cell membrane that occur during the detection of noxious stimuli (Cummins et al. 1998; Herzog et al. 2003). Nav1.7 also recovers slowly from inactivation (Klugbauer et al. 1995), which makes it ideally suited for the low-frequency firing characteristic of C-fiber DRG neurons. However, alternative hypotheses have been proposed, such as the idea that knockdown of Nav1.7 induces overexpression of endogenous enkephalin opioids by nociceptors that results in μ-opioid receptor dependent inhibition of pain (Minett et al. 2015).

Gain-of-function missense mutations in Nav1.7 have been found to cause clinically distinct painful neuropathies, depending on the location of the particular mutation within Nav1.7 (Vetter et al. 2017; Huang et al. 2017; Lampert et al. 2014). While many sub-variants have been described, three major Nav1.7 gain-of-function neuropathies have been extensively described: inherited erythromelalgia (IEM), paroxysmal extreme pain disorder (PEPD), and Nav1.7-dependent small fiber neuropathy (SFN) (Brouwer et al. 2014). IEM is in general caused by mutations that shift the activation $V_{1/2}$ toward more hyperpolarized voltages, allowing Nav1.7 to open with less robust membrane depolarization (Lampert et al. 2014). The clinical features of IEM generally include painful episodes that are triggered by a variety of stimuli including warm temperatures, exercise, humidity, and other various stresses. Commonly, the legs and/or arms of the affected individual will become reddened, concurrent with a powerful burning pain feeling that can be relieved to some extent by cooling (Fischer and Waxman 2010). PEPD, on the other hand, is generally caused by mutations in domains III or IV that shift steady-state inactivation to more depolarized voltages, making the channel more available to opening following membrane depolarization, and often with a small shift in the activation $V_{1/2}$ (also to more depolarized voltages) (Huang et al. 2017; Lampert et al. 2014). The clinical features of PEPD are extremely painful episodic sensations generally localized to the recta, ocular, or submandibular regions that are triggered by various normally non-noxious stimuli such as defecation, yawning, chewing, or emotional stresses (Fischer and Waxman 2010; Brouwer et al. 2014).

5 Nav1.7 Receptor Sites: Potential for Selective Targeting

The strong genetics linking Nav1.7 to human pain syndromes has prompted an intense search for selective Nav1.7 inhibitors as potential next generation analgesics (Bagal et al. 2015; de Lera Ruiz and Kraus 2015). However, the high sequence and structural conservation shared between the Nav1.1–Nav1.9 pore-forming α-subunits presents a significant challenge in the identification of subtype-selective Nav channel modulators (Figs. 2, 3, and 4). Except where otherwise noted, we have used the eel Nav1.4 channel-β1 subunit cryoEM structure to model human Nav1.7 and the ConSurf software (Landau et al. 2005) to map and display sequence conservation among the closest related channel subtypes, Nav1.1–Nav1.7 (Figs. 2, 3, and 4). Although a range of Nav channel blockers have been historically clinically useful therapeutics (Fozzard et al. 2011; Glaaser and Clancy 2006; Abdelsayed and Sokolov 2013), a truly molecular subtype-selective Nav channel drug has yet to emerge from ongoing discovery efforts. Here, we review the main receptor sites that are known on mammalian Nav channels with an emphasis on their potential for exploring the development of novel Nav1.7 subtype-selective channel modulators.

6 Inner Vestibule Nav Channel Antagonists

The inner vestibule or central cavity within the pore module of Nav channels houses an important pharmacological modulatory site (Fig. 1c). Clinically relevant antiarrhythmic, antiepileptic, and local anesthetic drugs such as disopyramide, phenytoin, and lidocaine are thought to suppress cardiac or neuronal firing by binding within the inner vestibule of Nav channels to promote or stabilize channel inactivation (Fozzard et al. 2011; Glaaser and Clancy 2006; Abdelsayed and Sokolov 2013). However, the inner vestibule site is also known for the diversity of chemistries that are able to access it, consistent with the poor molecular selectivity of most inhibitors that bind and block within this site (Pless et al. 2011; Ahern et al. 2008). The poor molecular selectivity of drugs that target the inner vestibule arises because the surrounding pore-lining residues are highly conserved between Nav channel subtypes (Fig. 2a, b). In order to maintain safety, antagonists that modulate Nav channels at the inner vestibule receptor site must therefore block via a functionally selective mechanism, whereby they require Nav channels to cycle through open, closed, and inactivated states in highly active neurons to achieve their therapeutic effect and safety window (Fozzard et al. 2011; Glaaser and Clancy 2006; Abdelsayed and Sokolov 2013). Thus, while these drugs lack molecular selectivity among the Nav1.1–1.9 channel subtypes due to the high sequence conservation found within the inner vestibule (Fig. 2a, b), Nav channel block is highly use-dependent.

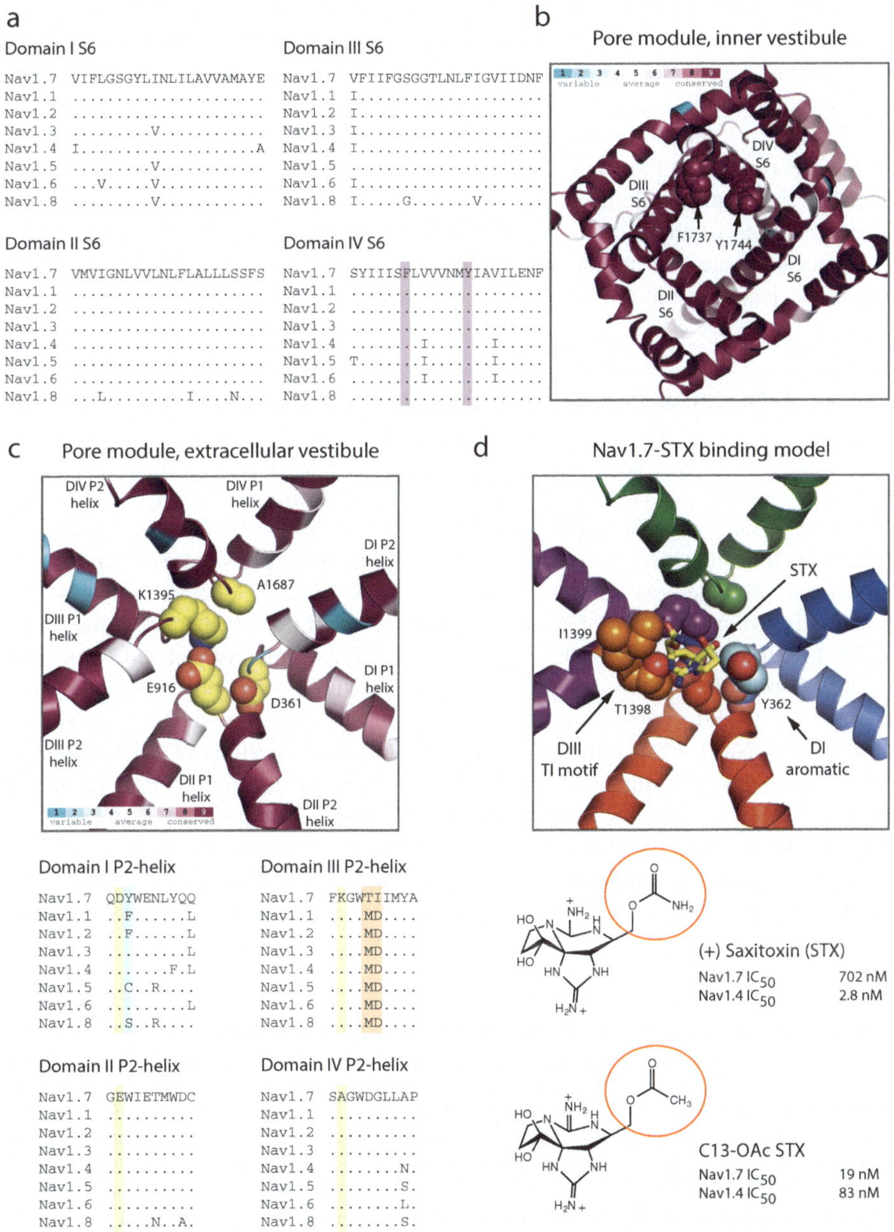

Fig. 2 The inner vestibule and extracellular vestibule receptor sites within the pore module of Nav channels. (**a**) Sequence alignment of the S6 regions comparing human Nav subtypes Nav1.1–Nav1.8. Highlighted in domain IV S6 are phenylalanine and tyrosine positions known to be important for local anesthetic binding; these are F1737 and Y1744 in human Nav1.7. (**b**) The inner vestibule receptor site. Homology model of human Nav1.7 was built based on the eel Nav1.4 structure (PDB 5XSY) and colored according to positional conservation comparing Nav1.1–Nav1.7

Despite the long-held view above, it is notable that molecularly selective inhibitors of Nav1.8 that appear to target the inner vestibule have been identified (Payne et al. 2015; Scanio et al. 2010). In this respect, however, Nav1.8 is more distinctive in its primary amino acid sequence within the inner vestibule compared to the Nav1.1–1.7 channel subtypes (Fig. 2a), and the same can be said for Nav1.9. Therefore, one can in theory envision the discovery of a Nav1.7-selective modulator based on the biophysical differences and minor sequence variations found within the inner vestibule of the Nav1.1–Nav1.7 channel subtypes (Fig. 2a, b); but we are presently unaware that any such inner vestibule antagonist with Nav1.7 molecular selectivity has been identified or advanced into preclinical studies.

7 Extracellular Vestibule Selectivity Filter Blockers

The small molecule guanidinium-containing toxins exemplified by tetrodotoxin (TTX) and saxitoxin (STX) are known to block Na^+ influx by binding directly into or near the selectivity filter from the extracellular side of the Nav channel α-subunit (Fig. 1c). In fact, since early studies on Nav channels, physiologists have long-postulated that the guanidinium moiety of TTX or STX directly engages and potentially competes for a Na^+ ion-binding site within the channel pore (Hille 1975; Lipkind and Fozzard 1994) (Fig. 2c, d). TTX, originally isolated from fugu pufferfish poison, has historically been used to divide subtypes within the Nav channel family into two groups: TTX-sensitive channel subtypes Nav1.1–Nav1.4 and Nav1.6–Nav1.7 are inhibited by nanomolar concentrations of TTX, whereas Nav1.8 and Nav1.9 require millimolar amounts to be blocked completely (Catterall et al. 2005a). Although Nav1.5 inhibition requires intermediate micromolar concentrations, TTX sensitivity can be substantially increased by replacing a cysteine in the domain I S5–S6 loop with a hydrophobic or aromatic residue (Lipkind and Fozzard 1994; Leffler et al. 2005), as this residue is a Phe or a Tyr in TTX-sensitive subtypes (Fig. 2c, d). These observations begin to highlight the molecular

Fig. 2 (continued) channel subtypes using the ConSurf software (Landau et al. 2005). The pore lining S6 residues are labeled and two key receptor site residues, F1737 and Y1744 (in human Nav1.7 numbering), are shown in sphere representation. (**c**) The extracellular vestibule receptor site. Sequence conservation mapping and homology modeling of human Nav1.7 were performed as in part (**b**). Residues of the DEKA motif in the selectivity filter are shown in yellow sphere representation (human Nav.17 numbering) for reference. Sequence alignment of the P2-helix regions comparing human Nav subtypes Nav1.1–Nav1.8 are shown, highlighting the aromatic side-chain in DI which is important for imparting TTX sensitivity and the MD/TI motif in DIII that influences STX potency. (**d**) Homology model of the human Nav1.7 extracellular vestibule receptor site highlighting the positions of the DIII TI motif (orange spheres) and the DI aromatic position (cyan spheres). Note, the structure of STX is shown in yellow stick representation and only placed approximately to indicate potential pose and region of contact within this receptor site. Shown below are chemical structures of STX and C13-OAc STX with potency data for Nav1.7 and Nav1.4 as reported by reference (Thomas-Tran and Du Bois 2016). The red circles highlight the differences between the two compounds

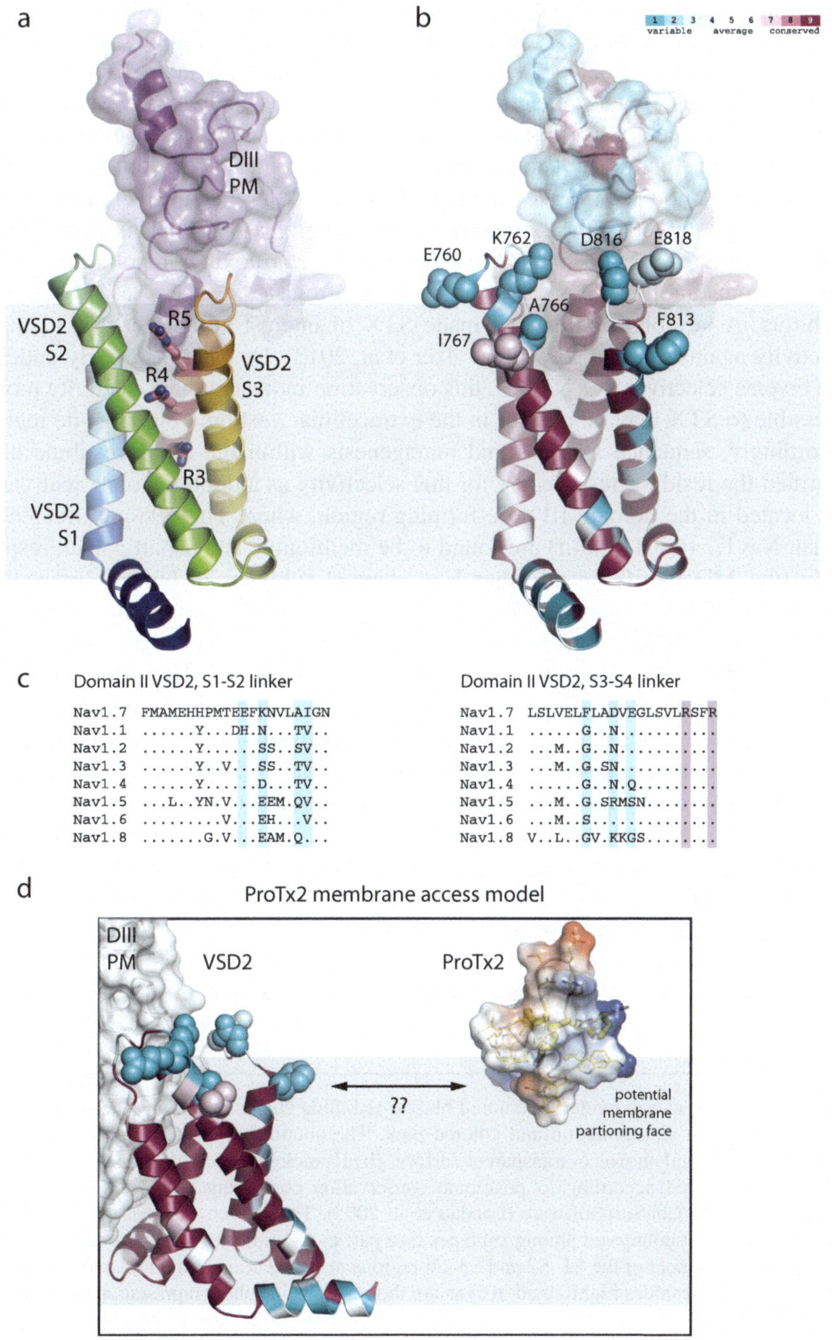

Fig. 3 The extracellular receptor site of voltage-sensor domain 2 (VSD2). (**a**) Homology model of human Nav1.7 was built based on the eel Nav1.4 structure (PDB 5XSY) and VSD2 is shown in

determinants of TTX sensitivity and suggest that it might be possible to identify or optimize subtype selective Nav channel blockers that bind to the extracellular selectivity filter pore site. Due to the complex molecular structure of TTX, however, the generation of extensive chemical derivatives remains challenging. Nevertheless, a natural TTX metabolite, 4,9-anhydro-TTX, has been shown to produce a shift in steady-state inactivation to more negative potentials in Nav1.6, but not in other TTX-sensitive subtypes (Rosker et al. 2007). Overall, these results begin to suggest that selectively targeting the extracellular vestibule in Nav channels is a possibility.

Recent progress has been made in using STX, a neurotoxin compound originally isolated from dinoflagellites, as a molecular scaffold to identify Nav1.7-selective inhibitors. A key observation that made STX of interest is the fact that it shows selectivity against human Nav1.7 (Walker et al. 2012). Despite this clearly undesirable inverse selectivity for Nav1.7, this observation indicated that selectivity motifs accessible to STX may be present in the extracellular vestibule binding site region. Accordingly, sequence analysis and mutagenesis within the outer vestibule have identified the residues responsible for this selectivity as a double amino acid variation located in the domain III pore-forming region, where Thr1398 and Ile1399 in human Nav1.7 (the TI motif) are found to be methionine and aspartic acid, respectively (the MD motif), in all other Nav channel subtypes in humans as well as non-primate Nav1.7 channels (Walker et al. 2012) (Fig. 2c, d). Detailed mutant-cycle analysis then led to the determination of the region of STX that likely directly interacts with these residues (Thomas-Tran and Du Bois 2016); this enabled a molecular model of the binding site region with sufficient resolution to allow structure-guided design efforts leading to the identification of a STX analog with true Nav1.7-subtype selectivity (Fig. 2d), although only ~5-fold selective relative to the human Nav1.4 channel. These results are exciting as further modifications of the STX scaffold may enable better interaction with the TI motif region present within human Nav1.7 and might allow for the identification of highly selective and therapeutically promising Nav1.7 blockers.

Given the intrinsic subtype selectivity of TTX and the progress made towards Nav1.7-directed targeting by STX analogs, it is important to note that subtype-selective peptide toxins that target the extracellular vestibule are also known, such

Fig. 3 (continued) carton rendering colored blue-to-red. Side-chains of S4 gating charge residues are shown in stick representation and colored pink. The abutting pore module (PM) of DIII is colored in purple and shown in transparent surface. (**b**) Homology model of human Nav1.7 from part (**a**) was colored according to positional conservation comparing Nav1.1–Nav1.7 channel subtypes using the ConSurf software (Landau et al. 2005). The side-chains near the extracellular side which are least conserved among subtypes (see part **c**) are indicated in sphere representation. (**c**) Sequence alignment of the S1–S2 and S3–S4 regions in VSD2 comparing human Nav subtypes Nav1.1–Nav1.8. Residues highlighted in cyan are those shown in sphere representation in part (**b**); two gating charges (R1 and R2) of S4 are also indicated in purple. (**d**) Transparent surface and electrostatic rendering of protoxin II (ProTx2; PDB 5O0U) are shown with an arrow to indicate its potential receptor site on VSD2 and highlight the proposed membrane access mechanism that is assumed for this class of peptides

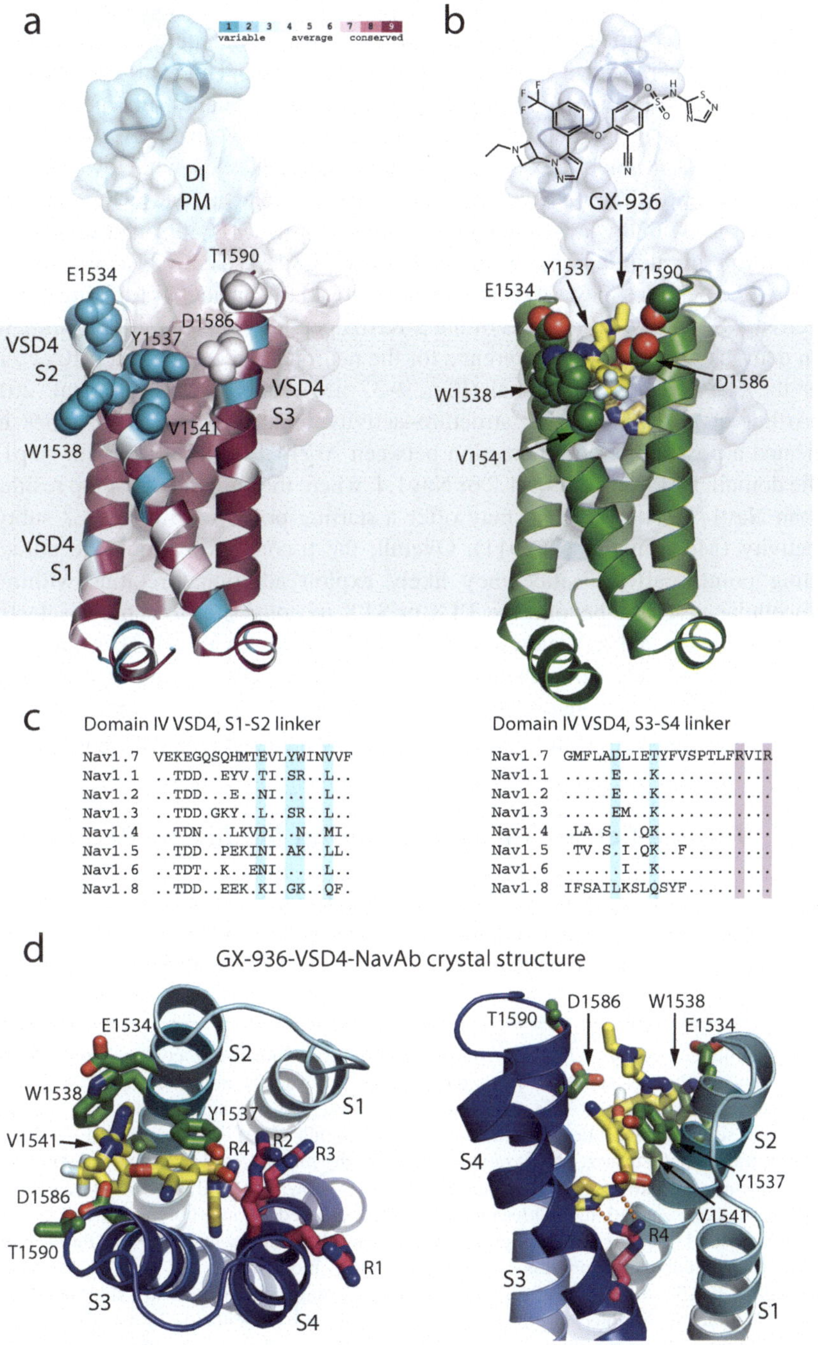

Fig. 4 The extracellular receptor site of voltage-sensor domain 4 (VSD4). (**a**) Homology model of human Nav1.7 was built based on the eel Nav1.4 structure (PDB 5XSY) and VSD4 is shown in

as μ-conotoxins (Wilson et al. 2011b; Knapp et al. 2012). μ-conotoxins (~20 amino acids) generally compete with TTX and STX for a binding site within the extracellular vestibule near the selectivity filter (French et al. 2010; Hui et al. 2002) (Fig. 1c). While some degree of molecular selectivity across Nav channel subtypes is known, no μ-conotoxin is particularly selective for Nav1.7. μ-PIIIA from *Conus purpurascens* is a notable example as it inhibits rat Nav1.4 with low nanomolar potency (IC_{50} 36 nM), rat Nav1.2 with high nanomolar potency (IC_{50} 620 nM), and rat Nav1.7 with low micromolar potency ($IC_{50} > 1$ μM) (Wilson et al. 2011b). In this light, as highlighted above for the inverse Nav1.7 selectivity noted for STX, μ-PIIIA-like peptides may serve as a viable scaffold to consider optimizing subtype-selectivity. Similarly, derivatives of the μ-KIIIA peptide from *Conus kinoshitai* have been demonstrated to have preference for the neuronal Nav1.2 channel (IC_{50} 5 nM) over the Nav1.4 muscle channel (IC_{50} 9–37 nM) subtype (Wilson et al. 2011b; McArthur et al. 2011). Here, structure–activity relationship (SAR) studies have indicated a possible direct interaction between Arg14 from μ-KIIIA with Asp1241 in the domain III P-loop of Nav1.2 or Nav1.4, where the equivalent P-loop residue in human Nav1.7 is Ile1410 and may offer a starting point to gain Nav1.7 subtype-selectivity (McArthur et al. 2011). Overall, the μ-conotoxins could be attractive starting point scaffolds since they likely exploit additional regions within the extracellular vestibule beyond the TTX or STX receptor sites to achieve selectivity (McArthur et al. 2011; Walewska et al. 2013; Brady et al. 2013). Nevertheless, μ-conotoxins are currently unsuitable to consider for oral dosing in their present form and generally lack state-dependence of block, which may or may not be ideal characteristics for achieving efficacy in therapeutically relevant pain settings.

8 Voltage-Sensor Targeting: Gating Modifying Peptides

Peptide toxins that target the VSDs of Nav channels have long been known because these neurotoxins played an essential role in early efforts to isolate, characterize, and ultimately clone the Nav channels (Beneski and Catterall 1980; Sharkey et al. 1984;

Fig. 4 (continued) carton rendering colored according to positional conservation comparing Nav1.1–Nav1.7 channel subtypes using the ConSurf software (Landau et al. 2005). The side-chains near the extracellular side which are least conserved among subtypes (see part **c**) are indicated in sphere representation. The abutting pore module (PM) of DI is colored similarly and shown in transparent surface. (**b**) Crystal structure of the human Nav1.7-VSD4-NavAb chimeric channel in complex with the aryl sulfonamide GX-936 is shown (PDB 5EK0), with GX-936 shown in yellow stick representation. Residues that are least conserved among subtypes (see part **c**) within ~5 Å of the bound GX-936 inhibitor are shown in green sphere representation. Note, the structure of GX-936 is shown above for reference, as is the homology model of the PM from DI. (**c**) Sequence alignment of the S1–S2 and S3–S4 regions in VSD4 comparing human Nav subtypes Nav1.1–Nav1.8. Residues highlighted in green are shown in sphere representation in part (**b**); two gating charges (R1 and R2) in S4 are indicated in purple. (**d**) Two close-in views of the GX-936 aryl sulfonamide (yellow sticks) in complex with the Nav1.7-VSD4-NavAb chimeric channel. Residues important for binding and potential selectivity are shown in stick representation and labeled

Rodriguez de la Vega and Possani 2005). These VSD-binding peptide toxins are generally thought to work principally as gating modifiers since they all target more peripheral binding sites on Nav channels (Fig. 1c), although effects on channel conductance or ion selectivity should not be ruled out *a priori*. Consistent with the functional specialization of the four VSDs within Nav eukaryotic channels (Chanda and Bezanilla 2002; Chanda et al. 2004; Capes et al. 2013; Lacroix et al. 2013; Pless et al. 2014), these toxins are mechanistically diverse; and there are many examples of VSD-binding peptide toxins that primarily inhibit channel activation, promote channel activation, or delay channel inactivation depending on their mechanism of action and primary interaction site with the α-subunit (Bosmans and Swartz 2010; Israel et al. 2017; Rodriguez de la Vega and Possani 2005).

Prototypical peptide toxins that target VSDs are often expected to be state-dependent modulators because they interact directly with the major gating apparatus of the channel (Pineda et al. 2014; Catterall et al. 2007; Rodriguez de la Vega and Possani 2005; Barhanin et al. 1983). However, the extent of state-dependent binding and modulation may depend on the peptide and conditions under study (Bosmans et al. 2008; Bosmans and Swartz 2010). Early studies with scorpion venoms capable of modulating Nav channels led to the identification of α-scorpion and β-scorpion toxins (Rodriguez de la Vega and Possani 2005; Barhanin et al. 1983; Catterall 1976, 1977; Couraud et al. 1978, 1982; Jaimovich et al. 1982; Wang and Strichartz 1982) (Fig. 1c). α-scorpion toxins target VSD4 where they limit S4 movement to inhibit fast inactivation and are often described as stabilizing VSD4 in a deactivated state (Campos et al. 2008; Rogers et al. 1996). Elegant chimeric studies using a Kv1.2 channel chassis has established the S3–S4 loop of VSD4 as a major binding determinant of the α-scorpion toxins (Bosmans et al. 2008), where more traditional approaches have additionally implicated a role for the S1–S2 loop (VSD4) and the S5–S6 loop (from domain I) (Thomsen and Catterall 1989; Rogers et al. 1996; Wang et al. 2011; Gur et al. 2011). The long-standing inference of a composite α-scorpion toxin receptor site on Nav channels is reasonable in light of the recent eukaryotic insect NavPas and eel Nav1.4 cryo-EM structures (Shen et al. 2017; Yan et al. 2017), as these channel structures suggest that the large α-scorpion toxins (~65 amino acids) could make multi-point contact with the pore-forming subunit to suppress S4 activation in VSD4 (Figs. 1c and 4a). Nevertheless, inhibiting channel fast inactivation of Nav1.7 is expected to phenocopy the genetics of GOF mutations, and this is in fact seen with the Nav1.7-selective α-like scorpion toxin OD1 that has proven to be a useful pharmacological tool to induce pain in animal models (Jalali et al. 2005; Maertens et al. 2006; Deuis et al. 2016). Notably, Hm1a, which is an unrelated toxin that can selectively inhibit fast inactivation in the Nav1.1 channel by targeting an apparently overlapping VSD4 receptor site, can also induce pain behavior in animal models (Osteen et al. 2016, 2017). Therefore, the utility of α-scorpion toxin-like molecules that suppress or delay fast inactivation is not regarded as a tractable approach to develop novel analgesics, but it is notable that high channel-subtype selectivity has been achieved from naturally isolated toxins.

In contrast to the modulation by α-scorpion toxins, classic β-scorpion toxins (65–70 amino acids) target VSD2 in Nav channels to shift the voltage dependence of activation and promote channel opening at more hyperpolarized potentials (Rodriguez de la Vega and Possani 2005; Vijverberg and Lazdunski 1984; Cestele et al. 2006). Moreover, in some cases, β-scorpion toxins have been described as trapping the S4 of VSD2 in its activated state (Leipold et al. 2012; Cestele et al. 1998, 2001). Chimeric studies using a Kv1.2 channel chassis have established that β-scorpion toxins appear to interact primarily with S3–S4 loop of VSD2 in Nav channels (Bosmans et al. 2008), while other studies have implicated additional determinants within the S1–S2 loop or domain III pore-loop in binding and block as well (Zhang et al. 2011, 2012b; Leipold et al. 2006). Mechanistically, although overly simplistic, it seems that β-scorpion toxins may require three major points of contact on the channel (i.e., VSD2 S1–S2, VSD2 S3–S4, DIII pore loop) to maintain the S4 of VSD2 in its activated conformation (Figs. 1c and 3a, b). Similarly, α-scorpion toxins likewise appear to require three points of contact on the channel (i.e., VSD4 S1–S2, VSD4 S3–S4, DI pore loop) to suppress the activation of S4 in VSD4 (Figs. 1c and 4a). Nevertheless, it is important to note that toxin pharmacology may be influenced depending upon the presence or absence β-subunit co-expression (Gilchrist et al. 2013; Das et al. 2016; Wilson et al. 2011a; Zhang et al. 2013), further highlighting the complex nature of the interactions that Nav channels have with their surroundings and modulators.

Two features of the VSD-binding toxin gating modulators are notable: (1) their binding sites are accessible from the extracellular side of the membrane and therefore potentially subject to modulation by small and large molecules (Fig. 1c); and (2) sequence conservation within the extracellular loops of VSDs is lower across all channel subtypes (Figs. 3b, c and 4a, c), making the identification of a Nav subtype selective modulator a potential reality. In this respect, peptide toxins that target the VSDs of Nav channels have generated significant interest because subtype-selective toxins have been identified (Flinspach et al. 2017; Deuis et al. 2017; Murray et al. 2015a; Shcherbatko et al. 2016). Although clinical efficacy has yet to be demonstrated in humans, these toxin molecules are being seriously considered as starting points to further improve their selectivity and potential drug-like properties. As numerous reports continue to surface describing the isolation and characterization of novel Nav1.7 VSD-targeting peptide toxin modulators (Chow et al. 2015; Klint et al. 2015a), here we will only aim to capture a snapshot of progress and challenges towards their potential therapeutic validation and clinical application.

8.1 The Pn3a Peptide

Pn3a (or μ-TRTX-Pn3a) is a peptide toxin isolated from the tarantula *Pamphobeteus nigricolor* that potently inhibits human Nav1.7 (IC_{50} 0.9 nM) and produces a rightward shift (21.3 mV) in Nav1.7 voltage dependence of activation while having a remarkable 40–1,000-fold selectivity over other Nav channel subtypes (Deuis et al. 2017). Pn3a shows similar potency across mouse, rat, and human Nav1.7 channels, without significant impact on the Kv, Cav, and nicotinic acetylcholine receptors tested. Pn3a exhibits a conserved inhibitor cystine knot (ICK) fold (~30 amino acids), which is common for peptides that modulate or inhibit Nav channels (Middleton et al. 2002; Cardoso et al. 2015), and a structural motif that may prove to be resistant to heat denaturation and proteolysis (Agwa et al. 2017a). Using a chimeric Kv2.1 channel chassis, Pn3a was shown to interact with the S3–S4 loops of VSD2 and VSD4 from Nav1.7. These receptor site-mapping studies are consistent with the observation that Pn3a inhibited Nav1.7 activation, presumably through interaction with VSD2, and also slowed fast inactivation and recovery from fast inactivation, presumably through interaction with VSD4 (Deuis et al. 2017).

Pn3a markedly inhibited the TTX-sensitive Nav currents in dissociated rat small diameter dorsal root ganglion neurons. Pn3a administered by intraperitoneal injection was able to dose-dependently reduce OD1-induced spontaneous pain behaviors in the absence of adverse effects, but Pn3a as a single agent displayed no analgesic activity in acute nociceptive or inflammatory pain in rodents when administered systemically despite its compelling apparent Nav1.7 selectivity, potency, and dual-VSD targeting mechanism (Deuis et al. 2017). Overall, these results identify Pn3a as a valuable scaffold to consider for the development of highly selective Nav1.7 targeting antagonists, but also highlight the challenges that may be associated with achieving broad analgesic efficacy in rodent models of pain using peripherally restricted selective Nav1.7 peptide toxin-based inhibitors.

8.2 ProTx2 and Derivatives

Protoxin II (or ProTx2) is a prototypical Nav1.7 targeting peptide toxin originally isolated from the tarantula *Thrixopelma pruriens* (Middleton et al. 2002). In contrast to α-scorpion or sea anemone toxins, spider toxins exemplified by ProTx2 typically do not affect the rate of inactivation but rather inhibit channel activity by shifting the voltage dependence of activation to more positive potentials and reducing conductance at all voltages. An early report on ProTx2 demonstrated marginal subtype-selectivity against Nav1.2, Nav1.5, Nav1.7, or Nav1.8, but noted recovery from inhibition varied widely, with Nav1.7 showing no recovery during a 20-min washout after ProTx2 exposure (Middleton et al. 2002). Subsequent reports have suggested that ProTx2 potently and selectively inhibits Nav1.7 with an IC_{50} of 0.3 nM compared to values of 26–146 nM for other Nav channel subtypes tested (Schmalhofer et al. 2008).

In contrast to the larger receptor sites assumed on Nav channels for the α-scorpion toxins (targeting VSD4) and β-scorpion toxins (targeting VSD2), which encompass respective S1–S2, S3–S4, and pore loops, mutagenesis and chimeric studies have demonstrated that the major determinants of ProTx2 binding appears to derive largely from a short stretch of residues in the S3–S4 loop of VSD2 (Bosmans et al. 2008; Bosmans and Swartz 2010; Schmalhofer et al. 2008; Sokolov et al. 2008). In this respect, it is notable that ProTx2 and related gating modifier peptide toxins (Milescu et al. 2009) have been suggested to partition into the membrane bilayer (Henriques et al. 2016; Smith et al. 2005), which may allow ProTx2 to gain considerable potency and selectivity for Nav1.7 (Lee and MacKinnon 2004; Agwa et al. 2017b; Klint et al. 2015b) (Fig. 3d). Here, the mechanism of ProTx2 antagonism is generally described as restricting S4 activation in VSD2 to produce a shift in channel activation to more positive potentials and a decrease in Na^+ current magnitude (Flinspach et al. 2017; Schmalhofer et al. 2008; Sokolov et al. 2008; Smith et al. 2005; Xiao et al. 2010). In addition to VSD2, it is important to note that potential multi-VSD targeting by ProTx2 on Nav1.7 has also been reported (Bosmans et al. 2008; Xiao et al. 2010).

The application of ProTx2 to desheathed cutaneous nerves has been shown to completely block the C-fiber compound action potential, but produces little effect on action potential propagation of an intact nerve (Schmalhofer et al. 2008). Accordingly, early reports using ProTx2 demonstrated no efficacy in rodent models of acute and inflammatory pain at tolerable doses, where lack of efficacy may result from an inability of ProTx2 to cross the blood-nerve barrier (Schmalhofer et al. 2008; Hackel et al. 2012). Intrathecal or perisciatic injections of ProTx2, however, did show efficacy in rodent pain models, furthering the notion that a blood-nerve barrier may be preventing access to the Nav1.7 target when the peptide is delivered systemically (Flinspach et al. 2017). Despite the apparent hurdles to achieving efficacy through intravenous injection or oral dosing, significant effort has been put forward to optimize the synthesis of ProTx2 (Park et al. 2012), and this has led to the understanding of important structure–activity relationships (Agwa et al. 2017a; Park et al. 2014). Structure–function studies have also extended beyond the ProTx2 parental sequence (Wright et al. 2017), including a rational approach to functionalize a non-Nav1.7 targeting toxin, whereby the Hhn2b toxin can be converted to a potent and selective Nav1.7 modulator through a few directed changes (Klint et al. 2015b).

Following the isolation of GpTx1 from the tarantula *Grammostola porteri*, a series of reports has described extensive characterization, synthesis, and structure – activity relationship studies that have produced a potent GpTx1 analog with improved Nav1.7 subtype-selectivity over ProTx2 (Murray et al. 2015a). Uniquely, this GpTx1 derivative shows low nanomolar potency on Nav1.7 (IC_{50} 1.6 nM) and >1,000-fold and >6,000-fold selectivity over the Nav1.4 and Nav1.5 muscle and cardiac channel subtypes, respectively (Murray et al. 2015a). In an effort to further explore the potential of GpTx1, homodimerization was tested and achieved through a bifunctional polyethylene glycol (PEG) linker, resulting in a compound with slightly increased potency for Nav1.7 (IC_{50} 0.6 nM) and a remarkably reduced off-rate where the recovery after washout had a $t_{1/2}$ of >45 min (Murray et al. 2015b). This observation warrants further consideration since multimerization of GpTx1 and related toxins may improve their intrinsic pharmacokinetic properties by increasing the molecular weight and/or hydrodynamic radius, although the in vivo stability and distribution of the bivalent-PEG-linked GpTx1 analog was not directly reported.

In a remarkable effort to improve circulating half-live and potentially alter biodistribution via an Fc receptor recycling mechanism, a GpTx1 derivative was conjugated site-specifically onto an engineered non-targeting monoclonal antibody (mAb) through PEG-based linkers (Biswas et al. 2017). This mAb-based conjugation of GpTx1 led to in vivo half-life extension from 0.6 to 80 h (~130-fold) relative to a nonconjugated GpTx1 peptide (Biswas et al. 2017). Moreover, differential biodistribution of the GpTx1-mAb conjugate to nerve fibers was observed in wild-type but not Nav1.7 knockout mice. Unfortunately, mAb conjugation of GpTx1 significantly impacted the in vitro potency against Nav1.7 (IC_{50} 297 nM); however, the conjugate was still able to inhibit TTX-sensitive currents in DRG neurons with an IC_{50} of 68 nM. Disappointingly, the GpTx1-mAb conjugate was unable to impact scratching behavior in a histamine-induced pruritis model, and results of other pain assays were not reported, perhaps suggesting the GpTx1-mAb conjugate was unable to obtain sufficient concentrations in the nerve to adequately block the Nav1.7 channel (Biswas et al. 2017). Nevertheless, this impressive body of work highlights the opportunities and challenges associated with the development of novel Nav1.7-targeted peptide-based therapeutics for the treatment of pain, since it may well be possible to achieve in vivo efficacy with GpTx1 or ProTx-like molecules (Flinspach et al. 2017). In this light, it is important to note that engineered channel constructs containing viable ProTx2 receptor sites for potential ligand discovery and optimization purposes have been established (Rajamani et al. 2017), and that new approaches to discover and develop macrocyclic peptides may produce novel lead scaffolds with improved pharmacokinetic, pharmacodynamic, and efficacy profiles (Hosseinzadeh et al. 2017; Over et al. 2016).

9 Trapping VSD4: Identification of the Subtype Selective Aryl Sulfonamides

A breakthrough study recently described the characterization of a novel class of subtype-selective small molecule Nav channel antagonist that interacts with a unique site on the α-subunit, which is distinct from known receptor sites targeted by nonselective small molecule modulators such as TTX or local anesthetics (McCormack et al. 2013) (Fig. 1c). This class of aryl sulfonamide inhibitors was originally identified in high-throughput screening efforts for Nav1.3 inhibitors that showed selectivity over the cardiac Nav1.5 channel subtype. Moreover, a truly exciting development in the search for Nav1.7-selective modulators was the discovery of closely related aryl sulfonamide small molecule inhibitors exemplified by PF-04856264 and GX-674 (McCormack et al. 2013; Ahuja et al. 2015), which are highly potent on human Nav1.7 (IC_{50} 28 nM and 0.1 nM, respectively) and exhibit up to 1,000-fold selectivity against other Nav channel subtypes including Nav1.3 and Nav1.5. As with other drug-like small molecule Nav channel antagonists, block of Nav1.7 by PF-04856264 and GX-674 is highly state dependent, where potent inhibition by GX-674 is observed when binding is equilibrated at −40 mV (IC_{50} 0.1 nM), a membrane voltage that promotes steady-state inactivation of Nav1.7, whereas GX-674 inhibition is ~2,400-fold weaker at −120 mV, a voltage that promotes a resting closed state of the channel (Ahuja et al. 2015). Uniquely, and in line with the profound molecular selectivity observed for PF-04856264 and GX-674, electrophysiological characterization of chimeric and point mutant Nav channels has firmly established an extracellular VSD4-based receptor site for the aryl sulfonamide class of inhibitors (McCormack et al. 2013; Ahuja et al. 2015).

High-resolution experimental structures of human Nav channels still remain elusive (Shen et al. 2017; Yan et al. 2017). As such, bacterial Nav (BacNav) channels have been exploited as proximal model systems (Payandeh and Minor 2015) to enable the visualization of bound small molecule modulators to provide insight into mechanism of action and potentially guide rational structure-based drug design (Clairfeuille et al. 2017). In fact, this strategy has provided the first experimental views of approved drugs and drug-like modulators bound to the pore module of representative BacNav and bacterial Cav (BacCav) channels (Bagneris et al. 2014; Tang et al. 2016). In order to extend this approach to truly enable structure-based drug design methods for the optimization and development of subtype selective Nav channel antagonists, the human Nav1.7 VSD4 sequence encompassing the presumed binding site of the aryl sulfonamide inhibitors was engineered onto a bacterial NavAb channel chassis (Payandeh et al. 2011, 2012), and the X-ray crystallographic structure of this Nav1.7-VSD4-NavAb chimeric channel was determined to 3.53 Å resolution in complex with GX-936, a potent and Nav1.7-subtype selective inhibitor (Fig. 4b, d) (Ahuja et al. 2015). This approach revealed the molecular details of how the aryl sulfonamide class of compounds antagonizes Nav channels and, for the first time, uncovered the structural basis for how small molecule inhibitors like GX-936 and PF-04856264 can achieve high molecular selectivity for VSD4 of the Nav1.7 channel.

GX-936 and related aryl sulfonamide modulators inhibit Nav1.7 by trapping the S4 voltage-sensor of VSD4 in the "up" or activated state. When bound within the extracellular vestibule of VSD4, the negatively charged sulfonamide moiety of GX-936 forms a direct salt-bridge contact with the forth arginine (R4) of the VSD4 S4 voltage-sensor (Fig. 4d). The importance of the observed sulfonamide-R4 interaction was confirmed by a ~2,700-fold potency shift for the single alanine mutation (R4A) studied in the context of the full-length Nav1.7 channel (GX-674 on R4A human Nav1.7, IC_{50} 270 nM). This strong and direct ionic interaction ultimately prevents the S4 voltage-sensor of VSD4 from returning to the deactivated (or "down") resting position following voltage-dependent channel activation. Notably, VSD4 is known to play a key role in the Nav channel inactivation process, as S4 activation in VSD4 has been shown to be necessary and sufficient to produce fast inactivation (Chanda and Bezanilla 2002; Capes et al. 2013). Therefore, by effectively trapping the VSD4 S4 voltage-sensor in the activated conformation, this class of aryl sulfonamide inhibitors is able to prevent the Nav channel from recovering from fast [and slow (Osteen et al. 2017; Theile et al. 2016)] inactivation (Ahuja et al. 2015; Capes et al. 2013). As the crystal structure of the GX-936-VSD4-NavAb channel complex clearly reveals, the aryl sulfonamide modulators can only bind to R4 of the S4 voltage-sensor when VSD4 is in the activated conformation (Fig. 4d), rationalizing why the binding affinity is so strongly state-dependent (McCormack et al. 2013; Ahuja et al. 2015).

As discussed above for peptide toxins targeting the VSDs of Nav channels, the extracellular surface of VSD4 is not absolutely conserved between all Nav channel subtypes, explaining why subtype-selective modulators can be achieved by small molecule compounds that are able to make intimate contacts with the more poorly conserved residues in VSD4. In this regard, the most critical selectivity motif that the aryl sulfonamide inhibitors like GX-936 and PF-04856264 target in human Nav1.7 VSD4 is Y1537/W1538 (the YW motif) (McCormack et al. 2013; Ahuja et al. 2015) (Fig. 4a–d). Specifically, the YW motif is present at this location of the S2 within VDS4 in only three of the nine Nav channel subtypes: Nav1.2, Nav1.6, and Nav1.7 (Fig. 4c). As aryl sulfonamides that can inhibit Nav1.7 make close contact with both side-chains of the YW motif (Fig. 4d), it is relatively straightforward to obtain inhibitors that are selective for these three channels over all other Nav channel subtypes. For example, GX-674 shows similar potencies on Nav1.2, Nav1.6, and Nav1.7 (IC_{50}'s all <1 nM), but it is at least 1,000-fold less potent on the other Nav channel subtypes that do not contain the YW motif (Ahuja et al. 2015). Importantly, additional residues within the aryl sulfonamide VSD4 binding site can in principle be used to make even more selective compounds (Ahuja et al. 2015), including: E1534 (S2), which is uniquely glutamate in Nav1.7 but asparagine in Nav1.2 and Nav1.6; V1541 (S2), which is uniquely valine in Nav1.7 but leucine in most other subtypes; D1586 (S3), which is aspartate in Nav1.4–Nav1.8 but glutamate in Nav1.1–Nav1.3; and T1590 (S3), which is uniquely threonine in Nav1.7 but lysine in most other subtypes (Fig. 4a–d). Likely by designing compounds that are able to interact with these specific residue side-chains, aryl sulfonamides with Nav1.7 selectivity ratios of >100 over Nav1.2 and Nav1.6 have been identified and reported, such as AMG8379 (Kornecook et al. 2017) and PF-06456384 (Storer et al. 2017).

With the recent discovery of the first truly Nav1.7-selective small molecule antagonists and elucidation of a high-resolution crystal structure of a representative aryl sulfonamide in complex within the VSD4 receptor site, significant efforts from multiple groups have since focused on optimizing the potency, selectivity, and drug-like properties of this class of molecules. Accordingly, a range of studies have begun to emerge that show promising results that aryl sulfonamide engagement of the VSD4 site of Nav1.7 in vivo leads to an analgesic effect in rodent models of acute and inflammatory pain (Focken et al. 2016; Roecker et al. 2017). In mice, engagement of the VSD4 receptor site in Nav1.7 by selective aryl sulfonamide inhibitors in vivo has been reported to reduce formalin-induced nociceptive pain behaviors (Pero et al. 2017; Wu et al. 2017b), histamine-induced scratching behavior (Marx et al. 2016; Weiss et al. 2017), capsaicin-induced licking in a nociception model of pain (Graceffa et al. 2017), and UVB radiation skin burn-induced thermal hyperalgesia (Kornecook et al. 2017). However, a notable lack of preclinical efficacy has also been reported for some aryl sulfonamide compounds (Storer et al. 2017; Wu et al. 2017c). Overall, results from these preclinical studies have warranted moving select aryl sulfonamides forward into human clinical trials (Jones et al. 2016).

10 Opportunities and Challenges in Nav1.7 Drug Discovery

Nav1.7 represents an extremely exciting target for the development of novel pain drugs. The challenge of identifying a selective Nav1.7 inhibitor with the proper drug-like properties to be a clinically useful pain drug, however, is immense. Each potential class of selective Nav1.7 inhibitor has its advantages and disadvantages, which we will consider here.

10.1 Pharmacokinetic Properties

Effective Nav1.7 inhibitors require exquisite potency, selectivity, and metabolic stability to enable a safe, low dose, and orally bioavailable pain medication. Selective outer vestibule inhibitors such as those based on STX will likely possess poor oral bioavailability as a result of the positively charged guanidinium moiety. The need to be injected either intravenously or subcutaneously may limit the clinical use of such compounds, possibly restricting them to in-hospital settings. Similar concerns exist in the case of peptide-based inhibitors that bind to VSD2, which will also likely suffer from poor oral bioavailability and rapid clearance. On the other hand, the aryl sulfonamide class of inhibitors has demonstrated that good oral bioavailability and metabolic stability is achievable. However, the negatively charged aryl sulfonamide group, which is required for binding the positively charged R4 on the S4 voltage-sensor of VSD4, in general makes such molecules highly protein-bound in the blood, limiting the free drug available for blocking Nav1.7 channels. The clinical candidate PF-05089771 is an example of a highly protein-bound aryl sulfonamide, where >99% of the compound is bound to plasma proteins and thus unavailable to block channels (Swain et al. 2017). Recent work showing that aryl sulfonamides with a zwitterionic character can have much lower levels of plasma protein binding while retaining potency (Roecker et al. 2017) suggests a strategy for making aryl sulfonamide compounds that may achieve better target coverage in vivo without requiring very high levels of total plasma concentration. Ultimately, finding compounds with the right balance of potency, selectivity, and metabolic stability will be required to deliver best-in-class Nav1.7-targeted pain medications.

10.2 State-Dependence of Inhibition

While state-dependence is a key attribute for nonselective Nav blockers, especially those used systemically, it is currently unclear whether state-dependence is a desirable feature for a selective Nav1.7 blocker. State-dependence can in principle allow for selective block of more active sensory fibers versus less active sensory fibers. Among the three classes of potential Nav1.7-selective blockers profiled in this chapter, only aryl sulfonamides show state-dependence that is expected to be relevant in vivo since extracellular vestibule inhibitors like STX in general lack significant state-dependence, and ProTx2-like peptides that bind to VSD2 show a preference for antagonizing closed Nav channels (Flinspach et al. 2017). The potency of aryl sulfonamides is highly voltage-dependent since these compounds only bind to the activated state of VSD4, which occurs primarily (or only) when Nav1.7 is in the inactivated state. The amount of time a channel spends in inactivated states depends on the membrane voltage, so aryl sulfonamides should inhibit Nav1.7 more effectively in more depolarized nerve fibers. This could in principle give rise to a safety factor between the desirable efficacy in chronic pain versus undesirable effects on acute nociception for the aryl sulfonamide class of inhibitors.

10.3 Efficiency of In Vivo Target Engagement

Even if a selective inhibitor can achieve high enough unbound concentrations in the plasma to cover the IC_{50} for Nav1.7 channel several-fold, it does not necessarily mean that it can access the Nav1.7 target within the micro-compartment in the nerve. A study of the in vivo efficacy of ProTx2, for example, showed no detectible effects of the peptide despite achieving high levels of coverage over the IC_{50} (>100-fold), whereas direct application of ProTx2 to the desheathed saphenous nerve was found to efficiently inhibit the C-fiber compound action potential (Schmalhofer et al. 2008). Accordingly, direct intrathecal or perisciatic injection of the ProTx2 peptide has shown in vivo efficacy (Flinspach et al. 2017). Additionally, a highly potent and selective aryl sulfonamide (PF-06456384) failed to show efficacy in the formalin-induced pain model despite achieving free plasma concentrations 62.5-fold over the mouse Nav1.7 IC_{50} (Storer et al. 2017). On the other hand, some aryl sulfonamide compounds are able to achieve much better target-engagement efficiencies. For example, compounds 5 and 9 from a recent medicinal chemistry effort were shown to inhibit formalin-induced flinching at free plasma concentrations only 11-fold and 3.5-fold over the mouse Nav1.7 IC_{50} (Roecker et al. 2017). Another example is compound AM8379, which showed efficacy in a histamine-induced itch model at free plasma concentrations of 5.3-fold over the mouse IC_{50} and showed efficacy in a UVB burn-induced heat hyperalgesia model at 23-fold over the mouse IC_{50} (Kornecook et al. 2017). Given that the efficiency of target engagement is clearly compound-dependent, careful examination of a potential inhibitor in Nav1.7-dependent target engagement models is required prior to considering human clinical trials.

10.4 Toxicity Considerations

Drug candidates based on any of the three aforementioned potentially Nav1.7-selective binding sites could in principle have off-target toxicity that will need to be evaluated preclinically and clinically. In addition, two types of on-target toxicity could occur with selective Nav1.7 inhibitors: those due to block of off-target Nav channels and those due to block of Nav1.7 itself. The effects on off-target Nav channels will depend in part on which receptor site a drug candidate targets. STX-like extracellular vestibule inhibitors may show equal potency on off-target Nav channels since only Nav1.7 contains the TI motif, whereas other Nav channels share an MD motif at this location. Peptide-based VSD2 binding inhibitors will likely have the weakest selectivity relative to Nav1.6 considering VSD2 sequence similarities, although it is not clear what toxicity effects may occur by blocking peripheral Nav1.6 channels since such peptide-based drugs will likely not be able to cross the blood–brain barrier. As described above, Nav1.7-selective aryl sulfonamide inhibitors will tend to be the least selective against Nav1.2 and Nav1.6 channel subtypes due to the common YW motif found in the S2 of VSD4. Thus, since aryl sulfonamides will likely penetrate the blood–brain barrier at least to some extent, CNS-mediated side effects might occur unless the drug is very selective against these specific channel subtypes. By contrast, true on-target side effects of aryl sulfonamide inhibitors should follow the expression pattern of Nav1.7. Effects on olfaction might be expected given the high level of expression of Nav1.7 in the olfactory epithelium. Autonomic side effects might also be possible given the high level expression of Nav1.7 in the sympathetic ganglia as well as some parasympathetic neurons such as those with cell bodies in the dorsal nucleus of the vagus nerve within the medulla. While autonomic defects have not been observed in CIP individuals with germline Nav1.7 mutations, it is possible that acute block of Nav1.7 might reveal autonomic side effects due to a lack of developmental compensation. Finally, block of Nav1.7 in the hypothalamus might, in principle, produce some deleterious outcomes such as effects on body weight (Branco et al. 2016).

11 Perspective and Future Outlook

Among voltage-gated sodium channels, Nav1.7 has risen to prominence in recent years due to genetics studies that have implicated this channel as playing an important role in human pain perception. Since the known physiology and human genetics indicates that Nav1.7-selective blockers may lead to effective analgesia without the potential for significant adverse side effects or addiction liabilities, Nav1.7 has become a highly scrutinized target by academic and industrial biomedical researches. Accordingly, an intensive search for highly selective and highly potent Nav1.7 antagonists has ensued. However, as outlined in preceding sections, selectivity and potency alone do not guarantee an efficacious or safe therapeutic in vivo in any preclinical or clinical model. In fact, there are many considerations and significant hurdles to overcome on the path to the discovery, optimization, and development of a potentially transformative new pain medicine. As we discuss above in the sections on state-dependence and toxicity, for example, there are many key considerations around Nav1.7 that still remain largely unexplored or unreported. Moreover, challenges in reproducing results (Murray et al. 2015a; Liu et al. 2016) of some reports describing potential Nav1.7-targeting lead molecules (Lee et al. 2014; Yang et al. 2013), and other contributions that conclude lack of single agent in vivo efficacy (Deuis et al. 2017; Storer et al. 2017; Wu et al. 2017c; Hockley et al. 2017) tend to dampen or confuse enthusiasm around Nav1.7. Nevertheless, many researchers still firmly believe that rigorous investigation of the therapeutic hypothesis that Nav1.7 may be a transformative target for the treatment of chronic and potentially neuropathic pain is worth the steady and careful scientific and clinical evaluation that such a challenging drug discovery effort truly requires.

References

Abdelsayed M, Sokolov S (2013) Voltage-gated sodium channels: pharmaceutical targets via anticonvulsants to treat epileptic syndromes. Channels (Austin) 7:146–152

Abriel H, Kass RS (2005) Regulation of the voltage-gated cardiac sodium channel Nav1.5 by interacting proteins. Trends Cardiovasc Med 15:35–40

Agwa AJ, Huang YH, Craik DJ, Henriques ST, Schroeder CI (2017a) Lengths of the C-terminus and interconnecting loops impact stability of spider-derived gating modifier toxins. Toxins (Basel) 9:248

Agwa AJ, Lawrence N, Deplazes E, Cheneval O, Chen RM, Craik DJ, Schroeder CI, Henriques ST (2017b) Spider peptide toxin HwTx-IV engineered to bind to lipid membranes has an increased inhibitory potency at human voltage-gated sodium channel $hNa_V1.7$. Biochim Biophys Acta 1859:835–844

Ahern CA, Eastwood AL, Dougherty DA, Horn R (2008) Electrostatic contributions of aromatic residues in the local anesthetic receptor of voltage-gated sodium channels. Circ Res 102:86–94

Ahern CA, Payandeh J, Bosmans F, Chanda B (2016) The hitchhiker's guide to the voltage-gated sodium channel galaxy. J Gen Physiol 147:1–24

Ahmad S, Dahllund L, Eriksson AB, Hellgren D, Karlsson U, Lund PE, Meijer IA, Meury L, Mills T, Moody A et al (2007) A stop codon mutation in SCN9A causes lack of pain sensation. Hum Mol Genet 16:2114–2121

Ahuja S, Mukund S, Deng L, Khakh K, Chang E, Ho H, Shriver S, Young C, Lin S, Johnson JP Jr et al (2015) Structural basis of Nav1.7 inhibition by an isoform-selective small-molecule antagonist. Science 350:aac5464

Alexandrou AJ, Brown AR, Chapman ML, Estacion M, Turner J, Mis MA, Wilbrey A, Payne EC, Gutteridge A, Cox PJ et al (2016) Subtype-selective small molecule inhibitors reveal a fundamental role for Nav1.7 in nociceptor electrogenesis, axonal conduction and presynaptic release. PLoS One 11:e0152405

Armstrong CM, Bezanilla F (1973) Currents related to movement of the gating particles of the sodium channels. Nature 242:459–461

Bagal SK, Marron BE, Owen RM, Storer RI, Swain NA (2015) Voltage gated sodium channels as drug discovery targets. Channels (Austin) 9:360–366

Bagneris C, DeCaen PG, Naylor CE, Pryde DC, Nobeli I, Clapham DE, Wallace BA (2014) Prokaryotic NavMs channel as a structural and functional model for eukaryotic sodium channel antagonism. Proc Natl Acad Sci U S A 111:8428–8433

Barhanin J, Pauron D, Lombet A, Norman RI, Vijverberg HP, Giglio JR, Lazdunski M (1983) Electrophysiological characterization, solubilization and purification of the *Tityus* γ toxin receptor associated with the gating component of the Na^+ channel from rat brain. EMBO J 2:915–920

Beneski DA, Catterall WA (1980) Covalent labeling of protein components of the sodium channel with a photoactivable derivative of scorpion toxin. Proc Natl Acad Sci U S A 77:639–643

Bezanilla F (2000) The voltage sensor in voltage-dependent ion channels. Physiol Rev 80:555–592

Biswas K, Nixey TE, Murray JK, Falsey JR, Yin L, Liu H, Gingras J, Hall BE, Herberich B, Holder JR et al (2017) Engineering antibody reactivity for efficient derivatization to generate $Na_V1.7$ inhibitory GpTx-1 peptide-antibody conjugates. ACS Chem Biol 12:2427–2435

Black JA, Dib-Hajj S, McNabola K, Jeste S, Rizzo MA, Kocsis JD, Waxman SG (1996) Spinal sensory neurons express multiple sodium channel α-subunit mRNAs. Brain Res Mol Brain Res 43:117–131

Black JA, Frezel N, Dib-Hajj SD, Waxman SG (2012) Expression of Nav1.7 in DRG neurons extends from peripheral terminals in the skin to central preterminal branches and terminals in the dorsal horn. Mol Pain 8:82

Bosmans F, Swartz KJ (2010) Targeting voltage sensors in sodium channels with spider toxins. Trends Pharmacol Sci 31:175–182

Bosmans F, Martin-Eauclaire MF, Swartz KJ (2008) Deconstructing voltage sensor function and pharmacology in sodium channels. Nature 456:202–208

Brady RM, Zhang M, Gable R, Norton RS, Baell JB (2013) De novo design and synthesis of a μ-conotoxin KIIIA peptidomimetic. Bioorg Med Chem Lett 23:4892–4895

Branco T, Tozer A, Magnus CJ, Sugino K, Tanaka S, Lee AK, Wood JN, Sternson SM (2016) Near-perfect synaptic integration by Nav1.7 in hypothalamic neurons regulates body weight. Cell 165:1749–1761

Brouwer BA, Merkies IS, Gerrits MM, Waxman SG, Hoeijmakers JG, Faber CG (2014) Painful neuropathies: the emerging role of sodium channelopathies. J Peripher Nerv Syst 19:53–65

Campos FV, Chanda B, Beirao PS, Bezanilla F (2008) α-Scorpion toxin impairs a conformational change that leads to fast inactivation of muscle sodium channels. J Gen Physiol 132:251–263

Cannon SC (2010) Voltage-sensor mutations in channelopathies of skeletal muscle. J Physiol 588:1887–1895

Capes DL, Goldschen-Ohm MP, Arcisio-Miranda M, Bezanilla F, Chanda B (2013) Domain IV voltage-sensor movement is both sufficient and rate limiting for fast inactivation in sodium channels. J Gen Physiol 142:101–112

Cardoso FC, Dekan Z, Rosengren KJ, Erickson A, Vetter I, Deuis JR, Herzig V, Alewood PF, King GF, Lewis RJ (2015) Identification and characterization of ProTx-III [μ-TRTX-Tp1a], a new voltage-gated sodium channel inhibitor from venom of the tarantula *Thrixopelma pruriens*. Mol Pharmacol 88:291–303

Catterall WA (1976) Purification of a toxic protein from scorpion venom which activates the action potential Na^+ ionophore. J Biol Chem 251:5528–5536

Catterall WA (1977) Membrane potential-dependent binding of scorpion toxin to the action potential Na^+ ionophore. Studies with a toxin derivative prepared by lactoperoxidase-catalyzed iodination. J Biol Chem 252:8660–8668

Catterall WA (2010) Ion channel voltage sensors: structure, function, and pathophysiology. Neuron 67:915–928

Catterall WA, Goldin AL, Waxman SG (2005a) International Union of Pharmacology. XLVII. Nomenclature and structure-function relationships of voltage-gated sodium channels. Pharmacol Rev 57:397–409

Catterall WA, Perez-Reyes E, Snutch TP, Striessnig J (2005b) International Union of Pharmacology. XLVIII. Nomenclature and structure-function relationships of voltage-gated calcium channels. Pharmacol Rev 57:411–425

Catterall WA, Cestele S, Yarov-Yarovoy V, Yu FH, Konoki K, Scheuer T (2007) Voltage-gated ion channels and gating modifier toxins. Toxicon 49:124–141

Cestele S, Catterall WA (2000) Molecular mechanisms of neurotoxin action on voltage-gated sodium channels. Biochimie 82:883–892

Cestele S, Qu Y, Rogers JC, Rochat H, Scheuer T, Catterall WA (1998) Voltage sensor-trapping: enhanced activation of sodium channels by β-scorpion toxin bound to the S3-S4 loop in domain II. Neuron 21:919–931

Cestele S, Scheuer T, Mantegazza M, Rochat H, Catterall WA (2001) Neutralization of gating charges in domain II of the sodium channel α subunit enhances voltage-sensor trapping by a β-scorpion toxin. J Gen Physiol 118:291–302

Cestele S, Yarov-Yarovoy V, Qu Y, Sampieri F, Scheuer T, Catterall WA (2006) Structure and function of the voltage sensor of sodium channels probed by a β-scorpion toxin. J Biol Chem 281:21332–21344

Cha A, Ruben PC, George AL Jr, Fujimoto E, Bezanilla F (1999) Voltage sensors in domains III and IV, but not I and II, are immobilized by Na^+ channel fast inactivation. Neuron 22:73–87

Chahine M, O'Leary ME (2011) Regulatory role of voltage-gated Na channel β subunits in sensory neurons. Front Pharmacol 2:70

Chanda B, Bezanilla F (2002) Tracking voltage-dependent conformational changes in skeletal muscle sodium channel during activation. J Gen Physiol 120:629–645

Chanda B, Asamoah OK, Bezanilla F (2004) Coupling interactions between voltage sensors of the sodium channel as revealed by site-specific measurements. J Gen Physiol 123:217–230

Chanda B, Asamoah OK, Blunck R, Roux B, Bezanilla F (2005) Gating charge displacement in voltage-gated ion channels involves limited transmembrane movement. Nature 436:852–856

Chow CY, Cristofori-Armstrong B, Undheim EA, King GF, Rash LD (2015) Three peptide modulators of the human voltage-gated sodium channel 1.7, an important analgesic target, from the venom of an Australian tarantula. Toxins (Basel) 7:2494–2513

Clairfeuille T, Xu H, Koth CM, Payandeh J (2017) Voltage-gated sodium channels viewed through a structural biology lens. Curr Opin Struct Biol 45:74–84

Cohen CJ, Bean BP, Colatsky TJ, Tsien RW (1981) Tetrodotoxin block of sodium channels in rabbit Purkinje fibers. Interactions between toxin binding and channel gating. J Gen Physiol 78:383–411

Cohen L, Ilan N, Gur M, Stuhmer W, Gordon D, Gurevitz M (2007) Design of a specific activator for skeletal muscle sodium channels uncovers channel architecture. J Biol Chem 282:29424–29430

Couraud F, Rochat H, Lissitzky S (1978) Binding of scorpion and sea anemone neurotoxins to a common site related to the action potential Na^+ ionophore in neuroblastoma cells. Biochem Biophys Res Commun 83:1525–1530

Couraud F, Jover E, Dubois JM, Rochat H (1982) Two types of scorpion receptor sites, one related to the activation, the other to the inactivation of the action potential sodium channel. Toxicon 20:9–16

Cox JJ, Reimann F, Nicholas AK, Thornton G, Roberts E, Springell K, Karbani G, Jafri H, Mannan J, Raashid Y et al (2006) An SCN9A channelopathy causes congenital inability to experience pain. Nature 444:894–898

Cox JJ, Sheynin J, Shorer Z, Reimann F, Nicholas AK, Zubovic L, Baralle M, Wraige E, Manor E, Levy J et al (2010) Congenital insensitivity to pain: novel SCN9A missense and in-frame deletion mutations. Hum Mutat 31:E1670–E1686

Cummins TR, Howe JR, Waxman SG (1998) Slow closed-state inactivation: a novel mechanism underlying ramp currents in cells expressing the hNE/PN1 sodium channel. J Neurosci 18:9607–9619

Das S, Gilchrist J, Bosmans F, Van Petegem F (2016) Binary architecture of the Nav1.2-β2 signaling complex. Elife. 2016;5. pii: e10960

de Lera Ruiz M, Kraus RL (2015) Voltage-gated sodium channels: structure, function, pharmacology, and clinical indications. J Med Chem 58:7093–7118

Deuis JR, Wingerd JS, Winter Z, Durek T, Dekan Z, Sousa SR, Zimmermann K, Hoffmann T, Weidner C, Nassar MA et al (2016) Analgesic effects of GpTx-1, PF-04856264 and CNV1014802 in a mouse model of $Na_V1.7$-mediated pain. Toxins (Basel). 201;8(3). pii: E78

Deuis JR, Dekan Z, Wingerd JS, Smith JJ, Munasinghe NR, Bhola RF, Imlach WL, Herzig V, Armstrong DA, Rosengren KJ et al (2017) Pharmacological characterisation of the highly $Na_V1.7$ selective spider venom peptide Pn3a. Sci Rep 7:40883

Dib-Hajj SD, Tyrrell L, Black JA, Waxman SG (1998) NaN, a novel voltage-gated Na channel, is expressed preferentially in peripheral sensory neurons and down-regulated after axotomy. Proc Natl Acad Sci U S A 95:8963–8968

Dib-Hajj SD, Cummins TR, Black JA, Waxman SG (2010) Sodium channels in normal and pathological pain. Annu Rev Neurosci 33:325–347

Dib-Hajj SD, Yang Y, Black JA, Waxman SG (2013) The $Na_V1.7$ sodium channel: from molecule to man. Nat Rev Neurosci 14:49–62

Doyle DA, Morais Cabral J, Pfuetzner RA, Kuo A, Gulbis JM, Cohen SL, Chait BT, MacKinnon R (1998) The structure of the potassium channel: molecular basis of K^+ conduction and selectivity. Science 280:69–77

Escayg A, Goldin AL (2010) Sodium channel SCN1A and epilepsy: mutations and mechanisms. Epilepsia 51:1650–1658

Fischer TZ, Waxman SG (2010) Familial pain syndromes from mutations of the $Na_V1.7$ sodium channel. Ann N Y Acad Sci 1184:196–207

Flinspach M, Xu Q, Piekarz AD, Fellows R, Hagan R, Gibbs A, Liu Y, Neff RA, Freedman J, Eckert WA et al (2017) Insensitivity to pain induced by a potent selective closed-state Nav1.7 inhibitor. Sci Rep 7:39662

Focken T, Liu S, Chahal N, Dauphinais M, Grimwood ME, Chowdhury S, Hemeon I, Bichler P, Bogucki D, Waldbrook M et al (2016) Discovery of aryl sulfonamides as isoform-selective inhibitors of NaV1.7 with efficacy in rodent pain models. ACS Med Chem Lett 7:277–282

Fozzard HA, Sheets MF, Hanck DA (2011) The sodium channel as a target for local anesthetic drugs. Front Pharmacol 2:68

French RJ, Yoshikami D, Sheets MF, Olivera BM (2010) The tetrodotoxin receptor of voltage-gated sodium channels – perspectives from interactions with μ-conotoxins. Mar Drugs 8:2153–2161

Fry M, Boegle AK, Maue RA (2007) Differentiated pattern of sodium channel expression in dissociated Purkinje neurons maintained in long-term culture. J Neurochem 101:737–748

Gilchrist J, Das S, Van Petegem F, Bosmans F (2013) Crystallographic insights into sodium-channel modulation by the β4 subunit. Proc Natl Acad Sci U S A 110:E5016–E5024

Glaaser IW, Clancy CE (2006) Cardiac Na^+ channels as therapeutic targets for antiarrhythmic agents. Handb Exp Pharmacol 171:99–121

Goldberg YP, MacFarlane J, MacDonald ML, Thompson J, Dube MP, Mattice M, Fraser R, Young C, Hossain S, Pape T et al (2007) Loss-of-function mutations in the Nav1.7 gene underlie congenital indifference to pain in multiple human populations. Clin Genet 71:311–319

Graceffa RF, Boezio AA, Able J, Altmann S, Berry LM, Boezio C, Butler JR, Chu-Moyer M, Cooke M, DiMauro EF et al (2017) Sulfonamides as selective $Na_V1.7$ inhibitors: optimizing potency, pharmacokinetics, and metabolic properties to obtain atropisomeric quinolinone (AM-0466) that affords robust in vivo activity. J Med Chem 60:5990–6017

Guo J, Zeng W, Chen Q, Lee C, Chen L, Yang Y, Cang C, Ren D, Jiang Y (2016) Structure of the voltage-gated two-pore channel TPC1 from Arabidopsis thaliana. Nature 531:196–201

Gur M, Kahn R, Karbat I, Regev N, Wang J, Catterall WA, Gordon D, Gurevitz M (2011) Elucidation of the molecular basis of selective recognition uncovers the interaction site for the core domain of scorpion α-toxins on sodium channels. J Biol Chem 286:35209–35217

Gutman GA, Chandy KG, Grissmer S, Lazdunski M, McKinnon D, Pardo LA, Robertson GA, Rudy B, Sanguinetti MC, Stuhmer W et al (2005) International Union of Pharmacology. LIII. Nomenclature and molecular relationships of voltage-gated potassium channels. Pharmacol Rev 57:473–508

Habib AM, Wood JN, Cox JJ (2015) Sodium channels and pain. Handb Exp Pharmacol 227:39–56

Hackel D, Krug SM, Sauer RS, Mousa SA, Bocker A, Pflucke D, Wrede EJ, Kistner K, Hoffmann T, Niedermirtl B et al (2012) Transient opening of the perineurial barrier for analgesic drug delivery. Proc Natl Acad Sci U S A 109:E2018–E2027

Henriques ST, Deplazes E, Lawrence N, Cheneval O, Chaousis S, Inserra M, Thongyoo P, King GF, Mark AE, Vetter I et al (2016) Interaction of tarantula venom peptide ProTx-II with lipid membranes is a prerequisite for its inhibition of human voltage-gated sodium channel $Na_V1.7$. J Biol Chem 291:17049–17065

Herzog RI, Cummins TR, Ghassemi F, Dib-Hajj SD, Waxman SG (2003) Distinct repriming and closed-state inactivation kinetics of Nav1.6 and Nav1.7 sodium channels in mouse spinal sensory neurons. J Physiol 551:741–750

Hille B (1975) The receptor for tetrodotoxin and saxitoxin. A structural hypothesis. Biophys J 15:615–619

Hille B (2001) Ion channels of excitable membranes, 3rd edn. Sinauer Associates, Sunderland

Ho C, O'Leary ME (2011) Single-cell analysis of sodium channel expression in dorsal root ganglion neurons. Mol Cell Neurosci 46:159–166

Hockley JR, Gonzalez-Cano R, McMurray S, Tejada-Giraldez MA, McGuire C, Torres A, Wilbrey AL, Cibert-Goton V, Nieto FR, Pitcher T et al (2017) Visceral and somatic pain modalities reveal NaV 1.7-independent visceral nociceptive pathways. J Physiol 595:2661–2679

Hodgkin AL, Huxley AF (1952) A quantitative description of membrane current and its application to conduction and excitation in nerve. J Physiol 117:500–544

Hosseinzadeh P, Bhardwaj G, Mulligan VK, Shortridge MD, Craven TW, Pardo-Avila F, Rettie SA, Kim DE, Silva D, Ibrahim YM et al (2017) Comprehensive computational design of ordered peptide macrocycles. Science 358:1461–1466

Huang W, Liu M, Yan SF, Yan N (2017) Structure-based assessment of disease-related mutations in human voltage-gated sodium channels. Protein Cell 8:401–438

Hui K, Lipkind G, Fozzard HA, French RJ (2002) Electrostatic and steric contributions to block of the skeletal muscle sodium channel by μ-conotoxin. J Gen Physiol 119:45–54

Israel MR, Tay B, Deuis JR, Vetter I (2017) Sodium channels and venom peptide pharmacology. Adv Pharmacol 79:67–116

Jaimovich E, Ildefonse M, Barhanin J, Rougier O, Lazdunski M (1982) Centruroides toxin, a selective blocker of surface Na^+ channels in skeletal muscle: voltage-clamp analysis and biochemical characterization of the receptor. Proc Natl Acad Sci U S A 79:3896–3900

Jalali A, Bosmans F, Amininasab M, Clynen E, Cuypers E, Zaremirakabadi A, Sarbolouki MN, Schoofs L, Vatanpour H, Tytgat J (2005) OD1, the first toxin isolated from the venom of the scorpion *Odonthobuthus doriae* active on voltage-gated Na^+ channels. FEBS Lett 579:4181–4186

Jones HM, Butt RP, Webster RW, Gurrell I, Dzygiel P, Flanagan N, Fraier D, Hay T, Iavarone LE, Luckwell J et al (2016) Clinical micro-dose studies to explore the human pharmacokinetics of four selective inhibitors of human Nav1.7 voltage-dependent sodium channels. Clin Pharmacokinet 55:875–887

Klint JK, Smith JJ, Vetter I, Rupasinghe DB, Er SY, Senff S, Herzig V, Mobli M, Lewis RJ, Bosmans F et al (2015a) Seven novel modulators of the analgesic target Na_V 1.7 uncovered using a high-throughput venom-based discovery approach. Br J Pharmacol 172:2445–2458

Klint JK, Chin YK, Mobli M (2015b) Rational engineering defines a molecular switch that is essential for activity of spider-venom peptides against the analgesics target $Na_V1.7$. Mol Pharmacol 88:1002–1010

Klugbauer N, Lacinova L, Flockerzi V, Hofmann F (1995) Structure and functional expression of a new member of the tetrodotoxin-sensitive voltage-activated sodium channel family from human neuroendocrine cells. EMBO J 14:1084–1090

Knapp O, McArthur JR, Adams DJ (2012) Conotoxins targeting neuronal voltage-gated sodium channel subtypes: potential analgesics? Toxins (Basel) 4:1236–1260

Kornecook TJ, Yin R, Altmann S, Be X, Berry V, Ilch CP, Jarosh M, Johnson D, Lee JH, Lehto SG et al (2017) Pharmacologic characterization of AMG8379, a potent and selective small molecule sulfonamide antagonist of the voltage-gated sodium channel $Na_V1.7$. J Pharmacol Exp Ther 362:146–160

Kurban M, Wajid M, Shimomura Y, Christiano AM (2010) A nonsense mutation in the SCN9A gene in congenital insensitivity to pain. Dermatology 221:179–183

La DS, Peterson EA, Bode C, Boezio AA, Bregman H, Chu-Moyer MY, Coats J, DiMauro EF, Dineen TA, Du B et al (2017) The discovery of benzoxazine sulfonamide inhibitors of $Na_V1.7$: tools that bridge efficacy and target engagement. Bioorg Med Chem Lett 27:3477–3485

Lacroix JJ, Campos FV, Frezza L, Bezanilla F (2013) Molecular bases for the asynchronous activation of sodium and potassium channels required for nerve impulse generation. Neuron 79:651–657

Lampert A, Eberhardt M, Waxman SG (2014) Altered sodium channel gating as molecular basis for pain: contribution of activation, inactivation, and resurgent currents. Handb Exp Pharmacol 221:91–110

Landau M, Mayrose I, Rosenberg Y, Glaser F, Martz E, Pupko T, Ben-Tal N (2005) ConSurf 2005: the projection of evolutionary conservation scores of residues on protein structures. Nucleic Acids Res 33:W299–W302

Lee SY, MacKinnon R (2004) A membrane-access mechanism of ion channel inhibition by voltage sensor toxins from spider venom. Nature 430:232–235

Lee CH, MacKinnon R (2017) Structures of the human HCN1 hyperpolarization-activated channel. Cell 168:111–120. e111

Lee JH, Park CK, Chen G, Han Q, Xie RG, Liu T, Ji RR, Lee SY (2014) A monoclonal antibody that targets a $Na_V1.7$ channel voltage sensor for pain and itch relief. Cell 157:1393–1404

Leffler A, Herzog RI, Dib-Hajj SD, Waxman SG, Cummins TR (2005) Pharmacological properties of neuronal TTX-resistant sodium channels and the role of a critical serine pore residue. Pflugers Arch 451:454–463

Leipold E, Hansel A, Borges A, Heinemann SH (2006) Subtype specificity of scorpion β-toxin Tz1 interaction with voltage-gated sodium channels is determined by the pore loop of domain 3. Mol Pharmacol 70:340–347

Leipold E, DeBie H, Zorn S, Borges A, Olivera BM, Terlau H, Heinemann SH (2007) μO-conotoxins inhibit Na_V channels by interfering with their voltage sensors in domain-2. Channels (Austin) 1:253–262

Leipold E, Borges A, Heinemann SH (2012) Scorpion β-toxin interference with Na_V channel voltage sensor gives rise to excitatory and depressant modes. J Gen Physiol 139:305–319

Leterrier C, Brachet A, Fache MP, Dargent B (2010) Voltage-gated sodium channel organization in neurons: protein interactions and trafficking pathways. Neurosci Lett 486:92–100

Lipkind GM, Fozzard HA (1994) A structural model of the tetrodotoxin and saxitoxin binding site of the Na^+ channel. Biophys J 66:1–13

Liu D, Tseng M, Epstein LF, Green L, Chan B, Soriano B, Lim D, Pan O, Murawsky CM, King CT et al (2016) Evaluation of recombinant monoclonal antibody SVmab1 binding to Na_V1.7 target sequences and block of human Na_V1.7 currents. F1000Res **5**:2764

Long SB, Campbell EB, Mackinnon R (2005) Crystal structure of a mammalian voltage-dependent shaker family K^+ channel. Science 309:897–903

Maertens C, Cuypers E, Amininasab M, Jalali A, Vatanpour H, Tytgat J (2006) Potent modulation of the voltage-gated sodium channel Nav1.7 by OD1, a toxin from the scorpion *Odonthobuthus doriae*. Mol Pharmacol 70:405–414

Mansouri M, Chafai Elalaoui S, Ouled Amar Bencheikh B, El Alloussi M, Dion PA, Sefiani A, Rouleau GA (2014) A novel nonsense mutation in SCN9A in a Moroccan child with congenital insensitivity to pain. Pediatr Neurol 51:741–744

Marx IE, Dineen TA, Able J, Bode C, Bregman H, Chu-Moyer M, DiMauro EF, Du B, Foti RS, Fremeau RT Jr et al (2016) Sulfonamides as selective Na_V1.7 inhibitors: optimizing potency and pharmacokinetics to enable in vivo target engagement. ACS Med Chem Lett 7:1062–1067

McArthur JR, Singh G, McMaster D, Winkfein R, Tieleman DP, French RJ (2011) Interactions of key charged residues contributing to selective block of neuronal sodium channels by μ-conotoxin KIIIA. Mol Pharmacol 80:573–584

McCormack K, Santos S, Chapman ML, Krafte DS, Marron BE, West CW, Krambis MJ, Antonio BM, Zellmer SG, Printzenhoff D et al (2013) Voltage sensor interaction site for selective small molecule inhibitors of voltage-gated sodium channels. Proc Natl Acad Sci U S A 110:E2724–E2732

McCusker EC, Bagneris C, Naylor CE, Cole AR, D'Avanzo N, Nichols CG, Wallace BA (2012) Structure of a bacterial voltage-gated sodium channel pore reveals mechanisms of opening and closing. Nat Commun 3:1102

Middleton RE, Warren VA, Kraus RL, Hwang JC, Liu CJ, Dai G, Brochu RM, Kohler MG, Gao YD, Garsky VM et al (2002) Two tarantula peptides inhibit activation of multiple sodium channels. Biochemistry 41:14734–14747

Milescu M, Bosmans F, Lee S, Alabi AA, Kim JI, Swartz KJ (2009) Interactions between lipids and voltage sensor paddles detected with tarantula toxins. Nat Struct Mol Biol 16:1080–1085

Minett MS, Pereira V, Sikandar S, Matsuyama A, Lolignier S, Kanellopoulos AH, Mancini F, Iannetti GD, Bogdanov YD, Santana-Varela S et al (2015) Endogenous opioids contribute to insensitivity to pain in humans and mice lacking sodium channel Nav1.7. Nat Commun 6:8967

Morinville A, Fundin B, Meury L, Jureus A, Sandberg K, Krupp J, Ahmad S, O'Donnell D (2007) Distribution of the voltage-gated sodium channel Na_V1.7 in the rat: expression in the autonomic and endocrine systems. J Comp Neurol 504:680–689

Murray JK, Ligutti J, Liu D, Zou A, Poppe L, Li H, Andrews KL, Moyer BD, McDonough SI, Favreau P et al (2015a) Engineering potent and selective analogues of GpTx-1, a tarantula venom peptide antagonist of the Na_V1.7 sodium channel. J Med Chem 58:2299–2314

Murray JK, Biswas K, Holder JR, Zou A, Ligutti J, Liu D, Poppe L, Andrews KL, Lin FF, Meng SY et al (2015b) Sustained inhibition of the $Na_V1.7$ sodium channel by engineered dimers of the domain II binding peptide GpTx-1. Bioorg Med Chem Lett 25:4866–4871

Nicole S, Fontaine B (2015) Skeletal muscle sodium channelopathies. Curr Opin Neurol 28:508–514

Noda M, Ikeda T, Suzuki H, Takeshima H, Takahashi T, Kuno M, Numa S (1986) Expression of functional sodium channels from cloned cDNA. Nature 322:826–828

O'Brien JE, Meisler MH (2013) Sodium channel SCN8A (Nav1.6): properties and de novo mutations in epileptic encephalopathy and intellectual disability. Front Genet 4:213

O'Malley HA, Isom LL (2015) Sodium channel beta subunits: emerging targets in channelopathies. Annu Rev Physiol 77:481–504

Osteen JD, Herzig V, Gilchrist J, Emrick JJ, Zhang C, Wang X, Castro J, Garcia-Caraballo S, Grundy L, Rychkov GY et al (2016) Selective spider toxins reveal a role for the Nav1.1 channel in mechanical pain. Nature 534:494–499

Osteen JD, Sampson K, Iyer V, Julius D, Bosmans F (2017) Pharmacology of the Nav1.1 domain IV voltage sensor reveals coupling between inactivation gating processes. Proc Natl Acad Sci U S A 114:6836–6841

Over B, Matsson P, Tyrchan C, Artursson P, Doak BC, Foley MA, Hilgendorf C, Johnston SE, Lee MD, Lewis RJ et al (2016) Structural and conformational determinants of macrocycle cell permeability. Nat Chem Biol 12:1065–1074

Papazian DM, Schwarz TL, Tempel BL, Jan YN, Jan LY (1987) Cloning of genomic and complementary DNA from Shaker, a putative potassium channel gene from Drosophila. Science 237:749–753

Park JH, Carlin KP, Wu G, Ilyin VI, Kyle DJ (2012) Cysteine racemization during the Fmoc solid phase peptide synthesis of the Nav1.7-selective peptide – protoxin II. J Pept Sci 18:442–448

Park JH, Carlin KP, Wu G, Ilyin VI, Musza LL, Blake PR, Kyle DJ (2014) Studies examining the relationship between the chemical structure of protoxin II and its activity on voltage gated sodium channels. J Med Chem 57:6623–6631

Payandeh J, Minor DL Jr (2015) Bacterial voltage-gated sodium channels (BacNavs) from the soil, sea, and salt lakes enlighten molecular mechanisms of electrical signaling and pharmacology in the brain and heart. J Mol Biol 427:3–30

Payandeh J, Scheuer T, Zheng N, Catterall WA (2011) The crystal structure of a voltage-gated sodium channel. Nature 475:353–358

Payandeh J, Gamal El-Din TM, Scheuer T, Zheng N, Catterall WA (2012) Crystal structure of a voltage-gated sodium channel in two potentially inactivated states. Nature 486:135–139

Payne CE, Brown AR, Theile JW, Loucif AJ, Alexandrou AJ, Fuller MD, Mahoney JH, Antonio BM, Gerlach AC, Printzenhoff DM et al (2015) A novel selective and orally bioavailable Nav 1.8 channel blocker, PF-01247324, attenuates nociception and sensory neuron excitability. Br J Pharmacol 172:2654–2670

Pero JE, Rossi MA, Lehman H, Kelly MJ 3rd, Mulhearn JJ, Wolkenberg SE, Cato MJ, Clements MK, Daley CJ, Filzen T et al (2017) Benzoxazolinone aryl sulfonamides as potent, selective Nav1.7 inhibitors with *in vivo* efficacy in a preclinical pain model. Bioorg Med Chem Lett 27:2683–2688

Pineda SS, Undheim EA, Rupasinghe DB, Ikonomopoulou MP, King GF (2014) Spider venomics: implications for drug discovery. Future Med Chem 6:1699–1714

Pitt GS, Lee SY (2016) Current view on regulation of voltage-gated sodium channels by calcium and auxiliary proteins. Protein Sci 25:1573–1584

Pless SA, Galpin JD, Frankel A, Ahern CA (2011) Molecular basis for class Ib anti-arrhythmic inhibition of cardiac sodium channels. Nat Commun 2:351

Pless SA, Elstone FD, Niciforovic AP, Galpin JD, Yang R, Kurata HT, Ahern CA (2014) Asymmetric functional contributions of acidic and aromatic side chains in sodium channel voltage-sensor domains. J Gen Physiol 143:645–656

Price N, Namdari R, Neville J, Proctor KJ, Kaber S, Vest J, Fetell M, Malamut R, Sherrington R, Pimstone SN et al (2017) Safety and efficacy of a topical sodium channel inhibitor (TV-45070) in patients with post herpetic neuralgia (PHN): a randomized, controlled, proof-of-concept, crossover study, with a subgroup analysis of the Nav1.7 R1150W genotype. Clin J Pain 33:310–318

Rajamani R, Wu S, Rodrigo I, Gao M, Low S, Megson L, Wensel D, Pieschl RL, Post-Munson DJ, Watson J et al (2017) A functional $Na_V1.7$-Na_VAb chimera with a reconstituted high-affinity ProTx-II binding site. Mol Pharmacol 92:310–317

Rodriguez de la Vega RC, Possani LD (2005) Overview of scorpion toxins specific for Na^+ channels and related peptides: biodiversity, structure-function relationships and evolution. Toxicon 46:831–844

Roecker AJ, Egbertson M, Jones KLG, Gomez R, Kraus RL, Li Y, Koser AJ, Urban MO, Klein R, Clements M et al (2017) Discovery of selective, orally bioavailable, N-linked arylsulfonamide Nav1.7 inhibitors with pain efficacy in mice. Bioorg Med Chem Lett 27:2087–2093

Rogers JC, Qu Y, Tanada TN, Scheuer T, Catterall WA (1996) Molecular determinants of high affinity binding of α-scorpion toxin and sea anemone toxin in the S3-S4 extracellular loop in domain IV of the Na^+ channel α subunit. J Biol Chem 271:15950–15962

Rosker C, Lohberger B, Hofer D, Steinecker B, Quasthoff S, Schreibmayer W (2007) The TTX metabolite 4,9-anhydro-TTX is a highly specific blocker of the $Na_{v1.6}$ voltage-dependent sodium channel. Am J Physiol Cell Physiol 293:C783–C789

Savio-Galimberti E, Argenziano M, Antzelevitch C (2017) Cardiac arrhythmias related to sodium channel dysfunction. Handb Exp Pharmacol. https://doi.org/10.1007/164_2017_43

Sawal HA, Harripaul R, Mikhailov A, Dad R, Ayub M, Jawad Hassan M, Vincent JB (2016) Biallelic truncating SCN9A mutation identified in four families with congenital insensitivity to pain from Pakistan. Clin Genet 90:563–565

Scanio MJ, Shi L, Drizin I, Gregg RJ, Atkinson RN, Thomas JB, Johnson MS, Chapman ML, Liu D, Krambis MJ et al (2010) Discovery and biological evaluation of potent, selective, orally bioavailable, pyrazine-based blockers of the $Na_V1.8$ sodium channel with efficacy in a model of neuropathic pain. Bioorg Med Chem 18:7816–7825

Schenkel LB, DiMauro EF, Nguyen HN, Chakka N, Du B, Foti RS, Guzman-Perez A, Jarosh M, La DS, Ligutti J et al (2017) Discovery of a biarylamide series of potent, state-dependent $Na_V1.7$ inhibitors. Bioorg Med Chem Lett 27:3817–3824

Schmalhofer WA, Calhoun J, Burrows R, Bailey T, Kohler MG, Weinglass AB, Kaczorowski GJ, Garcia ML, Koltzenburg M, Priest BT (2008) ProTx-II, a selective inhibitor of $Na_V1.7$ sodium channels, blocks action potential propagation in nociceptors. Mol Pharmacol 74:1476–1484

Sharkey RG, Beneski DA, Catterall WA (1984) Differential labeling of the α and β1 subunits of the sodium channel by photoreactive derivatives of scorpion toxin. Biochemistry 23:6078–6086

Shaya D, Findeisen F, Abderemane-Ali F, Arrigoni C, Wong S, Nurva SR, Loussouarn G, Minor DL Jr (2014) Structure of a prokaryotic sodium channel pore reveals essential gating elements and an outer ion binding site common to eukaryotic channels. J Mol Biol 426:467–483

Shcherbatko A, Rossi A, Foletti D, Zhu G, Bogin O, Galindo Casas M, Rickert M, Hasa-Moreno A, Bartsevich V, Crameri A et al (2016) Engineering highly potent and selective microproteins against Nav1.7 sodium channel for treatment of pain. J Biol Chem 291:13974–13986

Shen H, Zhou Q, Pan X, Li Z, Wu J, Yan N (2017) Structure of a eukaryotic voltage-gated sodium channel at near-atomic resolution. Science. 2017;355(6328). pii: eaal4326

Shorer Z, Wajsbrot E, Liran TH, Levy J, Parvari R (2014) A novel mutation in SCN9A in a child with congenital insensitivity to pain. Pediatr Neurol 50:73–76

Skerratt SE, West CW (2015) Ion channel therapeutics for pain. Channels (Austin) 9:344–351

Smith JJ, Alphy S, Seibert AL, Blumenthal KM (2005) Differential phospholipid binding by site 3 and site 4 toxins. Implications for structural variability between voltage-sensitive sodium channel domains. J Biol Chem 280:11127–11133

Sokolov S, Kraus RL, Scheuer T, Catterall WA (2008) Inhibition of sodium channel gating by trapping the domain II voltage sensor with protoxin II. Mol Pharmacol 73:1020–1028

Storer RI, Pike A, Swain NA, Alexandrou AJ, Bechle BM, Blakemore DC, Brown AD, Castle NA, Corbett MS, Flanagan NJ et al (2017) Highly potent and selective $Na_V1.7$ inhibitors for use as intravenous agents and chemical probes. Bioorg Med Chem Lett 27:4805–4811

Stuhmer W, Conti F, Suzuki H, Wang XD, Noda M, Yahagi N, Kubo H, Numa S (1989) Structural parts involved in activation and inactivation of the sodium channel. Nature 339:597–603

Sun J, MacKinnon R (2017) Cryo-EM structure of a KCNQ1/CaM complex reveals insights into congenital long QT syndrome. Cell 169:1042–1050. e1049

Sun S, Cohen CJ, Dehnhardt CM (2014) Inhibitors of voltage-gated sodium channel Nav1.7: patent applications since 2010. Pharm Pat Anal 3:509–521

Swain NA, Batchelor D, Beaudoin S, Bechle BM, Bradley PA, Brown AD, Brown B, Butcher KJ, Butt RP, Chapman ML et al (2017) Discovery of clinical candidate 4-[2-(5-amino-1H-pyrazol-4-yl)-4-chlorophenoxy]-5-chloro-2-fluoro-N-1,3-thiazol-4-ylbenzenesulfonamide (PF-05089771): design and optimization of diaryl ether aryl sulfonamides as selective inhibitors of $Na_V1.7$. J Med Chem 60:7029–7042

Tang L, Gamal El-Din TM, Swanson TM, Pryde DC, Scheuer T, Zheng N, Catterall WA (2016) Structural basis for inhibition of a voltage-gated Ca^{2+} channel by Ca^{2+} antagonist drugs. Nature 537:117–121

Theile JW, Fuller MD, Chapman ML (2016) The selective Nav1.7 inhibitor, PF-05089771, interacts equivalently with fast and slow inactivated Nav1.7 channels. Mol Pharmacol 90:540–548

Thomas-Tran R, Du Bois J (2016) Mutant cycle analysis with modified saxitoxins reveals specific interactions critical to attaining high-affinity inhibition of $hNa_V1.7$. Proc Natl Acad Sci U S A 113:5856–5861

Thomsen WJ, Catterall WA (1989) Localization of the receptor site for α-scorpion toxins by antibody mapping: implications for sodium channel topology. Proc Natl Acad Sci U S A 86:10161–10165

Ulbricht W (2005) Sodium channel inactivation: molecular determinants and modulation. Physiol Rev 85:1271–1301

Vargas E, Yarov-Yarovoy V, Khalili-Araghi F, Catterall WA, Klein ML, Tarek M, Lindahl E, Schulten K, Perozo E, Bezanilla F et al (2012) An emerging consensus on voltage-dependent gating from computational modeling and molecular dynamics simulations. J Gen Physiol 140:587–594

Vetter I, Deuis JR, Mueller A, Israel MR, Starobova H, Zhang A, Rash LD, Mobli M (2017) $Na_V1.7$ as a pain target – from gene to pharmacology. Pharmacol Ther 172:73–100

Vijverberg HP, Lazdunski M (1984) A new scorpion toxin with a very high affinity for sodium channels. An electrophysiological study. J Physiol Paris 79:275–279

Walewska A, Han TS, Zhang MM, Yoshikami D, Bulaj G, Rolka K (2013) Expanding chemical diversity of conotoxins: peptoid-peptide chimeras of the sodium channel blocker μ-KIIIA and its selenopeptide analogues. Eur J Med Chem 65:144–150

Walker JR, Novick PA, Parsons WH, McGregor M, Zablocki J, Pande VS, Du Bois J (2012) Marked difference in saxitoxin and tetrodotoxin affinity for the human nociceptive voltage-gated sodium channel Nav1.7. Proc Natl Acad Sci U S A 109:18102–18107

Wang GK, Strichartz G (1982) Simultaneous modifications of sodium channel gating by two scorpion toxins. Biophys J 40:175–179

Wang J, Yarov-Yarovoy V, Kahn R, Gordon D, Gurevitz M, Scheuer T, Catterall WA (2011) Mapping the receptor site for α-scorpion toxins on a Na^+ channel voltage sensor. Proc Natl Acad Sci U S A 108:15426–15431

Weiss J, Pyrski M, Jacobi E, Bufe B, Willnecker V, Schick B, Zizzari P, Gossage SJ, Greer CA, Leinders-Zufall T et al (2011) Loss-of-function mutations in sodium channel Nav1.7 cause anosmia. Nature 472:186–190

Weiss MM, Dineen TA, Marx IE, Altmann S, Boezio A, Bregman H, Chu-Moyer M, DiMauro EF, Feric Bojic E, Foti RS et al (2017) Sulfonamides as selective $Na_V1.7$ inhibitors: optimizing potency and pharmacokinetics while mitigating metabolic liabilities. J Med Chem 60:5969–5989

Whicher JR, MacKinnon R (2016) Structure of the voltage-gated K^+ channel Eag1 reveals an alternative voltage sensing mechanism. Science 353:664–669
Wilson MJ, Zhang MM, Azam L, Olivera BM, Bulaj G, Yoshikami D (2011a) Navbeta subunits modulate the inhibition of Nav1.8 by the analgesic gating modifier μO-conotoxin MrVIB. J Pharmacol Exp Ther 338:687–693
Wilson MJ, Yoshikami D, Azam L, Gajewiak J, Olivera BM, Bulaj G, Zhang MM (2011b) μ-Conotoxins that differentially block sodium channels $Na_V1.1$ through 1.8 identify those responsible for action potentials in sciatic nerve. Proc Natl Acad Sci U S A 108:10302–10307
Wright ZVF, McCarthy S, Dickman R, Reyes FE, Sanchez-Martinez S, Cryar A, Kilford I, Hall A, Takle AK, Topf M et al (2017) The role of disulfide bond replacements in analogues of the tarantula toxin ProTx-II and their effects on inhibition of the voltage-gated sodium ion channel Nav1.7. J Am Chem Soc 139:13063–13075
Wu J, Yan Z, Li Z, Yan C, Lu S, Dong M, Yan N (2015) Structure of the voltage-gated calcium channel Cav1.1 complex. Science 350:aad2395
Wu W, Li Z, Yang G, Teng M, Qin J, Hu Z, Hou L, Shen L, Dong H, Zhang Y et al (2017a) The discovery of tetrahydropyridine analogs as hNav1.7 selective inhibitors for analgesia. Bioorg Med Chem Lett 27:2210–2215
Wu YJ, Guernon J, McClure A, Luo G, Rajamani R, Ng A, Easton A, Newton A, Bourin C, Parker D et al (2017b) Discovery of non-zwitterionic aryl sulfonamides as Nav1.7 inhibitors with efficacy in preclinical behavioral models and translational measures of nociceptive neuron activation. Bioorg Med Chem 25:5490–5505
Wu YJ, Guernon J, Shi J, Ditta J, Robbins KJ, Rajamani R, Easton A, Newton A, Bourin C, Mosure K et al (2017c) Development of new benzenesulfonamides as potent and selective Nav1.7 inhibitors for the treatment of pain. J Med Chem 60:2513–2525
Xiao Y, Blumenthal K, Jackson JO 2nd, Liang S, Cummins TR (2010) The tarantula toxins ProTx-II and huwentoxin-IV differentially interact with human Nav1.7 voltage sensors to inhibit channel activation and inactivation. Mol Pharmacol 78:1124–1134
Yan Z, Zhou Q, Wang L, Wu J, Zhao Y, Huang G, Peng W, Shen H, Lei J, Yan N (2017) Structure of the Nav1.4-β1 complex from electric eel. Cell 170:470–482.e411
Yang S, Xiao Y, Kang D, Liu J, Li Y, Undheim EA, Klint JK, Rong M, Lai R, King GF (2013) Discovery of a selective $Na_V1.7$ inhibitor from centipede venom with analgesic efficacy exceeding morphine in rodent pain models. Proc Natl Acad Sci U S A 110:17534–17539
Yang SW, Ho GD, Tulshian D, Bercovici A, Tan Z, Hanisak J, Brumfield S, Matasi J, Sun X, Sakwa SA et al (2014) Bioavailable pyrrolo-benzo-1,4-diazines as $Na_V1.7$ sodium channel blockers for the treatment of pain. Bioorg Med Chem Lett 24:4958–4962
Yu FH, Catterall WA (2004) The VGL-chanome: a protein superfamily specialized for electrical signaling and ionic homeostasis. Sci STKE 2004:re15
Zhang JZ, Yarov-Yarovoy V, Scheuer T, Karbat I, Cohen L, Gordon D, Gurevitz M, Catterall WA (2011) Structure-function map of the receptor site for β-scorpion toxins in domain II of voltage-gated sodium channels. J Biol Chem 286:33641–33651
Zhang X, Ren W, DeCaen P, Yan C, Tao X, Tang L, Wang J, Hasegawa K, Kumasaka T, He J et al (2012a) Crystal structure of an orthologue of the NaChBac voltage-gated sodium channel. Nature 486:130–134
Zhang JZ, Yarov-Yarovoy V, Scheuer T, Karbat I, Cohen L, Gordon D, Gurevitz M, Catterall WA (2012b) Mapping the interaction site for a β-scorpion toxin in the pore module of domain III of voltage-gated Na^+ channels. J Biol Chem 287:30719–30728
Zhang MM, Wilson MJ, Azam L, Gajewiak J, Rivier JE, Bulaj G, Olivera BM, Yoshikami D (2013) Co-expression of $Na_V\beta$ subunits alters the kinetics of inhibition of voltage-gated sodium channels by pore-blocking μ-conotoxins. Br J Pharmacol 168:1597–1610

Part IV

Pathophysiology of Sodium Channels

Sodium Channelopathies of Skeletal Muscle

Stephen C. Cannon

Contents

Abstract

The $Na_V1.4$ sodium channel is highly expressed in skeletal muscle, where it carries almost all of the inward Na^+ current that generates the action potential, but is not present at significant levels in other tissues. Consequently, mutations of *SCN4A* encoding $Na_V1.4$ produce pure skeletal muscle phenotypes that now include six allelic disorders: sodium channel myotonia, paramyotonia congenita, hyperkalemic periodic paralysis, hypokalemic periodic paralysis, congenital myasthenia, and congenital myopathy with hypotonia. Mutation-specific alternations of $Na_V1.4$ function explain the mechanistic basis for the diverse phenotypes and identify opportunities for strategic intervention to modify the burden of disease.

S. C. Cannon (✉)
Department of Physiology, David Geffen School of Medicine at UCLA, Los Angeles, CA 90095, USA
e-mail: sccannon@mednet.ucla.edu

M. Chahine (ed.), *Voltage-gated Sodium Channels: Structure, Function and Channelopathies*, Handbook of Experimental Pharmacology 246,
https://doi.org/10.1007/164_2017_52

Keywords
Channelopathy · Gating pore · Myotonia · $Na_V1.4$ · Periodic paralysis · Sodium channel

1 The Na^+ Channel of Skeletal Muscle

The predominant sodium channel in skeletal muscle is a heterodimer of the pore-forming $Na_V1.4$ α subunit (Trimmer et al. 1989) and the non-covalently associated β_1 subunit (Isom et al. 1992). A separate TTX-insensitive isoform, $Na_V1.5$ which is the major α subunit in the heart (Gellens et al. 1992), conducts a minor component (~10%) of the total Na^+ current in adult skeletal muscle (Fu et al. 2011), but is the major isoform in fetal muscle and is upregulated in chronically denervated skeletal muscle (Rogart and Regan 1985; Yang et al. 1991). The human $Na_V1.4$ subunit is encoded by the *SCN4A* gene on chromosome 17q23 (George et al. 1992), and the β_1 subunit by *SCN1B* on chromosome 19q13.11 (McClatchey et al. 1993). Mutations of the β_1 subunit have been associated with epilepsy, ataxia, and cardiac arrhythmia (Calhoun and Isom 2014), but have not been linked to a skeletal muscle phenotype.

The sodium channel complex is expressed in skeletal muscle plasma membranes of both the transverse tubules and the sarcolemma (DiFranco and Vergara 2011; Jaimovich et al. 1976). Channels activate rapidly (<1 ms) in response to depolarization and conduct a large inward Na^+ current (~5 mA/cm^2) that drives the rapid upstroke of the action potential (dV/dt ~500 mV/ms). The channel density is about 100-fold higher at the endplate of the neuromuscular junction (Caldwell et al. 1986), which contributes to the high safety factor of synaptic transmission such that a muscle fiber action potential is elicited for each motoneuron action potential that invades the nerve terminal.

The functional consequences of $Na_V1.4$ mutations associated with muscle disorders have been studied primarily by heterologous expression in mammalian cell lines (HEK or tsA201 cells). Expression in Xenopus oocytes has also been used, but co-expression of the β_1 subunit is essential to suppress a gating mode with anomalously slow inactivation (Cannon et al. 1993b). A more limited number of studies have been performed on muscle biopsies (Lehmann-Horn et al. 1987), patient-derived cultured myotubes (Cannon et al. 1991), or knock-in mutant mouse models (Clausen et al. 2011). Fortunately, there has been good agreement in the conclusions drawn from this variety of experimental systems.

2 Clinical Phenotypes Associated with $Na_V1.4$ Mutations

Mutations of $Na_V1.4$ produce a variety of skeletal muscle phenotypes (Cannon 2015; Lehmann-Horn et al. 2004). Other excitable tissues, such as heart and brain, are not affected because the $Na_V1.4$ isoform is expressed at significant levels only in skeletal muscle (Trimmer et al. 1989). Muscle excitability may be either pathologically enhanced or reduced as a consequence of $Na_V1.4$ mutations.

Hyperexcitability presents clinically as muscle stiffness, termed *myotonia*, which results from an involuntary contraction that persists for several seconds after cessation of voluntary effort (Rüdel and Lehmann-Horn 1985). This after-contraction is caused by a burst of muscle action potentials that lasts for several seconds, independent from any excitation from the motoneuron (Brown and Harvey 1939). In patients with myotonia, the needle electromyogram shows sustained bursts of discharges that can be elicited by brief voluntary contraction or by direct mechanical percussion of the affected muscle. Symptomatically, the severity of myotonia fluctuates with the level of muscle activity. Myotonic stiffness is usually most pronounced with the first forceful movements after several minutes of rest, and then decreases over seconds to minutes with continued voluntary muscle activity (warm-up phenomenon). In some instances, myotonic stiffness may paradoxically worsen with repeated muscle effort and is termed *paramyotonia*. Myotonia may be aggravated by environmental trigger factors such as muscle cooling or ingestion of food with a high K^+ content. Muscle hyperexcitability with myotonia occurs with either gain-of-function defects in mutant $Na_V1.4$ subunits or loss-of-function mutations in the skeletal muscle chloride channel ClC-1 (Cannon 2015).

Reduced excitability of skeletal muscle arising from $Na_V1.4$ defects may produce transient attacks of weakness or a chronic state of permanent myopathic weakness. The most common manifestation is *periodic paralysis* with recurrent episodes of moderate to severe weakness resulting from sustained depolarization of the resting potential that inactivates $Na_V1.4$ and thereby reduces fiber excitability (Cannon 2015; Lehmann-Horn et al. 2004). The attacks of weakness are often triggered by environmental factors such as rest after vigorous exercise, diet (fasting, carbohydrate ingestion, high salt), or emotional stress. A typical episode has a gradual onset over minutes, lasts for several hours to a day or more, followed by spontaneous recovery. The severity of an attack may render a patient bedridden, unable to sit or raise a limb against gravity. Many patients with periodic paralysis also develop a late onset permanent weakness that begins around age 40 years and is slowly progressive. Another $Na_V1.4$-associated phenotype, with a much lower prevalence, is *myasthenia* characterized by rapid fatigue of muscle strength within seconds to minutes (Tsujino et al. 2003). Recovery of strength also occurs quickly over minutes, and the muscles of the face, eyelids, tongue, neck, and soft palate are most commonly affected. In exemplary cases, clinical electrophysiologic testing shows a decremental response of the compound muscle action potential elicited by repetitive nerve stimulation. This decremental response is the hallmark of a compromised safety factor of neuromuscular transition, but in this case the defect is with the generation of a muscle action potential from a normal postsynaptic endplate potential. A third phenotype is *congenital myopathy*, with fetal or neonatal hypokinesia, reduced muscle tone, and moderate to severe fixed myopathic weakness (Zaharieva et al. 2016).

3 Overview of $Na_V1.4$ Mutations

The sodium channelopathies of skeletal muscle are rare diseases, with combined prevalence of about 1 per 100,000 (Horga et al. 2013). The inheritance pattern is autosomal dominant and highly penetrant, except for the very rare syndrome of congenital myopathy which is autosomal recessive. Over 70 mutations of *SCN4A* have been identified in patients with skeletal muscle disorders (Huang et al. 2017; Lehmann-Horn and Jurkat-Rott 1999). Genotype–phenotype associations have emerged (Miller et al. 2004; Rüdel et al. 1993), wherein specific mutations are consistently found to cause a particular clinical syndrome amongst the 6 allelic disorders of the muscle sodium channelopathies (Fig. 1). A major focus of functional expression studies over the past two decades has been to determine the repertoire of biophysical defects of channel behavior that predispose affected muscle to a particular phenotype. Most mutations are missense changes that alter channel function, with only a handful of nonsense, frameshift, or splice site mutations found in the recessive congenital myopathy families (Zaharieva et al. 2016). A small subset of missense mutations in $Na_V1.4$ accounts for the majority of patients with myotonia or periodic paralysis (Horga et al. 2013; Matthews et al. 2008; Miller et al. 2004). Haplotype analysis of multiple families with the same missense mutation does not support the hypothesis of a common genetic ancestor (Wang et al. 1993), and the occurrence of de novo mutations has been verified in isolated probands.

Mammalian homologues $Na_V1.4$ skeletal muscle channelopathies in non-human species occurred spontaneously in American quarter horses and have been genetically engineered in mice. The equine mutation has been traced to a founder animal with myotonia and periodic paralysis, and the selective breeding practices in the horse industry lead to rapid dissemination of the trait that now affects 5% of all quarter horses (Rudolph et al. 1992). Three separate mouse lines with periodic paralysis or periodic paralysis plus myotonia have been generated by creating

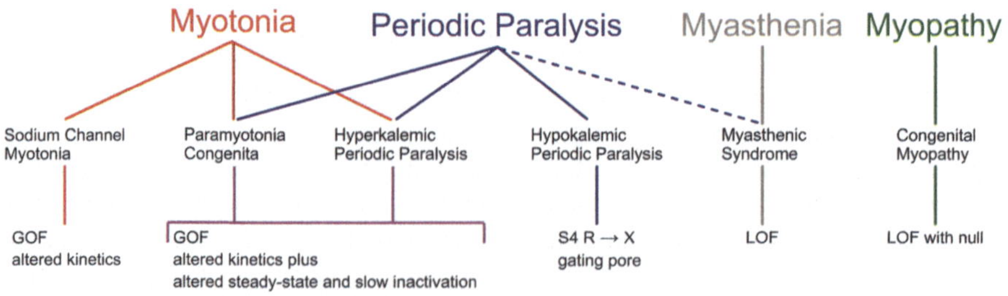

Fig. 1 Spectrum of clinical phenotypes, muscle diseases, and functional deficits for $Na_V1.4$ channelopathies. Paramyotonia congenita and hyperkalemic periodic paralysis have considerable overlap in clinical features and $Na_V1.4$ deficits. Sustained fluctuations in muscle strength may occur with myasthenia (*dashed line*), but the relation to periodic paralysis is uncertain. *GOF* gain-of-function, *LOF* loss-of-function

missense mutations of $Na_V1.4$ by site-directed knock-in (Hayward et al. 2008; Wu et al. 2011) or ENU mutagenesis (Corrochano et al. 2014).

3.1 Gain-of-Function Mutations Cause Myotonia and Hyperkalemic Periodic Paralysis

A subset of three muscle disorders with overlapping clinical symptoms have been associated with missense mutations that cause gain-of-function defects for $Na_V1.4$ (Fig. 1) (Cannon 2015; Lehmann-Horn et al. 2004). At one end of this spectrum, patients with *sodium channel myotonia* (SCM) have muscle stiffness and involuntary after-contractions without episodes of periodic paralysis. A diverse variety of clinical subtypes have been described, depending on the severity [e.g., myotonia permanens for severe dysfunction that may compromise breathing (Singh et al. 2014)] or on precipitating factors [e.g., potassium-aggravated myotonia (Heine et al. 1993)]. Myotonia that paradoxically worsens with repeated muscle activity or with muscle cooling is the hallmark of *paramyotonia congenita* (PMC), but patients may also have episodes of periodic paralysis. In *hyperkalemic periodic paralysis* (HyperPP) recurrent attacks of periodic paralysis, often in association with elevated serum $[K^+] > 5$ mM or precipitated by a K^+ challenge, are the predominant symptom. Patients with HyperPP frequently have myotonia, especially around the time of an episode of weakness. The clinical overlap of PMC and HyperPP is extensive, and family members with the same $Na_V1.4$ mutation may have a syndrome typical of PMC or HyperPP (Brancati et al. 2003; Kelly et al. 1997; McClatchey et al. 1992).

3.1.1 Gating Defects in Myotonia and HyperPP

Expression studies of $Na_V1.4$ mutant channels associated with SCM, PMC, and HyperPP have revealed a variety of gating defects, all of which result in gain-of-function changes that increase Na^+ influx. Most often, these changes disrupt inactivation gating of mutant channels (Fig. 2), and in some cases there is enhancement of activation. The inactivation defects include: (a) an increased persistent current that fails to inactivate over hundreds of milliseconds (Cannon et al. 1991; Cannon and Strittmatter 1993), (b) slower rate for entry to inactivation (Lerche et al. 1993; Yang et al. 1994), (c) faster rate of recovery from inactivation (Green et al. 1998), and (d) a depolarized shift in the voltage dependence of steady-state inactivation (Hayward et al. 1996; Mitrovic et al. 1995). These defects all disrupt conventional "fast" inactivation of $Na_V1.4$, which occurs over a time course of milliseconds and is important for shaping the repolarizing phase of the action potential and for limiting the propensity for repetitive firing during the refractory interval. The primary effect on fast inactivation gating is consistent with structure-function models of Na_V channels. The SCM/PMC/HyperPP mutations are clustered in the domain III-IV loop (the inactivation gate) (Stühmer et al. 1989; West et al. 1992), the cytoplasmic end of S5 and S6 segments at the inner mouth of the pore

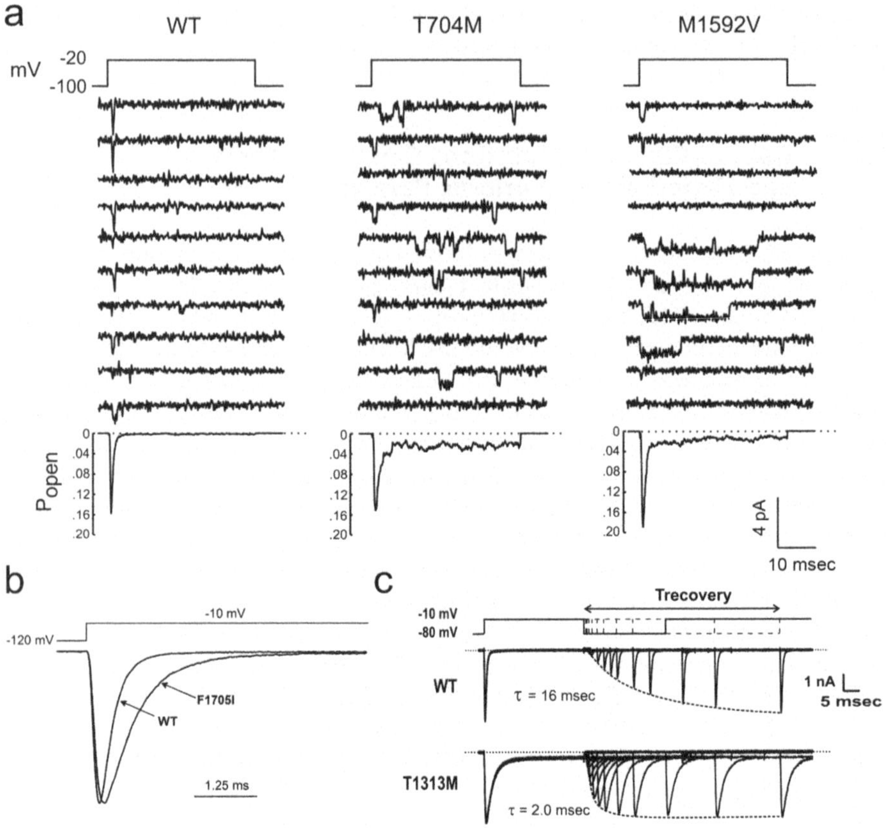

Fig. 2 Impairment of fast inactivation for $Na_V1.4$ mutations found in myotonia and HyperPP. (**a**) Cell-attached patch recordings from HEK cells show channel re-openings and prolonged open events for the two most commonly occurring HyperPP mutations (T704M and M1592V). Ensemble average (*bottom*) shows a small non-inactivating component [adapted from (Cannon and Strittmatter 1993)] (**b**) Amplitude normalized whole-cell current shows a slower rate of inactivation for the SCM mutation F1705I [adapted from (Wu et al. 2005)]. (**c**) Accelerated rate of recovery from fast inactivation for the PMC mutation T1313M [adapted from (Hayward et al. 1996)]

(the docking site for the inactivation gate), and the voltage sensor of domain IV (voltage sensor most tightly coupled to fast inactivation) (Capes et al. 2013).

All eukaryotic NaV channels also undergo slow inactivation, as revealed by a slow time course of recovery on the order of seconds. Channels become slow inactivated during prolonged intervals of depolarization lasting tens of seconds to minutes or during sustained high-frequency bursts of action potentials (Simoncini and Stuhmer 1987). Slow inactivation gating mechanisms are structurally distinct from those domains involved with fast inactivation. For example, mutations of the cytoplasmic III-IV loop may abolish fast inactivation but do not disrupt slow inactivation (Cummins and Sigworth 1996; Featherstone et al. 1996). Slow inactivation is impaired for several mutations associated with PMC/HyperPP, but not for SCM mutant channels (Hayward et al. 1997). The slow inactivation defects are manifest as a depolarized shift in voltage -dependence, an increased fraction of channels that do not slow inactivate, or a faster rate of recovery.

Activation gating may also be altered to produce gain-of-function changes in $Na_V1.4$ mutations associated with muscle channelopathies. The anomalous enhancement of activation for mutant channels occurs by a hyperpolarized shift in voltage dependence (Cummins et al. 1993) or a slower rate of deactivation (Featherstone et al. 1998).

3.1.2 Pathophysiologic Mechanism of Myotonia and HyperPP

Mutations of $Na_V1.4$ that produce gain-of-function defects with excessive inward Na^+ current at the channel level may cause either enhanced excitability (myotonia) or a transient loss of excitability (periodic paralysis) at the cellular level. Experimental evidence that a gain-of-function defect for $Na_V1.4$ is sufficient to cause these divergent effects on muscle excitability has been obtained for toxin-based models that disrupt $Na_V1.4$ inactivation (Cannon and Corey 1993) and for knock-in mutant mouse models (Corrochano et al. 2014; Hayward et al. 2008). The knock-in mouse homolog of M1592V has a robust HyperPP phenotype with myotonia by EMG, hindlimb stiffness, and reduced tetanic force in the setting of a 10 mM K^+ challenge in vitro (Hayward et al. 2008). Complementary studies using computer simulation of muscle excitability provide additional insights on pathomechanism by providing a mechanism to define the type and magnitude of gating defect that is necessary to cause myotonia or periodic paralysis (Cannon et al. 1993a). The model correlate of myotonia is sustained bursts of action potentials, whereas periodic paralysis is manifest as an anomalous stable depolarization of V_{rest} that inactivates the simulated Na^+ conductance and renders the model fiber inexcitable.

Simulations have demonstrated that some gain-of-function defects predispose to myotonia while others increase the susceptibility to periodic paralysis. For example, $Na_V1.4$ mutations found in SCM patients, who by definition have myotonia but no periodic paralysis, alter the *kinetics* of channel inactivation with a slower rate of onset and sometimes with faster recovery as well (Green et al. 1998; Yang et al. 1994). These changes increase the fraction of sodium channels that remain available for activation (i.e., non-inactivated) immediately after an action potential, which in turn increases the likelihood of repetitive after-discharges (Fig. 3, *middle*). Because the gain-of-function defect is transient, without a persistent Na^+ current in steady-state, the simulation will never produce the stable depolarization of V_{rest} required to produce periodic paralysis. The myotonic burst is sustained by the modest after-depolarization arising from use-dependent K^+ accumulation in the T-tubules, an extracellular space with restricted diffusion. Normally, $Na_V1.4$ channel availability is too low immediately after an action potential for this transient after-depolarization to trigger another spike, but for SCM mutant channels the increased availability may elicit an after-discharge. Experimental support for this mechanism has been obtained in animal models of myotonia, where muscle fiber

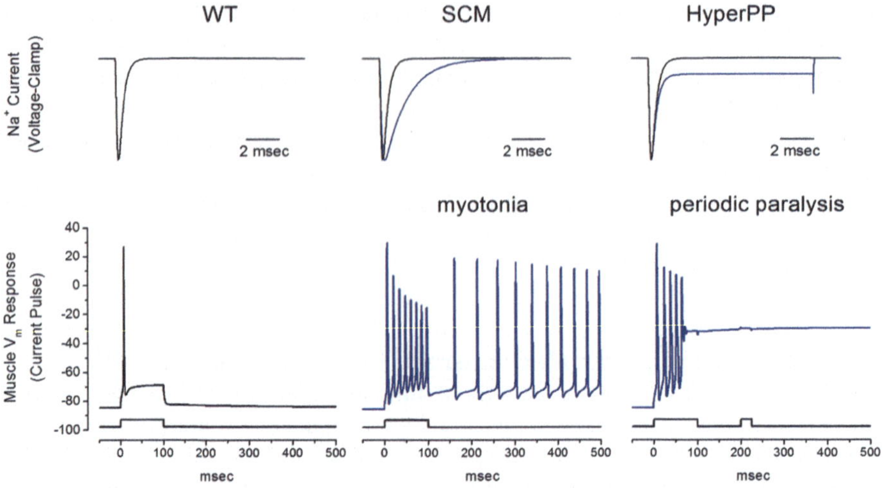

Fig. 3 Model simulation of myotonia and periodic paralysis resulting from gain-of-function defects in $Na_V1.4$. Top row shows simulated $Na_V1.4$ mutant currents (*blue lines*) typical for SCM (*middle*) and HyperPP (*right*). A two-compartment model for skeletal muscle (Cannon et al. 1993a), to simulate the sarcolemma and the t-tubule including K^+ accumulation, was used to simulate the response to current injection. The simulated muscle normally fires a single action potential and then accommodates (*left*, note the difference in time scale compared to the top row). A reduced rate for onset of fast inactivation as in SCM (*middle*) gives rise to a sustained burst of myotonic discharges that persists after termination of the stimulus. A small persistent Na^+ current to simulate HyperPP (*right*) also results in an initial myotonic burst, but then the membrane potential settles on an anomalously depolarized value, which inactivates the majority of $Na_V1.4$ channels and renders the fiber refractory from a second stimulus pulse

detubulation abolishes the transient after-depolarization and the burst of myotonic discharges (Cannon and Corey 1993).

In contrast, the gain-of-function defects for $Na_V1.4$ mutant channels associated with HyperPP often include an anomalous persistent Na^+ current from incomplete inactivation or a shift in the steady-state voltage dependence of gating (Cannon and Strittmatter 1993; Hayward et al. 1996). Voltage-clamp studies of wild-type $Na_V1.4$ show fast inactivation of Na^+ currents within a few milliseconds and a barely perceptible fraction of non-inactivating current that is about 0.2% of the transient peak amplitude. For HyperPP mutant channels, however, the persistent current is 1–4% of the transient peak (Cannon et al. 1991). While the absolute magnitude of this persistent current is small, the 5- to 20-fold relative increase has a large effect on simulated fiber excitability. Quiescent fibers have a normal V_{rest}, where $Na_V1.4$ channels are closed and the gain-of-function defect is silent. Brief stimulation may elicit a burst of myotonic discharges, and as K^+ accumulates in the T-tubules the moderate after-depolarization in conjunction with a persistent defect of $Na_V1.4$ inactivation results in a steady inward Na^+ current that keeps the fiber stably depolarized at about −45 mV (Fig. 3, *right*). From this anomalously depolarized V_{rest} the wild-type $Na_V1.4$ channels and most of the HyperPP mutant channels are inactivated, which renders the fiber chronically refractory and unable to generate an action potential. The mutant allele has a dominant-negative effect on fiber excitability, acting through voltage-dependent inactivation. This scenario also explains why exogenous administration of K^+ may elicit an attack of weakness in HyperPP. Model simulations show a similar mechanism occurs for HyperPP gain-of-function mutations that enhance activation from a left shift in voltage dependence. The increased overlap of steady-state inactivation and activation creates a persistent "window" current in the voltage range of −65 to −45 mV.

Slow inactivation of $Na_V1.4$ channels will attenuate the persistent current from a fast inactivation defect and thereby is predicted to reduce the susceptibility to periodic paralysis (Ruff 1994). Voltage-clamp studies revealed impairment of slow inactivation for the two most commonly occurring mutations in HyperPP (T704M and M1592V) and for a subset of other mutations in HyperPP and PMC (Cummins and Sigworth 1996; Hayward et al. 1997). About 90% of wild-type $Na_V1.4$ channels will slow inactivate at depolarized potentials, whereas the mutant channels have a combination of defects with larger non-inactivating fraction (20–40%), a right shift in voltage dependence, or a faster rate of recovery. Not all PMC/HyperPP mutant channels tested had a detectable impairment of slow inactivation, but every $Na_V1.4$ mutation with a defect of slow inactivation was associated with a phenotype where periodic paralysis was a prominent symptom (Hayward et al. 1999). The interpretation is that impaired slow inactivation markedly increases the risk of depolarization-induced attacks of periodic paralysis, but that because normal slow inactivation is only about 70–80% complete at −50 mV then a severe defect of fast inactivation alone is sufficient to produce stable depolarization with paralysis.

3.2 Anomalous Gating Pore Conduction in Hypokalemic Periodic Paralysis

Hypokalemic periodic paralysis presents with recurrent episodes of weakness in association with low serum K^+ (<3.0 mM). Attacks of weakness in HypoPP are often precipitated by events that promote a reduction of extracellular K^+ such as a concentrated oral load of carbohydrate (which shifts glucose and K^+ into muscle), or K^+ loss from vomiting or diarrhea (Cannon 2015; Lehmann-Horn et al. 2004). Myotonia does not occur in HypoPP, and is an exclusion criterion for the diagnosis. The genetic lesion in HypoPP is heterogeneous, with 60% of families having a missense mutation in the skeletal muscle calcium channel ($Ca_V1.1$ encoded by *CACNA1S*) and 20% with a missense mutation of *SCN4A* encoding $Na_V1.4$ (Sternberg et al. 2001). Remarkably, all 12 HypoPP mutations in $Na_V1.4$ and 8 of 9 mutations in $Ca_V1.1$ are missense substitutions at arginine residues in S4 segments of the voltage sensor domains (VSD) (Matthews et al. 2009). This convergence of mutation sites at homologous positions in the VSDs of $Ca_V1.1$ and $Na_V1.4$ has led to the notion that a common pathomechanism underlies the susceptibility to intermittent depolarization with paralysis (Cannon 2010). This hypothesis is consistent with the clinical experience that the signs and symptoms of HypoPP are indistinguishable for patients with $Ca_V1.1$ or $Na_V1.4$ mutations.

3.2.1 Gating Pore Current in HypoPP Mutant Channels

Functional expression studies of $Na_V1.4$ HypoPP mutant channels initially focused on voltage-dependent gating of the Na^+ conductance, since the mutations all occurred in VSDs. Moderate loss-of-function changes were identified, with enhanced slow inactivation [left shift and slower rate of recovery (Struyk et al. 2000)] or a left shift of fast inactivation with reduced current density (Jurkat-Rott et al. 2000). In addition, substantial decoupling of gating charge displacement to channel opening further contributes to a reduced peak Na^+ current in HypoPP mutant channels (Mi et al. 2014). These changes suggested a dichotomy to account for divergent clinical phenotypes from $Na_V1.4$ mutations – gain-of-function defects predispose to myotonia and HyperPP, whereas loss-of-function defects cause HypoPP – but there was no satisfactory explanation for how loss-of-function for $Na_V1.4$ would cause paradoxical depolarization and loss of excitability in low extracellular K^+.

Histidine scanning mutagenesis to ascertain voltage-dependent conformational changes in the S4 segments of Shaker K_V channels revealed a secondary pathway for translocation of protons across the membrane, either by a transporter mechanism (Starace et al. 1997) or conduction through an aqueous path (Starace and Bezanilla 2001, 2004). Substitution of the first arginine in S4 by smaller hydrophobic residues created a non-selective monovalent cation conductance, the so-called omega-current or gating pore current, as distinguished from the conventional "alpha" current through the canonical pore domain (Tombola et al. 2005). Homologous mutations at a pair of arginines in DIIS4 of $Na_V1.2$ also resulted in a gating pore current (Sokolov et al. 2005). This observation raised the possibility that all members of the voltage-gated ion channel superfamily might support a gating pore current when S4 in the VSD is mutated (Moreau et al. 2014), including by disease-associated mutations. The mechanistic interpretation was that voltage-dependent translocation of the S4 helix occurs through an aqueous crevice in the channel (Starace and Bezanilla 2004) that has a hydrophobic narrow waist – the gating charge transfer center (GCTC) – that normally impedes ion conduction (Tao et al. 2010; Wood et al. 2017). Mutation of an arginine in S4 may allow ion conduction through the gating pore, when the mutant residue is aligned with the GCTC. This scenario is consistent with the voltage dependence of the permissive state for ion conduction that has been observed for gating pore currents (Sokolov et al. 2005; Starace and Bezanilla 2004).

The HypoPP associated mutations of $Na_V1.4$ are in the outer arginines at the amino end of S4 segments (Matthews et al. 2009), and so a gating pore current would be expected to show inward rectification, with ion conduction occurring at hyperpolarized potentials that bias S4 in the "downward" closed channel conformation. When HypoPP $Na_V1.4$ mutant channels are expressed at very high levels in Xenopus oocytes, a TTX-insensitive inward rectifying current is detectable, consistent with a gating pore current (Fig. 4), and which is not seen for wild-type $Na_V1.4$ channels (Sokolov et al. 2007; Struyk and Cannon 2007; Struyk et al. 2008).

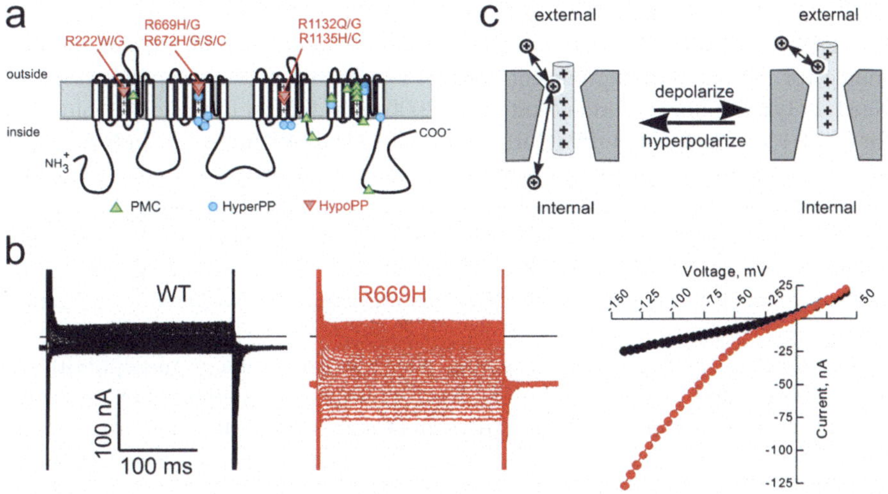

Fig. 4 Gating pore currents in $Na_V1.4$ HypoPP mutant channels. (**a**) Schematic representation of the $Na_V1.4$ α subunit showing the location of missense mutations associated with muscle syndromes that include periodic paralysis (PMC, HyperPP, HypoPP). The HypoPP mutations are all at arginine residences in S4 segments and none are in domain IV. (**b**) Currents recorded from oocytes expressing wild-type (WT, *black*) or R669H HypoPP mutant channels (*red*) in the presence of 1 μM TTX to block the $Na_V1.4$ pore. The steady-state I-V relation shows inward rectification for R669H but not WT channels. (**c**) Inward rectification is consistent with a gating pore current resulting from a voltage-dependent anomalous conduction pathway that allows cation permeation only when the mutant residue of S4 is within the gating charge transfer center

The two HypoPP mutations in DIIS4 that are arginine to histidine substitutions (R669H and R672H) both conduct proton-selective gating pore currents (Struyk and Cannon 2007; Struyk et al. 2008). Interestingly, the clinical phenotype is no different for proton-selective versus cation non-selective HypoPP mutations. Eight out of the 12 reported HypoPP mutations in $Na_V1.4$ have been screened, and all 8 had detectable gating pore currents. Moreover, the PMC associated mutation in DIVS4 (R1448C at the R1 position) did not have a detectable gating pore current (Francis et al. 2011), thereby showing specificity with gating pore currents observed only in HypoPP mutant channels. Histidine screening mutagenesis in S4 segments for all four domains of $Na_V1.4$ shows that the DIV VSD is more resistant to the creation of a gating pore leak, possibly because the GCTC is spread over a larger distance (Gosselin-Badaroudine et al. 2012). An inward rectifying current, consistent with a gating pore current, has also been detected in muscle fibers from R669H HypoPP mice, which verifies this anomalous current occurs in the context of mammalian skeletal muscle (Wu et al. 2011).

3.2.2 Pathophysiologic Mechanism for HypoPP

A confluence of experimental evidence and computer modeling supports the hypothesis that the gating pore current is the critical anomaly for HypoPP mutant channels in causing susceptibility to paradoxical depolarization and weakness in low K^+. Investigations with human HypoPP muscle biopsies implicated an unusual source for the depolarizing current, since TTX did not prevent depolarization of $Na_V1.4$ HypoPP fibers (Jurkat-Rott et al. 2000) nor did calcium channel blockers for $Ca_V1.1$ associated HypoPP (Ruff 1999). All 12 HypoPP mutations in $Na_V1.4$ are missense mutations at arginines in the outer ends of S4 segments in the VSD, and all 8 tested to date had detectable gating pore currents. Moreover, the $Ca_V1.1$ mutations in HypoPP are also R/X substitutions in S4 segments of the VSD, and voltage-clamp studies of $Ca_V1.1$-R528H human fibers (Jurkat-Rott et al. 2009; Ruff 1999) and knock-in mice (Wu et al. 2012) also revealed an inward rectifying gating pore current.

The mechanism by which the gating pore current produces susceptibility to depolarization and weakness in low K^+ has been tested and supported by computer simulation (Jurkat-Rott et al. 2010; Struyk and Cannon 2008). In quantitative terms, gating pore currents are very small, with the conductance being about 0.03% the peak Na^+ conductance through the conventional pore (Mi et al. 2014). In skeletal muscle at V_{rest}, the gating pore conductance is about 10 $\mu S/cm^2$ or 1% of the total resting conductance of the fiber. The gating pore current will promote depolarization, since the reversal potential is near 0 mV, regardless of whether the conductance is proton-selective or non-selective for monovalent cations. Based on these values, the impact of the anomalous gating pore current on a simulated resting fiber is modest in normal external K^+ (4.5 mM), with a predicted depolarization of about 3 mV. In low extracellular K^+, however, the impact can be large. As K^+ is reduced, the equilibrium or Nernst potential for K^+ becomes more negative, and initially the simulated fiber hyperpolarizes. In low K^+, the membrane voltage at which strong rectification occurs for the inward rectifier K channel (Kir) also becomes more negative. This shift of the rectification point for Kir will eventually reduce the outward K^+ current available to balance inward currents from ClC-1 chloride channels and the gating pore. (Recall in steady-state V_{rest} is always more positive than E_K so the current will be outward through Kir). When K^+ is further reduced and the Kir current becomes too small to balance the ClC-1 and gating pore inward currents, the fiber will depolarize until the delayed rectifier K channel (Kdr) opens and again establishes a net balance of 0 ionic current. In essence, the K^+ channel that dominates for setting V_{rest} switches from Kir (normally polarized V_{rest}) to Kdr (abnormally depolarized). At this anomalously depolarized V_{rest}, $Na_V1.4$ channels are inactivated, the fiber is inexcitable, and flaccid weakness ensues. Of course, this same mechanism applies to wild-type fibers, as has been demonstrated experimentally in mammalian fibers (Siegenbeek van Heukelom 1991), but in the absence of a gating pore current the critical $[K^+]$ for paradoxical depolarization is about 1 mM, far below the normal physiologic range. The anomalous gating pore current conducted by HypoPP mutant channels shifts the critical K^+ value up to about 3.0–3.5 mM, at the low end of the normal physiologic range. In addition, the loss-of-function changes in HypoPP mutant $Na_V1.4$ channels (enhanced inactivation and partial decoupling of charge movement to channel opening) will further contribute to the reduced excitability caused by depolarization of V_{rest}. The response of V_{rest} to changes in extracellular K^+ is summarized in the phase plot for simulated wild-type and HypoPP fibers in Fig. 5.

The chloride gradient across the sarcolemma has a strong influence over the susceptibility to depolarization and periodic paralysis for HypoPP fibers in low K^+. This modulation by Cl^- occurs because ClC-1 channels contribute about 85% of the total resting conductance in skeletal muscle (Palade and Barchi 1977). The reversal potential for Cl is about 3 mV depolarized from V_{rest} because influx via the Na^+-K^+-$2Cl^-$ cotransporter (NKCC) maintains the intracellular Cl^- slightly higher than the equilibrium value predicted for a passive electrochemical distribution (Aickin et al. 1989). Consequently, a steady inward Cl^- current (efflux of the Cl^- anion) through ClC-1 channels is the primary depolarizing current that is balanced at rest by the outward hyperpolarizing K^+ current conducted by Kir channels. Recordings from mammalian fibers and computer simulations (Gallaher et al. 2009) show that this combination of resting conductances gives rise to the possibility of two resting potentials (net ionic current is 0 for both cases) over a range of low external K^+ values (Fig. 5). The bistability of two resting potentials in skeletal muscle is not the result of an "N"-shaped steady-state I-V relation, with three critical membrane potentials at which the total current is 0 [e.g., the classical cardiac Purkinje cell model (Gadsby and Cranefield 1977)]. The large resting Cl^- conductance prevents such a scenario. Instead, a shift in the intracellular $[Cl^-]$ is the critical state variable that determines whether V_{rest} will lie on the hyperpolarized branch (low internal Cl^-, preserved excitability) or the depolarized branch (high internal Cl^-, loss of excitability). This insight led to the suggestion that maneuvers to manipulate the sarcolemma Cl^- gradient should have a strong modulating effect on the susceptibility to attacks of weakness in HypoPP (Geukes Foppen et al. 2002). Indeed, inhibition of the NKCC cotransporter with bumetanide, which will reduce intracellular $[Cl^-]$, completely prevents the loss of force from a 2 mM K^+ challenge in the $Na_V1.4$-R669H mouse model of HypoPP (Wu et al. 2013). Conversely, a hyperosmolar challenge, which stimulates NKCC and increases intracellular $[Cl^-]$, may trigger a loss of force or aggravate the response to a low K^+ challenge.

3.3 Loss-of-Function Mutations: Myasthenia and Congenital Myopathy

In myasthenia, voluntary muscle force rapidly fatigues, because of a loss in the safety factor of neuromuscular transmission. The synaptic defect is usually post-synaptic, as in myasthenia gravis with antibody-mediated loss of the nicotinic acetylcholine receptor and defacement of the endplate zone, or may be presynaptic as with congenital myasthenic syndromes (CMS) caused by genetic defects in acetylcholine synthesis or packaging into vesicles (Engel et al. 2003). A sodium channel based myasthenic syndrome was initially identified in a patient with fatigable weakness and the typical decremental response of the compound muscle action potential with repetitive nerve stimulation, but surprisingly the postsynaptic endplate potential was of normal amplitude (Tsujino et al. 2003). This combination

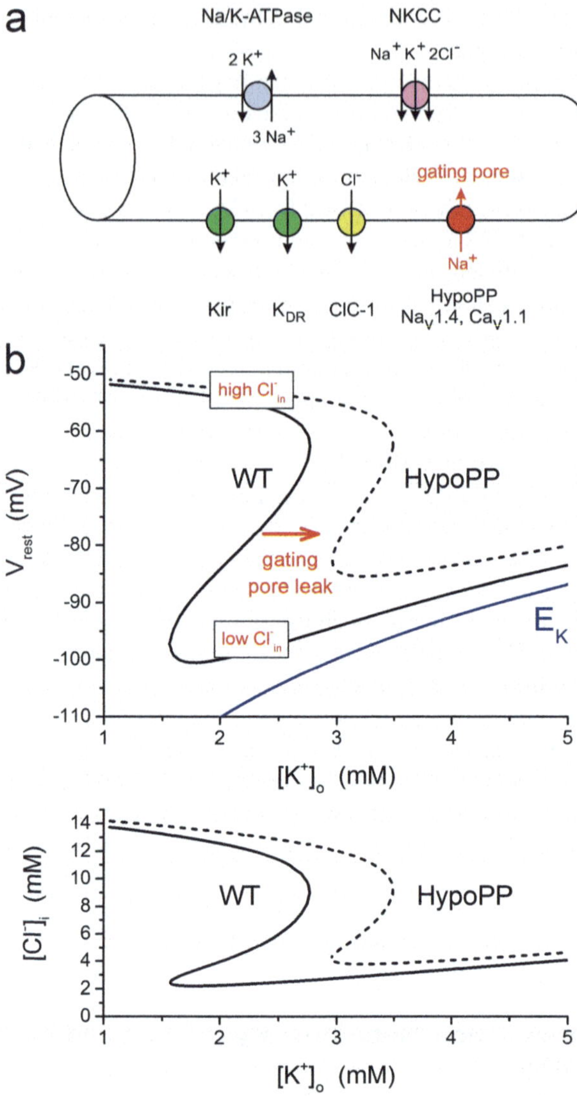

Fig. 5 Simulated paradoxical depolarization in low K^+ for a model HypoPP fiber. (**a**) Diagram of the pumps and transporters (*top* side) and ion channels (*bottom* side) to simulate the dependence of V_{rest} on extracellular K^+. *Arrows* indicate the direction of net transport for specific ions when the fiber is at the normal V_{rest}. The gating pore is present in HypoPP fibers only, has a reversal potential near 0 mV, and in this simulation is carried primarily Na^+ influx. (**b**) Phase plot of V_{rest} (*top*) and intracellular Cl^- (*bottom*) as the extracellular K^+ was varied from 1 to 5 mM. At each set value of K^+, the simulation searched for values of the membrane potential and internal Cl^- that simultaneously satisfied two constraints: that the sum of all ionic currents was zero (equilibrium point for Vm) and that the net Cl^- flux was also zero (mass balance for NKCC influx and ClC-1 efflux). For comparison, the Nernst potential for K^+ is shown in *blue*. For a simulated WT fiber (*solid line*), V_{rest} hyperpolarizes as K^+ is lowered from 5 mM until about 1.5 mM, at which point

of effects suggested a defect in muscle excitability downstream from the neuromuscular junction, and sequencing of *SCN4A* revealed a missense mutation in the voltage sensor of domain IV. Increased awareness that mutations of $Na_V1.4$ may occur in patients with symptomatic overlap between myasthenia and periodic paralysis led to the identification of additional isolated CMS cases, each of which was associated with recessive inheritance and homozygous missense mutations at arginines in the S4 of domain IV (Arnold et al. 2015; Habbout et al. 2016).

The congenital myopathies are a clinically and genetically diverse group of disorders with neonatal or early-onset weakness and characteristic changes on muscle biopsy. Over 20 different causative genes have been identified and the phenotype may vary from reduced fetal movements with neonatal fatality or mild disability with a normal lifespan. Whole exome sequencing from a large international consortium of congenital myopathy cases revealed 11 individuals with *SCN4A* mutations from six families (Zaharieva et al. 2016). The clinical features ranged from fetal loss or death within the first 2 days (7 cases) to reduced muscle tone at birth and moderately severe myopathy but retained ability for independent ambulation. The severe cases were all homozygous null mutations for $Na_V1.4$ whereas the less severe phenotypes had homozygous loss-of-function mutations or compound heterozygous mutations with a null allele and a loss-of-function mutation on the trans allele. Congenital myopathy patients who survived beyond childhood also had muscle fatigue and episodes of weakness, suggesting a partial overlap with myasthenia and periodic paralysis. All 12 parents were genetically confirmed to be heterozygous carriers, and all were asymptomatic with a normal neurological exam, thereby showing for the first time that a single null allele for *SCN4A* is well tolerated without a detectable clinical deficit.

3.3.1 Loss-of-Function Defects in Myasthenia and Myopathy

Expression studies of $Na_V1.4$ missense mutations associated with CMS or congenital myopathy show loss-of-function defects. The CMS mutations cause a pronounced leftward shift (-15 to -20 mV) for the voltage dependence of fast inactivation (Arnold et al. 2015; Tsujino et al. 2003), slower recovery from inactivation, and for R1454W a markedly faster onset of slow inactivation (>2 orders of magnitude) (Habbout et al. 2016). This combination of defects greatly enhances the use-dependent reduction of peak Na^+ current during a high frequency train (50 Hz) of brief depolarizations. The $Na_V1.4$ missense mutations found in congenital

Fig. 5 (continued) the fiber paradoxically depolarizes to -52 mV. This transition occurs because in very low K^+ the Kir conductance can no longer balance the inward Cl^- current. The fiber depolarizes, which must be accompanied with an increase of intracellular Cl^- (*bottom*) because of the high resting Cl^- conductance. Addition of the gating pore current in a simulated HypoPP fiber (*dashed line*) shifts the relation to the right such that the paradoxical depolarization now occurs at 3 mM which is in the low physiologic range. The system has the possibility of multiple equilibrium potentials at a single value of external K^+. For example, in the WT fiber this region of bistability for V_{rest} was with K^+ between 1.5 and 2.75 mM. The internal $[Cl^-]$ determines which value of V_{rest} the fiber will settle upon (hyperpolarized for low Cl^- and depolarized for high Cl^-)

myopathy either did not produce detectable Na^+ current (3 of 7), had substantially reduced Na^+ current density (3 of 4 expressing), or a depolarized shift of activation (2 of 4 expressing) (Zaharieva et al. 2016). Curiously, one missense mutation with an associated frameshift truncation (H1782Qfs65) was indistinguishable from wild-type in the HEK cell expression system. Otherwise, the congenital myopathy NaV1.4 mutations consistently displayed loss-of-function defects.

3.3.2 Pathophysiologic Mechanism of Myasthenic Weakness

The loss-of-function defects identified for $Na_V1.4$ in myasthenia will reduce the available pool of channels at the resting potential of −90 mV (large left shift of fast inactivation) and greatly accentuate the use-dependent reduction of peak Na^+ current during high-frequency discharges (slower recovery from inactivation, faster entry to slow inactivation). The mode of inheritance for the $Na_V1.4$ myasthenic syndrome is autosomal recessive (Arnold et al. 2015; Habbout et al. 2016), suggesting that a large use-dependent reduction of Na^+ current (an acute change) is required to cause intermittent failure of spike generation from the endplate depolarization at the NMJ or to reduce the action potential amplitude, and thereby cause fatigable weakness. The heterozygous parental carriers with one copy of the CMS mutation are asymptomatic, so a mild use-dependent loss of Na^+ current must be well tolerated. Chronically, even a complete null for one allele does not cause symptoms, as was detected in the parents of the congenital myopathy patients (Zaharieva et al. 2016).

An $Na_V1.4$ knock-out mouse model also supports the notion that an acute reduction for a substantial portion of the Na^+ current is required to produce myasthenia (Wu et al. 2016). Mice heterozygous for a deletion of exon 12 have a 50% reduction of Na^+ current density in muscle, but no obvious motor phenotype. So mice, like humans, tolerate a null allele for *SCN4A* without an appreciable deficit in behavior. Expression of the embryonic isoform, $Na_V1.5$, was not upregulated in adult mice to compensate for the reduced current density. Homozygous knock-out mice do not survive beyond the second postnatal day. In vitro contraction testing revealed a latent myasthenic phenotype in the heterozygous exon 2 deleted mice, with a sag in the muscle force plateau during tetanic stimulation and a decremental response of the compound muscle action potential during repetitive nerve stimulation in low-dose curare.

Taken together, the data for recessively inherited myasthenia or congenital myopathy imply a gene-dosage effect, and also a dependence on whether the mutant allele is a partial loss-of-function (hypomorphic) or a complete null. Figure 6 illustrates these interactions amongst the different alleles. The vertices of the triangle represent the three possible homozygous states (WT/WT, LOF/LOF, null/null), and the midpoints on the sides of the triangle represent all possible combinations of heterozygotes (WT/LOF, WT/null, LOF/null). The large yellow shaded area shows that three possible combinations will be asymptomatic (WT/WT, WT/LOF, WT/null). When both alleles are mutant, there is a spectrum of clinical phenotype severity ranging from the mildest being myasthenia (LOF/LOF), to congenital

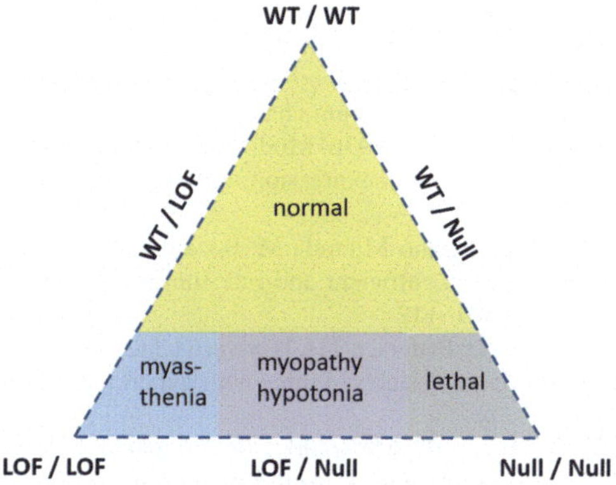

Fig. 6 Diagram illustrating the relationship between wild-type (WT), loss-of-function (LOF), and null alleles for $Na_V1.4$ in determining the clinical phenotype. The loss-of-function refers to mutant $Na_V1.4$ channels that express and conduct Na^+ current, albeit with reduced amplitude or duration, as distinct from the complete absence of function with a null allele. [Reproduced from (Cannon 2016)]

myopathy with survival to adulthood (LOF/null), and then the neonatal lethal incompatible with survival (null/null).

References

Aickin CC, Betz WJ, Harris GL (1989) Intracellular chloride and the mechanism for its accumulation in rat lumbrical muscle. J Physiol 411:437–455

Arnold WD, Feldman DH, Ramirez S, He L, Kassar D, Quick A, Klassen TL, Lara M, Nguyen J, Kissel JT, Lossin C, Maselli RA (2015) Defective fast inactivation recovery of Nav 1.4 in congenital myasthenic syndrome. Ann Neurol 77:840–850

Brancati F, Valente EM, Davies NP, Sarkozy A, Sweeney MG, LoMonaco M, Pizzuti A, Hanna MG, Dallapiccola B (2003) Severe infantile hyperkalaemic periodic paralysis and paramyotonia congenita: broadening the clinical spectrum associated with the T704M mutation in SCN4A. J Neurol Neurosurg Psychiatry 74:1339–1341

Brown GL, Harvey AM (1939) Congenital myotonia in the goat. Brain 62:341–363

Caldwell JH, Campbell DT, Beam KG (1986) Na channel distribution in vertebrate skeletal muscle. J Gen Physiol 87:907–932

Calhoun JD, Isom LL (2014) The role of non-pore-forming beta subunits in physiology and pathophysiology of voltage-gated sodium channels. Handb Exp Pharmacol 221:51–89

Cannon SC (2010) Voltage-sensor mutations in channelopathies of skeletal muscle. J Physiol 588:1887–1895

Cannon SC (2015) Channelopathies of skeletal muscle excitability. Compr Physiol 5:761–790

Cannon SC (2016) When all is lost...a severe myopathy with hypotonia from sodium channel mutations. Brain 139:642–644

Cannon SC, Corey DP (1993) Loss of Na+ channel inactivation by anemone toxin (ATX II) mimics the myotonic state in hyperkalaemic periodic paralysis. J Physiol 466:501–520

Cannon SC, Strittmatter SM (1993) Functional expression of sodium channel mutations identified in families with periodic paralysis. Neuron 10:317–326

Cannon SC, Brown RH Jr, Corey DP (1991) A sodium channel defect in hyperkalemic periodic paralysis: potassium-induced failure of inactivation. Neuron 6:619–626

Cannon SC, Brown RH Jr, Corey DP (1993a) Theoretical reconstruction of myotonia and paralysis caused by incomplete inactivation of sodium channels. Biophys J 65:270–288

Cannon SC, McClatchey AI, Gusella JF (1993b) Modification of the Na^+ current conducted by the rat skeletal muscle alpha subunit by coexpression with a human brain beta subunit. Pflugers Arch 423:155–157

Capes DL, Goldschen-Ohm MP, Arcisio-Miranda M, Bezanilla F, Chanda B (2013) Domain IV voltage-sensor movement is both sufficient and rate limiting for fast inactivation in sodium channels. J Gen Physiol 142:101–112

Clausen T, Nielsen OB, Clausen JD, Pedersen TH, Hayward LJ (2011) Na+,K+−pump stimulation improves contractility in isolated muscles of mice with hyperkalemic periodic paralysis. J Gen Physiol 138:117–130

Corrochano S, Mannikko R, Joyce PI, McGoldrick P, Wettstein J, Lassi G, Raja Rayan DL, Blanco G, Quinn C, Liavas A, Lionikas A, Amior N, Dick J, Healy EG, Stewart M, Carter S, Hutchinson M, Bentley L, Fratta P, Cortese A, Cox R, Brown SD, Tucci V, Wackerhage H, Amato AA, Greensmith L, Koltzenburg M, Hanna MG, Acevedo-Arozena A (2014) Novel mutations in human and mouse SCN4A implicate AMPK in myotonia and periodic paralysis. Brain 137:3171–3185

Cummins TR, Sigworth FJ (1996) Impaired slow inactivation in mutant sodium channels. Biophys J 71:227–236

Cummins TR, Zhou J, Sigworth FJ, Ukomadu C, Stephan M, Ptacek LJ, Agnew WS (1993) Functional consequences of a Na+ channel mutation causing hyperkalemic periodic paralysis. Neuron 10:667–678

DiFranco M, Vergara JL (2011) The Na conductance in the sarcolemma and the transverse tubular system membranes of mammalian skeletal muscle fibers. J Gen Physiol 138:393–419

Engel AG, Ohno K, Sine SM (2003) Neurological diseases: sleuthing molecular targets for neurological diseases at the neuromuscular junction. Nat Rev Neurosci 4:339–352

Featherstone DE, Richmond JE, Ruben PC (1996) Interaction between fast and slow inactivation in SkM1 sodium channels. Biophys J 71:3098–3109

Featherstone DE, Fujimoto E, Ruben PC (1998) A defect in skeletal muscle sodium channel deactivation exacerbates hyperexcitability in human paramyotonia congenita. J Physiol 506 (3):627–638

Francis DG, Rybalchenko V, Struyk A, Cannon SC (2011) Leaky sodium channels from voltage sensor mutations in periodic paralysis, but not paramyotonia. Neurology 76:1635–1641

Fu Y, Struyk A, Markin V, Cannon S (2011) Gating behaviour of sodium currents in adult mouse muscle recorded with an improved two-electrode voltage clamp. J Physiol 589:525–546

Gadsby DC, Cranefield PF (1977) Two levels of resting potential in cardiac Purkinje fibers. J Gen Physiol 70:725–746

Gallaher J, Bier M, Siegenbeek van Heukelom J (2009) The role of chloride transport in the control of the membrane potential in skeletal muscle – theory and experiment. Biophys Chem 143:18–25

Gellens ME, George AL Jr, Chen LQ, Chahine M, Horn R, Barchi RL, Kallen RG (1992) Primary structure and functional expression of the human cardiac tetrodotoxin-insensitive voltage-dependent sodium channel. Proc Natl Acad Sci 89:554–558

George AL Jr, Komisarof J, Kallen RG, Barchi RL (1992) Primary structure of the adult human skeletal muscle voltage-dependent sodium channel. Ann Neurol 31:131–137

Geukes Foppen RJ, van Mil HG, Siegenbeek van Heukelom J (2002) Effects of chloride transport on bistable behaviour of the membrane potential in mouse skeletal muscle. J Physiol 542:181–191

Gosselin-Badaroudine P, Delemotte L, Moreau A, Klein ML, Chahine M (2012) Gating pore currents and the resting state of Nav1.4 voltage sensor domains. Proc Natl Acad Sci U S A 109:19250–19255

Green D, George A, Cannon S (1998) Human sodium channel gating defects caused by missense mutations in S6 segments associated with myotonia: S804F and V1293I. J Physiol 510:685–694

Habbout K, Poulin H, Rivier F, Giuliano S, Sternberg D, Fontaine B, Eymard B, Morales RJ, Echenne B, King L, Hanna MG, Mannikko R, Chahine M, Nicole S, Bendahhou S (2016) A recessive Nav1.4 Mutation underlies congenital myasthenic syndrome with periodic paralysis. Neurology 86:161–169

Hayward LJ, Brown RH Jr, Cannon SC (1996) Inactivation defects caused by myotonia-associated mutations in the sodium channel III-IV linker. J Gen Physiol 107:559–576

Hayward LJ, Brown RH Jr, Cannon SC (1997) Slow inactivation differs among mutant Na channels associated with myotonia and periodic paralysis. Biophys J 72:1204–1219

Hayward LJ, Sandoval GM, Cannon SC (1999) Defective slow inactivation of sodium channels contributes to familial periodic paralysis. Neurology 52:1447–1453

Hayward LJ, Kim JS, Lee MY, Zhou H, Kim JW, Misra K, Salajegheh M, Wu FF, Matsuda C, Reid V, Cros D, Hoffman EP, Renaud JM, Cannon SC, Brown RH (2008) Targeted mutation of mouse skeletal muscle sodium channel produces myotonia and potassium-sensitive weakness. J Clin Invest 118:1437–1449

Heine R, Pika U, Lehmann-Horn F (1993) A novel SCN4A mutation causing myotonia aggravated by cold and potassium. Hum Mol Genet 2:1349–1353

Horga A, Raja Rayan DL, Matthews E, Sud R, Fialho D, Durran SC, Burge JA, Portaro S, Davis MB, Haworth A, Hanna MG (2013) Prevalence study of genetically defined skeletal muscle channelopathies in England. Neurology 80:1472–1475

Huang W, Liu M, Yan SF, Yan N (2017) Structure-based assessment of disease-related mutations in human voltage-gated sodium channels. Protein Cell 8:401–438

Isom LL, De Jongh KS, Patton DE, Reber BF, Offord J, Charbonneau H, Walsh K, Goldin AL, Catterall WA (1992) Primary structure and functional expression of the beta 1 subunit of the rat brain sodium channel. Science 256:839–842

Jaimovich E, Venosa RA, Shrager P, Horowicz P (1976) Density and distribution of tetrodotoxin receptors in normal and detubulated frog sartorius muscle. J Gen Physiol 67:399–416

Jurkat-Rott K, Mitrovic N, Hang C, Kouzmekine A, Iaizzo P, Herzog J, Lerche H, Nicole S, Vale-Santos J, Chauveau D, Fontaine B, Lehmann-Horn F (2000) Voltage-sensor sodium channel mutations cause hypokalemic periodic paralysis type 2 by enhanced inactivation and reduced current. Proc Natl Acad Sci U S A 97:9549–9554

Jurkat-Rott K, Weber MA, Fauler M, Guo XH, Holzherr BD, Paczulla A, Nordsborg N, Joechle W, Lehmann-Horn F (2009) K+-dependent paradoxical membrane depolarization and Na+ overload, major and reversible contributors to weakness by ion channel leaks. Proc Natl Acad Sci U S A 106:4036–4041

Jurkat-Rott K, Holzherr B, Fauler M, Lehmann-Horn F (2010) Sodium channelopathies of skeletal muscle result from gain or loss of function. Pflugers Arch 460:239–248

Kelly P, Yang WS, Costigan D, Farrell MA, Murphy S, Hardiman O (1997) Paramyotonia congenita and hyperkalemic periodic paralysis associated with a met 1592 Val substitution in the skeletal muscle sodium channel alpha subunit – a large kindred with a novel phenotype. Neuromuscul Disord 7:105–111

Lehmann-Horn F, Jurkat-Rott K (1999) Voltage-gated ion channels and hereditary disease. Physiol Rev 79:1317–1372

Lehmann-Horn F, Kuther G, Ricker K, Grafe P, Ballanyi K, Rüdel R (1987) Adynamia episodica hereditaria with myotonia: a non-inactivating sodium current and the effect of extracellular pH. Muscle Nerve 10:363–374

Lehmann-Horn F, Rüdel R, Jurkat-Rott K (2004) Nondystrophic myotonias and periodic paralyses. In: Engel AG, Franzini-Armstrong C (eds) Myology. McGraw-Hill, New York, pp 1257–1300

Lerche H, Heine R, Pika U, George AL Jr, Mitrovic N, Browatzki M, Weiss T, Rivet-Bastide M, Franke C, Lomonaco M, Ricker K, Lehmann-Horn F (1993) Human sodium channel myotonia:

slowed channel inactivation due to substitutions for a glycine within the III-IV linker. J Physiol 470:13–22
Matthews E, Tan SV, Fialho D, Sweeney MG, Sud R, Haworth A, Stanley E, Cea G, Davis MB, Hanna MG (2008) What causes paramyotonia in the United Kingdom? Common and new SCN4A mutations revealed. Neurology 70:50–53
Matthews E, Labrum R, Sweeney MG, Sud R, Haworth A, Chinnery PF, Meola G, Schorge S, Kullmann DM, Davis MB, Hanna MG (2009) Voltage sensor charge loss accounts for most cases of hypokalemic periodic paralysis. Neurology 72:1544–1547
McClatchey AI, McKenna-Yasek D, Cros D, Worthen HG, Kuncl RW, DeSilva SM, Cornblath DR, Gusella JF, Brown RH Jr (1992) Novel mutations in families with unusual and variable disorders of the skeletal muscle sodium channel. Nat Genet 2:148–152
McClatchey AI, Cannon SC, Slaugenhaupt SA, Gusella JF (1993) The cloning and expression of a sodium channel beta 1-subunit cDNA from human brain. Hum Mol Genet 2:745–749
Mi W, Rybalchenko V, Cannon SC (2014) Disrupted coupling of gating charge displacement to Na+ current activation for DIIS4 mutations in hypokalemic periodic paralysis. J Gen Physiol 144:137–145
Miller TM, Dias da Silva MR, Miller HA, Kwiecinski H, Mendell JR, Tawil R, McManis P, Griggs RC, Angelini C, Servidei S, Petajan J, Dalakas MC, Ranum LP, Fu YH, Ptacek LJ (2004) Correlating phenotype and genotype in the periodic paralyses. Neurology 63:1647–1655
Mitrovic N, George AL Jr, Lerche H, Wagner S, Fahlke C, Lehmann-Horn F (1995) Different effects on gating of three myotonia-causing mutations in the inactivation gate of the human muscle sodium channel. J Physiol 487:107–114
Moreau A, Gosselin-Badaroudine P, Chahine M (2014) Molecular biology and biophysical properties of ion channel gating pores. Q Rev Biophys 47:364–388
Palade PT, Barchi RL (1977) Characteristics of the chloride conductance in muscle fibers of the rat diaphragm. J Gen Physiol 69:325–342
Rogart RB, Regan LJ (1985) Two types of sodium channel with tetrodotoxin sensitivity and insensitivity detected in denervated mammalian skeletal muscle. Brain Res 329:314–318
Rüdel R, Lehmann-Horn F (1985) Membrane changes in cells from myotonia patients. Physiol Rev 65:310–356
Rüdel R, Ricker K, Lehmann-Horn F (1993) Genotype-phenotype correlations in human skeletal muscle sodium channel diseases. Arch Neurol 50:1241–1248
Rudolph JA, Spier SJ, Byrns G, Rojas CV, Bernoco D, Hoffman EP (1992) Periodic paralysis in quarter horses: a sodium channel mutation disseminated by selective breeding. Nat Genet 2:144–147
Ruff RL (1994) Slow Na+ channel inactivation must be disrupted to evoke prolonged depolarization-induced paralysis. Biophys J 66:542–545
Ruff RL (1999) Insulin acts in hypokalemic periodic paralysis by reducing inward rectifier K+ current. Neurology 53:1556–1563
Siegenbeek van Heukelom J (1991) Role of the anomalous rectifier in determining membrane potentials of mouse muscle fibres at low extracellular K+. J Physiol 434:549–560
Simoncini L, Stuhmer W (1987) Slow sodium channel inactivation in rat fast-twitch muscle. J Physiol 383:327–337
Singh RR, Tan SV, Hanna MG, Robb SA, Clarke A, Jungbluth H (2014) Mutations in SCN4A: a rare but treatable cause of recurrent life-threatening laryngospasm. Pediatrics 134:e1447–e1450
Sokolov S, Scheuer T, Catterall WA (2005) Ion permeation through a voltage-sensitive gating pore in brain sodium channels having voltage sensor mutations. Neuron 47:183–189
Sokolov S, Scheuer T, Catterall WA (2007) Gating pore current in an inherited ion channelopathy. Nature 446:76–78
Starace DM, Bezanilla F (2001) Histidine scanning mutagenesis of basic residues of the S4 segment of the shaker k+ channel. J Gen Physiol 117:469–490
Starace DM, Bezanilla F (2004) A proton pore in a potassium channel voltage sensor reveals a focused electric field. Nature 427:548–553

Starace DM, Stefani E, Bezanilla F (1997) Voltage-dependent proton transport by the voltage sensor of the shaker K+ channel. Neuron 19:1319–1327

Sternberg D, Maisonobe T, Jurkat-Rott K, Nicole S, Launay E, Chauveau D, Tabti N, Lehmann-Horn F, Hainque B, Fontaine B (2001) Hypokalaemic periodic paralysis type 2 caused by mutations at codon 672 in the muscle sodium channel gene SCN4A. Brain 124:1091–1099

Struyk AF, Cannon SC (2007) A Na^+ channel mutation linked to hypokalemic periodic paralysis exposes a proton-selective gating pore. J Gen Physiol 130:11–20

Struyk AF, Cannon SC (2008) Paradoxical depolarization of Ba^{2+}-treated muscle exposed to low extracellular K+: insights into resting potential abnormalities in hypokalemic paralysis. Muscle Nerve 37:326–337

Struyk AF, Scoggan KA, Bulman DE, Cannon SC (2000) The human skeletal muscle Na channel mutation R669H associated with hypokalemic periodic paralysis enhances slow inactivation. J Neurosci 20:8610–8617

Struyk AF, Markin VS, Francis D, Cannon SC (2008) Gating pore currents in DIIS4 mutations of NaV1.4 associated with periodic paralysis: saturation of ion flux and implications for disease pathogenesis. J Gen Physiol 132:447–464

Stühmer W, Conti F, Suzuki H, Wang XD, Noda M, Yahagi N, Kubo H, Numa S (1989) Structural parts involved in activation and inactivation of the sodium channel. Nature 339:597–603

Tao X, Lee A, Limapichat W, Dougherty DA, MacKinnon R (2010) A gating charge transfer center in voltage sensors. Science 328:67–73

Tombola F, Pathak MM, Isacoff EY (2005) Voltage-sensing arginines in a potassium channel permeate and occlude cation-selective pores. Neuron 45:379–388

Trimmer JS, Cooperman SS, Tomiko SA, Zhou J, Crean SM, Boyle MB, Kallen RG, Sheng Z, Barchi RL, Sigworth FJ, Goodman RH, Agnew WS, Mandel G (1989) Primary structure and functional expression of a mammalian skeletal muscle sodium channel. Neuron 3:33–49

Tsujino A, Maertens C, Ohno K, Shen XM, Fukuda T, Harper CM, Cannon SC, Engel AG (2003) Myasthenic syndrome caused by mutation of the SCN4A sodium channel. Proc Natl Acad Sci U S A 100:7377–7382

Wang J, Zhou J, Todorovic SM, Feero WG, Barany F, Conwit R, Hausmanowa-Petrusewicz I, Fidzianska A, Arahata K, Wessel HB, Sillen A, Marks HG, Hartlage P, Galloway G, Ricker K, Lehmann-Horn F, Hayakawa H, Hoffman EP (1993) Molecular genetic and genetic correlations in sodium channelopathies: lack of founder effect and evidence for a second gene. Am J Hum Genet 52:1074–1084

West JW, Patton DE, Scheuer T, Wang Y, Goldin AL, Catterall WA (1992) A cluster of hydrophobic amino acid residues required for fast Na^+-channel inactivation. Proc Natl Acad Sci 89:10910–10914

Wood ML, Freites JA, Tombola F, Tobias DJ (2017) Atomistic modeling of ion conduction through the voltage-sensing domain of the shaker K+ ion channel. J Phys Chem B 121:3804–3812

Wu FF, Gordon E, Hoffman EP, Cannon SC (2005) A C-terminal skeletal muscle sodium channel mutation associated with myotonia disrupts fast inactivation. J Physiol 565:371–380

Wu F, Mi W, Burns DK, Fu Y, Gray HF, Struyk AF, Cannon SC (2011) A sodium channel knockin mutant (NaV1.4-R669H) mouse model of hypokalemic periodic paralysis. J Clin Invest 121:4082–4094

Wu F, Mi W, Hernandez-Ochoa EO, Burns DK, Fu Y, Gray HF, Struyk AF, Schneider MF, Cannon SC (2012) A calcium channel mutant mouse model of hypokalemic periodic paralysis. J Clin Invest 122:4580–4591

Wu F, Mi W, Cannon SC (2013) Bumetanide prevents transient decreases in muscle force in murine hypokalemic periodic paralysis. Neurology 80:1110–1116

Wu F, Mi W, Fu Y, Struyk A, Cannon SC (2016) Mice with an NaV1.4 sodium channel null allele have latent myasthenia, without susceptibility to periodic paralysis. Brain 139:1688–1699

Yang JS, Sladky JT, Kallen RG, Barchi RL (1991) TTX-sensitive and TTX-insensitive sodium channel mRNA transcripts are independently regulated in adult skeletal muscle after denervation. Neuron 7:421–427

Yang N, Ji S, Zhou M, Ptacek LJ, Barchi RL, Horn R, George AL Jr (1994) Sodium channel mutations in paramyotonia congenita exhibit similar biophysical phenotypes in vitro. Proc Natl Acad Sci 91:12785–12789

Zaharieva I, Thor M, Oates E, Karnebeek C, Hendson G, Blom E, Witting N, Rasmussen M, Gabbett M, Ravenscroft G, Sframeli M, Sutterlin K, Sarkozy A, D'Argenzio L Hartley E Matthews M, Pitt J, Vissing M, Bellegaard C, Krarup A, Slordhal H, Halvorson C, Ye LH, Zhang N, Lokken U, Werlauf M, Abdelsayed MR, Davis L, Feng R, Phadke CA, Sewry JE, Morgan NG, Laing H, Vallance P, Ruben MG, Hanna S, Lewis EJ, Kamsteeg RM, Mutoni F (2016) Recessive loss-of-function SCN4A mutations associated with a novel phenotype of congenital myopathy. Brain 139:674–691

Cardiac Arrhythmias Related to Sodium Channel Dysfunction

Eleonora Savio-Galimberti, Mariana Argenziano, and Charles Antzelevitch

Contents

Abstract

The voltage-gated cardiac sodium channel ($Na_v1.5$) is a mega-complex comprised of a pore-forming α subunit and 4 ancillary β-subunits together with numerous protein partners. Genetic defects in the form of rare variants in one or more sodium channel-related genes can cause a loss- or gain-of-function of sodium channel current (I_{Na}) leading to the manifestation of various disease phenotypes, including Brugada syndrome, long QT syndrome, progressive cardiac conduction disease, sick sinus syndrome, multifocal ectopic Purkinje-related premature contractions, and atrial fibrillation. Some sodium channelopathies have also been shown to be responsible for sudden infant death syndrome (SIDS). Although these genetic defects often present as pure electrical diseases, recent studies point to a contribution of structural abnormalities to the electrocardiographic and arrhythmic manifestation in some cases, such as dilated cardiomyopathy. The same rare variants in *SCN5A* or related genes may present with different clinical phenotypes in different individuals and sometimes in members

E. Savio-Galimberti · M. Argenziano · C. Antzelevitch (✉)
Lankenau Institute for Medical Research, 100 E. Lancaster Avenue, Wynnewood, PA 19096, USA
e-mail: cantzelevitch@gmail.com

M. Chahine (ed.), *Voltage-gated Sodium Channels: Structure, Function and Channelopathies*, Handbook of Experimental Pharmacology 246,
https://doi.org/10.1007/164_2017_43

of the same family. Genetic background and epigenetic and environmental factors contribute to the expression of these overlap syndromes. Our goal in this chapter is to review and discuss what is known about the clinical phenotype and genotype of each cardiac sodium channelopathy, and to briefly discuss the underlying mechanisms.

Keywords
Atrial fibrillation · Brugada syndrome · Dilated cardiomyopathy · Early repolarization syndrome · Inherited cardiac arrhythmia syndromes · J wave syndromes · Long QT syndrome · Multifocal ectopic Purkinje-related premature contractions · Overlap syndromes · Progressive conduction disease · Sick sinus syndrome · Sudden infant death syndrome

1 Introduction

In the heart, voltage-gated sodium channels (Nav) are responsible for initiation and propagation of the action potential (AP). During the upstroke (or phase 0) of the AP, sodium channels open rapidly, generating an inward depolarizing current (I_{Na}), after which they quickly inactivate and enter into a nonconductive state. Inactivated sodium channels cannot respond to another stimulus and therefore cannot initiate a second AP (or after-depolarizations) until the channels recover from inactivation. Recovery from inactivation is time- and voltage-dependent. The sequential activation and inactivation of sodium channels work as a security mechanism, guaranteeing the directionality of the cardiac electrical activity through the myocardial syncytium, thus preventing the occurrence of pro-arrhythmic events that can trigger arrhythmias.

Nine different isoforms of voltage-gated sodium channel have been identified within the human body (Catterall et al. 2005). Although Nav 1.5 is the canonical cardiac sodium channel (Rogart et al. 1989; Gellens et al. 1992a, b; George et al. 1995), cardiac muscle expresses several other voltage-gated sodium channels, including neuronal sodium channels Nav 1.1, Nav 1.2, Nav 1.3, Nav 1.4, Nav 1.6, and Nav1.8 (Maier et al. 2004; Kaufmann et al. 2013; Westernbroek et al. 2013). Nav 1.8 (encoded by *SCN10A* gene) has been identified in human hearts (Facer et al. 2011; Yang et al. 2012) as well as in intracardiac neurons (Verkerk et al. 2012). *SCN10A* rare variants have been associated with alterations in the PR interval, QRS duration, and alterations in ventricular conduction (Chambers et al. 2010; Sotoodehnia et al. 2010a, b). Eukaryotic voltage-gated sodium channels share a similar structure that is highly preserved even when compared with the prokaryotic sodium channel. The channel is constituted by a single transmembrane copy of a protein of 2016 amino acids (~220 kDa) that comprises a cytoplasmic N-terminus, four homologous transmembrane domains (DI-DIV), and a cytoplasmic C-terminal domain (Fig. 1). Each one of the domains is similar to a prokaryotic subunit and is comprised of six α-helical transmembrane segments (termed S1 through S6) connected by extracellular and cytoplasmic loops. Segments S1 through S4 form the voltage-sensing domain and segments S5 and S6 comprise the pore-forming domain (Payandeh et al. 2011, 2012; Catterall 2014). The loop between S5 and S6

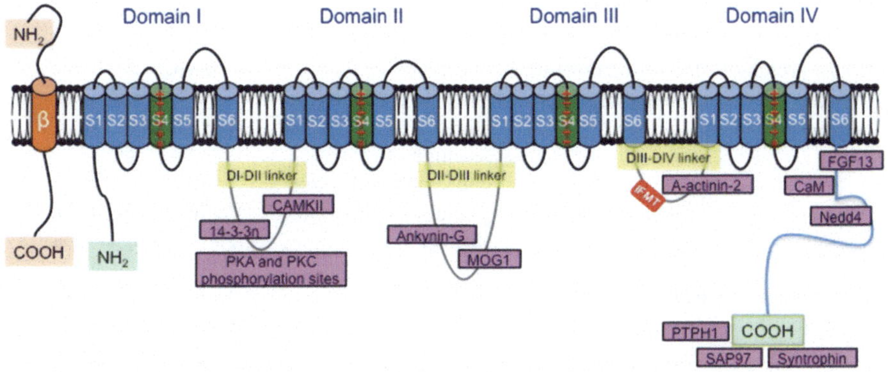

Fig. 1 Schematic representation of the α- and β-subunits of the VGSC. The four homologous domains (I–IV) of the α-subunit are represented; S5 and S6 are the pore-lining segments and S4 is the core of the voltage sensor. In the cytoplasmic linker between domains III and IV the IFMT (isoleucine, phenylalanine, methionine, and threonine) region is indicated. This is a critical part of the "inactivation particle" (inactivation gate), and substitution of amino acids in this region can disrupt the inactivation process of the channel. The "docking site" consists of multiple regions that include the cytoplasmic linker between S4 and S5 in domains III and IV, and the cytoplasmic end of the S6 segment in domain IV (*). Depending on the subtype of β-subunit considered they could interact (covalently or non-covalently) with the α-subunit. Some of the protein partners that can directly interact with the Nav1.5 are also shown in the figure (see also Table 1) (modified from Savio-Galimberti et al. 2012)

forms the selectivity filter. The S4 segment is heavily charged (arginine enriched region) and plays a central role in voltage sensing to increase channel permeability (activation of the channel) during depolarization of the cell.

The C-terminus and the linkers between domains contain interaction sites with which several protein partners ["sodium channel partners" or "Channel interactive proteins" (ChIP)] that regulate Nav 1.5 activity directly interact (Table 1 and Fig. 1). Other proteins that may indirectly interact with the Nav 1.5 include caveolin-3 (a scaffolding protein located within caveolar membranes) (Lu et al. 1999; Rybin et al. 2000; Yarbrough et al. 2002; Vatta et al. 2006), connexin-43 (Sato et al. 2011), telethonin (Valle et al. 1997; Mayans et al. 1998; Furukawa et al. 2001; Knoll et al. 2002; Haworth et al. 2004; Kojic et al. 2004; Mazzone et al. 2008), plakophilin-2 (Sato et al. 2009), ankyrin-B/ankyrin-2 (Jenkins and Bennett 2001; Garrido et al. 2003; Lemaillet et al. 2003; Mohler et al. 2004), glycerol-3-phosphate dehydrogenase 1-like protein (GPD1L), and Z-band-alternatively spliced-PDZ motif protein (ZASP) (Li et al. 2010; Remme 2013). The activity of the α-subunit of Nav 1.5 is also modulated by regulatory β-subunits (~30 kDa) of which there are four (β1-4). The stoichiometry between α- and β-subunits in the heart remains largely unknown. β-subunits can also act as cell adhesion molecules (CAMs) as well as modulate cell surface expression of Na_{V}s, enhancing sodium channel density and therefore cell excitability (Patino and Isom 2010; Savio-Galimberti et al. 2012). Mutations in the genes that encode several members of

Table 1 *SCN5A* protein partners

Nav 1.5 channel protein	Interacting proteins	Function	References
IQ motif (C-terminal)	Calmodulin (CaM)	Ubiquitous Ca-binding protein that may confer sensitivity to intracellular Ca levels	Tan et al. (2002); Shy et al. (2013)
PY motif (C-terminal)	Ubiquitin-protein Ligase Nedd 4-2	Ubiquitynation of Na channel	Van Bemmelen et al. (2004); Rougier et al. (2005)
PDZ domain-binding motif	PTPH1		Gavillet et al. (2006); Jespersen et al. (2006); Petitprez et al. (2011)
	SAP97		
	Syntrophins		
Other regions of C-terminus	Fibroblast growth factor homologous factor 13 (FGF13)	Delay fast inactivation of the channel	Dover et al. (2010); Wang et al. (2011)
Cytosolic DI–DII linker	14-3-3n; PKA and PKC phosphorylation sites; interaction site for CAMKII	Modulation of steady-state inactivation of the channel; Regulatory effects of Na channel availability and persistent current (late I_{Na}) magnitude	Allouis et al. (2006); Wagner et al. (2006); Ashpole et al. (2012)
DII–DIII linker	Ankyrin-G	Regulation of cell surface expression of Na channels	Lemaillet et al. (2003); Mohler et al. (2004); Kattygnarath et al. (2011)
	MOG1		
DIII–DIV linker	α-Actinin-2 (F-actin cross-linking protein family)	Increase of sodium current density with no effect on gating properties	Ziane et al. (2010)
Extracellular connecting loops between S5 and S6 segments	β1–β4 subunits	Modulation of Nav 1.5 channel density and kinetics	Malhotra et al. (2001); McEwen and Isom (2004); Ko et al. (2005); Meadows and Isom (2005); Medeiros-Domingo et al. (2007)

the ChIP group including *ANK2* (which encode Ankyrin-B/2 protein) and *SCN1B-3B* genes (that encode β1, β2, β3, and β4 subunits) have been associated with cardiac arrhythmia syndromes like long QT syndrome, structural heart disease (*ANK2*) (Swayne et al. 2017), Brugada syndrome (BrS; *SCN1B*, *SCN2B*, *SCN3B*) (Hu et al. 2009, 2010, 2012; Swayne et al. 2017), and atrial fibrillation (AF; *SCN1B*, *SCN2B*, *SCN3B*) (Olesen et al. 2011a, b, 2012a, b, c).

SCN5A mRNA transcription is regulated by several enhancers and repressors located near or within the promoter of the *SCN5A* gene (Arnolds et al. 2012; Van den Boogaard et al. 2012; Remme 2013). Transcriptional regulation of *SCN5A* can also be affected by gene-to-gene interaction. Van den Boogaard et al. (2012, 2014) reported that a common genetic variant within the intronic region of *SCN10A* modulates cardiac *SCN5A* expression (Van den Boogaard et al. 2012). Using high-resolution 4C-seq analysis of the *Scn10a-Scn5a* locus in murine heart tissue they showed that a cardiac enhancer located in *Scn10a*, encompassing *SCN10A* functional variant rs6801957, interacts with the promoter of *Scn5a*. An engineered transgenic mouse where they deleted the enhancer within *Scn10a* revealed that the enhancer was essential for Scn5a expression in cardiac tissue. Furthermore, in humans, the *SCN10A* variant rs6801957, which correlates with slowed conduction, was associated with decreased *SCN5A* expression (Van den Boogaard et al. 2012). These observations notwithstanding, the majority of *SCN10A* variants associated with BrS are exonic and not intronic as presumed in the *SCN5A-SCN10A* gene interaction hypothesis (Hu et al. 2014). Hu and coworkers presented evidence in support of the hypothesis that Nav1.5 encoded by *SCN5A* and $Na_v1.8$ encoded by *SCN10A* are physically associated in the cell membrane and that a mutation in *SCN10A* can lead to a major loss-of-function of Nav1.5 current, thus providing a mechanism to explain the association of *SCN10A* variants with BrS (Hu et al. 2014). Debate continues as to these two putative mechanisms.

2 *SCN5A* Mutations and Cardiac Arrhythmias

Gellens et al. (1992a, b) were the first to clone and characterize *SCN5A*. Three years later, George et al. (1995) mapped the *SCN5A* human gene to chromosome 3p21 by fluorescence in situ hybridization (FISH). In 1996 Wang et al. reported the genomic organization of the gene that contains 28 exons (Wang et al. 1996).

The first mutation in *SCN5A* was reported in a long QT syndrome (LQTS) type 3 by Mark Keating and his group in 1995 (Wang et al. 1995). Mutations in *SCN5A* (most of which are autosomal dominant) have been associated with a wide range of cardiac arrhythmia syndrome. These can be divided into three categories, based on minor allele frequency and gene location of the variant: rare *SCN5A* exonic variants, common exonic variants, and intronic noncoding variants.

2.1 Rare *SCN5A* Exonic Variants

2.1.1 Long QT Syndrome (LQTS)

Sixteen genes have been associated with LQTS to date (Table 2). Two forms have been identified: (1) Jervell and Lange-Nielsen syndrome (J-LN) is associated with deafness, and (2) the Romano-Ward syndrome (R-W) (Bennett et al. 1995; Schwartz et al. 2012). LQT1, 2, and 3 account for 90% of genotyped cases of LQTS. The prevalence of LQT3 among genotype-positive LQTS patients is 5–10%; LQT1 associated with loss-of-function mutations in *KCNQ1* gene accounts for 40–55%, and LQT2 associated with loss-of-function mutations in *KCNH2* gene accounts for 30–45% (Schwartz et al. 2012). Type 3 LQTS (LQT3) is associated with mutations in *SCN5A* giving rise to late or persistent sodium channel current (late I_{Na}) that effects the prolongation of the AP (Fig. 3) (Bennett et al. 1995;

Table 2 Genetic defects associated with the long QT syndrome

Chromosome		Gene	Ion channel	
LQT1	11	*KCNQ1, KvLQT1*	↓ I_{Ks}	90%
LQT2	7	*KCNH2, HERG*	↓ I_{Kr}	
LQT3	3	*SCN5A, Na$_v$1.5*	↑ Late I_{Na}	
LQT4	4	*Ankyrin-B, ANK2*	↑ Ca_i, ↑ Late I_{Na} ?	
LQT5	21	*KCNE1, minK*	↓ I_{Ks}	
LQT6	21	*KCNE2, MiRP1*	↓ I_{Kr}	
LQT7[a]	17	*KCNJ2, Kir2.1*	↓ I_{K1}	
LQT8[b]	6	*CACNA1C, Ca$_v$1.2*	↑ I_{Ca}	
LQT9	3	*CAV3, Caveolin-3*	↑ Late I_{Na}	
LQT10	11	*SCN4B, NavB4*	↑ Late I_{Na}	
LQT11	7	*AKAP9, Yotiao*	↓ I_{Ks}	
LQT12	20	*SNTA1, α-1 Syntrophin*	↑ Late I_{Na}	
LQT13	11	*KCNJ5, Kir3.4*	↓ $I_{K\text{-}ACh}$	
LQT14	14	CALM1, Calmodulin	↑ I_{Ca}, ↑ Late I_{Na}	
LQT15	2	CALM2, Calmodulin	↑ I_{Ca}, ↑ Late I_{Na}	
LQT16	19	CALM3, Calmodulin	↑ I_{Ca}, ↑ Late I_{Na}	

Augmentation of late I_{Na} is observed in variants associated with 8 of the 16 LQTS-susceptibility genes
[a]Andersen–Tawil syndrome
[b]Timothy syndrome

Schwartz et al. 2012). Variants in 8 of the 16 LQTS susceptibility genes have been shown to produce a gain-of-function in late I_{Na}, thus contributing to AP and QT prolongation (Table 2), which can be reversed with agents that block late I_{Na}, including ranolazine and mexiletine (Antzelevitch et al. 2014).

Clinical manifestation of LQTS includes syncopal episodes, frequently associated with cardiac arrest and leading to sudden cardiac death (SCD). The syncopal episodes are the consequence of an atypical polymorphic ventricular tachycardia known as torsade de pointes (TdP). SCD results when TdP degenerates into ventricular fibrillation. Most arrhythmic events in congenital LQT1 occur during physical or emotional stress, at rest or in association with sudden auditory stimulation in LQT2, and during sleep or rest in LQT3 patients (Priori et al. 2013). These phenotypic distinctions are consistent with the differential effects of catecholamines in the three genotypes (Fig. 2).

Ion channel dysfunctions associated with LQTS result in an inward shift in the balance of current leading to prolongation of the ventricular action potential and QT interval. The reduced repolarization reserve can lead to development of early afterdepolarizations (EADs). When EADs reach the threshold for activation of the inward calcium current, they generate triggered extrasystoles. Differences in the degree of AP prolongation among the three cell types that comprise the ventricular wall lead to development of transmural dispersion of repolarization (TDR), thus creating a vulnerable window across the ventricular wall and other regions of the ventricular myocardium, which can lead to development of reentrant arrhythmias. When an EAD-induced triggered response falls within this vulnerable window, the result is TdP (Antzelevitch 2007a, b) (Fig. 3).

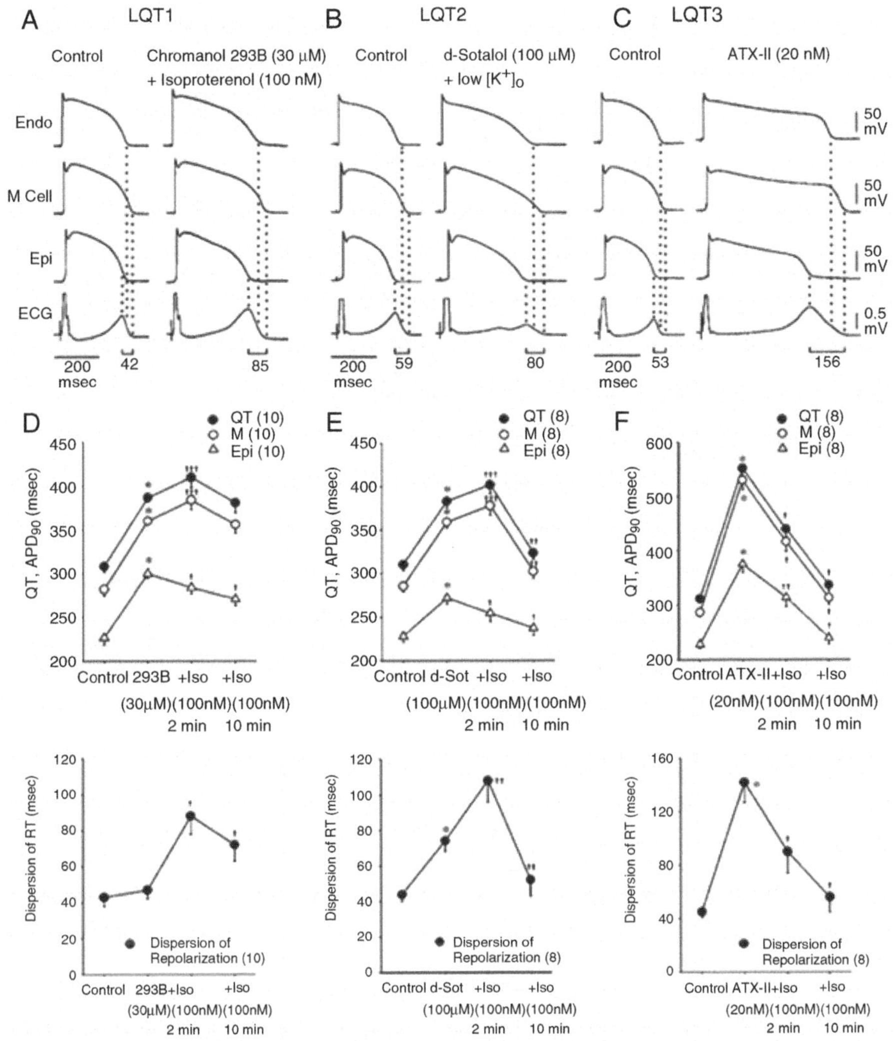

Fig. 2 Transmembrane action potentials (AP) and transmural electrocardiograms (ECG) in LQT1 (**a**), LQT2 (**b**), and LQT3 (**c**) models of long QT syndrome (LQTS) generated in arterially perfused canine left ventricular wedge preparations. Isoproterenol + chromanol 293B (an I_{Ks} blocker), d-sotalol + low $[K^+]_o$, and ATX-II – an agent that slows inactivation of late I_{Na} are used to mimic the LQT1, LQT2, and LQT3 syndromes, respectively. Panels (**a–c**) depict action potentials simultaneously recorded from endocardial (Endo), M, and epicardial (Epi) sites together with a transmural ECG. Basic cycle length = 2,000 ms. Transmural dispersion of repolarization (TDR) across the ventricular wall, defined as the difference in the repolarization time between M and Epi cells, is denoted below the ECG traces. Panels (**d–f**) show the effect of isoproterenol (Iso) in the LQT1, LQT2, and LQT3 models. In LQT1, Iso produces a persistent prolongation of the APD90 of the M cell and of the QT interval (at both 2 and 10 min), whereas the AP duration of 90(APD90) of the epicardial cell is always abbreviated, resulting in a persistent increase in TDR (**d**). In LQT2, Iso initially prolongs (2 min) and then abbreviates the QT interval and the APD90 of the M cell to the control level (10 min), whereas the APD90 of Epi cell is always abbreviated, resulting in a

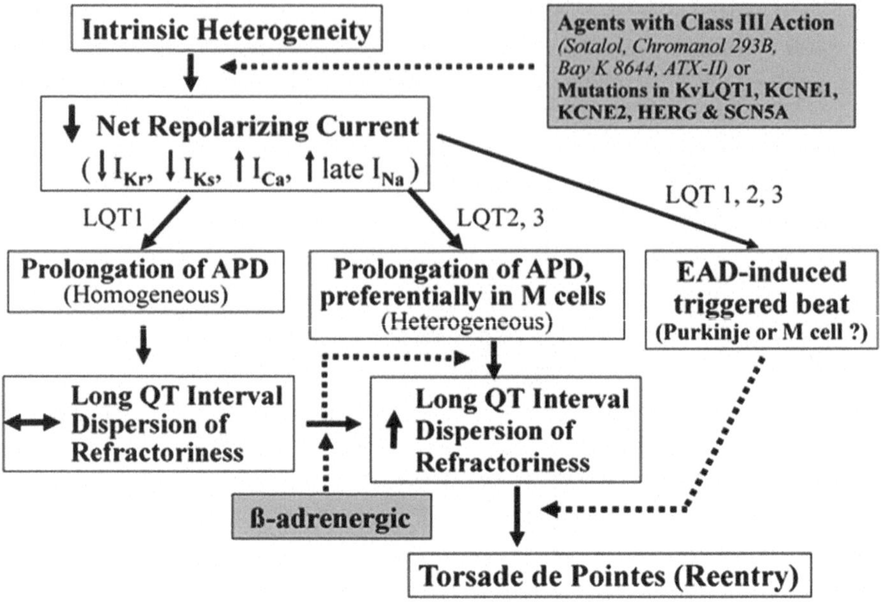

Fig. 3 Cellular and ionic mechanism underlying the development of Torsade de Pointes in the long QT syndrome. *APD* action potential duration, *EAD* early after-depolarization. Modified from Antzelevitch (2007a, b), with permission

2.1.2 J-Wave Syndromes: Brugada and Early Repolarization Syndrome

A prominent J wave is encountered in several life-threatening cardiac arrhythmia syndromes, including the Brugada (BrS) and early repolarization (ERS) syndromes. BrS and ERS differ with respect to the magnitude and lead location of abnormal J waves and are thought to represent a continuous spectrum of phenotypic expression termed J wave syndromes (JWSs). Both are associated with vulnerability to polymorphic ventricular tachycardia (VT) and ventricular fibrillation (VF) leading to SCD in young adults with apparently structurally normal hearts. Both syndromes characteristically display prominent J waves in the ECG that are thought to be a consequence of the presence of a transmural voltage-gradient caused by heterogeneous transmural distribution of I_{to} (Yan and Antzelevitch 1996). Both syndromes are predominantly observed in young males in their third and fourth decades of life, their electrical phenotype can be attenuated by quinidine, isoproterenol, milrinone, cilostazol, and tachypacing, and exacerbated by an increase in the vagal tone, and both can evolve into polymorphic ventricular tachycardia (VT)/ventricular

Fig. 2 (continued) transient increase in TDR (**e**). In LQT3, isoproterenol produced a persistent abbreviation of the QT interval and the APD90 of both M and Epi cells (at both 2 and 10 min), resulting in a persistent decrease in TDR (**f**). *RT* repolarization time $*p < 0.0005$ vs. control; $\dagger p < 0.0005$, $\dagger\dagger p < 0.005$, $\dagger\dagger\dagger p < 0.05$, vs. 293B, d-sotalol (d-Sot) or ATX-II. Modified from Shimizu and Antzelevitch (1997, 1998, 2000), with permission fibrillation (VF) with increased risk for sudden cardiac death (SCD) (Antzelevitch and Yan 2015; Antzelevitch et al. 2016). The region most affected by BrS is the anterior right ventricular outflow tract (RVOT), accounting for why J waves and ST-segment elevation are limited to the right precordial leads. The region most affected in ERS is the inferior wall of the left ventricle (LV), accounting for why the appearance of J waves or early repolarization in the inferior ECG leads is associated with the highest risk for development of arrhythmias and SCD. BrS and ERS have been linked to mutations in genes affecting ion channels leading to an outward shift in the balance of current during the early phase of the epicardial action potential (AP), thus causing accentuation of the AP notch, loss of the AP dome, and leading to the development of phase 2 reentry and polymorphic VT (Fig. 4).

BrS has been associated with variants in 19 different genes, whereas ERS has been associated with variants in seven genes (Table 3). Variants in 10 of the 19 BrS-susceptibility genes and 2 or the 7 ERS-susceptibility genes have been associated with a loss-of-function of I_{Na}. Loss-of-function mutations in *SCN5A* have been identified in 11–28% of probands with BrS in different regions of the world (Kapplinger et al. 2010).

BrS and ERS are strongly male dominant syndromes with a male:female ratio as high as 10:1 in the expression of the disease phenotype in the case of BrS (Antzelevitch and Yan 2010). Higher testosterone-mediated expression of I_{to} in the right ventricular epicardium is thought to be responsible (Antzelevitch 2003;

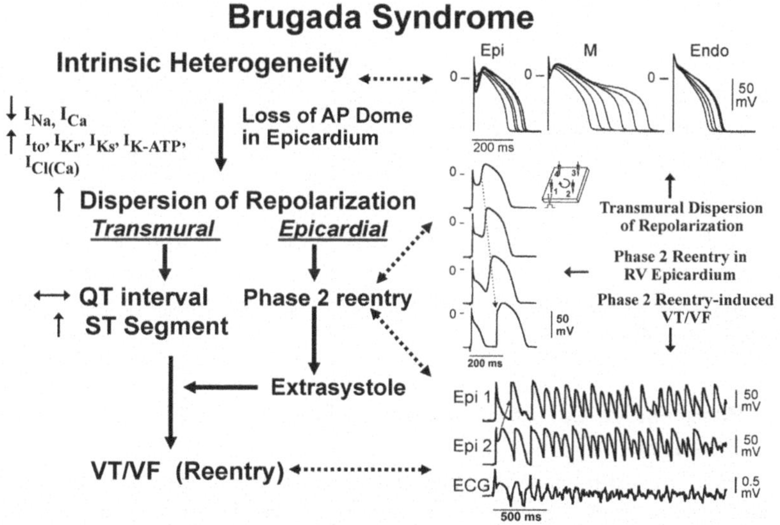

Fig. 4 Proposed mechanism for the Brugada syndrome. An outward shift in the balance of currents serves to amplify existing heterogeneities by causing loss of the action potential dome at some epicardial, but not endocardial sites. A vulnerable window develops as a result of the dispersion of repolarization and refractoriness within epicardium as well as across the wall. Epicardial dispersion leads to the development of phase 2 reentry, which provides the extrasystole that captures the vulnerable window and initiates VT/VF via a circus movement reentry mechanism. Modified from Antzelevitch (2001), with permission

Table 3 Genetics of Brugada and early repolarization syndromes

		Locus	Ion channel	Gene/protein
BrS	BrS1	3p21	$\downarrow I_{Na}$	*SCN5A*, $Na_V1.5$
	BrS2	3p24	$\downarrow I_{Na}$	*GPD1L*
	BrS3	12p13.3	$\downarrow I_{Ca}$	*CACNA1C*,$Ca_V1.2$
	BrS4	10p12.33	$\downarrow I_{Ca}$	*CACNB2b*, $Ca_v\beta2b$
	BrS5	19q13.1	$\downarrow I_{Na}$	*SCN1B*, $Na_V\beta1$
	BrS6	11q13-q14	$\downarrow I_{Ca}$	*KCNE3*. MiRP2
	BrS7	11q23.3	$\downarrow I_{Na}$	*SCN3B*, Navb3
	BrS8	12p11.23	$\uparrow I_{K\text{-}ATP}$	*KCNJ8*, Kir6.1
	BrS9	7q21.11	$\downarrow I_{Ca}$	*CACNA2D1*, $Ca_v\alpha2\delta1$
	BrS10	1p13.2	$\uparrow I_{to}$	*KCND3*, $K_v4.3$
	BrS11	17p13.1	$\downarrow I_{Na}$	*RANGRF*, MOG1
	BrS12	3p21.2-p14.3	$\downarrow I_{Na}$	*SLMAP*
	BrS13	12p12.1	$\uparrow I_{K\text{-}ATP}$	*ABCC9*, SUR2A
	BrS14	11q23	$\downarrow I_{Na}$	*SCN2B*, $Na_vß2$
	BrS15	12p11	$\downarrow I_{Na}$	*PKP2*, Plakophillin2
	BrS16	3q28	$\downarrow I_{Na}$	*FGF12*, FHAF1
	BrS17	3p22.2	$\downarrow I_{Na}$	*SCN10A*, $Na_v1.8$
	BrS18	6q	$\uparrow I_{Na}$	*HEY2 (transcriptional factor)*
	BrS19	1p3633	$\uparrow I_{to}$	*KCNAB2*, *Kv ß2*
ERS	ERS1	12p11.23	$\uparrow I_{K\text{-}ATP}$	*KCNJ8*, Kir6.1
	ERS2	12p13.3	$\downarrow I_{Ca}$	*CACNA1C*, $Ca_v1.2$
	ERS3	10p12.33	$\downarrow I_{Ca}$	*CACNB2b*, $Ca_vß2b$
	ERS4	7q21.11	$\downarrow I_{Ca}$	*CACNA2D1*, $Ca_v\alpha2\delta1$
	ERS5	12p12.1	$I_{K\text{-}ATP}$	*ABCC9*, SUR2A
	ERS6	3p21	$\downarrow I_{Na}$	*SCN5A*, $Na_v1.5$
	ERS7	3p22.2	$\downarrow I_{Na}$	*SCN10A*, $Na_v1.8$

Matsuo et al. 2003; Cordeiro et al. 2008; Ezaki et al. 2010). Barajas-Martinez and coworkers have also presented evidence in support of gender-related differences in transmural distribution of I_{Na} as the basis for the male predominance (Barajas-Martinez et al. 2009).

2.1.3 Progressive Cardiac Conduction Disease (PCCD or Lenegre-Lev Disease)

Familial PCCD is an inherited cardiac disease that may be associated with structural heart disease or may present as a primary electrical disease or channelopathy. In structurally normal hearts it is associated with genetic variants in the ion channel genes *SCN5A*, *SCN1B*, *SCN10A*, *TRPM4*, and *KCNK17*, as well as in genes coding for cardiac connexin proteins (Baruteau et al. 2015). Mutations in genes coding for cardiac transcriptional factors, including *NKX2.5* and *TBX5*, involved in the development of the cardiac conduction system and in cardiac morphogenesis, have been also been implicated in PCCD as well as in various congenital heart defects.

PCCD is clinically characterized by a progressive slow conduction through the His-Purkinje system, with right and/or left bundle branch block and widening of the QRS complex, leading to complete atrio-ventricular node block. PCCD can cause syncope and SCD and for this reason is the most frequent indication for implantation of permanent pacemakers globally (0.5 implantations/1,000 inhabitants/year in developed countries) (Scott et al. 1999). Scott and coworkers first reported a mutation in *SCN5A* that segregated with PCCD in an autosomal dominant manner in a French family in 1999 (Scott et al. 1999). Numerous loss-of-function *SCN5A* mutations, including splice-site, frameshift, nonsense, and missense mutations, have since been identified in association with PCCD (Barc and Bezzina 2014).

2.1.4 Sick Sinus Syndrome (SSS)

SSS is a disorder characterized by the dysfunction of the sinoatrial node. Patients affected by SSS exhibit sinus bradycardia, sinus arrest, and a reduced chronotropic response (Benson et al. 2003; Butters et al. 2010; Abe et al. 2014). Sinus node dysfunction has been associated with mutations in *SCN5A* (Benson et al. 2003; Makita et al. 2005), *HCN4* (Schulze-Bahr et al. 2003; Hategan et al. 2017), *CACNA1D* (Baig et al. 2011), and *GNB2* (Baig et al. 2011; Stallmeyer et al. 2017), which encodes the Gβ2 subunit of the heterotrimeric G-protein complex.

Like PCCD and J Wave syndrome, *SCN5A* mutations causing SSS are associated with loss-of-function of sodium channel current (Arnold et al. 2008). Although *SCN5A* is poorly expressed in central cells in the sinus node, mutations in this gene have a small impact on individual primary peacemaker cells. However, loss-of-function mutations in *SCN5A* reduced excitability in the cells located in the periphery of the sinus node thus slowing conduction and causing conduction block.

2.1.5 Sudden Infant Death Syndrome (SIDS)

SIDS is defined as sudden death of an infant <1-year old without any preceding symptoms. Based on the definition of SIDS, the historical era, demographics, and ethnicity of the population evaluated, SIDS affects ~2 infants per 1,000 live births, with a peak incidence between the ages of 2 and 5 months (Kinney and Thach 2009; Tester and Ackerman 2012). *SCN5A* mutations reported using candidate gene approaches account for 2–10% of SIDS cases (Ackerman et al. 2001). Although functional electrophysiological studies report a wide range of effects of *SCN5A* variants in SIDS cases, a gain-of-function mechanism is the most commonly associated (Schwartz et al. 2000). The first direct molecular link between SIDS and a cardiac arrhythmia was reported by Schwartz and coworkers and involved a gain-of-function missense mutation in *SCN5A* (S941N) that prolonged the QT interval via an increase in late I_{Na} (Schwartz et al. 2000).

2.1.6 Atrial Fibrillation (AF)

AF is the most frequent cardiac arrhythmia encountered and diagnosed in the clinic. It is characterized as a very rapid atrial activation and rapid and irregular ventricular rates (Brugada and Kaab 2008). Approximately 35% of AF cases have a positive family history suggesting a heritable basis for the arrhythmia (Wyse et al. 2014).

The Framingham Study showed that ~27% of individuals with AF have a first-degree relative with AF confirmed by ECG, and that familial AF is associated with a 40% increased risk of AF for other family members over a subsequent 8-year period, even after adjustment for established AF clinical risk factors (Lubitz et al. 2010). Linkage analysis and candidate gene approach have been used to identify mutations in *SCN5A* and its beta subunits (*SCN1B*, *SCN2B*, and *SCN3B*) in the AF population (McNair et al. 2004; Olson et al. 2005a, b; Laitinen-Forsblom et al. 2006; Ellinor et al. 2008; Makiyama et al. 2008a, b; Olesen et al. 2011a, b).

To date, rare variants in 32 genes have been associated with AF (Hayashi et al. 2017). Nineteen encode ion channel proteins and of these six affect sodium channel activity. Rare variants associated with AF exert a wide spectrum of effects on the biophysical properties of the sodium channel resulting in a loss-of-function with a consequent shortening of the APD (Ellinor et al. 2008; Watanabe et al. 2009a, b) as well as a gain-of-function (Makiyama et al. 2008a, b; Li et al. 2009). *SCN5A* and *SCN1B* variants are reported to show either a gain-of-function or a loss-of-function of I_{Na} (Olson et al. 2005a, b; Darbar et al. 2008; Makiyama et al. 2008a, b; Olesen et al. 2012a, b, 2012c; Hayashi et al. 2015), whereas rare variants in *SCN2B*, *SCN3B*, and *SCN4B* show loss-of-function effects (Watanabe et al. 2009a, b; Wang et al. 2010; Olesen et al. 2011a, b). Gain-of-function of I_{Na}, particularly of late I_{Na}, can promote ectopic activity and increase dispersion of repolarization and refractoriness, whereas loss-of-function can promote AF by abbreviating the refractory period and slowing of conduction, which provide the substrate for the development of reentrant arrhythmias. Rare variants in *SCN10A*, the gene that encodes $Na_v1.8$, have also been reported to be associated with AF (Savio-Galimberti et al. 2014; Jabbari et al. 2015).

2.1.7 Dilated Cardiomyopathy Disease (DCM)

DCM is a disorder characterized by ventricular dilatation and impaired systolic function, which usually results in the development of heart failure. Valvular heart disease, excess alcohol ingestion, hypertension, pregnancy, and infections are the most common underlying etiologic factors. Idiopatic DCM (IDCM) represents a subgroup of DCM patients where the etiology has not been determined, and where genetic, autoimmune, viral, and metabolic causes have been implicated as potential pathophysiological mechanisms of the disease. Approximately 40% of DCM has a positive family history for the disease displaying autosomal dominant inheritance, although X-linked, autosomal recessive, and mitochondrial inheritance have also been described as well, although less frequently. Mutations in numerous candidate genes (>40) have been identified in patients with IDCM. Rare genetic variants associated with DCM affect a range of diverse cellular structures and functions. Truncating variants in *titin* represent the single largest genetic cause of IDCM (Tayal et al. 2017). Two main forms of IDCM have been described: DCM with and without conduction system disease. Linkage analysis has been used to identify *SCN5A* mutations in the combined form. The first *SCN5A* mutation implicated in IDCM (D1275N) was independently discovered by McNair and coworkers and Olson and coworkers (McNair et al. 2004; Olson et al. 2005a, b) in a DCM family that was originally reported by Greenlee et al. in 1986 (Greenlee et al. 1986). Other *SCN5A* mutations (T220I, D1595H, 2550-2551 instTG, R814W) have been reported in IDCM patients (Olson et al. 2005a, b), in most cases these are loss-of-function mutations. The mechanisms by which sodium channel variants lead to DCM remain poorly understood.

2.1.8 Multifocal Ectopic Purkinje-Related Premature Contractions (MEPPC)

MEPPC is a cardiac arrhythmia recently linked to variants in *SCN5A* mutations. It was previously associated with both AF and DCM. Five families have been described with MEPPC. All presented with the same missense mutation in *SCN5A* (p.R222Q) involving the voltage sensor of the sodium channel in domain I causing an increase in the excitability of the channel (Laurent et al. 2012; Mann et al. 2012, Nair et al. 2012). The mutation is a rare variant, inherited as a dominant trait, with complete penetrance (McNair et al. 2011; Laurent et al. 2012; Mann et al. 2012; Nair et al. 2012).

2.1.9 Overlap Syndromes

SCN5A mutations can lead to a wide range of phenotypes. Although these entities may occur as isolated syndromes, in most cases an exhaustive investigation reveals their involvement in multiple "overlap syndromes" (Fig. 5). Loss-of-function *SCN5A* rare variants can be associated with multiple overlapping clinical manifestations including progressive conduction disease, BrS, AF, SSS, and DCM. Gain-of-function rare variants can be associated with LQTS, AF, and MEPCC. The particular phenotype expressed often depends on the genetic background, including modulating rare and common variants, as well as epigenetic and environmental factors.

2.2 Common *SCN5A* EXONIC Variants

Genetic screening of the exonic regions of *SCN5A* in patients with cardiac arrhythmias as well as in control populations often reveals the common occurrence of missense variants (polymorphisms) in the general population (common variants). The interest in these variants derives from the fact that they can potentially modulate the severity of disease phenotypes. This modulation is possible either because they affect the biophysics of the wild-type channel when they occur in the *trans* configuration (different allele from the one with the mutation) or because they can attenuate or exacerbate the biophysical behavior of the mutated channel when they occur in the *cis* configuration (on the same allele with the mutation). Viswanathan and coworkers were the first to introduce the concept that interaction between polymorphisms and mutations can exert important effects on the functional consequences of a mutation (Viswanathan et al. 2003). This has been referred as "trans-complementation effect" (Barc and Bezzina 2014), and there are several examples reported for *SCN5A* variants where this interaction has been confirmed (Viswanathan et al. 2003; Poelzing et al. 2006).

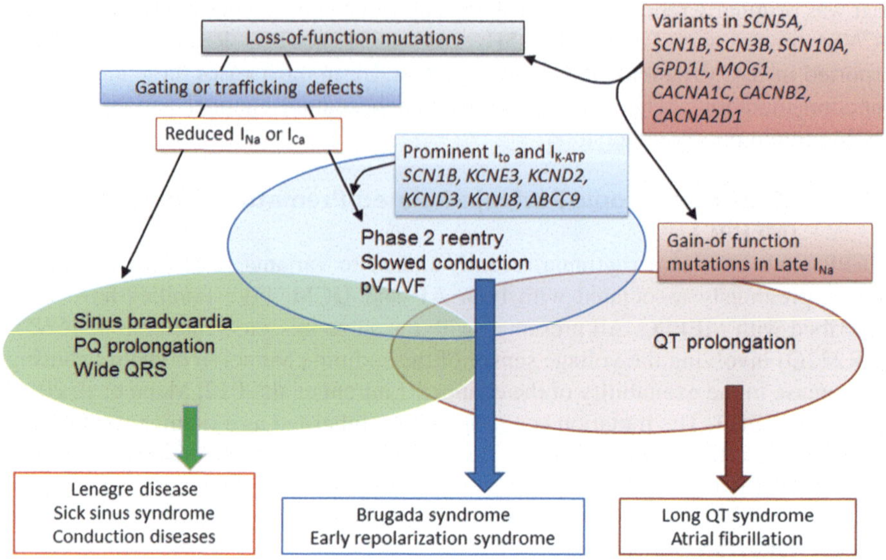

Fig. 5 Schematic showing overlap between Brugada and other inherited cardiac arrhythmia syndromes resulting from genetic defects secondary to loss-of-function of sodium (I_{Na}) and/or calcium (I_{Ca}) channel current. In the absence of prominent I_{to} or $I_{K\text{-}ATP}$, loss-of-function mutations in the inward currents result in various manifestations of conduction disease. In the presence of prominent I_{to} or $I_{K\text{-}ATP}$, loss-of-function mutations in inward currents cause conduction disease as well as the J wave syndromes (Brugada and Early repolarization syndromes). Early repolarization syndrome is believed to be caused by loss-of-function mutations of inward current in the presence of prominent I_{to} in certain regions of the left ventricle (LV), particularly the inferior wall of the LV. The genetic defects that contribute to BrS and ERS can also contribute to the development of long QT and conduction system disease, in some cases causing multiple expression of these overlap syndromes. In some cases, structural defects contribute to the phenotype. Modified from Antzelevitch et al. (2016), with permission

2.3 Common *SCN5A* Intronic Variants (*SCN5A-SCN10A* Interaction/Regulation)

Genetic screening studies (candidate gene approach, GWAS) conducted in control and diseased (BrS) populations have uncovered the role of genetic variation in the noncoding (intronic) regions of *SCN5A* in the regulation of cardiac electrophysiology. These variants (single nucleotide polymorphisms, SNP) may modulate the clinical phenotype by affecting the Nav 1.5 expression level as well as its biophysical properties of either the wild-type (WT) or the mutated channel. Bezzina and coworkers were the first to examine the first noncoding region in the *SCN5A* gene. They identified a haplotype including multiple polymorphisms in the promoter region of the *SCN5A* that are common in an Asian population (Bezzina et al. 2006).

The noncoding region surrounding *SCN5A* has also been studied in GWAS studies conducted in the general population. These studies set out to identify common variants that can modulate conduction and repolarization parameters on the ECGs (Kolder et al. 2012; Marsman et al. 2014). Several common variants in *SCN5A* and in and around *SCN10A* have been associated with changes in PR interval, QRS duration, and QT interval corrected (QTc) in the general population (Newton-Cheh et al. 2009; Pfeufer et al. 2009, 2010; Chambers et al. 2010; Holm et al. 2010; Sotoodehnia et al. 2010a, b; Smith et al. 2011).

Several groups have provided important information about the presence of regulatory elements (e.g., enhancers) within the noncoding regions of the genome. Also, much information has emerged concerning the potential interaction between regulatory regions between genes, like the interaction between *SCN10A* and *SCN5A* and its regulatory consequences in the context of cardiac arrhythmias like BrS (Arnolds et al. 2012; Van den Boogaard et al. 2012). Christoffels, Nobrega, Barnett, and Moscowitz groups have shown that the G to A nucleotide change identified at SNP rs6801957 is located in a consensus T-box transcription factor binding site within a cardiac enhancer located in the *SCN10A* gene. This change from G to A has at least three effects:

1. It reduces the T-box binding to the enhancer,
2. It affects the stimulation and repression by TBX5 and TBX3, respectively, of a reporter in in vitro assays, and
3. It reduces the activity of the enhancer in vivo (Van den Boogaard et al. 2012).

Moreover, the haplotype tagged by rs6801957 was associated with reduced *SCN5A* expression in human hearts (Van den Boogaard et al. 2012).

3 Summary

Consistent with the central role of *SCN5A* in cardiac electrophysiology, we describe a series of cardiac arrhythmia syndromes in which variants in this gene as well as the genes that encode its protein partners play a key role in the pathogenesis of disease. This pertains not only to the rare variants identified in coding (exonic) regions but also for the variants identified in the noncoding (intronic) regions; the latter generally involving intricate gene-to-gene interactions.

Conflict of Interest None

Financial Support Supported by NIH grant HL47678 (CA) and the Wistar and Martha Morris Fund (CA).

References

Abe K et al (2014) Sodium channelopathy underlying familial sick sinus syndrome with early onset and predominantly male characteristics. Circ Arrhythm Electrophysiol 7(3):511–517

Ackerman MJ, Siu BL, Sturner WQ et al (2001) Postmortem molecular analysis of SCN5A defects in sudden infant death syndrome. JAMA 286:2264–2269

Allouis M, Le Bouffant F, Wilders R, Peroz D, Schott JJ, Noireaud J, Le Marec H, Merot J, Escnade D, Baro I (2006) 14-3-3 is a regulator of the cardiac voltage-gated sodium channel Nav1.5. Circ Res 98:1538–1546

Antzelevitch C (2001) The Brugada syndrome: diagnostic criteria and cellular mechanisms. Eur Heart J 22(5):356–363

Antzelevitch C (2003) Androgens and male predominance of the Brugada syndrome phenotype. Pacing Clin Electrophysiol 26(7 Pt 1):1429–1431

Antzelevitch C (2007a) Ionic, molecular, and cellular bases of QT-interval prolongation and torsade de pointes. Europace 9(s4):iv4–iv15

Antzelevitch C (2007b) The role of spatial dispersion of repolarization in inherited and acquired sudden cardiac death syndromes. Am J Physiol Heart Circ Physiol 293(4):H2024–H2038

Antzelevitch C, Yan GX (2010) J wave syndromes. Heart Rhythm 7(4):549–558

Antzelevitch C, Yan GX (2015) J wave syndrome: Brugada and early repolarization syndromes. Heart Rhythm 12(8):1852–1866

Antzelevitch C, Nesterenko V, Shryock JC, Rajamani S, Song Y, Belardinelli L (2014) The role of late I_{Na} in development of cardiac arrhythmias. Handb Exp Pharmacol 221:137–168

Antzelevitch C, Yan GX, Ackerman MJ, Borggrefe M, Corrado D, Guo J, Gussak I, Hasdemir C, Horie M, Huikuri H, Ma C, Morita H, Nam GB, Sacher F, Shimizu W, Viskin S, Wilde AA (2016) J-wave syndromes expert consensus conference report: emerging concepts and gaps in knowledge. Heart Rhythm 13(10):e295–e324

Arnold SV, Morrow DA, Wang K, Lei Y, Mahoney EM, Scirica BM, Braunwald E, Cohen DJ, Investigators M-T (2008) Effects of ranolazine on disease-specific health status and quality of life among patients with acute coronary syndromes results from the MERLIN-TIMI 36 randomized trial. Circ Cardiovasc Qual Outcomes 1(2):107–115

Arnolds DE, Liu F, Fahrenbach JP, Kim GH, Schillinger KJ, Smemo S, McNally EM, Nobrega MA, Patel VV, Moskowitz IP (2012) TBX5 drives Scn5a expression to regulate cardiac conduction system function. J Clin Invest 122:2509–2518

Ashpole NM, Herren AW, Ginsburg KS, Brogan JD, Johnson DE, Cummins TR, Bers DM, Hudmon A (2012) Ca2+/calmodulin-dependent protein kinase II (CaMKII) regulates cardiac sodium channel Nav1.5 gating by multiple phosphorylation sites. J Biol Chem 287: 19856–19869

Baig SM, Koschak A, Lieb A, Gebhart M, Dafinger C, Nurnberg G, Ali A, Ahmad I, Sinnegger-Brauns MJ, Brandt N, Engel J, Mangoni ME, Farooq M, Khan HU, Nurnberg P, Striessnig J, Bolz HJ (2011) Loss of Ca(v)1.3 (CACNA1D) function in a human channelopathy with bradycardia and congenital deafness. Nat Neurosci 14(1):77–84

Barajas-Martinez H, Haufe V, Chamberland C, Blais Roy MJ, Fecteau MH, Cordeiro JM, Dumaine R (2009) Larger dispersion of I_{Na} in female dog ventricle as a mechanism for gender-specific incidence of cardiac arrhythmias. Cardiovasc Res 81(1):82–89

Barc J, Bezzina C (2014) Role of rare and common genetic variation in SCN5A in cardiac electrical function and arrhythmia. Card Electrophysiol Clin 6:665–677

Baruteau AE, Probst V, Abriel H (2015) Inherited progressive cardiac conduction disorders. Curr Opin Cardiol 30(1):33–39

Bennett PB, Yazawa K, Makita N, George AL Jr (1995) Molecular mechanism for an inherited cardiac arrhythmia. Nature 376(683–685):683

Benson DW, Wang DW, Dyment M, Knilans TK, Fish FA, Strieper MJ, Rhodes TH, George AL Jr (2003) Congenital sick sinus syndrome caused by recessive mutations in the cardiac sodium channel gene (SCN5A). J Clin Invest 112:1019–1028

Bezzina CR, Shimizu W, Yang P, Koopmann TT, Tanck MWT, Miyamoto Y, Kamakura S, Roden DM, Wilde AAM (2006) Common sodium channel promoter haplotype is asian subjects underlies variability in cardiac conduction. Circulation 113:338–344

van den Boogaard M, Barnett P, Christoffels VM (2014) From GWAS to function: genetic variation in sodium channel gene enhancer influences electrical patterning. Trends Cardiovasc Med 24(3):99–104

Brugada R, Kaab S (2008) Atrial fibrillation: from bench to bedside. In: Natale A, Jalife J (eds) Chapter 6 Humana Press, Totowa, NJ, pp 69–76

Butters TD, Aslanidi OV, Inada S, Boyett MR, Hancox JC, Lei M, Zhang H (2010) Mechanistic links between Na+ channel (SCN5A) mutations and impaired cardiac pacemaking in sick sinus syndrome. Circ Res 107(1):126–137

Catterall W (2014) Structure and function of voltage-gated sodium channels at atomic resolution. Exp Physiol 99(1):35–51

Catterall WA, Perez-Reyes E, Snutch TP, Striessnig J (2005) International Union of Pharmacology. XLVIII. Nomenclature and structure-function relationships of voltage-gated calcium channels. Pharmacol Rev 57(4):411–425

Chambers JC, Zhao J, Terraciano CMN, Bezzina CR, Zhang W, Kaba R, Navaratnarajah M, Lotlikar A, Sehmi JS, Kooner MK, Deng G, Siedlecka U, Parasramka S, El-Hamamsy I, Wass MN, Dekker LR, de Jong JS, Sternberg MJ, McKenna W, Severs NJ, de Silva R, Wilde AA, Anand P, Yacoub M, Scott J, Elliott P, Wood JN, Kooner JS (2010) Genetic variation in SCN10A influences cardiac conduction. Nat Genet 42:149–152

Cordeiro JM, Mazza M, Goodrow R, Ulahannan N, Antzelevitch C, Di Diego JM (2008) Functionally distinct sodium channels in ventricular epicardial and endocardial cells contribute to a greater sensitivity of the epicardium to electrical depression. Am J Physiol Heart Circ Physiol 295(1):H154–H162

Darbar D, Kannankeril PJ, Donahue BS, Kucera G, Stubblefield T, Haines JL, George AL Jr, Roden DM (2008) Cardiac sodium channel (*SCN5A*) variants associated with atrial fibrillation. Circulation 117(15):1927–1935

Dover K, Solinas S, D'Angelo E, Goldfarb M (2010) Long-term inactivation particle for voltage-gated sodium channels. J Physiol Lond 588:3695–3711

Ellinor PT, Nam EG, Shea MA et al (2008) Cardiac sodium channel mutations in atrial fibrillation. Heart Rhythm 5(1):99–105

Ezaki K, Nakagawa M, Taniguchi Y, Nagano Y, Teshima Y, Yufu K, Takahashi N, Nomura T, Satoh F, Mimata H, Saikawa T (2010) Gender differences in the ST segment: effect of androgen-deprivation therapy and possible role of testosterone. Circ J 74(11):2448–2454

Facer P, Punjabi PP, Abrari A, Kaba RA, Severs NJ, Chambers J, Kooner JS, Anand P (2011) Localisation of SCN10A gene product Na(v)1.8 and novel pain-related ion channels in human heart. Int Heart J 52(3):146–152

Furukawa T, Ono Y, Tsuchiya H, Katayama Y, Bang ML, Labeit D, Labeit S, Inagaki N, Gregorio CC (2001) Specific interaction of the potassium channel beta subunit minK with the sarcomeric protein T-cap suggests a T-tubule-myofibril linking system. J Mol Biol 313:775–784

Garrido JJ, Fernandes F, Moussif A, Fache MP, Giraud P, Dargent B (2003) Dynamic compartmentalization of the voltage-gated sodium channels in axons. Biol Cell 95:437–445

Gavillet B, Rougier JS, Domenighetti AA, Behar R, Boixel C, Ruchat P, Lehr HA, Pedrazzini T, Abriel H (2006) Cardiac sodium channel Nav1.5 is regulated by a multiprotein complex composed of syntrophins and dystrophin. Circ Res 99:407–414

Gellens ME, George AL Jr, Chen LQ, Chahine M, Horn R, Barchi RL, Kallen RG (1992a) Primary structure and functional expression of the human cardiac tetrodotoxin-insensitive voltage-dependent sodium channel. Proc Natl Acad Sci U S A 89(2):554–558

Gellens ME, George AL Jr, Chen LQ, Chahine M, Horn R, Barchi RL, Kallen RG (1992b) Primary structure and functional expression of the human cardiac tetrodotoxin-insensitive voltage-dependent sodium channel. Proc Natl Acad Sci U S A 89(2):554–558

George AL Jr, Varkony T, Drabkin HA, Han J, Knops JF, Finley WH, Brown GB, Ward DC, Haas M (1995) Assignment of the human heart tetrodotoxin-resistant voltage-gated Na+ channel alpha-subunit gene (SCN5A) to band 3p21. Cytogenet Cell Genet 68(1-2):67–70

Greenlee PR, Anderson JL, Lutz JR, Lindsay AE, Hagan AD (1986) Familial automaticity-conduction disorder with associated cardiomyopathy. West J Med 144:33–41

Hategan L, Csanyi B, Ordog B, Kakonyi K, Tringer A, Kiss O, Orosz A, Saghy L, Nagy I, Hegedus Z, Rudas L, Szell M, Varro A, Forster T, Sepp R (2017) A novel "splice site" HCN4 gene mutation, c.1737+1 G>T, causes familial bradycardia, reduced heart rate response, impaired chronotropic competence and increased short-term heart rate variability. Int J Cardiol 241:364–372

Haworth RS, Cuello F, Herron TJ, Franzen G, Kentish JC, Gautel M, Avkiran M (2004) Protein kinase D is a novel mediator of cardiac troponin I phosphorylation and regulates myofilament function. Circ Res 95:1091–1099

Hayashi K, Konno T, Tada H, Tani S, Liu L, Fujino N, Nohara A, Hodatsu A, Tsuda T, Tanaka Y, Kawashiri MA, Ino H, Makita N, Yamagishi M (2015) Functional characterization of rare variants implicated in susceptibility to lone atrial fibrillation. Circ Arrhythm Electrophysiol 8(5):1095–1104

Hayashi K, Tada H, Yamagishi M (2017) The genetics of atrial fibrillation. Curr Opin Cardiol 32(1):10–16

Holm H, Gudbjartsson D, Arnar DO et al (2010) Several common variants modulate heart rate, PR interval and QRS duration. Nat Genet 42(2):117–122

Hu D, Barajas-Martinez H, Burashnikov E, Springer M, Wu Y, Varro A, Pfeiffer R, Koopmann TT, Cordeiro JM, Guerchicoff A, Pollevick GD, Antzelevitch C (2009) A mutation in the beta 3 subunit of the cardiac sodium channel associated with Brugada ECG phenotype. Circ Cardiovasc Genet 2(3):270–278

Hu D, Barajas-Martinez H, Burashnikov E, Pfeiffer R, Schimpf R, Wolpert C, Borggrefe M, Antzelevitch C (2010) A novel mutation in SCN1BB linked to Brugada syndrome by modulating $Na_{v1.5}$ and $K_{V4.3}$ current. Heart Rhythm 7:S320

Hu D, Barajas-Martinez H, Medeiros-Domingo A, Crotti L, Tester DJ, Veltmann C, Schimpf R, Pfeiffer R, Dezi F, Liu Y, Burashnikov E, Giudicessi JR, Ye D, Wolpert C, Borggrefe M, Schwartz P, Ackerman MJ, Antzelevitch C (2012) Novel mutations in the sodium channel 2 subunit gene (SCN2B) associated with Brugada syndrome and atrial fibrillation. Circulation 126(21 Supplement):A16521

Hu D, Barajas-Martinez H, Pfeiffer R, Dezi F, Pfeiffer J, Buch T, Betzenhauser MJ, Belardinelli L, Kahlig KM, Rajamani S, DeAntonio HJ, Myerburg RJ, Ito H, Deshmukh P, Marieb M, Nam GB, Bhatia A, Hasdemir C, Haissaguerre M, Veltmann C, Schimpf R, Borggrefe M, Viskin S, Antzelevitch C (2014) Mutations in *SCN10A* are responsible for a large fraction of cases of Brugada syndrome. J Am Coll Cardiol 64(1):66–79

Jabbari J, Olesen MS, Yuan L, Nielsen JB, Liang B, Macri V, Christophersen IE, Nielsen N, Sajadieh A, Ellinor PT, Grunnet M, Haunso S, Holst AG, Svendsen JH, Jespersen T (2015) Common and rare variants in SCN10A modulate the risk of atrial fibrillation. Circ Cardiovasc Genet 8(1):64–73

Jenkins SM, Bennett V (2001) Ankyrin-G coordinates assembly of the spectrin-based membrane skeleton, voltage-gated sodium channels, and L1 CAMs at Purkinje neuron initial segments. J Cell Biol 155:739–746

Jespersen T, Gavillet B, van Bemmelen MX, Cordonier S, Ma T, Staub O, Abriel H (2006) Cardiac sodium channel Nav1.5 interacts with and is regulated by the protein tyrosine phosphatase PTPH1. Biochem Biophys Res Commun 348:1455–1462

Kapplinger JD, Tester DJ, Alders M, Benito B, Berthet M, Brugada J, Brugada P, Fressart V, Guerchicoff A, Harris-Kerr C, Kamakura S, Kyndt F, Koopmann TT, Miyamoto M, Pfeiffer R, Pollevick GD, Probst V, Zumhagen S, Vatta M, Towbin JA, Shimizu W, Schulze-Bahr E, Antzelevitch C, Salisbury BA, Guicheney P, Wilde AAM, Brugada R, Schott JJ, Ackerman MJ (2010) An international compendium of mutations in the SCN5A encoded cardiac sodium channel in patients referred for Brugada syndrome genetic testing. Heart Rhythm 7(1):33–46

Kaufmann SG, Westenbroek R, Maass AH, Lange V, Renner A, Wischmeyer E, Bonz A, Muck J, Ertl G, Catterall WA, Scheuer T, Maier SK (2013) Distribution and function of sodium channel subtypes in human atrial myocardium. J Mol Cell Cardiol 61:133–141
Kinney HC, Thach BT (2009) The sudden infant death syndrome. N Engl J Med 361(8):795–805
Knoll R, Hoshijima M, Hoffman MH, Person V, Lorenzen-Schmidt I, Bang ML, Hayashi T, Shiga N, Yasukawa H, Schaper W, McKenna W, Yokoyama M, Schork JN, Omens HJ, McCulloch DA, Kimura A, Gregorio CC, Poller W, Schaper J, Schultheiss HP, Chien KR (2002) The cardiac mechanical stretch sensor machinery involves a Z disc complex that is defective in a subset of human dilated cardiomyopathy. Cell 111:943–955
Ko SH, Lenkowski PW, Lee HC, Mounsey JP, Patel MK (2005) Modulation of Nav1.5 by beta 1 and beta 3 subunit co-expression in mammalian cells. Pflugers Arch 449:403–412
Kojic S, Medeot E, Guccione E, Krmac H, Zara I, Martinelli V, Valle G, Faulkner G (2004) The Ankrd2 protein, a link between the sarcomere and the nucleus in skeletal muscle. J Mol Biol 339:313–325
Kolder IC, Tanck MW, Bezzina CR (2012) Common genetic variation modulating cardiac ECG parameters and susceptibility to sudden cardiac death. J Mol Cell Cardiol 52:620–629
Laitinen-Forsblom PJ, Mäkynen P, Mäkynen H et al (2006) SCN5A mutation associated with cardiac conduction defect and atrial arrhythmias. J Cardiovasc Electrophysiol 17:480–485
Laurent G, Saal S, Amarouch MY, Beziau DM, Marsman RFJ, Faivre L, Barc J, Dina C, Bertaux G, Barthez O, Thauvin-Robinet C, Charron P, Fressart V, Maltret A, Villain E, Baron E, Merot J, Turpault R, Coudiere Y, Charpentier F, Schott J-J, Loussouarn G, Wilde AAM, Wolf J-E, Baro I, Kyndt F (2012) Multifocal ectopic Purkinje-related premature contractions: a new SCN5A-related cardiac channelopathy. J Am Coll Cardiol 60(2):144–156
Lemaillet G, Walker B, Lambert S (2003) Identification of a conserved ankyrin-binding motif in the family of sodium channel alpha subunits. J Biol Chem 278:27333–27339
Li Q, Huang H, Liv G, Lam K, Rutberg J, Green MS, Birnie DH, Lemery R, Chahine M, Gollob MH (2009) Gain-of-function mutation of Nav1.5 in atrial fibrillation enhances excitability and lowers the threshold for AP firing. Biochem Biophys Res Commun 380:132–137
Li Z, Ai T, Samani K, Xi Y, Tzeng HP, Xie M, Wu S, Ge S, Taylor MD, Dong JW, Cheng J, Ackerman MJ, Kimura A, Sinagra G, Brunelli L, Faulkner G, Vatta M (2010) A ZASP missense mutation, S196L, leads to cytoskeletal and electrical abnormalities in a mouse model of cardiomyopathy. Circ Arrhythm Electrophysiol 3:646–656
Lu T, Lee HC, Kabat JA, Shibata EF (1999) Modulation of rat cardiac sodium channel by the stimulatory G protein alpha subunit. J Physiol 518:371–384
Lubitz SA, Yin X, Fontes JD, Magnani JW, Rienstra M, Pai M, Villalon ML, Vasan RS, Pencina MJ, Levy D, Marson MG, Ellinor PT, Benjamin EJ (2010) Association between familial atrial fibrillation and risk of new-onset atrial fibrillation. JAMA 304(20):2263–2269
Maier SK, Westenbroek RE, McCormick KA, Curtis R, Scheuer T, Catterall WA (2004) Distinct subcellular localizationof different sodium channel alpha and beta subunits in single ventricular myocytes from mouse hearts. Circulation 109:1421–1427
Makita N, Sasaki K, Groenewegen WA, Yokota T, Yokoshiki H, Murakami T, Tsutsui H (2005) Congenital atrial standstill associated with coinheritance of a novel SCN5A mutation and connexin 40 polymorphisms. Heart Rhythm 2:1128–1134
Makiyama T, Akao M, Shizuta S, Doi T, Nishiyama K, Oka Y, Ohno S, Nishio Y, Tsuji K, Itoh H, Kimura T, Kita T, Horie M (2008a) A novel SCN5A gain-of-function mutation M1875T associated with familial atrial fibrillation. J Am Coll Cardiol 52(16):1326–1334
Makiyama T, Akao M, Shizuta S, Doi T, Nishiyama K, Oka Y, Ohno S, Nishio Y, Tsuji K, Itoh H, Kimura T, Kita T, Horie M (2008b) A novel SCN5A gain-of-function mutation M1875T associated with familial atrial fibrillation. J Am Coll Cardiol 52:1326–1334
Malhotra DJ, Chen C, Rivolta I, Abriel H, Malhotra R, Mattei LN, Brosius FC, Kass RS, Isom LL (2001) Characterization of sodium channel alpha and beta subunits in rat and mouse cardiac myocytes. Ciculation 103:1303–1310

Mann SA, Castro M, Ohanian M, Ohanian M, Guo G, Zodgekar P, Sheu A, Stockhammer K, Thompson T, Playford D, Subbiah R, Kuchar D, Aggarwal A, Vandenberg JI, Fatkin D (2012) R222Q SCN5A mutation is associated with reversible ventricular ectopy and dilated cardiomyopathy. J Am Coll Cardiol 60(16):1566–1573

Marsman RF, Tan HL, Bezzina CR (2014) Genetics of sudden cardiac death caused by ventricular arrhythmias. Nat Rev Cardiol 11(2):96–111

Matsuo K, Akahoshi M, Seto S, Yano K (2003) Disappearance of the Brugada-type electrocardiogram after surgical castration: a role for testosterone and an explanation for the male preponderance? Pacing Clin Electrophysiol 26(7 Pt 1):1151–1153

Mayans O, van der Ven PF, Wilm M, Mues A, Young P, Furst DO, Wilmanns M, Gautel M (1998) Structural basis for activation of the titin kinase domain during myofibrillogenesis. Nature 395: 863–869

Mazzone A, Strege PR, Tester DJ, Bernard CE, Faulkner G, de Giorgio R, Makielski JC, Stanghellini V, Gibbons SJ, Ackerman MJ, Farrugia G (2008) A mutation in telethonin alters Nav1.5 function. J Biol Chem 283:16537–16544

McEwen DP, Isom LL (2004) Heterophilic interactions of sodium channel beta 1 subunits with axonal and glial cell adhesion molecules. J Biol Chem 279:52744–52752

McNair WP, Ku L, Taylor MR, Fain PR, Dao D, Wolfer E, Mestroni L (2004) SCN5A mutations associated with dilated cardiomyopathy, conduction disorder, and arrhythmia. Circulation 110: 2163–2167

McNair WP, Sinagra G, Taylor MR et al (2011) SCN5A mutations associate with arrhythmic dilated cardiomyopathy and commonly localize to the voltage-sensing mechanism. J Am Coll Cardiol 57(21):2160–2168

Meadows LS, Isom LL (2005) Sodium channels as macromolecular complexes: implications for inherited arrhythmia syndromes. Cardiovasc Res 67:448–458

Medeiros-Domingo A, Kaku T, Tester DJ, Iturralde-Torres P, Itty A, Ye B, Valdivia C, Ueda K, Canizales-Quinteros S, Tusie-Luna MT, Makielski JC, Ackerman MJ (2007) SCN4B-encoded sodium chanel beta 4 subunit in congenital long-QT syndrome. Circulation 116:134–142

Mohler PJ, Rivolta I, Napolitano C, LeMaillet G, Lambert S, Priori SG, Bennett V (2004) Nav1.5 E1053K mutation causing Brugada syndrome blocks binding to ankyrin-G and expression of Nav1.5 on the surface of cardiomyocytes. Proc Natl Acad Sci U S A 101:17533–17538

Nair K, Pekhletski R, Harris L et al (2012) Escape capture bigeminy: phenotypic marker of cardiac sodium channel voltage sensor mutation R222Q. Heart Rhythm 9(10):1681–1688

Newton-Cheh C, Eijgelsheim M, Rice KM et al (2009) Common variants at ten loci influence QT interval duration in the QTGEN study. Nat Genet 41:399–406

Olesen MS, Jespersen T, Nielsen JB, Liang B, Moller DV, Hedley P, Christiansen M, Varro A, Olesen SP, Haunso S, Schmitt N, Svendsen JH (2011a) Mutations in sodium channel {beta}-subunit SCN3B are associated with early-onset lone atrial fibrillation. Cardiovasc Res 89(4): 786–793

Olesen MS, Jespersen T, Nielsen JB, Liang B, Moller DV, Hedley P, Christiansen M, Varro A, Olesen SP, Haunso S, Schmitt N, Svendsen JH (2011b) Mutations in sodium channel beta subunit SCN3B are associated with early-onset lone atrial fibrillation. Cardiovasc Res 89: 786–793

Olesen MS, Yuan L, Liang B, Holst AG, Nielsen N, Nielsen JB, Hedley PL, Christiansen M, Olesen SP, Haunso S, Schmitt N, Jespersen T, Svendsen JH (2012a) High prevalence of long QT syndrome-associated SCN5A variants in patients with early-onset lone atrial fibrillation. Circ Cardiovasc Genet 5(4):450–459

Olesen MS, Holst A, Svendsen JH, Haunso S, Tfelt-Hansen J (2012b) SCN1Bb R214Q found in 3 patients: 1 with Brugada syndrome and 2 with lone atrial fibrillation. Heart Rhythm 9: 770–773

Olesen MS, Holst AG, Svendsen JH, Haunso S, Tfelt-Hansen J (2012c) SCN1Bb R214Q found in 3 patients: 1 with Brugada syndrome and 2 with lone atrial fibrillation. Heart Rhythm 9(5): 770–773

Olson TM, Michels VV, Ballew JD, Reyna SP, Karst ML, Herron KJ, Horton SC, Rodeheffer RJ, Anderson JL (2005a) Sodium channel mutations and susceptibility to heart failure and atrial fibrillation. JAMA 293(4):447–454

Olson TM, Michels V, Ballew JD, Reyna SP, Karst ML, Herron KJ, Horton SC, Rodeheffer RJ, Anderson JL (2005b) Sodium channel mutations and susceptibility to heart failure and atrial fibrillation. JAMA 293:447–454

Patino GA, Isom L (2010) Electrophysiology and beyond: multiple roles of Na^+ channel beta subunits in development and disease. Neurosci Lett 486:53–59

Payandeh J, Scheuer T, Zheng N, Catterall WA (2011) The crystal structure of a voltage-gated sodium channel. Nature 475(7356):353–358

Payandeh J, Gamal El-Din TM, Scheuer T, Zheng N, Catterall WA (2012) Crystal structure of a voltage-gated sodium channel in two potentially inactivated states. Nature 486(7401):135–139

Petitprez S, Zmoos AF, Ogrodnik J, Balse E, Raad N, El-Haou S, Albesa M, Bittihn P, Luther S, Lehnart SE, Hatem SN, Coulombe A, Abriel H (2011) SAP97 and dystrophin macromolecular complexes determine two pools of cardiac sodium channels Nav1.5 in cardiomyocytes. Circ Res 108:294–304

Pfeufer A, Sanna S, Arking DE, Müller M et al (2009) Common variants at ten loci modulate the QT interval duration in the QTSCD study. Nat Genet 41(4):407–414

Pfeufer A, van Noord C, Marciante KD et al (2010) Genome-wide association study of PR interval. Nat Genet 42(2):153–159

Poelzing S, Forleo C, Samodell M, Dudash L, Sorrentino S, Anaclerio M, Troccoli R, Iacoviello M, Romito R, Guida P, Chahine M, Pitzalis M, Deschenes I (2006) SCN5A polymorphism restores trafficking of a Brugada syndrome mutation on a separate gene. Circulation 114:368–376

Priori SG, Wilde AA, Horie M, Cho Y, Behr ER, Berul C, Blom N, Brugada J, Chiang CE, Huikuri H, Kannankeril P, Krahn A, Leenhardt A, Moss A, Schwartz PJ, Shimizu W, Tomaselli G, Tracy C (2013) HRS/EHRA/APHRS expert consensus statement on the diagnosis and management of patients with inherited primary arrhythmia syndromes: document endorsed by HRS, EHRA, and APHRS in May 2013 and by ACCF, AHA, PACES, and AEPC in June 2013. Heart Rhythm 10(12):1932–1963

Remme C (2013) Cardiac sodium channelopathy associated with SCN5A mutations: electrophysiological, molecular and genetic aspects. J Physiol 591(17):4099–4116

Rogart RB, Cribbs LL, Muglia LK, Kephart DD, Kaiser MW (1989) Molecular cloning of a putative tetrodotoxin-resistant rat heart Na+ channel isoform. Proc Natl Acad Sci U S A 86(20): 8170–8174

Rougier JS, van Bemmelen MX, Bruce MC, Jespersen T, Gavilet B, Apotheloz E, Cordonier S, Staub O, Rotin D, Abriel H (2005) Molecular determinants of voltage-gated sodium channel regulation by the Nedd4/Nedd4-like proteins. Am J Phys Cell Phys 288:C692–C701

Rybin VO, Xu X, Lisanti MP, Steinberg SF (2000) Differential targeting of beta adrenergic receptor subtypes and adenylyl cyclase to cardiomyocyte caveolae. J Biol Chem 275: 41447–41457

Sato PY, Musa H, Coombs W, Guerrero-Serna G, Patinio GA, Taffet SM, Isom LL, Delmar M (2009) Loss of plakophillin-2 expression leads to decreased sodium current and slower conduction velocity in cultured cardiac myocytes. Circ Res 105:523–526

Sato PY, Coombs W, Lin X, Nekrasova O, Green KJ, Isom LL, Taffet SM, Delmar M (2011) Interactions between ankyrin-G, plakophilin-2, and connexin43 at the cardiac intercalated disc. Circ Res 109:193–201

Savio-Galimberti E, Gollob MH, Darbar D (2012) Voltage-gated sodium channels: biophysics, pharmacology, and related channelopathies. Front Pharmacol 3:1–19

Savio-Galimberti E, Weeke P, Muhammad R, Blair M, Ansari S, Short L, Atack TC, Kor K, Vanoye CG, Olesen MS, Yang T, George AL Jr, Roden DM, Darbar D (2014) SCN10A/Nav1.8 modulation of peak and late sodium currents in patients with early onset atrial fibrillation. Cardiovasc Res 104(2):355–363

Schulze-Bahr E, Neu A, Friederich P, Kaupp UB, Breithardt G, Pongs O, Isbrandt D (2003) Pacemaker channel dysfunction in a patient with sinus node disease. J Clin Invest 111(10): 1537–1545

Schwartz PJ, Priori SG, Dumaine R, Napolitano C, Antzelevitch C, Stramba-Badiale M, Richard T, Berti MR, Bloise R (2000) A molecular link between the sudden infant death syndrome and the long-QT syndrome. N Engl J Med 343:262–267

Schwartz PJ, Crotti L, Insolia R (2012) Long QT syndrome: from genetics to management. Circ Arrhythm Electrophysiol 5(4):868–877

Scott JJ, Alshinawi C, Kyndt F, Probst V, Hoorntje TM, Hulsbeek M, Wilde AA, Escande D, Mannens MM, Le MH (1999) Cardiac conduction defects associated with mutations in SCN5A. Nat Genet 23:20–21

Shimizu W, Antzelevitch C (1997) Sodium channel block with mexiletine is effective in reducing dispersion of repolarization and preventing torsade de pointes in LQT2 and LQT3 models of the long-QT syndrome. Circulation 96(6):2038–2047

Shimizu W, Antzelevitch C (1998) Cellular basis for the ECG features of the LQT1 form of the long QT syndrome: effects of beta-adrenergic agonists and antagonists and sodium channel blockers on transmural dispersion of repolarization and torsade de pointes. Circulation 98(21): 2314–2322

Shimizu W, Antzelevitch C (2000) Differential effects of beta-adrenergic agonists and antagonists in LQT1, LQT2 and LQT3 models of the long QT syndrome. J Am Coll Cardiol 35:778–786

Shy D, Gillet L, Abriel H (2013) Cardiac sodium channel Nav 1.5 distribution in myocytes via interacting proteins: the multiple pool model. Biochim Biophys Acta 1833:886–894

Smith JG, Magnani J, Palmer C, Meng YA, Soliman EZ, Musani SK et al (2011) Genome-wide association studies of the PR interval in African Americans. PLoS Genet 7(2):e1001304

Sotoodehnia N, Isaacs A, de Bakker PI, Dorr M, Newton-Cheh C, Nolte IM, van der Harst HP, Muller M, Eijgelsheim M, Alonso A, Hicks AA, Padmanabhan S, Hayward C, Smith AV, Polasek O, Giovannone S, Fu J, Magnani JW, Marciante KD, Pfeufer A, Gharib SA, Teumer A, Li M, Bis JC, Rivadeneira F, Aspelund T, Kottgen A, Johnson T, Rice K, Sie MP, Wang YA, Klopp N, Fuchsberger C, Wild SH, Mateo L, Estrada IK, Volker U, Wright AF, Asselbergs FW, Qu J, Chakravarti A, Sinner MF, Kors JA, Petersmann A, Harris TB, Soliman EZ, Munroe PB, Psaty BM, Oostra BA, Cupples LA, Perz S, de Boer RA, Uitterlinden AG, Volzke H, Spector TD, Liu FY, Boerwinkle E, Dominiczak AF, Rotter JI, van Herpen G, Levy D, Wichmann HE, Van Gilst WH, Witteman JC, Kroemer HK, Kao WH, Heckbert SR, Meitinger T, Hofman A, Campbell H, Folsom AR, Van Veldhuisen DJ, Schwienbacher C, O'Donnell CJ, Volpato CB, Caulfield MJ, Connell JM, Launer L, Lu X, Franke L, Fehrmann RS, Te MG, Groen HJ, Weersma RK, van den Berg LH, Wijmenga C, Ophoff RA, Navis G, Rudan I, Snieder H, Wilson JF, Pramstaller PP, Siscovick DS, Wang TJ, Gudnason V, van Duijn CM, Felix SB, Fishman GI, Jamshidi Y, Stricker BHC, Samani NJ, Kaab S, Arking DE (2010a) Common variants in 22 loci are associated with QRS duration and cardiac ventricular conduction. Nat Genet 42(12):1068–1076

Sotoodehnia N, Isaacs A, de Bakker PIW et al (2010b) Common variants in 22 loci are associated with QRS duration and cardiac ventricular contraction. Nat Genet 42:1068–1076

Stallmeyer B, Kuss J, Kotthoff S, Zumhagen S, Vowinkel K, Rinne S, Matschke LA, Friedrich C, Schulze-Bahr E, Rust S, Seebohm G, Decher N, Schulze-Bahr E (2017) A mutation in the G-protein gene GNB2 causes familial sinus node and Atrioventricular conduction dysfunction. Circ Res 120(10):e33–e44

Swayne LA, Murphy NP, Asuri S, Chen L, Xu X, McIntosh S, Wang C, Lancione PJ, Roberts JD, Kerr C, Sanatani S, Sherwin E, Kline CF, Zhang M, Mohler PJ, Arbour LT (2017) Novel variant in the ANK2 membrane-binding domain is associated with Ankyrin-B syndrome and structural heart disease in a first nations population with a high rate of long QT syndrome. Circ Cardiovasc Genet 10(e001537):1–11

Tan HL, Kupershmidt S, Zhang R, Stepanovic S, Roden DM (2002) A calcium sensor in the sodium channel modulates cardiac excitability. Nature 415:442–447

Tayal U, Prasad S, Cook SA (2017) Genetics and genomics of dilated cardiomyopathy and systolic heart failure. Genome Med 9(1):20

Tester DJ, Ackerman MJ (2012) The molecular autopsy: should the evaluation continue after the funeral? Pediatr Cardiol 33(3):461–470

Valle G, Faulkner G, De Antoni A, Pacchioni B, Pallavicini A, Pandolfo D, Tiso N, Toppo S, Trevisan S, Lanfranchi G (1997) Telethonin, a novel sarcomeric protein of heart and skeletal muscle. FEBS Lett 415:163–168

Van Bemmelen MX, Rougier JS, Gavillet B, Apotheloz F, Daidie D, Tateyama M, Rivolta I, Thomas MA, Kass RS, Staub O, Abriel H (2004) Cardiac voltage-gated sodium channel Nav1.5 is regulated by Nedd4-2 mediated ubiquitination. Circ Res 95:284–291

Van den Boogaard M, Wong LY, Tessadori F, Bakker ML, Dreizenhnter LK, Wakker V, Bezzina CR, 't Hoen PA, Bakkers J, Barnett P, Christoffels VM (2012) Genetic variations in T-box binding element functionally affects SCN5A/SCN10A enhancer. J Clin Invest 122:2519–2530

Vatta M, Ackerman MJ, Ye B, Makielski JC, Ughanze EE, Taylor EW, Tester DJ, Balijepalli RC, Foell JD, Li Z, Kamp TJ, Towbin JA (2006) Mutant caveolin-3 induces persistent late sodium current and is associated with long-QT syndrome. Circulation 114(20):2104–2112

Verkerk AO, Remme CA, Schumacher CA, Scicluna BP, Wolswinkel R, de Jonge B, Bezzina CR, Veldkamp MW (2012) Functional Nav1.8 channels in intracardiac neurons: the link between SCN10A and cardiac electrophysiology. Circ Res 111(3):333–343

Viswanathan PC, Benson DW, Balser JR (2003) A common SCN5A polymorphism modulates the biophysical effects of an SCN5A mutation. J Clin Invest 111:341–346

Wagner S, Dybkova N, Rasenack EC, Jacobshagen C, Fabritz L, Kirchhof P, Maier SK, Zhang T, Hasenfuss G, Brown JH, Bers DM, Maier LS (2006) Ca/calmodulin-dependent protein kinase II regulates cardiac Na channels. J Clin Invest 116:3127–3128

Wang Q, Shen J, Splawski I, Atkinson D, Li Z, Robinson JL, Moss AJ, Towbin JA, Keating MT (1995) SCN5A mutations associated with an inherited cardiac arrhythmia, long QT syndrome. Cell 80:805–811

Wang Q, Li Z, Shen J, Keating MT (1996) Genomic organization of the human SCN5A gene encoding the cardiac sodium channel. Genomics 34:9–16

Wang C, Wang C, Hoch EG, Pitt GS (2011) Identification of novel interaction sites that determine specificity between fibroblast growth factor homologous factors and voltage-gated sodium channels. J Biol Chem 286:24253–24263

Wang P, Yang Q, Wu X, Yang Y, Shi L, Wang C, Wu G, Xia Y, Yang B, Zhang R, Xu C, Cheng X, Li S, Zhao Y, Fu F, Liao Y, Fang F, Chen Q, Tu X, Wang QK (2010) Functional dominant-negative mutation of sodium channel subunit gene SCN3B associated with atrial fibrillation in a Chinese GeneID population. Biochem Biophys Res Commun 398(1):98–104

Watanabe H, Darbar D, Kaiser DW, Jiramongkolchai K, Chopra S, Donahue BS, Kannankeril PJ, Roden DM (2009a) Mutations in sodium channel beta 1 and beta 2 subunits associated with atrial fibrillation. Circ Arrhythm Electrophysiol 2:268–275

Watanabe H, Darbar D, Kaiser DW, Jirasirirojanakorn K, Chopra S, Donahue BS, Kannankeril P, Roden DM (2009b) Mutations in sodium channel b1 and b2 subunits associated with atrial fibrillation. Circ Arrhythm Electrophysiol 2(3):268–275

Westernbroek RE, Bischoff S, Fu Y, Maier SK, Catterall WA, Scheuer T (2013) Localization of sodium channel subtypes in mouse ventricular myocytes using quantitative immunocytochemistry. J Mol Cell Cardiol 64:69–78

Wyse DG, Van Gelder IC, Ellinor PT, Go AS, Kalman JM, Narayan SM, Nattel S, Schotten U, Rienstra M (2014) Lone atrial fibrillation: does it exist? A "white paper" of the journal of the American College of Cardiology. J Am Coll Cardiol 63(17):1715–1723

Yan GX, Antzelevitch C (1996) Cellular basis for the electrocardiographic J wave. Circulation 93(2):372–379

Yang T, Atack TC, Stroud DM, Zhang W, Hall L, Roden DM (2012) Blocking Scn10a channels in heart reduces late sodium current and is antiarrhythmic. Circ Res 111(3):322–332

Yarbrough TL, Lu T, Lee HC, Shibata EF (2002) Localization of cardiac sodium channels in caveolin-rich membrane domains: regulation of sodium current amplitude. Circ Res 90: 443–449

Ziane R, Huang H, Moghadaszadeh B, Beggs AH, Levesque G, Chahine M (2010) Cell membrane expression of cardiac sodium channel Nav1.5 modulated by alpha-actinin-2 interaction. Biochemistry 49:166–178

Translational Model Systems for Complex Sodium Channel Pathophysiology in Pain

Katrin Schrenk-Siemens, Corinna Rösseler, and Angelika Lampert

Contents

Abstract

Chronic pain patients are often left with insufficient treatment as the pathophysiology especially of neuropathic pain remains enigmatic. Recently, genetic variations in the genes of the voltage-gated sodium channels (Navs) were linked to inherited neuropathic pain syndromes, opening a research pathway to foster our understanding of the pathophysiology of neuropathic pain. More than 10 years ago, the rare, inherited pain syndrome erythromelalgia was linked to mutations in the subtype Nav1.7, and since then a plethora of mutations and genetic variations in this and other Nav genes were identified. Often the biophysical changes induced by the genetic alteration offer a straightforward explanation for the clinical symptoms, but mutations in some channels, especially Nav1.9,

K. Schrenk-Siemens
Institute of Pharmacology, Heidelberg University, Heidelberg, Germany

C. Rösseler · A. Lampert (✉)
Institute of Physiology, Uniklinik RWTH Aachen University, Aachen, Germany
e-mail: alampert@ukaachen.de

M. Chahine (ed.), *Voltage-gated Sodium Channels: Structure, Function and Channelopathies*, Handbook of Experimental Pharmacology 246,
https://doi.org/10.1007/164_2017_91

paint a more complex picture. Although efforts were undertaken to significantly advance our knowledge, translation from heterologous or animal model systems to humans remains a challenge. Here we present recent advances in translation using stem cell-derived human sensory neurons and their potential application for identification of better, effective, and more precise treatment for the individual pain patient.

Keywords
CIP · Erythromelalgia · Genetic pain syndrome · iPSCs · Nav1.7 · Nav1.8 · Nav1.9 · PEPD · SFN · Stem cell derived sensory neurons

1 Congenital Pain Syndromes

From the evolutionary aspect pain is important to protect the human being from noxious stimuli. However, when pain becomes chronic (often caused by genetic diseases), it requires (sometimes lifelong) medical attention. About every fifth person in the world suffers from chronic pain (WHO-Guidelines 2008). To date often no satisfactory medication for chronic pain is available, which underlines the ongoing need for drug development. Recent model systems allow patient-specific genetic disease modeling, which can provide adapted medication and illustrates a way to personalized medicine. Voltage-gated sodium channels (Navs) are in the focus of recent analgesic drug development, as several mutations were identified in patients suffering from inherited pain syndromes. Navs are, among others, located on sensory neurons of the peripheral nervous system. Nociceptors are specialized sensory neurons, which detect potentially harmful (noxious) stimuli from the periphery or internal organs and transduce the information via long neurites to the spinal cord. Several Nav subtypes were identified as potentially disease causing when mutated: Nav1.7, Nav1.8, and Nav1.9, all of which are expressed in nociceptors (Leipold et al. 2013; Lampert et al. 2014).

1.1 Nav1.7 in Human Pain Syndromes

The first mutation to be linked to a congenital pain syndrome was described in the *SCN9A* gene, which codes for Nav1.7 (Yang et al. 2004). In the years to follow, several distinct inherited pain syndromes were identified to result from mutations in Nav1.7: paroxysmal extreme pain disorder [PEPD (Fertleman et al. 2006)], inherited erythromelalgia (IEM), and small fiber neuropathy [SFN (Faber et al. 2012a)]. All are characterized by gain-of-function mutations of Nav1.7, and patients suffer from intense burning pain. PEPD has a very early onset, sometimes already in utero, and is characterized by physical assaults of severe pain in proximal parts of the body such as the face and the anogenital region. Potential triggers of the attacks are cold, mechanical stress, or intestinal peristalsis, and attacks are often associated with reddening of the skin. IEM may also start during childhood but mostly manifests when patients are young adults. Patients show typically intense, burning pain and erythema in the extremities triggered by exposure to warmth or mild exercise. In contrast, the onset of SFN is usually later in life and correlates with a reduction of small nerve fibers in the skin epidermis (Cazzato and Lauria 2017), which seems to be contradictory to the burning pain sensation reported by the patients. Chronic insensitivity to pain (CIP) can be caused by homozygous functional knockout mutations of Nav1.7, rendering the channel protein nonfunctional. These patients do not feel pain and suffer from self-inflicted injuries. Interestingly, gain-of-function mutations of Nav1.9 can have similar phenotypes, which cannot be explained intuitively.

To date several dozens of pain-linked Nav mutations were described, and numerous reviews have summarized correlations between the clinical manifestations and the biophysical gating effects of the respective mutations (Drenth and Waxman 2007; Lampert et al. 2010, 2014; Dabby 2012; Hoeijmakers et al. 2012; Dib-Hajj et al. 2013; Waxman 2013; Bennett and Woods 2014; Brouwer et al. 2014; Habib et al. 2015; Vetter et al. 2017). It is quite clear that humans carrying nonfunctional Nav1.7 suffer from CIP, stressing the importance of Nav1.7 for human pain perception (Cox et al. 2006). A loss of Nav1.7 may result in the inability to produce action potentials in nociceptors, thereby a lack of signal transduction to the CNS, and thus the inability to sense pain. Still, the exact mechanism how a loss of Nav1.7 leads to insensitivity to pain remains to be determined, especially as reliable knockout mouse models are scarce. Recently, a Nav1.7 CIP patient was able to feel pain for the first time in her life after administration of naloxone, an opioid antagonist (Minett et al. 2015). This raises the question whether the endogenous opioid system may play an important role in Nav1.7 knockdown-mediated human loss of pain.

A characteristic feature of PEPD mutations is a depolarized, and thus impaired, fast channel inactivation, and every mutation tested so far showed enhanced resurgent currents. These are depolarizing sodium currents, which occur during the falling phase of the action potential and may thus support hyperexcitability (Cannon and Bean 2010). Resurgent currents are thought to occur due to an open channel block and can physiologically be found in some small sensory neurons and may be enhanced under pathological conditions (Jarecki et al. 2010; Theile et al. 2011; Eberhardt et al. 2014; Sittl et al. 2012).

IEM can be clearly linked to a facilitated channel activation of Nav1.7. Under physiological conditions, Navs open when the membrane potential reaches a certain voltage threshold, which occurs due to a depolarizing stimulus. Sodium ions enter the cell, and an action potential is initiated. In case of IEM, the activation threshold is shifted to more hyperpolarized potentials by most of the mutations. Thus, a smaller difference in the membrane potential is needed to open the channels for induction of action potentials, and therefore smaller stimuli are sufficient to elicit a nociceptive signal. To date more than 30 mutations in the Nav1.7 channel gene have been characterized, leading to IEM. 27 of them were shown to have a hyperpolarized activation threshold, when investigated by overexpression in cell lines. Unfortunately, up to now this shift was only shown in heterologous expression systems, as human sensory neurons are not easily available.

Electrophysiological analyses of SFN mutations paint a more complex picture. Some of these mutations show an impaired slow inactivation, while it remains unchanged in others. Increased resurgent currents were reported, but also a shift of fast inactivation to more depolarized potentials, which could render the expressing neuron hyperexcitable. When overexpressed in rodent sensory neurons, all SFN-linked Nav1.7 mutations produce an increased firing rate. Thus, a clear link to a biophysical characteristic, as it was proposed for IEM or PEPD, cannot be claimed for SFN. It also remains to be discovered, how a functional hyperexcitability may lead to degeneration of the small nerve fibers in the skin. Some data point toward a dysfunctional calcium homeostasis (Rolyan et al. 2016), but an impairment of skin-nerve signaling could also be involved.

Genetic screens of neuropathic pain patients led to the identification of mutations and genetic variations in other Nav channels, such as the already mentioned Nav1.8 and Nav1.9 subtypes, as well as Nav1.1 and Nav1.6. All of these may also play an important role in peripheral nociception. Nav1.8 and Nav1.9 will be discussed in the following section.

1.2 Nav1.8 and Nav1.9 in Pain (Less) Disorders

Animal and human studies have shown a role for Nav1.8 (encoded by the *SCN10A* gene) in pain symptoms. Mutations described in this gene are mostly linked to SFN (Faber et al. 2012b), but some are also described for IEM (Kist et al. 2016). Nav1.8 is important for the fast upstroke of the action potential and thus may have an influence on cellular excitability. Additionally, Nav1.8 knockout mice show impairments in the electrocardiogram [ECG (Yang et al. 2012)], suggesting a more general role for Nav1.8 in regulation of physiological functions.

Nav1.9 (encoded by the *SCN11A* gene), like Nav1.8, is resistant to the pufferfish toxin tetrodotoxin (TTX). Mutations in Nav1.9 were linked to gain of pain as well as to loss of pain. The channel is special as it is slowly gating and produces large persistent currents. Its open probability can be enhanced by G proteins, which are also part of the second messenger pathways of inflammatory processes that occur during tissue damage (Vanoye et al. 2013).

Patients suffering from pain were also reported to carry gain-of-function mutations of Nav1.9 (Zhang et al. 2013; Huang et al. 2014; Leipold et al. 2015; Han et al. 2017). The link between a gain of function and pain is intuitive, as the enhanced persistent current of Nav1.9 may support higher frequency firing. On the other hand, the L811P mutation, a very strong gain-of-function mutation, was described in 2013 in a patient suffering from CIP (Leipold et al. 2013). Nav1.9 channels with this mutation showed a slowed deactivation and strongly hyperpolarized voltage dependence of activation, suggesting that cells expressing this mutated Nav1.9 would be hyperexcitable. This apparent discrepancy between channel gain of function and clinical picture (loss-of-pain) was solved when the authors showed that the strong increase in activity of Nav1.9 led to a depolarized resting membrane potential and thus, after initial strong firing, to a depolarization block (Leipold et al. 2015). A similar mechanism was described for the L1302F mutation, which was identified in a patient with with impaired pain sensation, for which an U-shaped relationship between the resting membrane potential and the neuronal action potential threshold was reported (Phatarakijnirund et al. 2016; Huang et al. 2017). Also here, a strong gain-of-function phenotype determined the clinical symptoms characterized by a loss of pain. Electrophysiological studies showed that this is most likely due to the inactivation of voltage-gated ion channels by depolarization of the resting membrane potential. Patients carrying one of these two mutations suffer not only from CIP but also from hyperflexibility, gastrointestinal symptoms, itch, and bone defects, suggesting a wide range of physiological functions for Nav1.9.

Nav1.9 has been notoriously hard to express in heterologous cell systems. Nav1.8 knockout sensory neurons of mice were used successfully to express the TTX-resistant Nav1.9 for electrophysiological recordings (e.g., Leipold et al. 2013), and only very few groups managed to express this channel heterologously in neuronal cell lines (Leipold et al. 2013; Vanoye et al. 2013). Larger currents can be achieved by using a chimera of Nav1.9 and Nav1.4 (Goral et al. 2015), and only recently Lin et al. managed with major efforts to create a stable Nav1.9 cell line using human embryonic kidney (HEK293) cells, which is also ambenable to drug testing (Lin et al. 2016).

In order to investigate the impact of a Nav mutation, heterologous expression systems are valuable, and in some cases, as for many IEM mutations, 100% of clinical penetrance of the mutations can be observed, clearly showing the link between the mutation and the disease. Other genetic variants give a more complex picture and the identification of pain-associated SNPs stresses the importance of the patient's specific genetic background when assessing the potential disease causing effects.

The pathological importance of genetic modifications is not easy to interpret, especially when focusing on single genetic variations in isolation, i.e., expressing one mutation, SNP, or genetic variation in a non-neuronal, non-patient-specific cellular background. Patient sensory neurons contain all relevant gene products of the specific person, and the combination of genetic variations may impact cell function and disease.

2 Translation from Dish to Rodent to Human

Over the past five decades, animal models have become an invaluable tool to study cellular, molecular, and mechanistic aspects of somatosensation and pain. However, which aspects of this knowledge can be translated to devise new analgesic strategies has become a major focus of discussion. In particular, the discrepancy between the number of successfully conducted preclinical studies and the number of clinical trials that had to be terminated or did not reveal any measurable improvement for the patients sparked a discussion among scientists and clinicians about the reasons for these failures and the possibilities to develop alternative, new approaches for future studies (Mogil 2009; Burma et al. 2017).

One of the possible reasons why compounds that did well in animal models failed to show beneficial effects in clinical trials may stem from the fact that too little is known about differences of fundamental processes that trigger pain responses and maladaptations in humans compared to lower vertebrates (Dib-Hajj 2014). Besides the psychological component of pain perception, which can be very different in human beings according to social upbringing, personal stress levels, and other factors – and which is currently not possible to model in animals – working with humans or tissues derived from humans has been very difficult in the past. The latter is due to the impeded accessibility of dorsal root ganglia or trigeminal ganglia, the native sensory tissue that houses the cell bodies of nociceptive neurons. Therefore, studying the molecular and functional make up of sensory neurons, the primary transducers of nociceptive and innocuous stimuli that can evoke pain, was hampered by limited access to human tissue – particularly in conditions that would allow functional characterization.

2.1 Heterologous Expression of Human Proteins

As native human sensory tissue is difficult to obtain, an easy way to investigate the function of a human protein is introducing an expression vector harboring the genetic information of the target molecule into a cell line such as HEK 293 cells. In many instances heterologous expression of the protein of interest is unproblematic, and the function of the introduced protein can be analyzed using different techniques, such as the patch-clamp technique. Many of the results described in the first part of this chapter were gathered in this way. In some cases, however, the target protein cannot be adequately and functionally expressed in conventionally used expression systems. Nevertheless, heterologous expression in cellular systems has been widely used in the past to gain information about human proteins and to compare them to their rodent counterparts. In the case of Nav channel physiology, researchers also used rodent dorsal root ganglion neurons as target cells to express and investigate the human channels in a more native environment. Using this method, differences between human and rodent Nav1.8 have been described (Han et al. 2015). The researchers took DRG neurons from a Nav1.8 knockout mouse line and introduced either human or rat Nav1.8. Thereby, they found that human Nav1.8 displays larger persistent, ramp, and window currents than their rat counterparts, which also affected action potential characteristics (Han et al. 2015).

The data clearly demonstrate species-specific differences between rodent and human Nav1.8, which have to be taken into account when inferring therapeutic possibilities based on rodent Nav1.8. Heterologous cell systems are important research tools for identifying a starting point for investigating the function of human channels. Nevertheless, the properties in native human neurons were still shown to be different, which argues for the importance of a natural environment of the channel to reveal its full physiological function. Therefore, new strategies to utilize human sensory neurons and native expression levels of the receptors and ion channels under investigation are of imminent importance.

2.2 Human DRGs as a Model System

Differences as well as similarities between human and rodent sensory neurons have already been established. Some literature is available, showing that marker profiles can be obtained from and electrophysiological recordings can be performed with sensory tissue collected from postmortem patients and were compared to rodent neurons. Those differences include, for example, a nearly complete absence of P_2X_2 in human nociceptors, while the orthologous purinergic receptor in rodents has been proposed to play a prominent role in models of chronic pain (Serrano et al. 2012).

A comparative study assessing the presence of specific nociceptive markers in tissue slices of mouse and human DRGs using in situ hybridization was recently performed in our laboratory (Rostock et al. 2017). The expression of sensory markers such as *TRPV1*, *RET*, Nav1.7, Nav1.8, and Nav1.9 was analyzed in relation to the expression of *TRKA*, an accepted marker for peptidergic nociceptors: we observed a higher percentage of *TRKA* positive neurons expressing *TRPV1*, *RET*, Nav1.8, and Nav1.9 in human versus mouse DRG sections. No difference in the expression of Nav1.7 was observed. Another striking difference is the almost universal expression of the neurofilament heavy subunit (NF200) in human sensory neurons compared to a rather confined expression of the filament in myelinated medium to large diameter mouse sensory neurons that give rise to Aβ and Aδ fibers (Vega et al. 1994; Rostock et al. 2017). Already from these few studies one can learn that the molecular makeup of human sensory neurons cannot automatically be inferred from expression data that has been obtained from rodent cells. Additionally, further studies need to be performed to assess if these differences do have significant implications on the functional level of the sensory neurons.

Unfortunately, so far functional data from cultured human DRG neurons is scarce. Two recent studies assessed some fundamental electrophysiological properties of cultured small-diameter human nociceptors and found the majority of cells to exhibit a "shoulder" on the falling phase of the action potential that has been proposed to be the result of calcium as well as sodium influx (Davidson et al. 2014; Han et al. 2015). This electrophysiological feature has also been found in small-diameter rodent nociceptors (Ritter and Mendell 1992; Gold et al. 1996; McCarthy and Lawson 1997). The "shoulder" phenomenon in rodents acts to prolong action potential duration (Blair and Bean 2002) and can be correlated with the presence of the isolectin-B4 (IB4), a marker widely used in rodent models to label non-peptidergic, small-diameter nociceptors (Stucky and Lewin 1999). However, Davidson et al. were not able to show binding of IB4 to cultured human DRGs (Davidson et al. 2014), which possibly is another species difference, although other groups did observe IB4 binding to human sensory neurons (Shi et al. 2008; Pan et al. 2012).

In summary, there clearly is a need to analyze human sensory neurons in more detail on molecular as well as functional levels for several reasons: (1) only when we know how human nociceptors operate, we will be able to find suitable means of interfering with pathways involved in the onset and perpetuation of maladaptive forms of pain; (2) in light of the growing interest to use human pluripotent stem cells to derive sensory neurons from human subjects including pain patients, we need to be able to compare the derived neurons with their in vivo equivalents.

2.3 Human Microneurography

In order to assess the functionality of human nerve fibers, the microneurography technique has been used for decades to record electrical activity from nerve fibers in living humans (Vallbo and Hagbarth 1968). In particular, peripheral nerves innervating the skin can be accessed by inserting a fine (100–200 μm) tungsten electrode in close vicinity to a human nerve. A well-placed electrode is able to discriminate impulses from a single nerve fiber of interest in relation to an adjacent reference electrode. As the method can be exerted in fully awake human beings and does not harm or destruct the nerve, it provides a valuable tool to study alterations in the electrical activity of specific single nerve fibers and their response to stimuli in their receptive field. By these means researchers were able to demonstrate that patients suffering from neuropathic pain show greater spontaneous activity in C-fibers (Kleggetveit et al. 2012; Serra et al. 2012). The method even allows to specify the type of C-fiber involved; in the case of neuropathic pain, silent nociceptors that are mechano-insensitive (CMi-nociceptors) are thought to be responsible for the increase in spontaneous activity (Kleggetveit et al. 2012). Additionally, a faster recovery of activity-dependent slowing of conduction could also be observed in chronic pain patients, which indicates alterations of axonal ion channels. Despite the possibilities offered by microneurography, such as the identification of fiber types involved in particular pathological pain condition, the technique only rarely allows to deduce molecular mechanisms underlying the observed electrical properties. To gain mechanistic insight, model systems are indispensable. Nevertheless, the method can be used to test defined (and "prescreened") compounds for their ability to, for example, reduce or even stop spontaneous activity in diseased sensory neurons, thereby helping to identify analgesic substances (Serra 2010; Wehrfritz et al. 2011; Kankel et al. 2012; Schwarz et al. 2017).

Taken together, the already available data and the stagnancy regarding the successful implementation of preclinical data show the need to reevaluate the model systems currently used and to assess the possibility/feasibility of incorporating emerging new model systems.

With the isolation of human embryonic stem cells (hESCs) in 1998 (Thomson et al. 1998) and remarkable technological advances that allow the reprogramming of somatic cells into induced pluripotent stem cells [iPSCs (Takahashi et al. 2007)], we have now tools at hand that could help to invigorate pain research. A crucial step forward is the development of differentiation procedures by several researches to generate human sensory neuron-like cells from hESCs and hiPSCs in the dish (Valensi-Kurtz et al. 2010; Chambers et al. 2012; Blanchard et al. 2015; Schrenk-

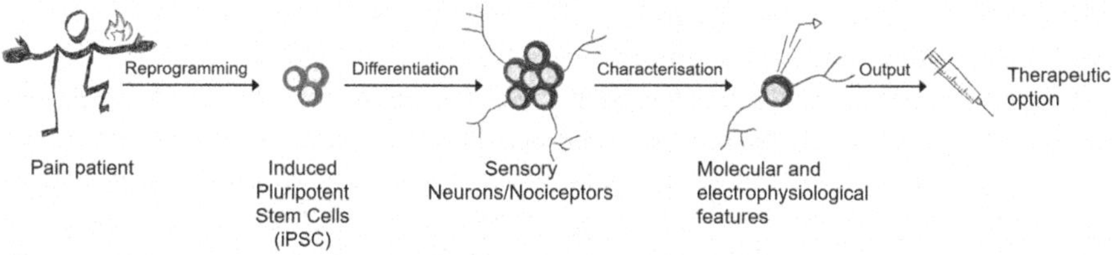

Fig. 1 Model system using human nociceptor-like cells. By obtaining a tissue sample from a pain patient, these cells can be reprogrammed into induced pluripotent stem cell (iPSCs). iPSCs can be differentiated into sensory neurons, which carry the genetic background of the patient. The pathophysiology of the nociceptors is analyzed using molecular and electrophysiological techniques, such as staining for markers, patch-clamp, and multielectrode recordings. Drugs can be tested on iPSC-derived sensory neurons, and personalized treatment may be identified which could reverse patient-specific pathophysiological changes.

Siemens et al. 2015; Wainger et al. 2015). These cells may offer the possibility to help elucidating and understanding possible differences in the cellular makeup and functionality of human and rodent sensory neurons and thereby refine our way of developing means to interfere with pain and the maladaptation of it (Fig. 1).

2.4 Human Pluripotent Stem Cell-Derived Nociceptors

Currently, there are two different strategies, how researchers succeeded in generating human nociceptor-like cells in a dish: One is by interfering with specific cell intrinsic pathways (Chambers et al. 2012), the other via overexpression of transcription factors relevant for sensory neuron development (Blanchard et al. 2015; Wainger et al. 2015).

In the first approach, Chambers et al. could show that the combination of five small-molecule pathway inhibitors was sufficient to efficiently trigger the differentiation of human pluripotent stem cells into neurons in a rather short time frame compared to classical multistep differentiation procedures. The resulting neurons showed functional and molecular hallmarks of nociceptive neurons such as the expression of TRKA. Also the upregulation of several ion channels such as SCN9A (Nav1.7), SCN10A (Nav1.8), and SCN11A (Nav1.9), the purinergic receptor P2RX3, and the TRP channels TRPV1 and TRPM8 indicated that the derived cells have a nociceptive-like phenotype. A later study compared the mRNA transcript profile of the differentiated sensory neurons after 30 days of culture with those from native human DRG cells and found them to be highly comparable. On the level of ion channel expression 84% of human ion channel genes were also expressed in human-differentiated sensory neurons (Young et al. 2014). Especially the presence of Navs in these model cells (Eberhardt et al. 2015) makes this method attractive to use as an in vitro model system for the analysis of the signaling pathways involved in pain perception and transduction. The universality of the protocol makes it possible to derive nociceptor-like cells also from hiPSCs that could be isolated from patients suffering from pain conditions triggered by mutations in the Nav genes such as Nav1.7-dependent inherited erythromelalgia (Cao et al. 2016).

Using overexpression of transcription factors, known to be involved in sensory neuron development of the mouse, has been the basis of two recently published differentiation strategies (Blanchard et al. 2015; Wainger et al. 2015). Both studies used virus-mediated infection of human fibroblasts, thereby introducing genes for reprogramming the cells into sensory-like neurons, including nociceptors. Consensus of both studies is that human fibroblasts are sufficient as starting material for nociceptor generation, and no reprogramming into iPSCs is needed.

Blanchard et al. (2015) used the basic helix-loop-helix transcription factors neurogenin 1 (NGN1) or neurogenin 2 (NGN2) in combination with BRN3A (POU4F1) to drive the generation of sensory neurons (Blanchard et al. 2015). They made use of an inducible expression system and could thereby regulate the transcription factor expression by doxycycline treatment. The conversion of fibroblasts resulted in the generation of neurons showing molecular and key electrophysiological features of all three main subtypes (mechano-, noci-, and proprioceptors) of sensory neurons. Investigating the derived nociceptor-like cells in more detail, the authors could show the expression of receptor ion channels TRPV1, TRPM8, and TRPA1 which are responsible for the detection of heat, cold, and noxious chemicals, respectively. Evaluating the electrophysiological properties of the derived nociceptive-like cells showed characteristics of functional neurons including the presence of TTX-resistant currents, which refers to the presence of Nav1.8 and/or Nav1.9.

Wainger et al. (2015) used a similar approach: also here virus-mediated overexpression of transcription factors was harnessed to induce the conversion of human fibroblast cells into sensory neuron-like cells (Wainger et al. 2015). In this case, five different transcription factors, some of which have been connected to nociceptor generation in vivo, were continuously overexpressed: ASCL1, MYT1L, ISL2, NGN1, and KLF7. The resulting neurons showed features of nociceptors on a functional level such as firing broad action potentials and the presence of TTX-resistant currents, which refers to the presence of Nav1.8 and/or Nav1.9.

As a result of the progress that was made during the last few years, it is now possible to generate human nociceptive-like cells from human iPSCs. Maturation of the so derived sensory neurons varies and expression of the TTX-resistant subtype Nav1.5 hints toward immaturity in some sensory neurons derived via the small-molecule approach (Eberhardt et al. 2015). Nevertheless, reliable expression of slowly gating TTXr currents which resemble Nav1.8 suggests that these neurons are still useful model systems to investigate the function of several sodium channels. Additionally, species differences may also occur here, and it is possible that expression of Nav1.5 in human DRGs is more pronounced compared to rodent sensory neurons. Although the published differentiation strategies differ in their outcome with regard to the homogeneity of the derived sensory neuron subtype as well as the level of maturity (Eberhardt et al. 2015) and comparability to the in vivo human situation, several research labs implemented the protocols already in their pain research, and first results have been published. One of these studies that used stem cell-derived nociceptive-like cells deals with IEM (Cao et al. 2016). Human iPSC lines were generated from four different IEM patients, harboring different Nav1.7 mutations as well as four different non-IEM patients without any Nav1.7 mutations as control group. The pluripotent stem cells were differentiated using the small-molecule inhibitor approach (Chambers et al. 2012) for a total of 9 weeks before electrophysiological experiments were performed. The main goal of the study was to test the contribution of Nav1.7 on the firing pattern of sensory neurons in IEM subjects and the usefulness of two selective Nav1.7 blockers (clinical compound PF-05089771 and the in vitro tool PF-05153462) in abrogating pathological firing patterns in Nav1.7 mutant cells.

Without any treatment, the IEM-derived nociceptive-like neurons showed a significant higher proportion of spontaneously active cells compared to the non-IEM donors, although one IEM donor and one non-IEM donor showed spontaneous firing to a very similar degree. This might point toward some intrinsic heterogeneity among the derived cell lines. Another difference could be observed regarding the rheobase (which indicates the minimal current injection that is required to evoke an action potential), which was on average lower in the IEM-derived neurons. The investigators tested both Nav1.7 blockers on the generated nociceptive-like cells and saw a reduction of spontaneous firing in those IEM cells that showed elevated spontaneous activity. Investigating the involvement of Nav1.7 on reducing rheobase, the scientists found that treatment with both blockers showed an increase in rheobase for mutated as well as control Nav1.7, while the magnitude of the increase was significantly greater in cells derived from IEM patients.

Heat can trigger pain attacks in IEM patients. A modest, innocuous elevation of temperature was already sufficient to increase the excitability of IEM-derived neurons in vitro. Application of PF-0515462 was able to reverse the effect of temperature on rheobase in IEM patient-derived nociceptive-like neurons, suggesting that mutated Nav1.7 channel contributes to the increased heat sensitivity in IEM patient-derived sensory neuron-like cells.

Although Cao et al. only had five IEM patients of which four allowed the isolation of hiPSCs, the researchers also conducted a small in vivo study by treating the patients themselves with the Nav1.7 blocker PF-05089771 or a placebo. Onset of pain was triggered by using a controlled heat stimulus, and the subjects rated their pain using a pain intensity numerical rating scale. A reduction in the magnitude of pain perception by some of the individual patients could be observed in the presence of the drug in comparison to subjects who received placebo, although only to a limited extent and in a very narrow time window.

Although it is quite obvious that the number of patients involved in this study is too small to draw a final conclusion about the usefulness of Nav1.7 channel blockers, the study demonstrates how the derivation of hiPSCs from IEM patients, their differentiation into nociceptive-like cells, in vitro characterization, and subsequent treatment of the patients themselves can help to understand the molecular base of a disease as well as testing compounds that can interfere with the pathological mechanisms.

3 Concluding Remarks

The long-lasting assumption that data generated in animal models in the field of pain research can be easily translated to human patients has been a misconception. One indicator for this is the strong imbalance between the high number of successfully conducted preclinical animal studies and the low number of successfully conducted clinical trials thereof.

With the sensational achievement of reprogramming human somatic cells into pluripotent stem cells in 2007 and the development of differentiation strategies to generate neurons showing hallmarks of peripheral sensory neurons, we have now the opportunity to implement some aspects of translational nociceptive research. However, we are still in need of data from native human DRG neurons concerning their molecular make up and their functional features for comparative purposes and to assess if the derived sensory neurons faithfully recapitulate features of their native counterparts. It is still early days, but some recent studies already illustrate the potential stem cell-derived nociceptors could have on drug research in the future.

Personalized medicine could become one of the avenues to pursue by generation of patient-derived iPSCs to study basic aspects of the disease, perform individual drug treatments on the cells, and translate the findings back to the patient itself, with the hope to find a better suitable, personalized therapy.

References

Bennett DL, Woods CG (2014) Painful and painless channelopathies. Lancet Neurol 13(6):587–599

Blair NT, Bean BP (2002) Roles of tetrodotoxin (TTX)-sensitive Na+ current, TTX-resistant Na+ current, and Ca2+ current in the action potentials of nociceptive sensory neurons. J Neurosci 22(23):10277–10290

Blanchard JW, Eade KT, Szucs A, Lo Sardo V, Tsunemoto RK, Williams D et al (2015) Selective conversion of fibroblasts into peripheral sensory neurons. Nat Neurosci 18(1):25–35

Brouwer BA, Merkies IS, Gerrits MM, Waxman SG, Hoeijmakers JG, Faber CG (2014) Painful neuropathies: the emerging role of sodium channelopathies. J Peripher Nerv Syst 19(2):53–65

Burma NE, Leduc-Pessah H, Fan CY, Trang T (2017) Animal models of chronic pain: advances and challenges for clinical translation. J Neurosci Res 95(6):1242–1256

Cannon SC, Bean BP (2010) Sodium channels gone wild: resurgent current from neuronal and muscle channelopathies. J Clin Invest 120(1):80–83

Cao L, McDonnell A, Nitzsche A, Alexandrou A, Saintot PP, Loucif AJ et al (2016) Pharmacological reversal of a pain phenotype in iPSC-derived sensory neurons and patients with inherited erythromelalgia. Sci Transl Med 8(335):335ra356

Cazzato D, Lauria G (2017) Small fibre neuropathy. Curr Opin Neurol 30(5):490–499

Chambers SM, Qi Y, Mica Y, Lee G, Zhang XJ, Niu L et al (2012) Combined small-molecule inhibition accelerates developmental timing and converts human pluripotent stem cells into nociceptors. Nat Biotechnol 30(7):715–720

Cox JJ, Reimann F, Nicholas AK, Thornton G, Roberts E, Springell K et al (2006) An SCN9A channelopathy causes congenital inability to experience pain. Nature 444(7121):894–898

Dabby R (2012) Pain disorders and erythromelalgia caused by voltage-gated sodium channel mutations. Curr Neurol Neurosci Rep 12(1):76–83

Davidson S, Copits BA, Zhang J, Page G, Ghetti A, Gereau RW (2014) Human sensory neurons: membrane properties and sensitization by inflammatory mediators. Pain 155(9):1861–1870
Dib-Hajj SD (2014) Human pain in a dish: native DRG neurons and differentiated pluripotent stem cells. Pain 155(9):1681–1682
Dib-Hajj SD, Yang Y, Black JA, Waxman SG (2013) The Na(V)1.7 sodium channel: from molecule to man. Nat Rev Neurosci 14(1):49–62
Drenth JP, Waxman SG (2007) Mutations in sodium-channel gene SCN9A cause a spectrum of human genetic pain disorders. J Clin Invest 117(12):3603–3609
Eberhardt M, Nakajima J, Klinger AB, Neacsu C, Huhne K, O'Reilly AO et al (2014) Inherited pain: sodium channel Nav1.7 A1632T mutation causes erythromelalgia due to a shift of fast inactivation. J Biol Chem 289(4):1971–1980
Eberhardt E, Havlicek S, Schmidt D, Link AS, Neacsu C, Kohl Z et al (2015) Pattern of functional TTX-resistant sodium channels reveals a developmental stage of human iPSC- and ESC-derived nociceptors. Stem Cell Rep 5(3):305–313
Faber CG, Hoeijmakers JG, Ahn HS, Cheng X, Han C, Choi JS et al (2012a) Gain of function Nanu1.7 mutations in idiopathic small fiber neuropathy. Ann Neurol 71(1):26–39
Faber CG, Lauria G, Merkies IS, Cheng X, Han C, Ahn HS et al (2012b) Gain-of-function Nav1.8 mutations in painful neuropathy. Proc Natl Acad Sci U S A 109(47):19444–19449
Fertleman CR, Baker MD, Parker KA, Moffatt S, Elmslie FV, Abrahamsen B et al (2006) SCN9A mutations in paroxysmal extreme pain disorder: allelic variants underlie distinct channel defects and phenotypes. Neuron 52(5):767–774
Gold MS, Dastmalchi S, Levine JD (1996) Co-expression of nociceptor properties in dorsal root ganglion neurons from the adult rat in vitro. Neuroscience 71(1):265–275
Goral RO, Leipold E, Nematian-Ardestani E, Heinemann SH (2015) Heterologous expression of NaV1.9 chimeras in various cell systems. Arch Eur J Physiol 467(12):2423–2435
Habib AM, Wood JN, Cox JJ (2015) Sodium channels and pain. Handb Exp Pharmacol 227:39–56
Han C, Estacion M, Huang J, Vasylyev D, Zhao P, Dib-Hajj SD et al (2015) Human Na(v)1.8: enhanced persistent and ramp currents contribute to distinct firing properties of human DRG neurons. J Neurophysiol 113(9):3172–3185
Han C, Yang Y, Te Morsche RH, Drenth JP, Politei JM, Waxman SG et al (2017) Familial gain-of-function Nav1.9 mutation in a painful channelopathy. J Neurol Neurosurg Psychiatry 88(3):233–240
Hoeijmakers JG, Faber CG, Lauria G, Merkies IS, Waxman SG (2012) Small-fibre neuropathies – advances in diagnosis, pathophysiology and management. Nat Rev Neurol 8(7):369–379
Huang J, Han C, Estacion M, Vasylyev D, Hoeijmakers JG, Gerrits MM et al (2014) Gain-of-function mutations in sodium channel Na(v)1.9 in painful neuropathy. Brain J Neurol 137(Pt 6):1627–1642
Huang J, Vanoye CG, Cutts A, Goldberg YP, Dib-Hajj SD, Cohen CJ et al (2017) Sodium channel NaV1.9 mutations associated with insensitivity to pain dampen neuronal excitability. J Clin Invest 127(7):2805–2814
Jarecki BW, Piekarz AD, Jackson JO 2nd, Cummins TR (2010) Human voltage-gated sodium channel mutations that cause inherited neuronal and muscle channelopathies increase resurgent sodium currents. J Clin Invest 120(1):369–378
Kankel J, Obreja O, Kleggetveit IP, Schmidt R, Jorum E, Schmelz M et al (2012) Differential effects of low dose lidocaine on C-fiber classes in humans. J Pain 13(12):1232–1241
Kist AM, Sagafos D, Rush AM, Neacsu C, Eberhardt E, Schmidt R et al (2016) SCN10A mutation in a patient with erythromelalgia enhances C-fiber activity dependent slowing. PLoS One 11(9): e0161789
Kleggetveit IP, Namer B, Schmidt R, Helas T, Ruckel M, Orstavik K et al (2012) High spontaneous activity of C-nociceptors in painful polyneuropathy. Pain 153(10):2040–2047
Lampert A, O'Reilly AO, Reeh P, Leffler A (2010) Sodium channelopathies and pain. Arch Eur J Physiol 460(2):249–263
Lampert A, Eberhardt M, Waxman SG (2014) Altered sodium channel gating as molecular basis for pain: contribution of activation, inactivation, and resurgent currents. Handb Exp Pharmacol 221:91–110

Leipold E, Liebmann L, Korenke GC, Heinrich T, Giesselmann S, Baets J et al (2013) A de novo gain-of-function mutation in SCN11A causes loss of pain perception. Nat Genet 45(11):1399–1404

Leipold E, Hanson-Kahn A, Frick M, Gong P, Bernstein JA, Voigt M et al (2015) Cold-aggravated pain in humans caused by a hyperactive NaV1.9 channel mutant. Nat Commun 6:10049

Lin Z, Santos S, Padilla K, Printzenhoff D, Castle NA (2016) Biophysical and pharmacological characterization of Nav1.9 voltage dependent sodium channels stably expressed in HEK-293 cells. PLoS One 11(8):e0161450

McCarthy PW, Lawson SN (1997) Differing action potential shapes in rat dorsal root ganglion neurones related to their substance P and calcitonin gene-related peptide immunoreactivity. J Comp Neurol 388(4):541–549

Minett MS, Pereira V, Sikandar S, Matsuyama A, Lolignier S, Kanellopoulos AH et al (2015) Endogenous opioids contribute to insensitivity to pain in humans and mice lacking sodium channel Nav1.7. Nat Commun 6:8967

Mogil JS (2009) Animal models of pain: progress and challenges. Nat Rev Neurosci 10(4):283–294

Pan A, Wu H, Li M, Lu D, He X, Yi X et al (2012) Prenatal expression of purinergic receptor P2X3 in human dorsal root ganglion. Purinergic Signal 8(2):245–254

Phatarakijnirund V, Mumm S, McAlister WH, Novack DV, Wenkert D, Clements KL et al (2016) Congenital insensitivity to pain: fracturing without apparent skeletal pathobiology caused by an autosomal dominant, second mutation in SCN11A encoding voltage-gated sodium channel 1.9. Bone 84:289–298

Ritter AM, Mendell LM (1992) Somal membrane properties of physiologically identified sensory neurons in the rat: effects of nerve growth factor. J Neurophysiol 68(6):2033–2041

Rolyan H, Liu S, Hoeijmakers JG, Faber CG, Merkies IS, Lauria G et al (2016) A painful neuropathy-associated Nav1.7 mutant leads to time-dependent degeneration of small-diameter axons associated with intracellular Ca2+ dysregulation and decrease in ATP levels. Mol Pain 12:1744806916674472

Rostock C, Schrenk-Siemens K, Pohle J, Siemens J (2017) Human vs. mouse nociceptors – similarities and differences. Neuroscience. https://doi.org/10.1016/j.neuroscience.2017.11.047

Schrenk-Siemens K, Wende H, Prato V, Song K, Rostock C, Loewer A et al (2015) PIEZO2 is required for mechanotransduction in human stem cell-derived touch receptors. Nat Neurosci 18(1):10–16

Schwarz MG, Namer B, Reeh PW, Fischer MJM (2017) TRPA1 and TRPV1 antagonists do not inhibit human acidosis-induced pain. J Pain 18(5):526–534

Serra J (2010) Microneurography: an opportunity for translational drug development in neuropathic pain. Neurosci Lett 470(3):155–157

Serra J, Bostock H, Sola R, Aleu J, Garcia E, Cokic B et al (2012) Microneurographic identification of spontaneous activity in C-nociceptors in neuropathic pain states in humans and rats. Pain 153(1):42–55

Serrano A, Mo G, Grant R, Pare M, O'Donnell D, Yu XH et al (2012) Differential expression and pharmacology of native P2X receptors in rat and primate sensory neurons. J Neurosci 32(34):11890–11896

Shi TJ, Liu SX, Hammarberg H, Watanabe M, Xu ZQ, Hokfelt T (2008) Phospholipase C{beta}3 in mouse and human dorsal root ganglia and spinal cord is a possible target for treatment of neuropathic pain. Proc Natl Acad Sci U S A 105(50):20004–20008

Sittl R, Lampert A, Huth T, Schuy ET, Link AS, Fleckenstein J et al (2012) Anticancer drug oxaliplatin induces acute cooling-aggravated neuropathy via sodium channel subtype Na(V)1.6-resurgent and persistent current. Proc Natl Acad Sci U S A 109(17):6704–6709

Stucky CL, Lewin GR (1999) Isolectin B(4)-positive and -negative nociceptors are functionally distinct. J Neurosci 19(15):6497–6505

Takahashi K, Okita K, Nakagawa M, Yamanaka S (2007) Induction of pluripotent stem cells from fibroblast cultures. Nat Protoc 2(12):3081–3089

Theile JW, Jarecki BW, Piekarz AD, Cummins TR (2011) Nav1.7 mutations associated with paroxysmal extreme pain disorder, but not erythromelalgia, enhance Navbeta4 peptide-mediated resurgent sodium currents. J Physiol 589(Pt 3):597–608

Thomson JA, Itskovitz-Eldor J, Shapiro SS, Waknitz MA, Swiergiel JJ, Marshall VS et al (1998) Embryonic stem cell lines derived from human blastocysts. Science 282(5391):1145–1147

Valensi-Kurtz M, Lefler S, Cohen MA, Aharonowiz M, Cohen-Kupiec R, Sheinin A et al (2010) Enriched population of PNS neurons derived from human embryonic stem cells as a platform for studying peripheral neuropathies. PLoS One 5(2):e9290

Vallbo AB, Hagbarth KE (1968) Activity from skin mechanoreceptors recorded percutaneously in awake human subjects. Exp Neurol 21(3):270–289

Vanoye CG, Kunic JD, Ehring GR, George AL Jr (2013) Mechanism of sodium channel NaV1.9 potentiation by G-protein signaling. J Gen Physiol 141(2):193–202

Vega JA, Humara JM, Naves FJ, Esteban I, Del Valle ME (1994) Immunoreactivity for phosphorylated 200-kDa neurofilament subunit is heterogeneously expressed in human sympathetic and primary sensory neurons. Anat Embryol 190(5):453–459

Vetter I, Deuis JR, Mueller A, Israel MR, Starobova H, Zhang A et al (2017) NaV1.7 as a pain target – from gene to pharmacology. Pharmacol Ther 172:73–100

Wainger BJ, Buttermore ED, Oliveira JT, Mellin C, Lee S, Saber WA et al (2015) Modeling pain in vitro using nociceptor neurons reprogrammed from fibroblasts. Nat Neurosci 18(1):17–24

Waxman SG (2013) Painful Na-channelopathies: an expanding universe. Trends Mol Med 19(7):406–409

Wehrfritz A, Namer B, Ihmsen H, Mueller C, Filitz J, Koppert W et al (2011) Differential effects on sensory functions and measures of epidermal nerve fiber density after application of a lidocaine patch (5%) on healthy human skin. Eur J Pain 15(9):907–912

WHO-Guidelines (2008) Scoping document for WHO treatment guideline on non-malignant pain in adults. In: Adopted in WHO Steering Group on Pain Guidelines, 14 Oct 2008

Yang Y, Wang Y, Li S, Xu Z, Li H, Ma L et al (2004) Mutations in SCN9A, encoding a sodium channel alpha subunit, in patients with primary erythermalgia. J Med Genet 41(3):171–174

Yang T, Atack TC, Stroud DM, Zhang W, Hall L, Roden DM (2012) Blocking Scn10a channels in heart reduces late sodium current and is antiarrhythmic. Circ Res 111(3):322–332

Young GT, Gutteridge A, Fox H, Wilbrey AL, Cao L, Cho LT et al (2014) Characterizing human stem cell-derived sensory neurons at the single-cell level reveals their ion channel expression and utility in pain research. Mol Ther 22(8):1530–1543

Zhang XY, Wen J, Yang W, Wang C, Gao L, Zheng LH et al (2013) Gain-of-function mutations in SCN11A cause familial episodic pain. Am J Hum Genet 93(5):957–966

Gating Pore Currents in Sodium Channels

J. R. Groome, A. Moreau, and L. Delemotte

Contents

Abstract

Voltage-gated sodium channels belong to the superfamily of voltage-gated cation channels. Their structure is based on domains comprising a voltage sensor domain (S1–S4 segments) and a pore domain (S5–S6 segments). Mutations in

J. R. Groome (✉)
Department of Biological Sciences, Idaho State University, Pocatello, ID 83209, USA
e-mail: groojame@isu.edu

A. Moreau
Institut NeuroMyogene, ENS de Lyon, Site MONOD, Lyon, France

L. Delemotte
Science for Life Laboratory, Department of Physics, KTH Royal Institute of Technology, Box 1031, 171 21 Solna, Sweden

M. Chahine (ed.), *Voltage-gated Sodium Channels: Structure, Function and Channelopathies*, Handbook of Experimental Pharmacology 246,
https://doi.org/10.1007/164_2017_54

positively charged residues of the S4 segments may allow protons or cations to pass directly through the gating pore constriction of the voltage sensor domain; these anomalous currents are referred to as gating pore or omega (ω) currents. In the skeletal muscle disorder hypokalemic periodic paralysis, and in arrhythmic dilated cardiomyopathy, inherited mutations of S4 arginine residues promote omega currents that have been shown to be a contributing factor in the pathogenesis of these sodium channel disorders. Characterization of gating pore currents in these channelopathies and with artificial mutations has been possible by measuring the voltage-dependence and selectivity of these leak currents. The basis of gating pore currents and the structural basis of S4 movement through the gating pore has also been studied extensively with molecular dynamics. These simulations have provided valuable insight into the nature of S4 translocation and the physical basis for the effects of mutations that promote permeation of protons or cations through the gating pore.

Keywords
Arrhythmic dilated cardiomyopathy · Gating pore · Hypokalemic periodic paralysis · Molecular dynamics · Omega current · Sodium channel

1 Part I: Voltage Sensing Domains and Gating Pores

1.1 The Voltage Sensor Domain

Excitable cells function to propagate bioelectric signals through the generation of action potentials (AP). The cell membrane acts as a barrier to selectively modulate unbalanced ionic concentration gradients. Voltage-gated ion channel proteins in that membrane ensure the rapid and selective flux of ions during AP signaling through highly regulated opening and closing events. Their existence was proposed as early as 1952 where Hodgkin and Huxley stated that their recording of the AP in squid giant axon would rely on "molecules with a charge or dipole moment". From the deduced amino acid sequence of the first cloned voltage-gated ion (Na^+) channel (Noda et al. 1984), the hydrophobicity plot allowed the authors to propose that these large proteins were composed of 24 transmembrane segments organized into four domains.

It is now well accepted that voltage-gated ion channels feature two main functional structures, a channel pore domain (PD) that mediates ionic flux, and a voltage sensor domain (VSD) that modulates the PD. The PD has been used as a screening structure to identify the voltage-gated-like superfamily (VGL-Chanome) consisting of 143 members of voltage-gated and related cation channels (Yu and Catterall 2004). Many of the VGL-Chanome proteins are regulated by voltage, and a VSD-featuring superfamily can be built (Fig. 1a, b; Moreau et al. 2014).

The VSD is made of four transmembrane segments (S1–S4) that confer the protein its voltage sensitivity and thus modulate the PD function. VSDs can be divided into two main components: the positively charged S4 segments that contain

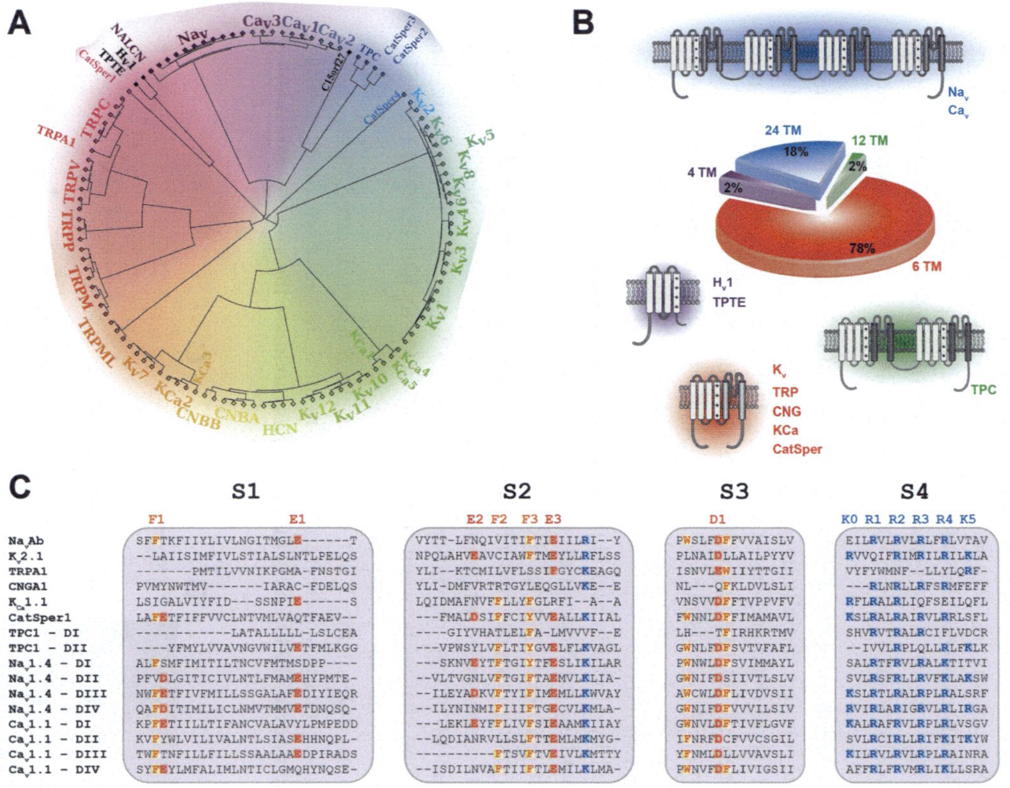

Fig. 1 The voltage-gated-like protein superfamily. (**a**) Phylogenic representation of proteins featuring a VSD. (**b**) Distribution of VSD-containing proteins into four families according to the number of transmembrane segments (TM): 24, 12, 6, or 4 TM. Main differences with the VGL-Chanome consist in the lack of two-pore K^+ channels and Kir channels while several proteins without PD such as Hv1 channels or TPTE appear in the superfamily. (**c**) Sequence alignment of S1–S4 segments of VSD-featuring proteins. Highly conserved amino acids are highlighted in the alignment (*top row*) and are displayed and in the schematic example of a VSD (*bottom left*). Most highly conserved amino acids comprise the positively charged S4 residues, negative and aromatic residues in S1–S3 segments, including the GCTC (Adapted from Moreau et al. 2014. Copyright Front Pharmacol)

3–7 positively charged arginine or lysine residues, and the S1–S3 segments that contain highly conserved, negatively charged residues which stabilize the highly charged S4 in the hydrophobic membrane environment (Fig. 1c). While the S4 segment is clearly mobile, the surrounding S1–S3 helices that stabilize that movement are considered as a more rigid structure (Banjeree and MacKinnon 2008; Tao et al. 2010). The gating charge transfer center (GCTC) formed primarily by two negatively charged residues and one aromatic residue (S2 and S3) is critical for this stabilization. Several studies demonstrate that S1–S3 segments are slightly more mobile than initially expected (Payandeh et al. 2012; Lin et al. 2010; Campos et al. 2007). Flexibility in S1–S3 may be requisite for optimal voltage gating; the creation of metal bridges between S2 and S3 segments significantly delays channel opening (Lin et al. 2010).

From an evolutionary point of view, the VSD would have emerged from a single transmembrane segment initially designed to sense the cellular volume through the membrane curvature (Kumanovics et al. 2002). The ancestral segment would most likely have been an S3-like segment since it is the most preserved. Given the existence of proteins consisting as VSD or PD only, these two structures may have evolved separately and were later assembled to construct voltage-gated ion channels as known today.

As predicted by Hodgkin and Huxley in 1952, the VSD possesses a "charged particle" shown later by Noda et al. (1984) to comprise the positively charged S4 segments, hence coined "gating charges". Movement of positive charge in S4 underlies the gating currents first recorded by Armstrong and Bezanilla (1973). Gating currents are recorded while ionic flux through the PD is blocked, and reveal the relationship of charge movement to voltage (Q/V, Fig. 2) representing the probability to find charges between stable resting and activated states of the VSD, according to the membrane potential.

The functional role of the VSD was investigated using directed mutagenesis where S4 positive charges were replaced by neutral, negative, or other positive charges (Stuhmer et al. 1989). Several mutations modulate the voltage-dependence

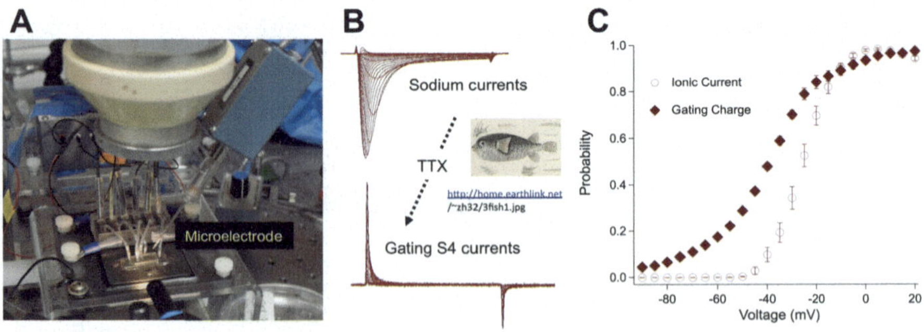

Fig. 2 Measurement of S4 movement. (**a**) Cut-open oocyte recording configuration. (**b**) Traces of sodium ionic currents, and gating (S4) currents after block by tetrodotoxin (TTX). (**c**) Probability of channel opening (ionic current, I/V) and gating charge movement (Q/V)

of activation, leading to early or late opening. It is now largely accepted that upon changes in membrane potential, the S4 segment undergoes a movement that initiates the modulation of the PD conformation. The first direct evidence of S4 movement was provided using cysteine scanning mutagenesis (Yang and Horn 1995; Yang et al. 1996); methane thiosulfonate (MTS) reagents immobilize the S4 segment as its movement reaches intracellular or extracellular space, leaving the channel non-functional.

S4 movement has been the subject of many debates and is still actively studied. The first proposed model of S4 movement has been called "sliding helix" or "helical screw" (Fig. 3a; Catterall 1986; Guy and Seetharamulu 1986; Yang et al. 1996). It predicts a large rotation and translation of the S4 segment during depolarization. The paddle model was proposed with the publication of the first crystal structure of a voltage-gated channel (KvAP, Jiang et al. 2003). That structure predicts an α-helically hairpin-like paddle motif located near the intracellular surface in the resting conformation. Under depolarized conditions, this paddle would move to the outer surface, carrying the gating charge. The paddle model has been contested with directed mutagenesis studies that reveal that at rest, the S4

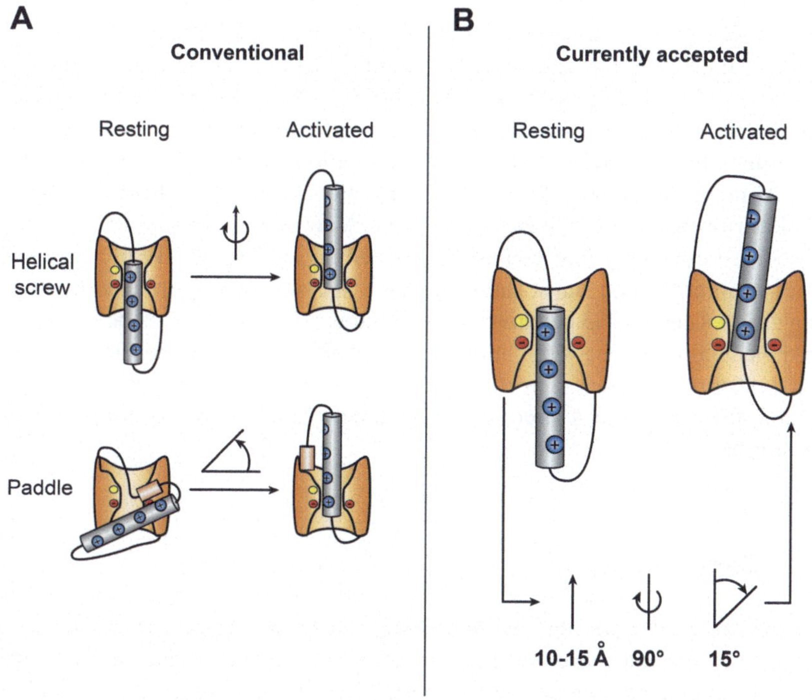

Fig. 3 Models of S4 segment motion during depolarization. In (**a**) the two traditional models of S4 motion during a depolarization are depicted (helical screw and paddle models). (**b**) Present modifications to the conventional model incorporate a slight rotation and tilt (Adapted from Moreau and Chahine 2015. Copyright Medicine Sciences)

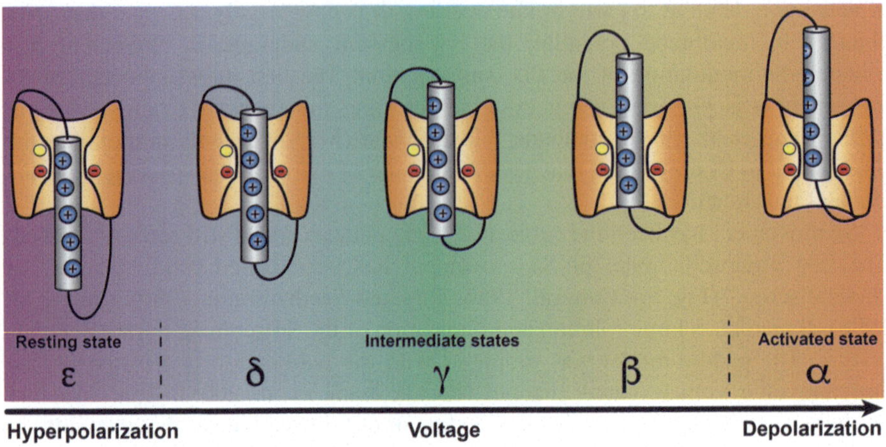

Fig. 4 Intermediate states of S4 movement. S4 segment movement from its hyperpolarized-favored position at rest (*left*) to a depolarized-favored position with activation (*right*) is facilitated by transient salt bridge interactions comprising S4 positive charges and S1–S3 negative countercharges (Adapted from Moreau and Chahine 2015. Copyright Medicine Sciences)

outermost charge is accessible from the extracellular side (Yang and Horn 1995; Posson et al. 2005). The currently accepted model of S4 movement reconciles helical screw and sliding helix models (Fig. 3b). At hyperpolarized voltages, the S4 segment is in its resting conformation with the outermost positive charge interacting with the GCTC. During its activation course, the S4 undergoes a 10–15 Å translation, a 60°–90° clockwise rotation, and a 15° tilt (Delemotte et al. 2011; Henrion et al. 2012; Jensen et al. 2012; Vargas et al. 2012).

This latter model presents activation proceeding in a ratchet-like motion through a series of intermediate states stabilized by a network of salt bridges between S4 gating charges and S1–S3 negative countercharges (Fig. 4). This model is predicted from molecular dynamics simulations and supported by the crystal structure of the *Ciona intestinalis* voltage sensitive protein (Li et al. 2014) in both resting and activated states, allowing a direct comparison between these two conformations and S4 position.

1.2 Gating Pores

While VSDs usually are impermeant to ionic flux, "gating pores" constitute such a permeation pathway directly through the VSD structure. The first ion current flowing through the gating pore was reported in a study of VSD function and more particularly S4 motion during voltage changes in the K^+ *Shaker* channel (Starace and Bezanilla 2004). When the first S4 arginine was replaced by histidine (R362H) and recorded in a non-conducting background, a large H^+ inward current was recorded at hyperpolarized voltages, suggesting a permeation pathway different from the physiological α-pore. These results were confirmed using agitoxin II (AgTxII), a toxin known to obstruct the physiological pore, as H^+ current amplitude was not affected by the addition of AgTxII. Since histidine can be protonated and deprotonated at physiological pH, the H^+ current amplitude recorded with the *Shaker* R362H mutant was sensitive to the H^+ driving force, leading the authors to propose that histidine can serve as a shuttle for H^+. The description of this novel permeation pathway was expanded when S4 arginines were replaced with amino acids such as alanine, cysteine, serine, or valine (Tombola et al. 2005), resulting in a cationic permeation pathway through the gating pore. The spatial location of that pathway was confirmed using MTS reagents that blocked the current when the first S4 arginine was replaced by cysteine. This novel permeation pathway was called the gating pore or omega (Ω) pore, distinct from the physiological α-pore of voltage-gated ion channels.

Due to their location, the biophysical properties of gating pore currents are intimately related to the biophysical properties of VSDs. Gating pores are created with mutations of VSD residues that participate in maintaining the hydrophobic septum in the VSD (Starace and Bezanilla 2004; Sokolov et al. 2007; Struyk and Cannon 2007; Gosselin-Badaroudine et al. 2012a; Groome et al. 2014; Moreau et al. 2015a, b). As such, mutations of outermost S4 arginines create gating pores opened at hyperpolarized voltages, supporting the thought that arginines interact with the GCTC at hyperpolarized voltages (i.e., when the VSD is in its resting position, Fig. 5). Conversely, mutations of innermost S4 arginines create gating pore currents at depolarized potentials (i.e., when the VSD is in an activated state). In native channels, the side-chain of the arginine has been proposed to occlude the permeation pathway (Tombola et al. 2005; Hong et al. 2013). Gating pore current selectivity is determined by the particular mutation. When S4 arginines are replaced with histidine, the gating pore current appears to be proton-selective (Starace and Bezanilla 2004; Gosselin-Badaroudine et al. 2012a, b). In contrast, when S4 arginines are replaced with other amino acids, gating pore current is cation-selective and non-specific.

The voltage-dependence of gating pore current is determined by the probability of localization of the mutated amino acid in the GCTC, as given by the Q/V curve. The relationship between gating pore currents and VSD voltage-dependence reveals notable differences between the four VSDs in Na_v channels (Gosselin-Badaroudine et al. 2012b). For example, for mutated arginines in the middle of the S4 segment, the maximum open probability of the gating pore is found at half maximal voltage of the Q/V relationship, where the probability to find the VSD in its resting versus activated state is equal. Transitions between VSD states are thus maximal and these should exactly correspond to the state where gating pore is most conductive (Starace and Bezanilla 2004). Longer duration depolarizations elicit immobilization of the S4 gating charge (Cha et al. 1999). This phenomenon may explain that for mutations of inner S4 arginines, a gating pore current can be recorded at hyperpolarized potentials after long depolarization (Sokolov et al. 2008; Fan et al. 2013; Groome et al. 2014; Moreau et al. 2015b). Finally, there is only a single published example of mutations outside the S4 segment that open a

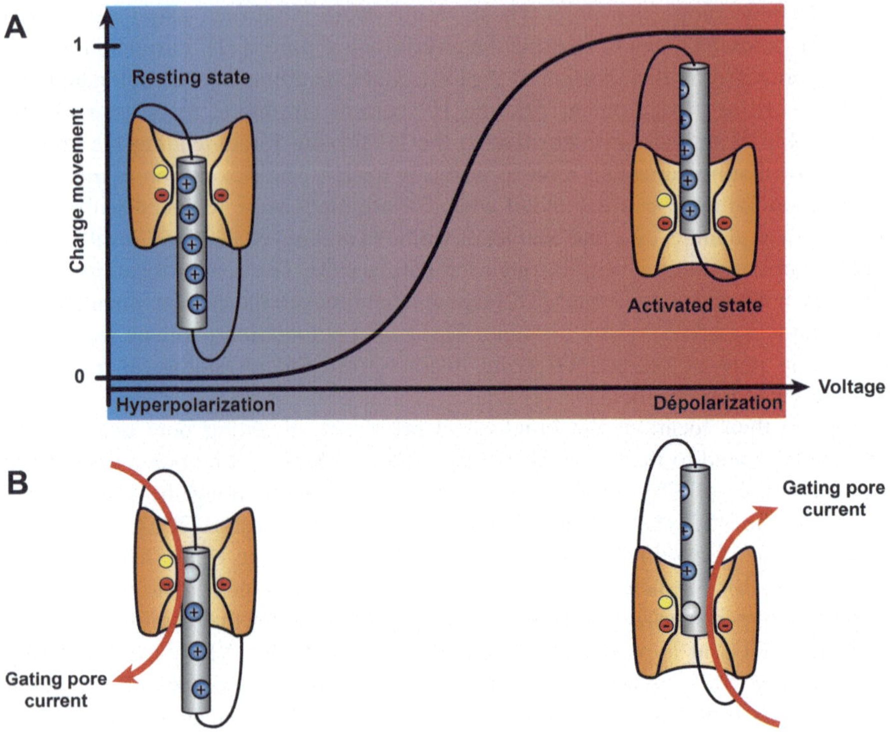

Fig. 5 Effect of S4 mutations on the omega current. (**a**) S4 translocation from hyperpolarized to depolarized-favored positions. (**b**) S4 mutations of outer arginines result in inward Na^+ omega current at resting membrane potential, whereas S4 mutations of inner arginines result in a large outward K^+ current during the depolarizing phase of the action potential (Adapted from Moreau and Chahine 2015. Copyright Medicine Sciences)

gating pore (Campos et al. 2007) for residues thought to maintain a portion of the hydrophobic septum.

The VSD is a protein structure non-permeable to ions without intrinsic ionic selectivity. Consequently, except for arginine to histidine S4 mutations, gating pores have been described as cation-selective (Tombola et al. 2005; Sokolov et al. 2007; Struyk et al. 2008; Francis et al. 2011; Groome et al. 2014; Moreau et al. 2015a). The selection for cations would be due to the deficit in positive charge with replacement of the native arginine (Delemotte et al. 2010; Khalili-Araghi et al. 2012) and further determined by the propensity of the permeating ion to dehydrate. Consequently, larger ions with a greater propensity to dehydrate are expected to be more permeant up to an exclusion size of 3.5 Å. For proton-selective gating pores, H^+ selectivity is explained by histidine protonation and deprotonation at physiological pH, with permeation most likely occurring through a Grotthus hopping mechanism (Starace and Bezanilla 2004).

1.3 Gating Pore Currents and Action Potentials

Gating pores have also been identified for mutations in patients suffering from skeletal muscle or cardiac disorders as described in Part II. The pathological nature of gating pores thus resides in the ion flux through this unusual permeation pathway. Since gating pores do not exhibit inactivation, ion flux due to omega current during an action potential is defined by driving force. Gating pores activated under hyperpolarized conditions are permeant when cells reach their resting membrane potential and until a depolarization is initiated; a Na^+ inward current will be the predominant current flowing through such gating pores while an outward K^+ current would be minimal. The case of gating pores activated under depolarized conditions is slightly more complicated since S4 immobilization also acts as a determinant. Indeed, during depolarized conditions, such pore will be mainly permeant to K^+, with lesser Na^+ current. During the falling phase of the action potential and membrane hyperpolarization, driving forces change instantly while the gating pore will not close if that S4 is immobilized. Consequently, for such pores a predominant inward Na^+ current will temporarily flow in the cell under hyperpolarized conditions.

2 Part II: Gating Pores and Sodium Channelopathies

2.1 Skeletal Muscle Channelopathies: Mutation-Based Phenotype and Gating Defects

A number of neuromuscular disorders have been linked to inherited mutations in voltage-gated ion channels of the outer sarcoplasmic or intracellular membrane (Morrill et al. 1998; Lehmann-Horn and Jurkat-Rott 1999; Lehmann-Horn et al. 2004; Jurkat-Rott and Lehmann-Horn 2005; Vicart et al. 2005; Venance et al. 2006; Jurkat-Rott et al. 2006, 2010; Cannon 2007, 2010; Matthews et al. 2010). These "channelopathies" have been investigated using heterologous expression of mutant channels and in some instances with knock-in mutations to characterize, respectively, the gating and behavioral defects associated with these mutations. For skeletal muscle sodium channelopathies, mutations found in patients with paramyotonia congenita, potassium aggravated myotonia, congenital myasthenic syndrome, or with periodic paralysis have been expressed in myotubes, mammalian cells, and/or in *Xenopus* oocytes for functional characterization. Alterations of gating properties observed in mutant channels for most sodium channelopathies are consistent with the observed clinical defects, and their description is reviewed elsewhere (Cannon 2010, 2015).

Interestingly, a number of mutations of the skeletal muscle sodium channelopathies have homologs in cardiac syndromes (Loussouarn et al. 2016). Here we focus on the impact of skeletal muscle Na_v mutations in hypokalemic periodic paralysis (HypoPP) type II, with mutations in domains I to III (Fig. 6 and Table 1), and follow with a description of cardiac muscle Na_v mutations in arrhythmic dilated

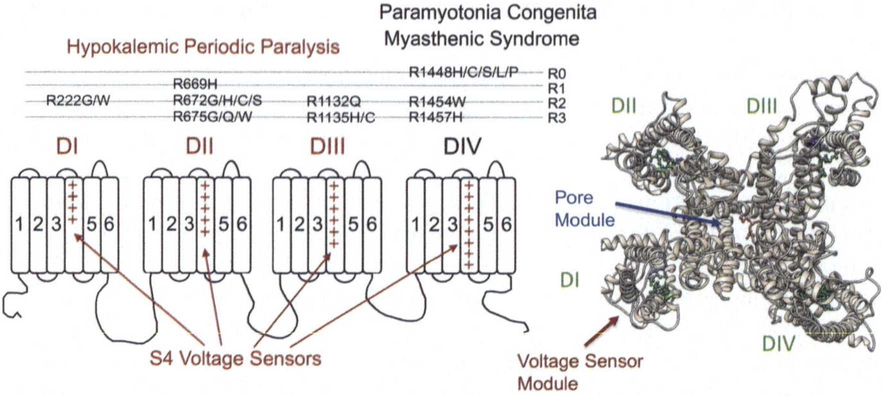

Fig. 6 Channelopathy mutations in $Na_v1.4$. Diagram of $hNa_v1.4$ at *left*, with positions of arginine residues (+) in S4 segments, and mutations for HypoPP type II (domains I–III), PC or CMS (domain IV). Homology model of $hNa_v1.4$ based on cryo-EM structure 5X0M.pdb from *Periplaneta*, at *right*. Domains comprise a peripheral voltage sensor module (S1–S4, with S4 residues in *green*) and a central pore module (S5–S6). The IFMT inactivation motif (*orange*) of the DIII–DIV linker is shown

cardiomyopathy in domains I and II (Table 2). For each of these sodium channelopathies, gating pore current appears to be a significant contributing factor to disease pathogenesis.

For HypoPP, a consistent theme of S4 segment mutations is observed for sodium (*SCN4A*; $Na_v1.4$) or calcium (*CACNA1S*; Cav1.1) channels (Matthews et al. 2009), with mutations in potassium channels *KCNJ2* (Kir2.1) and *KCNJ18* (Kir2.6) associated with periodic paralysis in Andersen-Tawil syndrome or (thyrotoxic) non-familial periodic paralysis (Plaster et al. 2001; Ryan et al. 2010; Tristani-Firouzi and Etheridge 2010; Burge and Hanna 2012; Lin and Huang 2012; Cheng et al. 2011, 2013). Most inherited HypoPP mutations in patients are found in *CACNA1S* (60%, HypoPP type I) with about 10% localized to *SCN4A* (HypoPP type II; Platt and Griggs 2009). These latter mutations have been identified for S4 residues in positions R1–R3, and only rarely elsewhere with overlap syndromes of myotonia (Sugiura et al. 2003; Webb and Cannon 2008). Characterization of HypoPP type II mutations focused initially on several mutations in DII-S4 (Jurkat-Rott et al. 2000; Struyk et al. 2000; Bendahhou et al. 2001; Kuzmenkin et al. 2002) with generalized effects of reduced sodium channel current, and enhanced fast or slow inactivation (Table 1).

2.2 Hypokalemic Periodic Paralysis: Role of the Omega Current

Skeletal muscle fibers maintain a normal, negative membrane potential in response to perturbation, through mechanisms that include inwardly rectifying potassium (Kir) channels and Na/K/Cl transport (Geukes Foppen et al. 2002;

Table 1 Mutations associated with gating pore currents in hypokalemic periodic paralysis type II (skeletal muscle sodium channel, *SCN4A*)

Mutation	Location	Reference genetics	Reference function	Reference gating pore
Domain I				
R222G	DIS4-R2	Holzherr et al. (2010)	Holzherr et al. (2010)[a,b]	Holzherr et al. (2010) and Jurkat-Rott et al. (2012)
Domain II				
R669H	DIIS4-R1	Bulman et al. (1999)	Struyk et al. (2000)[b,c]	Struyk and Cannon (2007)
			Kuzmenkin et al. (2002)[a,b,c,d,e]	Sokolov et al. (2007) and Mi et al. (2014)
R672H	DIIS4-R2	Jurkat-Rott et al. (2000)	Jurkat-Rott et al. (2000)[a,d,e,f,g]	Struyk et al. (2008)
			Kuzmenkin et al. (2002)[a,b,c,e,h]	Sokolov et al. (2007, 2010)
R672G	DIIS4-R2	Jurkat-Rott et al. (2000)	Jurkat-Rott et al. (2000)[a,b,d,e,i,j]	Struyk et al. (2008)
			Kuzmenkin et al. (2002)[a,b,c,d,e,h]	Sokolov et al. (2007)
		Sternberg et al. (2001)	Sokolov et al. (2007)[a,h]	Sokolov et al. (2010) and Mi et al. (2014)
R672S	DIIS4-R2	Kim et al. (2004)	Bendahhou et al. (2001)[a,b,c,d]	Struyk et al. (2008)
R675G	DIIS4-R3	Vicart et al. (2004) and Song et al. (2012)	Sokolov et al. (2008)	Sokolov et al. (2008)
			None	
R675Q	DIIS4-R3	Vicart et al. (2004) and Song et al. (2012)	Sokolov et al. (2008)[k,l,m]	Sokolov et al. (2008)
R675W	DIIS4-R3	Vicart et al. (2004)	Sokolov et al. (2008)[ek]	Sokolov et al. (2008)
Domain III				
R1132Q	DIIIS4-R2	Carle et al. (2006)	Carle et al. (2006)[a,b,f,h]	Francis et al. (2011)
R1135H	DIIIS4-R3	Matthews et al. (2009)	Groome et al. (2014)[a,b,c,d,g]	Groome et al. (2014)
R1135C	DIIIS4-R3	Groome et al. (2014)	Groome et al. (2014)[a,b,c,d]	Groome et al. (2014)

[a]Enhanced FI – left shift
[b]Enhanced SI – left shift
[c]Slowed recovery from SI
[d]Slowed recovery from FI
[e]Decreased current density
[f]Slowed entry into FI inactivation
[g]Slowed activation
[h]Decreased activation – right shift or decreased voltage sensitivity
[i]Accelerated inactivation

(continued)

Table 1 (continued)
[j]Accelerated activation
[k]Slowed entry into SI
[l]Accelerated recovery from SI
[m]Decreased voltage sensitivity, SI
[n]Increased persistent or window current

Table 2 Mutations associated with gating pore currents in dilated cardiomyopathy (cardiac sodium channel, *SCN5A*)

Mutation	Location	Reference genetics	Reference function	Reference gating pore
Domain I				
R219H	DIS4-R1	Gosselin-Badaroudine et al. (2012a)	Gosselin-Badaroudine et al. (2012a)	Gosselin-Badaroudine et al. (2012a)
				None
R222Q	DIS4-R2	Hershberger et al. (2008), Laurent et al. (2012), and Mann et al. (2012)	Cheng et al. (2010)[a,b] and Laurent et al. (2012)	Moreau et al. (2015a)[a,b,c,d]
				Mann et al. (2012)[a,b,d]
				Nair et al. (2012)[a,b]
R225P	DIS4-R3	Beckermann et al. (2014)	Beckermann et al. (2014)	Moreau et al. (2015b)[d,e,f,g]
R225W	DIS4-R3	Bezzina et al. (2003)	Bezzina et al. (2003)	Moreau et al. (2015a)[f,h,i]
Domain II				
R814W	DIIS4-R3	Nguyen et al. (2008) and Olson et al. (2005)	Nguyen et al. (2008)[b,d,j,k,l,m,n,o]	Moreau et al. (2015b)

[a]Enhanced FI – left shift
[b]Enhanced activation
[c]Accelerated entry into FI
[d]Enhanced persistent or window current
[e]Slowed entry into FI inactivation
[f]Decreased activation – right shift or decreased voltage sensitivity
[g]Increased current density
[h]Decreased FI – right shift
[i]Decreased current density
[j]Slowed recovery from FI
[k]Enhanced SI – more complete
[l]Accelerated entry into SI
[m]Slowed recovery from SI
[n]Slowed activation
[o]Slowed deactivation

Struyk and Cannon 2008). With a severe drop in extracellular potassium, Kir channels become increasingly ineffective at stabilizing membrane potential (Struyk and Cannon 2008), with muscle fibers increasing in tendency to exhibit bi-stable membrane potentials, P_1 and P_2 (Jurkat-Rott et al. 2009). The potential at P_2 is a "paradoxical" depolarization in response to reduced potassium, and while rare in normal muscle fibers (i.e., reduction of serum potassium to 1 mM), *CACNA1S* or *SCN4A* HypoPP S4 mutations provide a depolarizing influence on membrane potential with a modest decrement in serum potassium observed in patients.

The depolarization of patient muscle fibers in response to hypokalemia is not explained by effects of S4 mutations on inactivation (Ruff 2000). A mechanism for paradoxical depolarization was suggested following the identification of proton- and cation-selective gating pore currents in *Shaker* K^+ channels as described earlier. The possibility that channelopathy-linked mutations might promote omega current was confirmed with the identification of gating pore currents for HypoPP type II $Na_v1.4$ S4 mutations in DII-S4 (Sokolov et al. 2007, 2008; Struyk and Cannon 2007; Struyk et al. 2008). Gating pore currents for HypoPP type I Cav1.1/R1242G (DIV-R3) have been studied using transfected myotubes (Fan et al. 2013), and a recent report describes robust gating current measurements in Cav1.1/R528H (DII-R1) using an enhanced expression system (Wu and Cannon 2017).

Gating pore currents have been identified in mutations of HypoPP type II S4 arginine residues in domains I–III (Table 1), but are not associated with myotonia (R1448C) in domain IV (Francis et al. 2011) whose voltage sensor module contains a larger hydrophobic septum (Gosselin-Badaroudine et al. 2012b). These gating pore currents are determined by blocking ionic conductance with tetrodotoxin, measuring total leak current, and normalizing that leak against the ohmic relationship. In DII-S4, an inwardly directed proton current is caused by histidine mutations at DII-S4 R1 and R2 in $rNa_v1.4$ (analogous to HypoPP mutations R669H and R672H in the human isoform $hNa_v1.4$) and is observed at hyperpolarized voltages under physiological conditions for the proton driving force (Sokolov et al. 2007; Struyk and Cannon 2007; Struyk et al. 2008; Wu et al. 2011). Other substitutions at DII-S4 R2 (R672G/S/C) provide cationic permeation through the gating pore (Fig. 7), as does R1132Q (DIII-S4 R2, Francis et al. 2011), or R222G (DI-S4 R2, Holzherr et al. 2010; Jurkat-Rott et al. 2012).

Estimates of the inward current through the gating pore in DII-S4 mutations suggest that omega currents contribute less than 1% of the central pore conductance (Struyk and Cannon 2007; Sokolov et al. 2008; Struyk et al. 2008). Nevertheless, this contribution may result in paradoxical depolarization of skeletal muscle fibers and resulting paralysis with serum K^+ decrement within the range typically observed in patients, and is considered as a significant contributing factor in the pathogenesis of the disease (Cannon 2010; Jurkat-Rott et al. 2012). Additional support for gating pore current as a contributor to the pathogenesis of HypoPP is the correlation of mutation-induced cation leak with muscular sodium overload and edema (Jurkat-Rott et al. 2009; Nagel et al. 2014; Weber et al. 2016).

Several *SCN4A* S4 mutations at R3 produce gating pore currents in response to membrane depolarization including R675Q/G/W (Sokolov et al. 2008) in DII-S4, and rat R1128H/C (analogous to HypoPP mutations R1135H/C in $hNa_v1.4$) in DIII-S4 (Groome et al. 2014). Omega currents observed with hyperpolarization of R675Q/G/W channels recovering from slow inactivation or R1128H/C channels

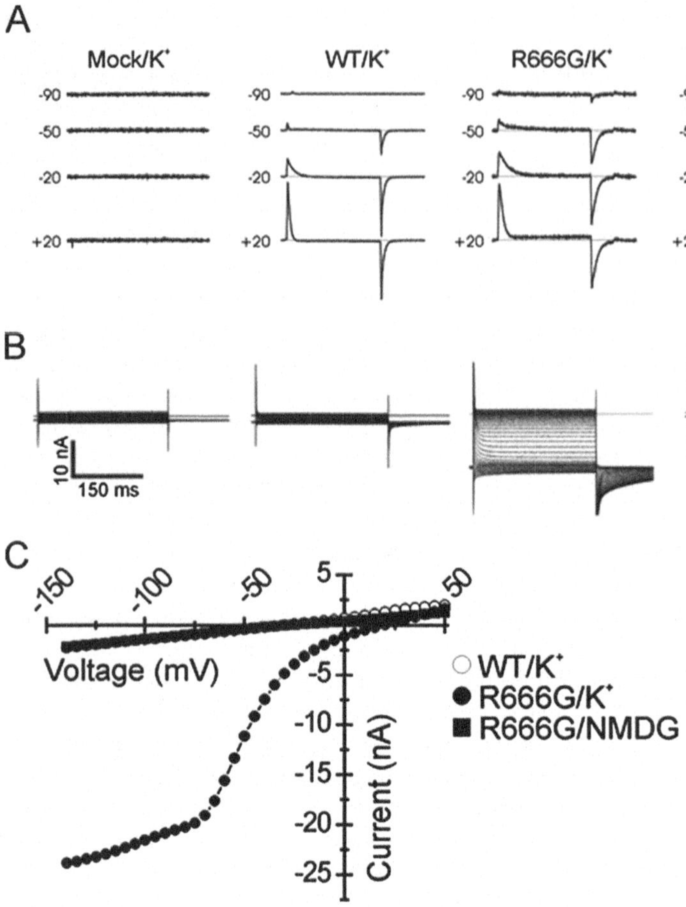

Fig. 7 Classic features of omega current. (**a**) Gating current amplitudes are similar for wild type and hypokalemic periodic paralysis mutation rR666G (R672G in human isoform) (**b**) Leak current is observed only in the mutant channel, and over a limited range of membrane potential (**c**) Adapted from Struyk et al. (2008). Copyright Journal of General Physiology

recovering from fast inactivation are larger in amplitude than those elicited with depolarization. These findings suggest that gating pore currents for R3 mutants might contribute more to a prolonged recovery from slow- or fast inactivation than through possible effects on activation through depolarization-activated currents.

Mi et al. (2014) found that DII-S4 mutations R669H and R672G each significantly decrease the coupling of gating charge displacement to peak ionic current, suggesting a mechanism for action potential disruption in muscle fibers harboring those mutations by uncoupling activation from pore opening. Charge remobilization in DIII-S4 mutations rR1128H/C is disrupted to promote prolonged recovery (Groome et al. 2014). It will be interesting to see if effects of HypoPP mutations on charge movement and omega current are causally linked to explain observed gating defects in this disorder.

2.3 Cardiac Channelopathies

Cardiac arrhythmia and failure comprise a complex set of pathologies that include a contribution from inheritance of mutations in ion channels including the cardiac sodium channel gene *SCN5A* (Chockalingham and Wilde 2012; Bezzina et al. 2015). A wide array of mutations in this gene have been identified in patients with Brugada syndrome exhibiting loss of function phenotype, or with long QT syndrome type 3 (LQT3) presenting with gain of function phenotype; these syndromes have been extensively characterized with electrophysiological analyses of the effects of *SCN5A* mutations (reviewed by Jones and Ruben 2008). The finding of proton- and cation-selective gating pores associated with HypoPP prompted the hypothesis that gain of function mutations in the VSD of *SCN5A* associated with LQT3 might also be associated with a gating pore current (supplemental Table in Sokolov et al. 2007).

2.4 Cardiac Channelopathies: Role of the Omega Current

Interestingly, the first demonstration of gating pore current with cardiac channelopathies has come from the study of several mutations in patients with combined cardiac arrhythmia and dilated cardiomyopathy (DCM, Table 2). Surveys of mutations in patients with arrhythmic DCM have revealed a common locus of mutations to the voltage-sensing regions S3 and S4 (McNair et al. 2011; Gosselin-Badaroudine et al. 2014). In domain I, an R219H proton current was characterized using a TTX-sensitive mutation of $hNa_V1.5$; the mutation exhibits a hyperpolarization-activated leak current amplified under increasingly acidic conditions (Gosselin-Badaroudine et al. 2012a). The proposed proton-wire mechanism of R219H in patients might promote gain of function arrhythmia with myocardial depolarization caused by increased inward current; acidification of myocytes could disrupt (sodium and) calcium homeostasis (Fabiato and Fabiato 1978) and gap junction (connexon43) properties (Bukauskas and Peracchia 1997; Bukauskas et al. 2001) to contribute as pathogenic mechanisms for the DCM phenotype.

SCN5A DI-S4 mutations R222Q (R2) and R225W (R3) in other patients with a mixed arrhythmia and DCM phenotype exhibit opposing biophysical effects; R222Q promotes gain of function (increased current density and enhanced window current, Mann et al. 2012), while R225W promotes loss of function (decreased current density and depolarizing shift of activation, Bezzina et al. 2003). Investigation of these mutations in heterologous expression and with molecular dynamics reveals that while they produce very different effects on gating properties, each produce a depolarization-activated gating pore current (Fig. 8, Moreau et al. 2015a) with greatest permeation to cesium, and significant for both sodium and potassium.

Additional mutations in patients with mixed arrhythmias and DCM include R225P (DI-S4 R3, Beckermann et al. 2014) and R814W (DII-S4 R3, Nguyen et al. 2008) that affect the voltage-dependence of channel activation (hyperpolarizing shift) and kinetics (slowed), in addition to an enhanced window current similar to that

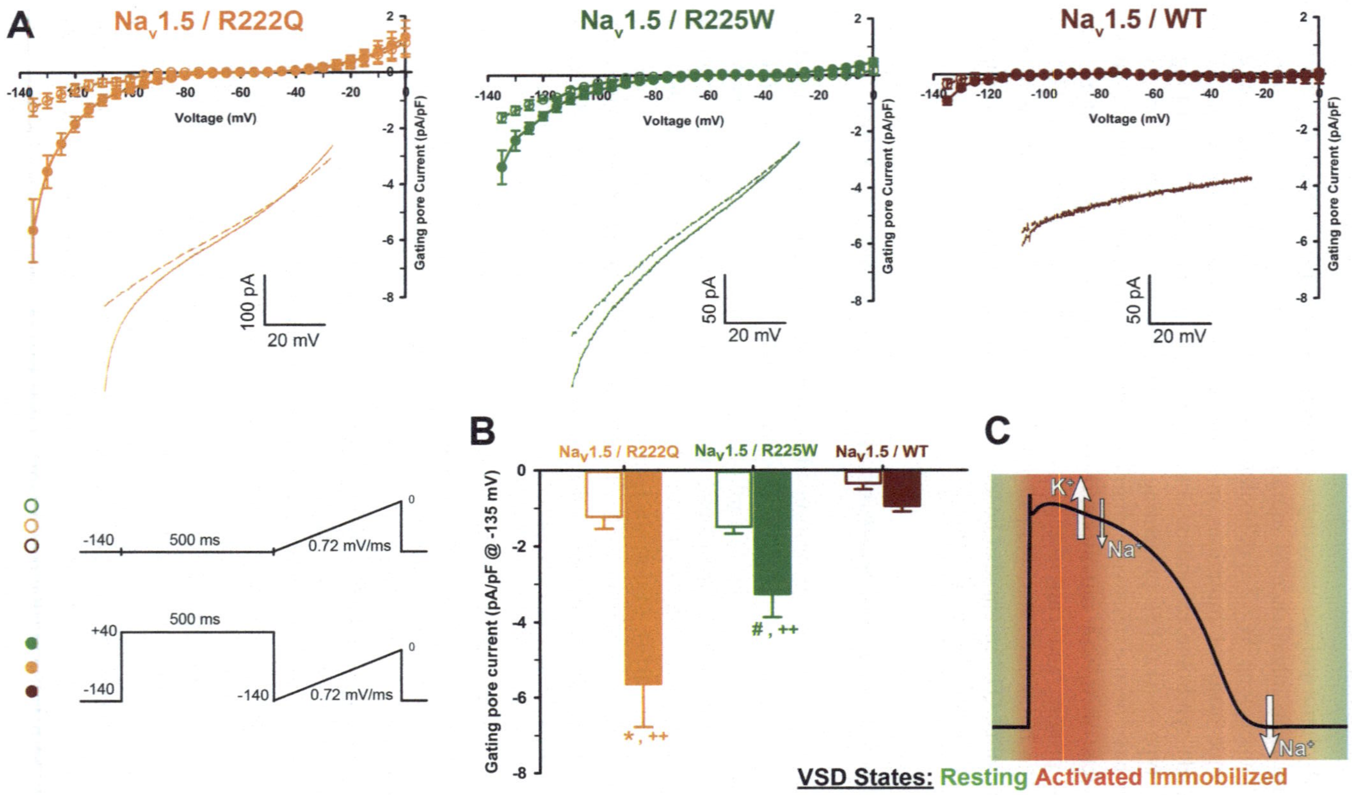

Fig. 8 Gating pore currents in *SCN5A* DCM mutations. (**a**, **b**) Omega currents in R222Q and R225W. Produced following prolonged depolarization are larger in amplitude than those elicited from a hyperpolarized potential. (**c**) Proposed gating pore flux from resting potential and during an immobilized state. Adapted from Moreau et al. (2015b) Copyright Journal of General Physiology

observed with R222Q and R225W mutations. These mutations produce a gating pore conductance similar to that shown for R222Q and R225W (Moreau et al. 2015b). Thus, a number of cardiac mutations in patients with the arrhythmic/DCM phenotype produce a divergent set of biophysical defects with a common gating pore conductance, and are located at several loci in S4 segments of two domains. For each of the aforementioned S4 mutations at R2 or R3 in *SCN5A*, gating pore currents are observed during hyperpolarization following long duration depolarization. Together, these findings suggest that for these mutations, the channel gating pore conducts cationic current during both activation and recovery to promote the arrhythmic phenotype (Moreau et al. 2015a, b).

2.5 Pharmacology of Sodium Channelopathies Associated with Gating Pore Current

The carbonic anhydrase inhibitors acetazolamide and dichlorphenamide have been utilized in treatment of periodic paralysis with some efficacy in reducing the severity and frequency of attacks in HypoPP patients (Venance et al. 2006; Sansone et al. 2008, 2016; Tricarico and Camerino 2011). These agents may produce benefit by activation of Ca^{2+} sensitive K^+ channels (Tricarico et al. 2004, 2006); acetazolamide, but not dichlorphenamide, may limit vacuolar myopathy (Tricarico et al. 2008). Acetazolamide is significantly more effective at treating symptoms of patients with HypoPP type I versus type II (Matthews et al. 2011) and is least effective in patients with glycine mutations in HypoPP. While acetazolamide does not block the gating pore conductance in the $Na_v1.4$ mutant R1132Q (Francis et al. 2011), it is speculated that proton permeation elicited by histidine mutations in $Na_v1.4$ is a more attractive target to CA inhibitors (Matthews et al. 2010). Pharmacological targeting of Na/K/Cl transport by bumetanide has proven effective in alleviating muscle weakness in transgenic mice harboring histidine HypoPP mutations (Wu et al. 2013a, b).

Block of the gating pore current for DII-S4 mutations by divalent cations was reported by Sokolov et al. (2007); dose-response of gating pore block by divalent or trivalent cations is reported at several hundred μM (Sokolov et al. 2010). A study by Struyk et al. (2008) suggests that divalent cations diminish gating pore conductance as an indirect effect on the leak conductance. Guanidine derivatives have also been tested on gating pore conductance according to the proposed occlusion of the gating pore by guanidyl side-chain of arginine in native channels as that residue enters the gating pore (Sokolov et al. 2010). Of the compounds tested, 1-(2,4-xylyl) guanidine carbonate was shown to partially block the gating pore current in rR666G (analogous to R672G in $hNa_v1.4$) carried by either sodium ions or guanidine (sulfate). These findings suggest that while present treatments that affect skeletal muscle fibers "downstream" of gating pore conductance are most effective, development of small molecules to specifically target the gating pore itself does hold promise for eventual pharmaceutical application (for review see Moreau et al. 2014).

3 Part III: Computational Approaches to Investigate Gating Pore Current

3.1 Action Potential Modeling

The significance of channelopathy mutations at the cellular level may be investigated by determining the impact of gating defects in mathematical simulations of action potential initiation. Skeletal and cardiac muscle fiber muscle models employ the electrical correlates of the lipid bilayer and ion channels of the cell membrane (and) transverse tubules for which modified Hodgkin-Huxley parameters are utilized in a series of differential equations to simulate individual or repetitive action potential firing (Cannon et al. 1993; Filatov et al. 2005), and as used to describe the effects of mutation-induced gating defects in myotonia or paralysis (Hayward et al. 1997; Richmond et al. 1997; Featherstone et al. 1998; Fan et al. 2013; Groome et al. 2014; Mankodi et al. 2015).

Mathematical modeling has been used to investigate the role of omega current in promoting a bi-stable membrane potential in muscle fibers of patients with HypoPP. Here, the relative importance of the balance of the inward leak current (I_{Leak}) with outward current from Kir and delayed rectifier (KDR) channels has been demonstrated empirically with barium-treated muscle fibers (Struyk and Cannon 2008). Simulations predict that the small gating pore conductance (I_{GP}) elicits a decrease in potassium reduction necessary to promote a membrane potential stabilized at the balance between I_{Leak} and I_{KDR} at P_2, instead of a balance between I_{Leak} and I_{Kir} at P_1. These parameters are included in simulations for native or mutant muscle fibers in response to decreased K^+ (Cannon 2010; Jurkat-Rott et al. 2012; Fan et al. 2013).

3.2 All-Atom Molecular Dynamics

Another type of mathematical modeling that has extensively been used to investigate the impact of mutations in voltage-gated ion channels, and in particular of mutations responsible for the appearance of gating pore currents, is all-atom molecular dynamics (MD) simulation (Leach 2001). In such a model, atoms are represented as particles with a defined mass and charge and their interactions are tabulated in a so-called force field. More precisely, the interaction force field can be divided into two parts: bonded interactions (typically, bonds, angles, and dihedral angles are described by simple equations representing the oscillations around their equilibrium values) and non-bonded ones (divided typically into electrostatic and van der Waals interactions). The force field parameters have been derived and refined according to various recipes over the years and are typically now considered accurate enough to represent biological systems such as proteins, nucleic acids, and model membranes. Typical force fields used for biophysical systems today are CHARMM (MacKerell et al. 2002), AMBER (Wang et al. 2006), and OPLS (Kaminski et al. 2001).

The systems of interest typically consist of an ion channel embedded in a model membrane patch (such as a POPC bilayer) and plunged in a salt solution. Such systems represent usually a few hundred thousand atoms, and it is thus impossible to solve the equations of motion describing such a system analytically. A numerical approach is then taken where time is discretized to time steps less than the period of the shortest oscillation occurring in the system (i.e., bond oscillations involving light atoms such as hydrogens) (Allen and Tildesley 1987). The equations of motion are then solved iteratively and the process is repeated to reach the time scale of interest. With a time step of ~1 fs (10^{-15} s), 10^9 steps are needed to reach the microsecond time scale; high performance computers are generally used to complete the simulations.

Given the timescales accessible to MD simulations, folding of membrane proteins cannot be observed and modelers rely on the availability of experimentally determined protein structures. Whereas initial structures were obtained exclusively by X-ray crystallography, recent progress in cryo-electron microscopy has provided the community with a large number of structures for additional members of the voltage-gated ion channel superfamily (Egelman 2016). Starting from these structures, MD simulations then allow one to observe the time evolution of an ion channel at equilibrium, or submitted to an external stimulus such as a transmembrane potential. Beyond the dynamics, these simulations can allow proteins to cross free energy barriers and reach transition states that have not been characterized experimentally. If sampling is adequate, it is even possible to reconstruct the full free energy landscape along the important degree of freedom of the system through the Boltzmann relationship:

$$\Delta G(x) \propto -kT \log P(x) \quad (1)$$

where $\Delta G(x)$ represents the free energy difference along x the desired degree of freedom, k is the Boltzmann constant, T the temperature of the system, and $P(x)$ the probability distribution along x. Using MD simulations, it is also possible to study the effect on the structure and dynamics of the system in response to minor perturbations in temperature, the binding of a small molecule, or, of particular interest here, the effect of a point mutation.

For much of the early work the effect of S4 mutations, the only high resolution structure of a voltage-gated ion channel that was available was one in the Kv1 family, the rat Kv1.2 channel. Even though no mutation with clinical relevance was reported in the S4 segment of the human isoform of this channel, it was used as a prototype to observe the molecular level effect of S4 mutations. This was deemed reasonable because of the sequence and structure similarity between the members of the large family of VGCCs (Fig. 1).

First, "artificial mutants" for which specific S4 residues were substituted by uncharged equivalents were considered, mimicking a mutation to glutamine. MD simulations and solvent accessibility experiments revealed that Kv channel VSDs are filled with water molecules extending from the intra- and extra-cellular media and which solvate the gating S4 charges and negatively charged salt bridge counterparts (Treptow and Tarek 2006; Jogini and Roux 2007; Krepkiy et al. 2009; Freites et al. 2006; Larsson et al. 1996). The central part of the VSD is dehydrated, thanks to the presence of a hydrophobic plug comprised mainly by a conserved Phe residue on S2 (Tao et al. 2010; Delemotte et al. 2011; Khalili-Araghi et al. 2010; Schwaiger et al. 2012). During activation, as the S4 segment translocates across the bilayer, the salt bridge network rearranges: whereas in the activated states of the VSD inner S4 charges are close to the constriction, in the resting state, outer S4 charges take their place. It is therefore expected that mutation of the charges involved in this constriction will affect VSD stability.

A set of simulations of the Kv1.2 full channel was conducted, in the open/activated and the closed/resting state, respectively, for which these specific residues were substituted by uncharged equivalents (Fig. 9; Delemotte et al. 2010). Mutations to an uncharged residue always disrupted the salt bridge in which the residue is involved, but, as expected, only the mutation of residues involved in the constriction region (K5/R6 double mutant and K5 single mutant in the activated state and R1 single mutant in the resting state) opened a hydrophilic pathway through the VSD, identified as the omega pore (Fig. 5). When submitted to a highly depolarized transmembrane voltage, the activated state double (K5/R6) conducted K^+ ions in the outward direction. The activated K5 single mutant, on the other hand,

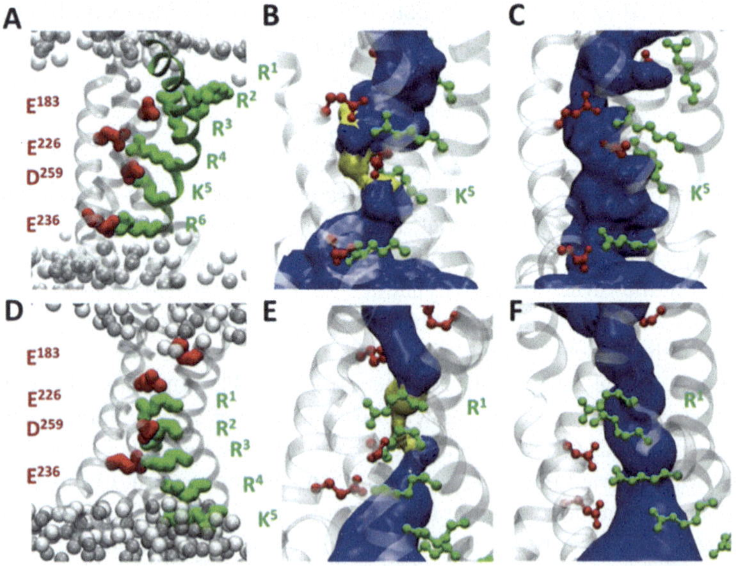

Fig. 9 Topologies of the activated (*top panels*) and resting (*lower panels*) Kv1.2 VSD conformations. (**a**, **d**) Location of the S4 gating charges (*green*) and the salt bridges they form with negatively charged residues (*red*). (**b**, **e**) Solvent-accessible volume (*blue*) within the VSD of the WT channel, for which the most constricted regions (pore radius <1.15 Å) are depicted in *yellow*. (**c**, **f**) K5 and R1 respective mutants, in which omega pores are formed (Adapted from Delemotte et al. 2010), copyright Biophysical Journal

showed no conduction event over the time scale of the simulation. In the resting state single R1 mutant, a large hyperpolarized transmembrane voltage generated inwards K^+ conduction. The ion permeation pathway is similar to the pathway taken by the gating charges during VSD activation and deactivation (Delemotte et al. 2010). The simulations conducted at the time, however, were too short to calculate meaningful conductance values.

Tens of K^+ and Cl^- conduction events were later observed in simulations of the 100 ns timescale of a resting-state Kv1.2 VSD R294S/E226D variant (equivalent to *Shaker* R326S/E283D). These simulations identified the putative selectivity filter for cations as a region at the top of the extracellular VSD crevice containing important acidic side chains (Khalili-Araghi et al. 2012). Long (10 μs timescale) simulations of the R296S Kv1.2–Kv2.1 paddle-chimera VSD (equivalent to R365S in *Shaker*) in a resting-state conformation revealed the appearance of K^+ currents and identified the K^+ permeation pathway (Jensen et al. 2012). A recent paper complements these investigations by reporting μs-long simulations of the *Shaker* VSD variant R362S/E283D/S357C/M356D (the so-called big omega-current or BOM mutant (Tombola et al. 2006)) in the resting-state. The authors rationalize that permeation of guanidinium is more favorable than that of K^+ likely because the ion permeation pathway is lined by binding sites that have evolved to bind arginine side-chains (Wood et al. 2017).

For channels of unknown structure, homology modeling has been useful to understand the effect of S4 mutations. In particular, bacterial sodium channel crystal structures have been used as templates to gain an insight in the effect of genetic mutations of human voltage-gated ion channels. For example, in $Na_v1.4$, single histidine (H) substitutions of gating charges were used to probe the hyperpolarized-favored positions of the S4 helices in each domain (Gosselin-Badaroudine et al. 2012b). Indeed, the appearance of a leak current indicated that the mutated residue was accessible from both sides of the membrane, thus enabling placement of DI/R1, DII/R1, and DIII/R2 in the constriction in the VSD. The case of DIV was complicated by the fact that the single R1H, R2H, and R3H mutations did not produce proton currents. Molecular dynamics simulations of a homology model of the resting states of $Na_v1.4$ built using a Na_vAb sodium channel X-ray structure as a template provided plausible explanations for the observed phenomena. In particular, the hydrophobic septum of DIV was shown to be longer than that of the other VSDs, preventing water molecules from hydrating the center of the VSD, and thus breaking the proton conduction pathway. A recent cryo-EM structure of a eukaryotic Na_v (*Periplaneta*) confirms this notion (Shen et al. 2017).

Homology modeling and molecular dynamics also provide an insight into the molecular basis of two $Na_v1.5$ DI-S4 mutants (R222Q and R225W), responsible for atypical clinical phenotypes involving complex arrhythmias and dilated cardiomyopathy (Moreau et al. 2015a) that both produce a depolarization-activated gating pore current as described in Part II. Molecular dynamics simulations of an activated state of the $Na_v1.5$ DI-S4 R222Q mutant and an intermediate state of the $Na_v1.5$ DI-S4 R225W mutant reveal that these two mutations produce very different effects on the structure of the VSD gating pore (Fig. 10). The pore created by the R225W

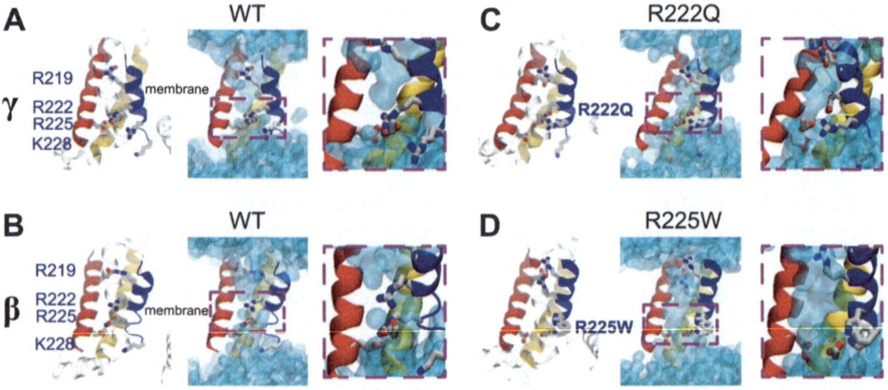

Fig. 10 Structural models of the DI VSDs of $Na_V1.5$ WT, R222Q and R225W mutants. Structural models of the relaxed DI VSD of the WT in the second intermediate γ (**a**) and first intermediate β (**b**) states. (**c**) Structural model of the relaxed DI VSD of the R222Q mutant in the γ state. (**d**) Structural model of the relaxed DI VSD of the R225W mutant in the β state. For all structural models, the VSD protein backbone is represented as a ribbon (*left panel*, S1 in *yellow*, S2 in *red*, S3 in *transparent cyan*, and S4 in *blue*). The gating charges of S4 and the counter charges of S2 and S3 are shown as sticks (carbon in *gray*, nitrogen in *blue*, and oxygen in *red*; hydrogens are omitted for clarity). In the middle panel, the water-accessible volume is shown as a transparent *cyan* surface. For each configuration, a higher magnification of the GCTC (*dotted pink box*) is shown (Adapted from Moreau et al. 2015a. Copyright Journal of General Physiology)

mutant is much wider, due to the preferential orientation of the tryptophan residue in the hydrophobic core of the lipid bilayer. This is likely at the origin of the higher gating pore conductance values observed in R225W.

References

Allen MP, Tildesley DJ (1987) Computer simulation of liquids. Clarendon Press, Oxford

Armstrong CM, Bezanilla F (1973) Currents related to movement of the gating particles of the sodium channels. Nature 242(5398):459–461

Banjeree A, MacKinnon R (2008) Inferred motions of the S3a helix during voltage-dependent K^+ channel gating. J Mol Biol 381(3):569–580. https://doi.org/10.1016/jmb.2008.06.010.

Beckermann TM, McLeod K, Murday V, Potet F, George AL Jr (2014) Novel SCN5A mutation in amiodarone-responsive multifocal ventricular ectopy-associated cardiomyopathy. Heart Rhythm 11(8):1446–1453. https://doi.org/10.1016/j.hrthm.2014.04.042

Bendahhou S, Cummins TR, Griggs RC, Fu Y-H, Ptacek LJ (2001) Sodium channel inactivation defects are associated with acetazolamide-exacerbated hypokalemic periodic paralysis. Ann Neurol 50(3):417–420

Bezzina CR, Rook MB, Groenewegen A, Herfst LJ, van der Wal AC, Lam J, Jongsma HJ, Wilde AM, Mannens MMAM (2003) Compound heterozygosity for mutations (W156X and R225W) in SCN5A associated with severe cardiac conduction disturbances and degenerative changes in the conduction system. Circ Res 92:159–168

Bezzina CR, Lahoruchi N, Priori SG (2015) Genetics of sudden cardiac death. Circ Res 116:1919–1936. https://doi.org/10.1161/CIRCRESAHA. 116.304030

Bukauskas FF, Peracchia C (1997) Two distinct gating mechanisms in gap junction channels: C02-sensitive and voltage-sensitive. Biophys J 72:2137–2142

Bukauskas FF, Bukauskiene A, Bennett MVL, Verselis V (2001) Gating properties of gap junction channels assembled from connexin43 and connexin43 fused with green fluorescent protein. Biophys J 81:137–152

Bulman DE, Scoggan KA, van Oene MD, Nicolle MW, Hahn AF, Tollar LL, Ebers GC (1999) A novel sodium channel mutation in a family with hypokalemic periodic paralysis. Neurology 53:1932–1936

Burge JA, Hanna MG (2012) Novel insights into the pathomechanisms of skeletal muscle channelopathies. Curr Neurol Neurosci Rep 12(1):62–69. https://doi.org/10.1007/s11910-011-0238-3

Campos FV, Chanda B, Roux B, Bezanilla F (2007) Two atomic constraints unambiguously position the S4 segment relative to S1 and S2 segments in the closed state of the Shaker K channel. Proc Natl Acad Sci U S A 104(19):7904–7909

Cannon SC (2007) Physiologic principles underlying ion channelopathies. Neurotherapeutics 4 (2):174–183

Cannon SC (2010) Voltage-sensor mutations in channelopathies of skeletal muscle. J Physiol 588 (11):1887–1895. https://doi.org/10.1113/jphysiol.2010.186874

Cannon SC (2015) Channelopathies of skeletal muscle excitability. Compr Physiol 5(2):761–790. https://doi.org/10.1002/cphy.c140062

Cannon SC, Brown RH Jr, Corey DP (1993) Theoretical reconstruction of myotonia and paralysis caused by incomplete inactivation of sodium channels. Biophys J 65:270–288

Carle T, Lhuillier L, Luce S, Sternberg D, Devuyst O, Fontaine B, Tabti N (2006) Gating defects of a novel Na^+ channel mutant causing hypokalemic periodic paralysis. Biochem Biophys Res Commun 348:653–661. https://doi.org/10.1016/j.bbrc.2006.07.101

Catterall WA (1986) Molecular properties of voltage-sensitive sodium channels. Annu Rev Biochem 55:953–985

Cha A, Ruben PC, George AL Jr, Fujimoto E, Bezanilla F (1999) Voltage sensors in domains III and IV, but not I and II, are immobilized by Na^+ channel fast inactivation. Neuron 22(1):73–87

Cheng J, Morales A, Siegfried JD, Li D, Norton N, Song J, Gonzalez-Quintana J, Makielski JC, Hershberger RE (2010) SCN5A rare variants in familial dilated cardiomyopathy decrease peak sodium current depending on the common polymorphism H558R and common splice variant Q1077del. Clin Transl Sci 3:287–294. https://doi.org/10.1111/j.1752-8062.01000249.x.

Cheng C-J, Lin S-H, Lo Y-F, Yang S-S, Hsu Y-J, Cannon SC, Huang C-L (2011) Identification and functional characterization of Kir2.6 mutations associated with non-familial hypokalemic periodic paralysis. J Biol Chem 286(31):27425–27435. https://doi.org/10.1074/jbc.M111.249656

Cheng C-J, Kuo E, Huang C-L (2013) Extracellular potassium homeostasis: insights from hypokalemic periodic paralysis. Semin Nephrol 33(3):237–247. https://doi.org/10.1016/j.semnephrol.2013.04.004

Chockalingham P, Wilde A (2012) The multifaceted cardiac sodium channel and its clinical implications. Heart 98(17):1318–1324. https://doi.org/10.1136/heartjnl-2012-301784

Delemotte L, Treptow W, Klein ML, Tarek M (2010) Effect of sensor domain mutations on the properties of voltage-gated ion channels: molecular dynamics studies of the potassium channel Kv1.2. Biophys J 99(9):L72–L74. https://doi.org/10.1016/bpj.2010.08.069.

Delemotte L, Tarek M, Klein ML, Amaral C, Treptow W (2011) Intermediate states of the Kv1.2 voltage sensor from atomistic molecular dynamics simulations. Proc Natl Acad Sci U S A 108 (15):6109–6114. https://doi.org/10.1073/pnas.1102724108

Egelman EH (2016) The current revolution in Cryo-EM. Biophys J 110:1008–1012. https://doi.org/10.1016/j.bpj.2016.02.001

Fabiato A, Fabiato F (1978) Effects of pH on the myofilaments and the sarcoplasmic reticulum of skinned cells from cardiac and skeletal muscles. J Physiol 276:233–255

Fan C, Lehmann-Horn F, Weber MA, Bednarz M, Groome JR, Jonsson MK, Jurkat-Rott K (2013) Transient compartment-like syndrome and normokalemic periodic paralysis due to a $Ca_{(v)}1.1$ mutation. Brain 136(12):3775–3786. https://doi.org/10.1039/brain/awt300.
Featherstone DE, Fujimoto E, Ruben PC (1998) A defect in sodium channel deactivation exacerbates hyperexcitability in human paramyotonia congenita. J Physiol 506(3):627–638
Filatov GN, Pinter MJ, Rich MM (2005) Resting potential-dependent regulation of the voltage sensitivity of sodium channel gating in rat skeletal muscle in vivo. J Gen Physiol 126 (2):161–172
Francis DG, Rybalchencko V, Struyk AF, Cannon SC (2011) Leaky sodium channels from voltage sensor mutations in periodic paralysis, but not paramyotonia. Neurology 76(19):1635–1641. https://doi.org/10.1212/WNL.0b013e318219fb57
Freites JA, Tobias DJ, White SH (2006) A voltage-sensor water pore. Biophys J 91:L90–L92
Geukes Foppen RJ, van Mil HGJ, van Heukelom JS (2002) Effects of chloride transport on bistable behavior of the membrane potential in mouse skeletal muscle. J Physiol 542(1):181–191
Gosselin-Badaroudine P, Keller DI, Huang H, Pouliot V, Chatelier A, Osswald S, Brink M, Chahine M (2012a) A proton leak current through the cardiac sodium channel is linked to mixed arrhythmia and the dilated cardiomyopathy phenotype. PLoS One 7(5):e38331. https://doi.org/10.1371/journal.pone.0038331
Gosselin-Badaroudine P, Delemotte L, Moreau A, Klein ML, Chahine M (2012b) Gating pore currents and the resting state of $rNa_v1.4$ voltage sensor domains. Proc Natl Acad Sci U S A 109 (47):19250–19255. https://doi.org/10.1073/pnas.1217990
Gosselin-Badaroudine P, Moreau A, Chahine M (2014) $Na_v1.5$ mutations linked to dilated cardiomyopathy phenotypes. Is the gating pore current the missing link? Channels 8(1):90–94. https://doi.org/10.4161/chan.27179.
Groome JR, Lehmann-Horn F, Fan C, Wolf M, Winston V, Merlini L, Jurkat-Rott K (2014) $Na_v1.4$ mutations cause hypokalemic periodic paralysis by disrupting IIIS4 movement during recovery. Brain 137(4):998–1008. https://doi.org/10.1093/brain/awu015
Guy HR, Seetharamulu P (1986) Molecular model of the action potential sodium channel. Proc Natl Acad Sci U S A 83(2):508–512
Hayward LJ, Brown RH, Cannon SC (1997) Slow inactivation differs among mutant Na channels associated with myotonia and periodic paralysis. Biophys J 72:1204–1219. https://doi.org/10.1016/S0006-3495(97)78768-X
Henrion U, Renhorn J, Borjesson SI, Neslon SI, Nelson EM, Schwaiger CS, Bjelkmar P, Wallner B, Lindhal E, Elinder F (2012) Tracking a complete voltage sensor with metal-ion bridges. Proc Natl Acad Sci U S A 109(22):8552–8557. https://doi.org/10.1073/pnas.116938109.
Hershberger RE, Parks SB, Kushner JD, Li D, Ludwigsen S, Jakobs P, Nauman D, Burgess D, Partain J, Litt M (2008) Coding sequence mutations identified in MYH7, TNNT2, SCN5A, CSRP3, LBD3, and TCAP from 313 patients with familial or idiopathic dilated cardiomyopathy. Clin Transl Sci 1:21–26. https://doi.org/10.1111/j.1752-8062.2008.00017.x
Hodgkin AL, Huxley AF (1952) A quantitative description of membrane current and its application to conduction and excitation in nerve. J Physiol 117:500–544
Holzherr BD, Groome JR, Fauler M, Nied E, Lehmann-Horn F, Jurkat-Rott K (2010) Characterization of a novel $hNa_V1.4$ mutation causing hypokalemic periodic paralysis. Biophys Soc Pos-LB201
Hong L, Pathak MM, Kim IH, Ta D, Tombola F (2013) Voltage-sensing domain of voltage-gated proton channel Hv1 shares mechanism of block with pore domains. Neuron 77:274–287. https://doi.org/10.1016/j.neuron.2012.11.013
Jensen MO, Jogini V, Borhani DW, Leffler AE, Dror RO, Shaw DE (2012) Mechanism of voltage gating in potassium channels. Science 336(6078):229–233. https://doi.org/10.1126/science.1216533
Jiang Y, Lee A, Ruta V, Cadene M, Chait BT, MacKinnon R (2003) X-ray structure of a voltage-dependent K^+ channel. Nature 423(6935):33–41

Jogini V, Roux B (2007) Dynamics of the Kv1.2 voltage-gated K^+ channel in a membrane environment. Biophys J 93:3070–3082

Jones DK, Ruben PC (2008) Biophysical defects in voltage-gated sodium channels associated with long QT and Brugada syndromes. Channels 2(2):70–80

Jurkat-Rott K, Lehmann-Horn F (2005) Muscle channelopathies and critical points in functional and genetic studies. J Clin Invest 115(8):2000–2009. https://doi.org/10.1172/JCI25525

Jurkat-Rott K, Mitrovic N, Hang C, Kouzmekine A, Iaizzo P, Herzog J, Lerche H, Nicole S, Vale-Santos J, Chaveau D, Fontaine B, Lehmann-Horn F (2000) Voltage-sensor sodium channel mutations cause hypokalemic periodic paralysis type 2 by enhanced inactivation and reduced current. Proc Natl Acad Sci U S A 97(17):9549–9554

Jurkat-Rott K, Fauler M, Lehmann-Horn F (2006) Ion channels and ion transporters of the transverse tubular system of skeletal muscle. J Muscle Res Cell Motil 27(5-7):275–290

Jurkat-Rott K, Weber M-A, Fauler M, Guo X-H, Holzherr B, Paczulla A, Nordsborg N, Joechle W, Lehmann-Horn F (2009) K^+-dependent paradoxical membrane depolarization and Na^+ overload, major and reversible contributors to weakness by ion channel leaks. Proc Natl Acad Sci U S A 106(10):4036–4041. https://doi.org/10.1073/pnas.0811277106

Jurkat-Rott K, Holzherr B, Fauler M, Lehmann-Horn F (2010) Sodium channelopathies of skeletal muscle result from gain or loss of function. Pflugers Arch 460(2):239–248. https://doi.org/10.1007/s00424-010-0814-4

Jurkat-Rott K, Groome J, Lehmann-Horn F (2012) Pathophysiological role of omega pore current in channelopathies. Front Pharmacol 3(112):1–19. https://doi.org/10.3389/fphar.2012.00112.

Kaminski GA, Friesner RA, Tirado-Rives J, Jorgensen WL (2001) Evaluation and reparametrization of the OPLS-AA force field for proteins via comparison with accurate quantum chemical calculations on peptides. J Phys Chem B 105:6474–6487. https://doi.org/10.1021/jp003919d

Khalili-Araghi F, Jogini V, Yarov-Yarovoy V, Tajkhorshid E, Roux B, Schulten K (2010) Calculation of the gating charge for the Kv1.2 voltage activated potassium channel. Biophys J 98(10):2189–2198. https://doi.org/10.1016/j.bpj.2010.02.056

Khalili-Araghi F, Tajkhorsid E, Roux B, Schulten K (2012) Molecular dynamics investigation of the ω-current in the Kv1.2 voltage sensor domains. Biophys J 102(2):258–267. https://doi.org/10.1016/j.bpj.2011.10.057

Kim M-K, Lee S-H, Park M-S, Kim B-C, Cho K-H, Lee M-C, Kim J-H, Kim S-M (2004) Mutation screening in Korean hypokalemic periodic paralysis: a novel *SCN4A* Arg672Cys mutation. Neuromuscul Disord 14(11):727–731

Krepkiy D, Mihailescu M, Freites JA, Schow EV, Worcester DL, Gawrisch K, Tobias DJ, White SH, Swartz KJ (2009) Structure and hydration of membranes embedded with voltage-sensing domains. Nature 462:473–479. https://doi.org/10.1038/nature08542

Kumanovics A, Levin G, Blount P (2002) Family ties of gating pores: evolution of the sensor module. FASEB J 16(12):1632–1639

Kuzmenkin A, Muncan V, Jurkat-Rott K, Hang C, Lerche H, Lehmann-Horn F, Mitrovic N (2002) Enhanced inactivation and pH sensitivity of $Na^{(+)}$ channel mutations causing hypokalemic periodic paralysis type II. Brain 125(4):835–843

Larsson HP, Baker OS, Dhillon DS, Isacoff EY (1996) Transmembrane movement of the Shaker K^+ channel S4. Neuron 16:387–397

Laurent G, Saal S, Amarouch MY, Beziau DM, Marsman RFJ, Faivre L, Barc J, Dina C, Bertaux G, Barthez O, Thauvin-Roubinet C, Charron P, Fressart V, Maltret A, Villain E, Baron E, Merot J, Turpault R, Coudiere Y, Charpentier F, Schott J-J, Loussouarn G, Wilde AAM, Wolf J-E, Baro I, Kyndt F, Probst V (2012) Multifocal ectopic Purkinje-related premature contractions: a new SCN5A-related cardiac channelopathy. J Am Coll Cardiol 60:144–156. https://doi.org/10.1016/j.jacc.2012.02.052.

Leach AR (2001) Molecular modeling – principles and applications, 2nd edn. Pearson Education, Harlow

Lehmann-Horn F, Jurkat-Rott K (1999) Voltage-gated ion channels and hereditary disease. Physiol Rev 79(4):1317–1372

Lehmann-Horn F, Rudel R, Jurkat-Rott K (2004) Non-dystrophic myotonias and periodic paralyses. In: Engel AG, Franzini-Armstrong C (eds) Myology, 3rd edn. McGraw-Hill, New York, pp 1257–1300

Li Q, Wanderling S, Paduch M, Medovoy D, Singharoy A, McGeevy R, Villalba-Galea C, Hulse RE, Roux B, Schulten K, Kossiakoff A, Perozo E (2014) Structural mechanism of voltage-dependent gating in an isolated voltage-sensing domain. Nat Struct Mol Biol 21 (3):244–252. https://doi.org/10.1038/nsmb.2768

Lin S-H, Huang C-L (2012) Mechanism of thyrotoxic periodic paralysis. J Am Soc Nehprhol 23 (6):985–988. https://doi.org/10.1681/ASN.2012010046

Lin M-C, Abramson J, Papazian DM (2010) Transfer of ion binding site from ether-a-go-go to Shaker: Mg^{2+} binds to resting state to modulate channel opening. J Gen Physiol 135 (5):415–431. https://doi.org/10.1085/jgp.200910320

Loussouarn G, Sternberg D, Nicole S, Marionneau C, Bouffant FL, Toumaniantz G, Barc J, Malak OA, Fressart V, Pereon Y, Baro I, Charpentier F (2016) Physiological and pathophysiological insights of $Na_v1.4$ and $Na_v1.5$ comparison. Front Pharmacol 6(314):1–20. https://doi.org/10.3389/fphar.2015.00314.

MacKerell AD, Brooks B, Brooks CL, Nilsson L, Roux B, Won Y, Karplus M (2002) CHARMM: the energy function and its parameterization. In: Encyclopedia of computational chemistry. Wiley, Chichester

Mankodi A, Grunseich C, Skov M, Cook L, Aue G, Purev E, Bakar D, Lehky T, Jurkat-Rott K, Pedersen TH, Childs RW (2015) Divalent cation-responsive myotonia and muscle paralysis in skeletal muscle channelopathy. Neuromuscul Disord 25(11):908–912. https://doi.org/10.1016/j.nmd.2015.08.007

Mann SA, Castro ML, Ohanian M, Guo G, Zodgekar P, Sheu A, Stockhammer K, Thompson T, Playford D, Subbiah R, Kuchar D, Aggarwal A, Vandenberg JI, Fatkin D (2012) R222Q SCN5A mutation is associated with reversible ventricular ectopy and dilated cardiomyopathy. J Am Coll Cardiol 60(16):1566–1573. https://doi.org/10.1016/j.jacc.2012.05.050

Matthews E, Labrum R, Sweeney MG, Sud R, Haworth A, Chinnery PF, Meola G, Schorge S, Kullman DM, Davis MB, Hanna MG (2009) Voltage sensor loss accounts for most cases of hypokalemic periodic paralysis. Neurology 72:1544–1547. https://doi.org/10.1212/01.wnl.0000342387.65477.46.

Matthews E, Fialho D, Tan SV, Venance SL, Cannon SC, Sternberg D, Fontaine B, Amato AA, Barohn RJ, Griggs RC, Hanna MG (2010) The non-dystrophic myotonias: molecular pathogenesis, diagnosis and treatment. Brain 133(1):9–22. https://doi.org/10.1093/brain/awp294

Matthews E, Portaro S, Ke Q, Sud R, Haworth A, Davis MB, Griggs RC, Hanna MG (2011) Acetazolamide efficacy in hypokalemic periodic paralysis and the predictive role of genotype. Neurology 77(22):1960–1964. https://doi.org/10.1212/WNL.0b013e31823a0cb6

McNair WP, Sinagra G, Taylor MRG, Lenarda AD, Ferguso DA, Salcedo EF, Slavov D, Zhu X, Caldwell JH, Mestroni L (2011) SCN5A mutations associate with arrhythmic dilated cardiomyopathy and commonly localize to the voltage sensing mechanism. J Am Coll Cardiol 57 (21):2160–2168. https://doi.org/10.1016/jacc.2010.09.084.

Mi W, Rybalchenko V, Cannon SC (2014) Disrupted coupling of gating charge displacement to Na^+ current activation for DIIS4 mutations in hypokalemic periodic paralysis. J Gen Physiol 144 (2):137–145. https://doi.org/10.1085/jgp.201411199

Moreau A, Chahine M (2015) Omega pore, and alternative ion channel permeation pathway involved in the development of several channelopathies. Med Sci 31(8–9):735–741. https://doi.org/10.1051/medsci/20153108011.

Moreau A, Gosselin-Badaroudine P, Chahine M (2014) Biophysics, pathophysiology, and pharmacology of ion channel gating pores. Front Pharmacol 5(53):1–19. https://doi.org/10.3389/fphar.2014.00053.

Moreau A, Gosselin-Badaroudine P, Delemotte L, Klein ML, Chahine M (2015a) Gating pore currents are defects in common with two $Na_v1.5$ patients with mixed arrythmias and dilated cardiomyopathy. J Gen Physiol 145(2):93–106. https://doi.org/10.1085/jgp.201411304

Moreau A, Gosselin-Badaroudine P, Boutjdir M, Chahine M (2015b) Mutations in the voltage sensors of domains I and II of $Na_V1.5$ that are associated with arrhythmias and dilated cardiomyopathy generate gating pore currents. Front Pharmacol 6(301):1–12. https://doi.org/10.3389/fphar.2015.00301.

Morrill JA, Brown RH Jr, Cannon SC (1998) Gating of the L-type Ca channel in human skeletal myotubes: an activation defect caused by hypokalemic periodic paralysis mutation R528H. J Neurosci 18(24):10320–10334

Nagel AM, Lehmann-Horn F, Weber M-A, Jurkat-Rott K, Wolf MB, Radbruch A, Umathum R, Semmler W (2014) In vivo 35Cl MR imaging in humans: a feasibility study. Radiology 271 (2):585–595. https://doi.org/10.1148/radiol.1313151617

Nair K, Pekhletski R, Harris L, Care M, Morel C, Farid T, Backx PH, Szabo E, Nanthakumar K (2012) Escape capture bigeminy: phenotypic marker of cardiac soidum channel voltage sensor mutation R222Q. Heart Rhythm 9:1681–1688. https://doi.org/10.1016/j.hrthm.2012.06.029

Nguyen TP, Wang DW, Rhodes TH, George AL Jr (2008) Divergent biophysical defects caused by mutant sodium channels in dilated cardiomyopathy with arrhythmia. Circ Res 102(3):364–371. https://doi.org/10.1161/CIRCRESAHA.107.164673

Noda MS, Shizimu S, Tanabe T, Takai T, Kayano T, Ikeda T, Takahashi H, Nakayami Y, Kamaoka N, Minamino N, Kangawa K, Matsuo K, Raferty H, Hirose M, Inayama T, Hayashida H, Miyata T, Numa S (1984) Primary structure of *Electrophorus electricus* sodium channel deduced from cDNA sequence. Nature 312:121–127

Olson TM, Michels VV, Ballew JD, Reyna SP, Karst ML, Herron KJ, Horton SC, Rodeheffer RJ, Anderson JL (2005) Sodium channel mutations and susceptibility to heart failure and atrial fibrillation. JAMA 293(4):447–454. https://doi.org/10.1001/jama.293.4.447

Payandeh J, El-Din G, Scheuer T, Zheng N, Catterall WA (2012) Crystal structure of a voltage-gated sodium channel in two potentially inactivated states. Nature 486(7401):135–140. https://doi.org/10.1038/nature10238

Plaster NM, Tawil R, Tristani-Firouzi M, Canun S, Bendahhou S, Tsunoda A, Donaldson MR, Iannaccone ST, Brunt E, Barohn R, Clark J, Deymeer F, George AL Jr, Fish FA, Hahn A, Nitu A, Ozdemir C, Serdaroglu P, Subramony SH, Wolfe G, Fu Y-H, Ptacek LJ (2001) Mutations in Kir2.1 cause the development and episodic electrical phenotypes of Anderson's syndrome. Cell 105(4):511–519

Platt D, Griggs R (2009) Skeletal muscle channelopathies: new insights into the periodic paralyses and non-dystrophic myotonias. Curr Opin Neurol 22(5):524–531. https://doi.org/10.1097/WCO.0b013e32832efa9f.

Posson DJ, Ge P, Miller C, Bezanilla F, Selvin PR (2005) Small vertical movement of a K^+ voltage-sensor measured with luminescence energy transfer. Nature 436(7052):848–851. https://doi.org/10.1038/nature03819

Richmond JE, VanDeCarr D, Featherstone DE, George AL Jr, Ruben PC (1997) Defective fast inactivation recovery and deactivation account for sodium channel myotonia in the I1160V mutant. Biophys J 73(4):1896–1903

Ruff RL (2000) Skeletal muscle sodium current is reduced in hypokalemic periodic paralysis. Proc Natl Acad Sci U S A 97(18):9832–9833

Ryan DP, da Silva MRD, Soong TW, Fontaine B, Donaldson MR, Kung AWC, Jongjaroenprasert W, Liang MC, Khoo DHC, Cheah JS, Ho SC, Bernstein HS, Macie RMB, Brown RH Jr, Ptacek LJ (2010) Mutations in potassium channel Kir2.6 cause susceptibility to thyrotoxic hypokalemic periodic paralysis. Cell 140(1):88–98. https://doi.org/10.1016/j.cell2009.12.024

Sansone V, Meola G, Links TP, Panzeri M, Rose MR (2008) Treatment for periodic paralysis. Cochrane Database Syst Rev 1:CD005045. https://doi.org/10.1002/14651858.CD005045.pub2

Sansone VA, Burge J, McDermott MP, Smith PC, Herr B, Tawil R, Pandya S, Kissel J, Ciafaioni E, Shieh P, Ralph JW, Amato A, Cannon SC, Trivedi J, Barohn R, Crum B, Misumoto H, Pestronk A, Meola G, Griggs R (2016) Randomized, placebo-controlled trials of dichlorophenamide in periodic paralysis. Neurology 86(15):1408–1416. https://doi.org/10.1212/WNL.0000000000002416

Schwaiger CS, Börjesson SI, Hess B, Wallner B, Elinder F, Lindahl E (2012) The free energy barrier for arginine gating charge translation is altered by mutations in the voltage sensor domain. PLoS One 7:e45880. https://doi.org/10.1371/journal.pone.0045880

Shen H, Zhou Q, Pan X, Li Z, Wu J, Yan N (2017) Structure of a eukaryotic voltage-gated sodium channel at near-atomic resolution. Science 355:eaal4326. https://doi.org/10.1126/science.aal4326

Sokolov S, Scheuer T, Catterall WA (2007) Gating pore currents in an inherited channelopathy. Nature 446(7131):76–78

Sokolov S, Scheuer T, Catterall WA (2008) Depolarization-activated gating pore current conducted by mutant sodium channels in potassium-sensitive normokalemic periodic paralysis. Proc Natl Acad Sci U S A 105(50):19980–19985. https://doi.org/10.1073/pnas.0810562105

Sokolov S, Scheuer T, Catterall WA (2010) Ion permeation and block of the gating pore in the voltage sensor of $Na_v1.4$ channels with hypokalemic periodic paralysis mutations. J Gen Physiol 136(2):225–236. https://doi.org/10.1085/jgp.201010414

Song Y-W, Kim S-J, Heo T-H, Kim M-H, Kim J-B (2012) Normokalemic periodic paralysis is not a distinct disease. Muscle Nerve 46(6):914–916. https://doi.org/10.1002/mus.23441

Starace DM, Bezanilla F (2004) A proton pore in a potassium channel voltage sensor reveals a focused electric field. Nature 427(6974):548–553

Sternberg D, Maisonobe T, Jurkat-Rott K, Nicole S, Launay E, Chauveau D, Tabti N, Lehmann-Horn F, Hainque B, Fontaine B (2001) Hypokalemic periodic paralysis type 2 caused by mutations at codon 672 in the muscle sodium channel gene *SCN4A*. Brain 124 (6):1091–1099

Struyk AF, Cannon SC (2007) A Na^+ channel mutation linked to hypokalemic periodic paralysis exposes a proton-selective gating pore. J Gen Physiol 130(1):11–20

Struyk AF, Cannon SC (2008) Paradoxical depolarization of Ba^{2+} treated muscle exposed to low extracellular K^+: insights into resting potential abnormalities in hypokalemic periodic paralysis. Muscle Nerve 37(3):326–337

Struyk AF, Scoggan KA, Bulman DE, Cannon SC (2000) The human skeletal muscle Na channel mutation R669H associated with hypokalemic periodic paralysis enhances slow inactivation. J Neurosci 20(23):8010–8017

Struyk AF, Markin VS, Francis D, Cannon SC (2008) Gating pore currents in DIIS4 mutations of $Na_v1.4$ associated with periodic paralysis: saturation of ion flux and implications of disease pathogenesis. J Gen Physiol 132(4):447–464. https://doi.org/10.1085/jgp.200809967

Stuhmer W, Conti F, Suzuki H, Wang X, Noda N, Yahagi N, Kubo H, Numa S (1989) Structural parts involved in activation and inactivation of the sodium channel. Nature 339(6226):597–603

Sugiura Y, Makita N, Li L, Noble PJ, Kimura J, Kumagai Y, Soeda T, Yamamoto T (2003) Cold induces shifts of voltage dependence in mutant *SCN4A*, causing hypokalemic periodic paralysis. Neurology 61(7):914–918

Tao X, Lee A, Limapichat W, Dougherty DA, MacKinnon R (2010) A gating charge transfer center in voltage sensors. Science 328:67–73. https://doi.org/10.1126/science.1185954

Tombola F, Pathak MM, Isacoff EY (2005) Voltage-sensing arginines in a potassium channel permeate and occlude cation-selective pores. Neuron 45(3):379–388

Tombola F, Pathak MM, Gorostiza P, Isacoff EY (2006) The twisted ion-permeation pathway of a resting voltage-sensing domain. Nature 445:546–549. https://doi.org/10.1038/nature05396

Treptow W, Tarek M (2006) Environment of the gating charges in the Kv1.2 Shaker potassium channel. Biophys J 90:L64–L66

Tricarico D, Camerino DC (2011) Recent advances in the pathogenesis and drug action in periodic paralysis and related channelopathies. Front Pharmacol 2(8):1–8. https://doi.org/10.3389/fphar.2011.00008.

Tricarico D, Barbieri M, Mele A, Carbonara G, Camerino DC (2004) Carbonic anhydrase inhibitors are specific openers of skeletal muscle BK channel of K^+ deficient rats. FASEB J 18(6):760–761

Tricarico D, Mele A, Camerino DC (2006) Carbonic anhydrase inhibitors ameliorate the symptoms of hypokalemic periodic paralysis in rats by opening the muscular Ca^{2+}-activated-K^{+} channels. Neuromuscul Disord 16(1):39–45

Tricarico D, Lovaglio S, Mele A, Rotondo G, Mancinelli E, Meola G, Camerino DC (2008) Acetazolamide prevents vacuolar myopathy in skeletal muscle $K^{(+)}$-depleted rats. Br J Pharmacol 154(1):183–190. https://doi.org/10.1038/bjp.2008.42

Tristani-Firouzi M, Etheridge SP (2010) Kir2.1 channelopathies: the Anderson-Tawil syndrome. Pflugers Arch 460(2):289–294. https://doi.org/10.1007/s00424-010-0820-6

Vargas E, Yarov-Yarovoy V, Khalili-Araghi F, Catterall WA, Klein ML, Tarek M, Lindhal E, Schulten K, Perozo E, Bezanilla F, Roux B (2012) An emerging consensus on voltage-dependent gating from computational modeling and molecular dynamics simulations. J Gen Physiol 140(6):587–594. https://doi.org/10.1085/jgp.201210873

Venance SL, Cannon SC, Fialho D, Fontaine B, Hanna MG, Ptacek LJ, Tristani-Fiorouzi M, Tawil R, Griggs RC (2006) The primary periodic paralyses: diagnosis, pathogenesis and treatment. Brain 129(1):8–17

Vicart S, Sternberg D, Fournier E, Ochsner F, Laforet P, Kuntzer T, Eymard B, Hainque B, Fontaine B (2004) New mutations at *SCN4A* cause a potassium-sensitive normokalemic periodic paralysis. Neurology 63(11):2120–2127

Vicart S, Sternberg D, Fontaine B, Meola G (2005) Human skeletal muscle sodium channelopathies. Neurol Sci 26(4):194–202

Wang J, Wang W, Kollman PA, Case DA (2006) Automatic atom type and bond type perception in molecular mechanical calculations. J Mol Graph Model 25:247–260. https://doi.org/10.1016/j.jmgm.2005.12.005

Webb J, Cannon SC (2008) Cold-induced defects of sodium channel gating in atypical periodic paralysis plus myotonia. Neurology 70(10):755–761

Weber MA, Nagel AM, Marschar AM, Glemser P, Jurkat-Rott K, Wolf MB, Ladd ME, Schlemmer HP, Kauczor HU, Lehmann-Horn F (2016) 7-T (35)Cl and (23)Na MR imaging for detection of mutation-dependent alterations in muscular edema and fat fraction with sodium and chloride concentrations in periodic paralysis. Radiology 280(3):848–859. https://doi.org/10.1148/radiol.2016151617

Wood ML, Freites JA, Tombola F, Tobias DJ (2017) Atomistic modeling of ion conduction through the voltage-sensing domain of the Shaker K^{+} ion channel. J Phys Chem B 121:3804–3812. https://doi.org/10.1021/acs.jpcb.6b12639

Wu F, Cannon SC (2017) Stac3 facilitated expression of Cav1.1 in Xenopus oocytes to assess functional consequences of HypoPP mutant Cav1.1-R528H. Biophys Abstr 112(3):245a

Wu F, Mi W, Burns DK, Fu Y, Gray HF, Struyk AF, Cannon SC (2011) A sodium channel knockin mutant (Na_v1.4-R669H) mouse model of hypokalemic periodic paralysis. J Clin Invest 121 (10):4082–4094. https://doi.org/10.1172/JCI57398

Wu F, Mi W, Cannon SC (2013a) Bumetanide prevents transient decreases in muscle force in murine hypokalemic periodic paralysis. Neurology 80(12):1110–1116. https://doi.org/10.1212/WNL.0b013e3182886a0e

Wu F, Mi W, Cannon SC (2013b) Beneficial effects of bumetamide in a Cav1.1-R528H mouse model of hypokalemic periodic paralysis. Brain 136(12):3766–3774. https://doi.org/10.1093/brain/awt280

Yang N, Horn R (1995) Evidence for voltage-dependent S4 movement in sodium channels. Neuron 15(1):213–218

Yang N, George AL Jr, Horn R (1996) Molecular basis of charge movement in voltage-gated sodium channels. Neuron 16(1):113–122

Yu FH, Catterall WA (2004) The VGL-chanome: a protein superfamily specialized for electrical signaling and ionic homeostasis. Sci STKE 2004:re15. https://doi.org/10.1126/stke.2532004re15

Calculating the Consequences of Left-Shifted Nav Channel Activity in Sick Excitable Cells

Bela Joos, Benjamin M. Barlow, and Catherine E. Morris

Contents

Abstract

Two features common to diverse sick excitable cells are "leaky" Nav channels and bleb damage-damaged membranes. The bleb damage, we have argued, causes a channel kinetics based "leakiness." Recombinant (node of Ranvier type) Nav1.6 channels voltage-clamped in mechanically-blebbed cell-attached patches undergo a damage intensity dependent kinetic change. Specifically, they experience a coupled hyperpolarizing (left) shift of the activation and inactivation processes. The biophysical observations on Nav1.6 currents formed the

B. Joos (✉) · B. M. Barlow
Department of Physics, University of Ottawa, Ottawa, ON, Canada
e-mail: bjoos@uottawa.ca

C. E. Morris
Neurosciences, Ottawa Hospital Research Institute, Ottawa, ON, Canada
e-mail: cmorris@uottawa.ca

M. Chahine (ed.), *Voltage-gated Sodium Channels: Structure, Function and Channelopathies*, Handbook of Experimental Pharmacology 246,
https://doi.org/10.1007/164_2017_63

basis of Nav-Coupled Left Shift (Nav-CLS) theory. Node of Ranvier excitability can be modeled with Nav-CLS imposed at varying LS intensities and with varying fractions of total nodal membrane affected. Mild damage from which sick excitable cells might recover is of most interest pathologically. Accordingly, Na^+/K^+ ATPase (pump) activity was included in the modeling. As we described more fully in our other recent reviews, Nav-CLS in nodes with pumps proves sufficient to predict many of the pathological excitability phenomena reported for sick excitable cells. This review explains how the model came about and outlines how we have used it. Briefly, we direct the reader to studies in which Nav-CLS is being implemented in larger scale models of damaged excitable tissue. For those who might find it useful for teaching or research purposes, we coded the Nav-CLS/node of Ranvier model (with pumps) in NEURON. We include, here, the resulting "Regimes" plot of classes of excitability dysfunction.

Keywords
Bleb · Ectopic · Excitability · Hyperpolarizing shift · Leaky sodium channels · Left shift · Membrane damage · Mild injury · Modeling

1 Introduction

Excitable cells express voltage-gated sodium channels (Nav) or in some cases voltage-gated calcium channels that enable them to create and propagate the voltage spikes known as action potentials (APs). For reasons that long seemed obscure, many pathological conditions that involve excitable cells included, in the list of malfunctioning parts, "leaky sodium channels." In recent reviews we have labeled such cells "sick excitable cells" and have argued that the acquired sodium channelopathies noted under these conditions almost certainly arise from the widely reported but otherwise rather overlooked bleb damage inherent to the various disease states. Specifically, it would be bleb damage to the Nav-bearing membranes that is critical for sick excitable cells (Morris et al. 2012a, b; Morris and Joós 2016). The damage can result from mechanical trauma (and the attendant intra-tissue shear stresses), ischemia, inflammation, excitotoxicity, or other conditions. Blebbing of the Nav-bearing membrane would occur progressively as, progressively, the filamentous web of actin-spectrin cytoskeleton became detached from the inner bilayer leaflet.

In ways that are only partially understood, the adherent cortical cytoskeleton contributes both to trans-leaflet asymmetry of healthy membranes and to their dynamic lateral non-random heterogeneous organization (Sheetz et al. 2006). Ischemia and inflammation lead to excessive cytoplasmic calcium, thereby hyperactivating enzymes that hydrolyze the cytoskeleton. Shear forces, by contrast, would act directly to produce detachments. The living healthy plasma membrane features not self-organized bilayer, but rather, a bilayer whose molecular structure is cell-mediated. Presumably there is a continuum between the high-entropy "self-organized" state of the outright membrane blebs observed sloughing off sick

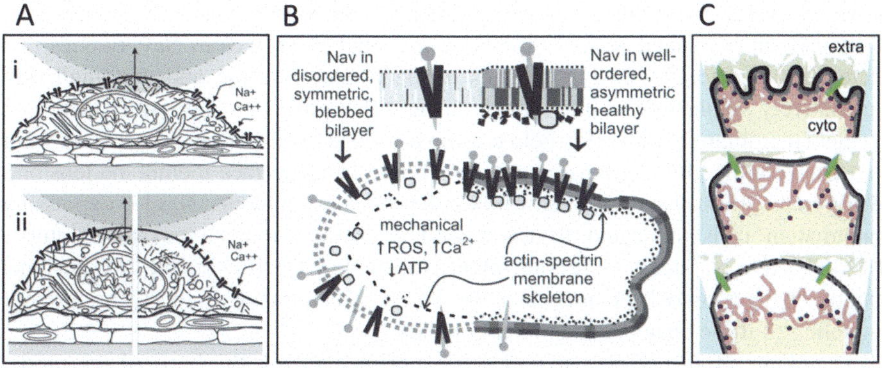

Fig. 1 Cartoon depictions of bleb damage developing in Nav-bearing plasma membranes. (**A**) Depiction of an atomic force microscopy experiment that tests the mechanical state of an excitable cell's surface/subsurface region before and after a bleb-inducing insult (Zou et al. 2013) (ii). A neuron cultured on astrocytes is depicted with a ~20 μm sphere whose maximum displacement is outlined. Excitotoxicity was mimicked with glutamate and Nav channel agonists that dissipate [Ca^{2+}] and [Na^+] gradients. Neurons inflate on exposure to hypotonic medium and to excitotoxic agonists. (ii) Early and late expected states of an enlarged neuron exposed to excitotoxic agonists is depicted: at left, once the initial channel-mediated Na^+ influx and osmotically-obligated H_2O has hydrostatically inflated the neuron (countered to some extent by Ca^{2+}-mediated actomyosin contractility), and at right, after Ca^{2+}-toxicity has damaged the previously adherent (and contractile) neuronal membrane skeleton, allowing the plasma membrane to bleb pathologically. The bilayer in (**A**) is depicted by a simple line but the adjacent cartoon, (**B**), highlights some key changes that would be occurring in Nav channel bearing membranes as it went from healthy to blebbed (from Morris and Joós 2016). (**C**) depicts a cell-attached patch (blue (gray in print versions) is pipette walls) with extracellular, cytoplasmic, and membrane disruptions expected after a gigaohm-seal is gently formed (top) then after milder (middle) and more severe (bottom) bleb damage has occurred due to pipette aspiration. Bilayer structure denatures relative to its intact plasma membrane state, but membrane proteins (green arrow-tipped lines (light gray in print versions)) do not denature. In the case of Nav channels, a fully functional, inherently mechanosensitive voltage-gated channel finds itself in a bilayer whose mechanical state (see discussions of a bilayer's lateral pressure profile in Finol-Urdaneta et al. 2010) undergoes a major change between the healthy and fully blebbed state.

excitable cells, and the light-microscopically invisible incipient state of blebbing. As depicted in Figure 1, atomic force microscopy probing cortical neurons subjected to excitotoxic stimuli (Zou et al. 2013) suggests that during the course of blebbing, where conditions cause osmotic swelling, elevated far-field tension could be expected in the blebbed Nav-bearing membrane.

Experiments on recombinant Nav channels heterologously expressed in *Xenopus* oocytes established connection between mechanically imposed membrane damage and changes in the kinetics of the Nav channels. Channel activity was monitored as macroscopic current in cell-attached patches. The effect of membrane damage first became evident for Nav1.4 channels (Tabarean et al. 1999; Shcherbatko et al. 1999), but the irreversibly changed Nav channel kinetics observed in those experiments were not, at the time, attributed to the bleb damage per se. Subsequently,

Wang et al. (2009), doing similar experiments with Nav1.6, made the connection. They also distinguished between the large irreversible changes in Nav kinetics that were attributed to irreversible changes in membrane structure and the small (but qualitatively similar) reversible kinetic changes that arise from reversibly increased far-field membrane tension (far-field tension is used here to make it clear that we refer to a Hookean-like bilayer tension and not an interfacial surface membrane tension). Wang et al. (2009) also performed whole cell trauma experiments; Nav1.6-expressing mammalian cells subjected to traumatic-type stretch were found to exhibit a TTX-sensitive Na^+-leak. Here, our intent is to review the irreversible Nav channel kinetic changes associated mechanically induced bleb-damage, and to review our modeling of those changes in cellular contexts.

The Wang et al. (2009) patch clamp experiments form the basis of the model discussed in this review, i.e., coupled left shift (or CLS), whose properties and predictions were first presented in Boucher et al. (2012) then further explored in Yu et al. (2012) and by Lachance et al. (2014). As Morris and Joós (2016) have repeatedly emphasized, the biophysical link between blebbing and Nav channelopathy revealed by the work of Wang et al. (2009) would benefit from further experimental lines of inquiry. The issue that specifically needs to be addressed is whether it is indeed blebbed Nav-bearing membrane that should be the therapeutic target for addressing pathologies arising from abnormal Nav kinetics in sick excitable cells.

2 Experimental Basis of the Nav-CLS Model

The predominant Nav channels at nodes of Ranvier and in the distal regions of axon initial segments are Nav1.6 channels. Because Nav channel blockers are protective in cell and tissue models of mechanical, ischemic, inflammatory and other injuries, it is understood that these white matter Nav channels become lethally leaky under pathological conditions. Fast-mode gating dominates in Nav1.6 and their response to progressive bleb damage turns out to be straightforward. Reversibly applied pipette suction ("aspiration") yields irreversible and approximately equal hyperpolarizing shifts of the equilibrium conductance $g(V)$ and availability (V) as a function of the membrane voltage V. In Wang et al. (2009) (Fig. 2; Fig. 2B in Wang et al.), those quantities are shown for "pre-stretch" membrane patches and the "post-stretch" membranes. On average the cumulative applied suction produced ~20 mV of hyperpolarizing (or "left") shift. The rising curves were obtained by stepping from $V_{\mathrm{hold}} = -110$ mV to the test potential. The g(V) curve is obtained from the normalized peak current observed at the potentials along the X-axis. At left are Nav availability curves. From V_{hold} the voltage was stepped for 210 ms to the voltages on the X-axis (210 ms is more than sufficient for fast inactivation to equilibrate) and then for 10 ms to the test voltage (0 mV) to assess availability based on peak current.

Direct evidence that both activation and fast inactivation kinetics simply "left-shift" due to membrane damage is shown in Fig. 3 (see legend) with the most

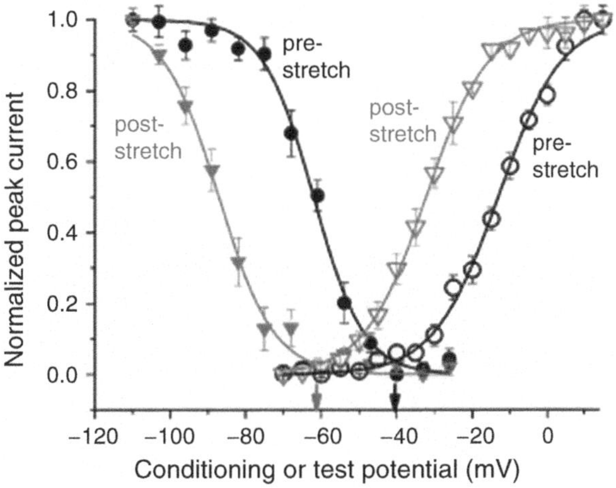

Fig. 2 Coupled-left shift (Nav-CLS). After traumatic membrane stretch, the kinetics of recombinant Nav1.6 channels expressed in oocytes are irreversibly changed. They undergo irreversible hyperpolarizing shifts of activation and steady-state fast inactivation (availability) as explained in the text. From Wang et al. (2009) where experimental details can be found

straightforward observations presented in the bottom section (*v*). Here, $I_{\mathrm{Na}}(t)$ time courses before and after stretch at several voltages are shown for one particular cell-attached oocyte patch. Fortuitously, for this patch stretch damage produced a left-shift of exactly 20 mV. Since Nav currents were assessed at 5 mV intervals, this meant that the 20 mV left shift corresponded precisely to a membrane potential tested before and after application of membrane-traumatizing suction (stretch). Note how currents post-stretch overlap completely (over the whole activation and inactivation time course) with amplitude-normalized pre-stretch currents obtained at voltages 20 mV more hyperpolarized. This shows that after irreversible damage, both activation and inactivation are irreversibly accelerated to the same extent. Part A (i, ii) of Fig. 3 (a different patch) illustrates pre- and post-stretch currents at 0 mV in a patch whose inactivation time constant (fits not shown) shifted leftward by 16 mV. Parts iii and iv show traces for a given patch that is stepped to −15 mV pre- and post-stretch; when the currents are amplitude normalized, the slower (pre-stretch) current can then be rescaled in the time domain till it completely overlaps. This double-rescaling is another direct way in which coupled-left shift manifests itself. Figure 3B i, ii shows data from two other patches in which the observed irreversible left-shift was, fortuitously, an almost exact multiple of 5 mV (i.e., 10 and 35 mV). In the Hodgkin-Huxley formulation, fast inactivation is explicitly described as an independently voltage dependent process (i.e., it is not kinetically coupled to activation, unlike in real Nav channels; see Banderali et al. 2010 and see discussions in Morris and Joós 2016) and so the "double-rescaling" procedure does not work. However, for our modeling purposes, the consequences of this Hodgkin-Huxley departure from reality are trivial.

In some experiments irreversible stretch induced damage was taken to a point at which no further irreversible left shift occurred (this was closer to 30 mV than the ~20 of Fig. 2; approaching the "saturation" point makes membrane rupture very likely, so this was avoided usually; the ~20 mV average coupled left shift of Fig. 2 therefore underestimates left shift for these channels in fully blebbed membrane).

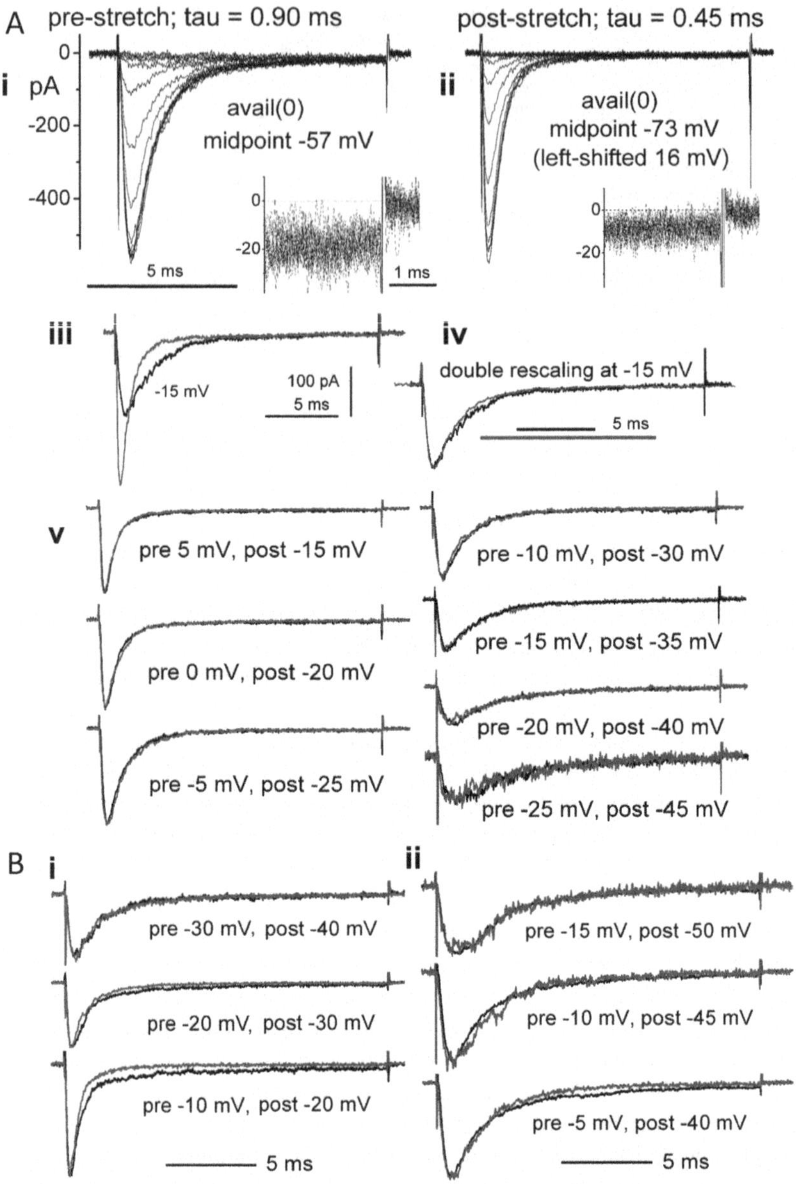

Fig. 3 Cell-attached patch clamp currents from Nav1.6 channels before and after traumatic (bleb-inducing) membrane stretch. The Nav CLS theory is based on data such as that in (**A**) i-v and in (**B**) i,ii, as explained in the text. From Wang et al. (2009) where experimental details can be found

Once "saturation" was achieved in these few experiments, it was possible to test effects of reversibly increased membrane tension. In other words, mechanosensitive (MS) gating changes could be tested. They amounted to at most a few mV of reversible coupled left shift (this has been explored more fully in Nav1.5 channels; see Morris and Juranka 2007; Banderali et al. 2010 and the references therein). The fact that these reversible MS changes are qualitatively identical to the larger irreversible changes associated with blebbing suggests the following: from the point of view of the Nav channel's kinetic behavior, a bleb damaged membrane "looks and feels" like a bilayer that has been made to thin due to imposed stretch (for more detail and discussion see Morris and Joós 2016; Morris et al. 2012a, b). We have no information on whether blebbing Nav-bearing membrane in sick excitable cells is ever at a sufficiently elevated membrane tension to produce appreciable MS kinetic changes. We suggest, however, that even if some blebs are at elevated tension (as was suggested for the earliest stages of excitotoxicity (Fig. 1A ii)), this would be almost certainly be irrelevant since bleb damage itself impacts Nav channel gating more strongly in a kinetically identical manner.

3 The Coupled Left-Shift Model (CLS)

The experiments of Wang et al. (2009) justify a mathematically simple model of damage to nodal Nav channel function in which the kinetics are simply left-shifted by a constant voltage which we call LS and that will become a measure of the damage (Boucher et al. 2012; Morris et al. 2012a, b). As both activation and inactivation shift by the same amount, the model was called the *coupled left-shift* (CLS) model. (For discussions of the likely molecular basis of this coupling, see Banderali et al. 2010 and Morris and Joós 2016.) The voltage V_m in all expressions of the kinetics of the intact Navs is simply replaced by (V_m + LS) for the Navs in the damaged regions of the membrane.

An action potential (AP) represents a spike in the plasma membrane potential. The plasma membrane's lipid bilayer is assigned a capacitance per unit area of 1 μF/cm^2. The healthy node of Ranvier is, most of the time, in a quiescent state of readiness for firing (at its resting potential, V_{rest}) with the membrane conductance dominated by non-specific leak conductance (g_L). For computational purpose the reversal potential of g_L is set at a potential somewhat more depolarized than E_{K}. There is a low internal (axoplasmic) [Na^+] and a high external [Na^+], and vice versa for [K^+]. Depolarization of the membrane potential $V_m = V_{\text{in}} - V_{\text{out}}$ occurs when, for example a brief stimulus such as an injected current triggers Nav channels to open and support a Na^+ influx. With a brief delay, this depolarization next triggers the opening of voltage-gated K^+ channels (Kvs). The resulting outflow of K^+ drives V_m back in the hyperpolarizing direction and eventually V_m returns to V_{rest}. Homeostasis is either assumed in the CLS models that will be discussed (fixed values for the Na^+ and K^+ reversal potentials) or is explicitly included as a process by mathematically depicting the Na^+/K^+ ATPase (pumps) that continuously remove 3 Na^+ from the cell and bring in 2 K^+ per ATP consumed.

To explore qualitatively how LS might affect the excitability of nodes of Ranvier, we use the Hodgkin-Huxley (HH) model (Hodgkin and Huxley 1952) to model the Navs, Kvs and leaks, but similar calculations could be carried out using other kinetic models. Pumps are included in much of our modeling, and for this Michaelis-Menten kinetics are used in the manner outlined by Läuger (1991). Adding the pumps acknowledges the impact of extended AP firing on excitable systems as they contend with ion homeostasis of their finite internal and external volumes.

Using the convention that ion current is positive when cations are outflowing, the differential equation governing the rate of change of V_m (written for simplicity V) is given by:

$$C\frac{\mathrm{d}V}{\mathrm{d}t} = -I_{\mathrm{Na}} - I_{\mathrm{K}} - I_{\mathrm{L}} + I_{\mathrm{stim}}, \tag{1}$$

Note that positive ion current hyperpolarizes V_m whereas negative ion current (e.g., an inflow of Na^+ ions) depolarizes V_m. In Eq. (1) I_{Na} flows through Navs and I_{K} through Kvs. I_L is the leak current and I_{stim} the injected external stimulus that, typically, is used to depolarize the membrane enough to elicit an AP.

Dividing both sides of Eq. (1) gives dV/dt = the sum of the 4 currents divided by C; solving this differential equation yields $V(t)$, which, when all is well, takes the form of an AP. Since there is no analytical solution for dV/dt, numerical methods are used.

Expressions for each ion current and their kinetics are:

$$I_{\mathrm{Na}} = g_{\mathrm{Na}}(V - E_{\mathrm{Na}}), \text{ where } g_{\mathrm{Na}} = \bar{g}_{\mathrm{Na}} m^3 h \tag{2}$$

$$I_{\mathrm{K}} = g_{\mathrm{K}}(V - E_{\mathrm{K}}), \text{ where } g_{\mathrm{K}} = \bar{g}_{\mathrm{K}} n^4. \tag{3}$$

$\bar{g}_{\mathrm{Na}}$ and $\bar{g}_{\mathrm{K}}$ are the maximal conductances of the Nav and Kv channels, respectively; E_{Na} and E_{K} are the Na^+ and K^+ reversal potentials, respectively. The non-dimensional gating variables $j = m$, h and n evolve with time according to:

$$\frac{\mathrm{d}j}{\mathrm{d}t} = \alpha_j(1 - j) - \beta_j j \tag{4}$$

Each gating variable j has voltage dependent forward (α_j) and backward (β_j) rates. The variable m monitors the activation of the Nav channel whereas h describes the availability of the channel such that m^3h is the fraction of the full conductance or the open probability for the channel population. Nav channels inactivate quickly (~1 ms).

$$\alpha_m = 0.1\frac{(V + 40)}{1 - \exp[-(V + 40)/10]}, \tag{5}$$

$$\beta_m = 4\exp[-(V + 65)/18], \tag{6}$$

$$\alpha_h = 0.07\exp[-(V + 65)/20], \tag{7}$$

$$\beta_h = \frac{1}{1 + \exp[-(V + 35)/10]}. \tag{8}$$

The Kvs, which exhibit no inactivation (at least, not on the time scale of APs) have a steady-state open probability given by n^4. The forward and backward rates of change for this variable are given by

$$\alpha_n = 0.01 \frac{(V + 55)}{1 - \exp\left[-\frac{V+55}{10}\right]} \text{ and } \beta_n = 0.125\exp\left[-\frac{V + 65}{80}\right] \tag{9}$$

Details of the kinetics can be seen in Sterratt et al. (2011).

A damaged node may have a population of Navs with a distribution of LS values. The kinetics of a Nav with a left shift of LS is modeled simply by replacing in Eqs. (5)–(9) V by V + LS. If we suppose that the node is comprised of N fractions f_i of Navs with left-shifts LS$_i$ then with activation variables m_i (i = 1, 2,. . .N) and inactivation variables h_i (i = 1, 2,. . .N), the total Nav current will be given by Yu et al. (2012):

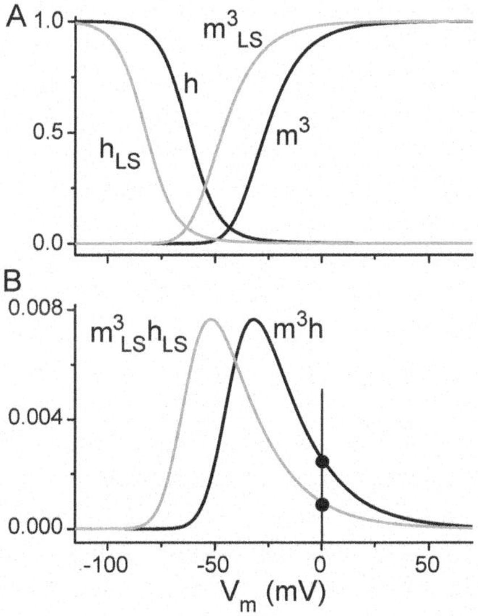

Fig. 4 Coupled left-shift (CLS) with LS = 20 mV in the Hodgkin-Huxley formulation. (**A**) Equilibrium values of activation (m^3 (V)) and inactivation ($h(V)$) variables for intact membrane (*black*) and after a 20 mV left-shift (*gray*). (**B**) The steady-state open probability, m^3h, of the intact and 20 mV left-shifted condition. Current flowing through the displaced window conductance would constitute a "Nav-leak." Note that the "window conductance" magnitude is greater at voltages nearer to normal V_{rest} and less at 0 mV (vertical line and circles). For further explanation, see Boucher et al. (2012)

$$I_{Na} = g_{Na}(V_m - E_{Na}), \text{ where } g_{Na} = \bar{g}_{Na}\sum\nolimits_{i=1}^{N} f_i m_i^3 h_i \text{ and } \sum\nolimits_{i=1}^{N} f_i = 1 \quad (10)$$

Figure 4 shows the impact of LS on the kinetics of a channel. The equilibrium ($t \rightarrow \infty$) values of the kinetic variables m and h shift by LS (Fig. 4A), leading to a shift in the equilibrium open probability of the channel (Fig. 4B). Consequently, the so-called window conductance operates at voltages too near the normal V_{rest} (−65.5 mV in Boucher et al. 2012). This constitutes "leaky Nav channels." A further outcome is a reduced steady-state I_{Na} at depolarized voltages (e.g., at 0 mV, as highlighted in the figure). We stress this point: when fast-mode Nav channel activity is responsible for "Nav leak," a decrease in steady-state I_{Na} near 0 mV is expected, not an increase.

The consequences of such shifts are examined next. The increased leakage near V_{rest} will tax the node requiring increased energy input to maintain the node in a resting homeostatic state. We will examine these effects when we introduce pumps.

4 CLS in a Node with Two Nav Populations (Intact and Damaged) and No Pumps

To gain insight into the effect of LS on nodal excitability we consider the simple case of two sub-populations of Nav channels, a left-shifted one with $f_1 = \text{AC}$ and an intact one with $f_2 = 1-\text{AC}$ (AC stands for "affected channels"). For parameters given in Boucher et al. (2012) Fig. 5A shows the impact of LS with perfect homeostasis assumed (i.e., Nernst potentials kept constant). Mild damage renders the node more excitable ("hypersensitive," *regime a*), lowers the threshold for initiating APs, and increases the frequency of firing. As LS increases, the node will begin to fire ectopically (i.e., without stimulus). In this idealized (perfect homeostasis) situation, the regime plot shows a large "zone" of tonic firing (*regime b*) and clearly, when a node is firing tonically, this is definitely not mild injury. Inspection of the regime plot shows that the mild damage regime (hypersensitive but not ectopically firing, *regime a*) corresponds either to any sufficiently small LS (zone along the X-axis) or to large LS/small AC (zone along the Y-axis) (see Fig. 5A). As LS and AC values push the system into the tonic firing regimes (*b* and *c*), the node also experiences increasingly greater difficulty responding to an imposed stimulus (see traces for *b* and *c*): from the *V*(*t*) traces, zone *b* fires tonically but zone *c* exhibits depolarizing block. *Regime d* no longer shows tonic firing and is inexcitable. Note that the set of *V*(*t*) traces illustrated are all for AC = 1 (all channels affected). AC = 1 would presumably correspond to the patch clamp experiments of Fig. 3. To provide some perspective, recall that there we saw Nav1.6 channels in damaged membranes exhibiting shifts of 16 and 20 mV. In the CLS model (for AC = 1), depolarizing block occurs around LS = 17 mV and total inexcitability near 19 mV. Experimentalists need to bear in mind the following: mild damage, though pathologically critical, would likely be extraordinarily difficult to measure biophysically. However, if CLS underlies pathological Nav leaks, mild and severe damage represent quantitatively but not

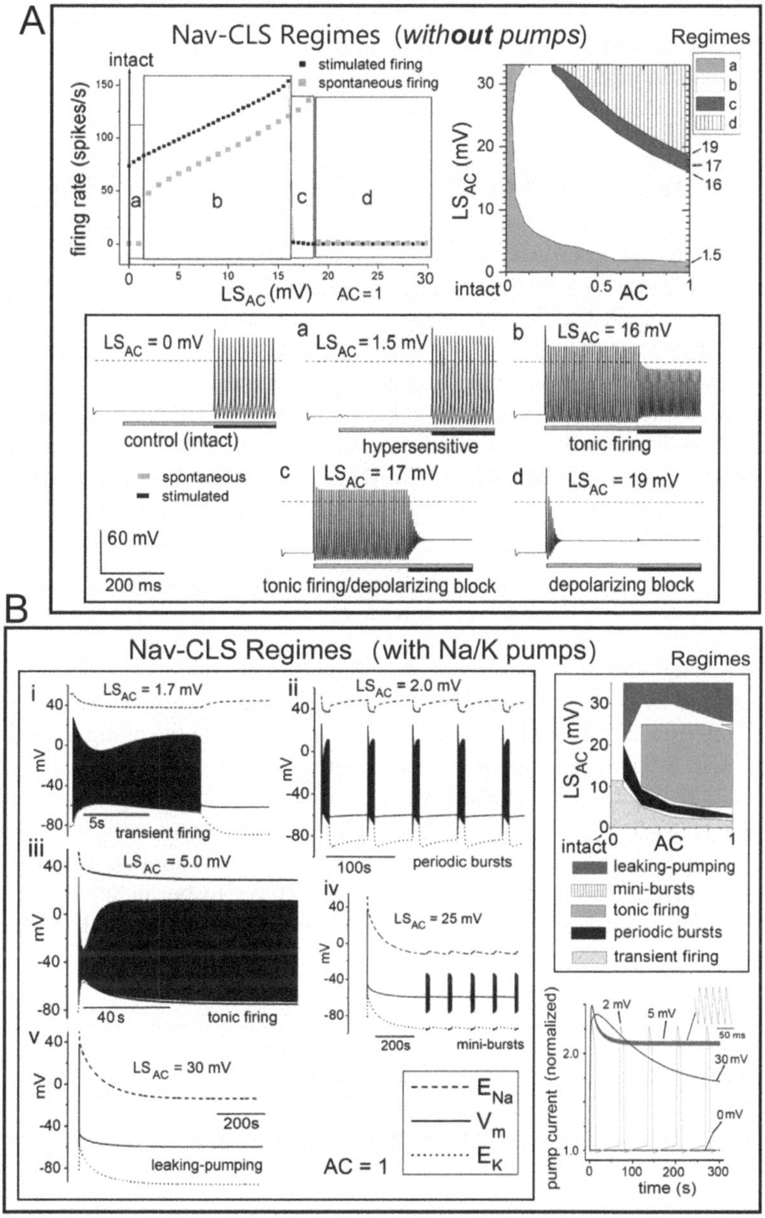

Fig. 5 Nav-CLS excitability regimes without and with pumps. (**A**) (without pumps) Computed excitability with Nav-CLS injury, as labeled. Regime plots are alluded to in the text. In voltage plots, dashed lines indicate 0 mV. (**B**) (with pumps) Na/K pumps are included with a surface s to inner volume vol_i ratio, 20 cm^2/μL, small enough to observe neuropathic behaviors over conveniently short simulation times. This ratio was used in Boucher et al. (2012), Yu et al. (2012), and Lachance et al. (2014). (Parameter Tables in Boucher et al. 2012 and Yu et al. 2012 have a typo: "m^3" instead of liters, L, for volumes. For all simulations here, $vol_i = vol_o = 3 \times 10^{-15}$ L $= 3 \times 10^{-9}$ μL). Figure modified from Boucher et al. (2012), as published in Morris and Joós (2016)

qualitatively different states. For channel neuropharmacology, it would be valid to study channels in severely damaged membrane even when the aim was eventually to target channels in mildly damaged cells.

5 Excitability and CLS Damage in a Node with Pumps

When ion fluxes are described for finite volumes of nodal axoplasm and the external volume, then whether a node is firing or quiescent, the ion gradients dissipate. When they have fuel, Na^+/K^+ ATPase pumps continuously remove 3 Na^+ from the cell and bring in 2 K^+ per ATP consumed to restore those gradients. In the Michaelis-Menten model of pump action one ATP must be hydrolyzed for each cycle of the pump. Three Na^+ ions bind the pump protein from inside and move to the external side and two K^+ ions bind from the external side and move to the inner side. Accordingly, I_{pump} depends on both $[Na^+]_i$ and $[K^+]_o$. This is the basis of the pump formulation given by Läuger (1991) and also used by Käger et al. (2000):

$$I_{\mathrm{pump}} = I_{\mathrm{max\ pump}}\left(1 + \frac{\mathrm{K}_{M_{\mathrm{K}}}}{[\mathrm{K}^+]_o}\right)^{-2} \times \left(1 + \frac{\mathrm{K}_{M_{\mathrm{Na}}}}{[\mathrm{Na}^+]_i}\right)^{-3} \tag{11}$$

The Michaelis-Menten coefficients $\mathrm{K}_{M_{\mathrm{Na}}}$ and $\mathrm{K}_{M_{\mathrm{K}}}$ measure the efficiency of the process on the two sides of the pumps. $I_{\mathrm{max\ pump}}$ is the maximal current generated by the pump (to a first approximation it would correspond to the pump density), and the Na^+ and K^+ currents flowing through the pump are $I_{\mathrm{Napump}} = 3I_{\mathrm{pump}}$ and $I_{\mathrm{Kpump}} = -2I_{\mathrm{pump}}$. To model the time evolution, these currents will have to be added to Eq. (1) $I_{\mathrm{Napump}} + I_{\mathrm{Kpump}} = I_{\mathrm{pump}}$. To maintain homeostasis these will not suffice because to ensure $V =$ const under rest (i.e., no stimulation) conditions, pump leak currents associated with the two ions have to be included I_{Naleak} and I_{Kleak} leading to a modified Eq. (1) governing the time evolution of the voltage:

$$C\frac{\mathrm{d}V_m}{\mathrm{d}t} = -I_{\mathrm{Na}} - I_{\mathrm{K}} - I_{\mathrm{Naleak}} - I_{\mathrm{Kleak}} - I_{\mathrm{leak}} - I_{\mathrm{pump}}, \tag{12}$$

where $I_{\mathrm{Naleak}} = g_{\mathrm{Naleak}}(V - E_{\mathrm{Na}})$ and $I_{\mathrm{Kleak}} = g_{\mathrm{Kleak}}(V - E_{\mathrm{K}})$ (g_{Naleak} and g_{Kleak} are constants).

The ion fluxes across the membrane of the node of Ranvier lead to concentration changes:

$$\frac{\mathrm{d}[\mathrm{Na}^+]_i}{\mathrm{d}t} = -\frac{(I_{\mathrm{Na}} + I_{\mathrm{Napump}} + I_{\mathrm{Naleak}})A}{F\mathrm{Vol}_i} \tag{13}$$

$$\frac{\mathrm{d}[\mathrm{Na}^+]_o}{\mathrm{d}t} = \frac{(I_{\mathrm{Na}} + I_{\mathrm{Napump}} + I_{\mathrm{Naleak}})A}{F\mathrm{Vol}_o} \tag{14}$$

$$\frac{\mathrm{d}[\mathrm{K}^+]_i}{\mathrm{d}t} = -\frac{\left(I_{\mathrm{K}} + I_{\mathrm{Kpump}} + I_{\mathrm{Kleak}}\right)A}{F\mathrm{Vol}_i} \tag{15}$$

$$\frac{\mathrm{d}[\mathrm{K}^+]_o}{\mathrm{d}t} = \frac{\left(I_{\mathrm{K}} + I_{\mathrm{Kpump}} + I_{\mathrm{Kleak}}\right)A}{F\mathrm{Vol}_o} \tag{16}$$

where F is the Faraday constant, A is the surface area of the nodal membrane, and Vol_i (or Vol_o) is the intracellular (or extracellular) volume of the node of Ranvier under study (we take $\mathrm{Vol}_i = \mathrm{Vol}_o$ for simplicity). Ion concentration dynamics alter the reversal potentials E_{Na} and E_{K} appearing in Eqs. (2) and (3). They obey the Nernst equations:

$$E_{\mathrm{Na}} = -\frac{RT}{F}\ln\frac{[\mathrm{Na}^+]_i}{[\mathrm{Na}^+]_o} \quad \text{and} \quad E_{\mathrm{K}} = -\frac{RT}{F}\ln\frac{[\mathrm{K}^+]_i}{[\mathrm{K}^+]_o} \tag{17}$$

The key parameter determining the time evolution of the firing node is the surface area (SA) to intracellular volume (vol_i) ratio $r = \mathrm{SA}/\mathrm{vol}_i$. For a node of Ranvier of length 1 μm and radius 1 μm this would give $r = 0.5 \times 10^6\ \mathrm{m}^{-1}$, but, in reality the relevant intracellular volume includes not just the slice under the nodal membrane but the full internodal volume, which for an internode, say, 1 mm, would yield $r = 0.5 \times 10^3\ \mathrm{m}^{-1}$, i.e. a three orders of magnitude smaller. In Boucher et al. (2012) and subsequent papers (Yu et al. 2012; Lachance et al. 2014) r was set at $2 \times 10^6\ \mathrm{m}^{-1}$ to allow the different regimes of excitability to be investigated with reasonable computational resources. (As an aside, Boucher et al. (2012) erroneously gave the intracellular volume as $2 \times 10^{-15}\ \mathrm{m}^3$ instead of 2×10^{-15} L.) Consequently, the pathological changes noted in Boucher et al. (2012), Yu et al. (2012), and Lachance et al. (2014) reveal themselves more quickly than would be expected in reality. Importantly, however, the choice of volumes affects only the time scales of these slow changes and not the behaviors per se. For the resulting computations, Fig. 5B shows different patho-excitability regimes for an excitable system that pumps ions in and out of finite volumes at it attempts to maintain ion homeostasis (Boucher et al. 2012). The difficulty of maintaining ion homeostasis increases as values of LS and AC increase. With E_{Na} and E_{K} free to vary, the excitability regimes are different as seen by comparing Fig. 5A, B. In Fig. 5B, with AC = 1, increasing LS yields a range of behaviors. Mild injury elicits transient firing (just after the injury is imposed) but the mildly injured system then relaxes into a new quiescent (albeit hyperexcitable) state as homeostatic pumping catches up. With deeper injury there is ectopic firing, first in the form of periodic bursts then tonic firing, then mini-bursts or subthreshold bursts, and finally a quiescent state that is unexcitable.

6 CLS-Induced Pathological Activity for Realistically Complex Membrane Damage

The previous sections showed different regimes of excitability for nodes with a fraction of the Nav population suffering an LS. Even though a rich range of behavior was observed for mildly damaged nodes, real nodes would probably experience more complex damage patterns. In general, the regimes described above should still be relevant, but important pathological nuances could be overlooked. As an example of somewhat more realistic complexity, consider distributions of Nav population that combine both mild damage (LS = 2 mV) and heavy damage (LS > 20 mV). One such situation was examined in Yu et al. (2012). Nodes were given three Nav populations (LS_i = [26.5, 2.0, 0] mV and f_i = [0.2, 0.08, 0.72]) where f_I with i = 1, 2, or 3 is the fraction of the node with left shift LS_i (the Hodgkin-Huxley curves are given in Fig. 6B) and the resulting "window

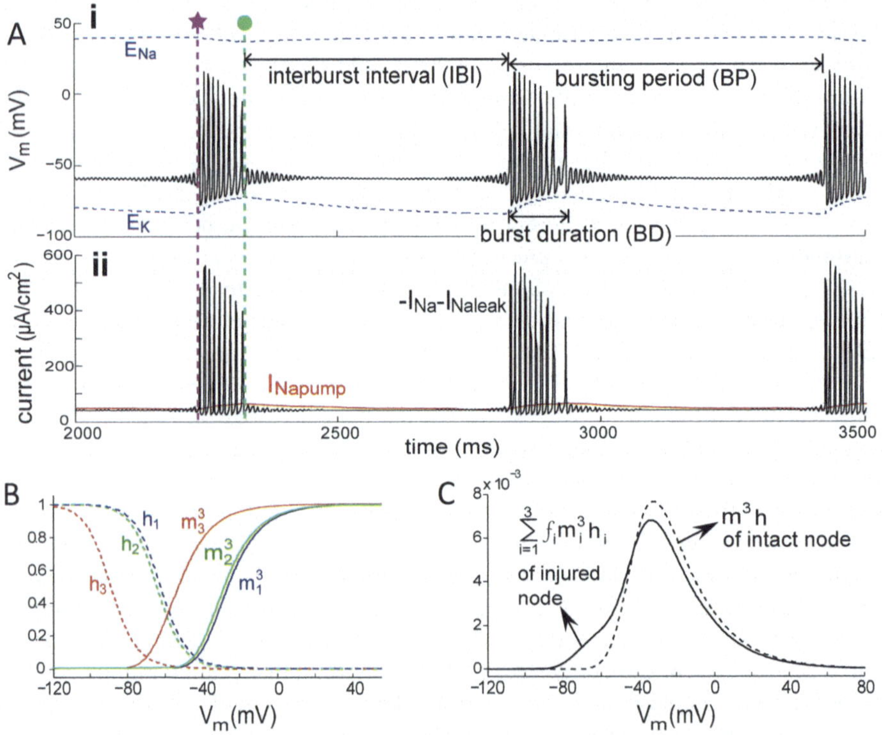

Fig. 6 Transitions between STO and burst behaviors. (**A**). Upper: V (black solid line) at an injured node and the varying E_{Ion} (E_{Na} and E_{K}; blue dotted lines). Three g_{Na} populations were used: LS_i = [0, 2, 26.5] mV and f_i = [0.72, 0.08, 0.2] with $Vol_i = Vol_o = 10^{-15}$ m^3. Initiation and termination times of a burst of spikes (pink star, green dot, respectively) are used in Fig. 7. Lower: corresponding Na^+ currents, as labeled. (**B**). Equilibrium values of the Nav kinetic constants for the three LS values in (**A**). (**C**) Window current for node producing bursts in (**A**) compared with intact node (from Yu et al. 2012)

conductance" is shown in Fig. 6C. We will discuss this case in some detail as it produces a pattern of behavior (i.e., $V_m(t)$) as per Fig. 6A i that is prevalent in neuropathic firing, i.e., irregularly spaced bursts of APs followed and preceded by subthreshold oscillations (STOs). Also shown is the changing Nav channel I_{Na} and a small pump I_{Na} in Fig. 6A ii. This simulation starts with the Na^+ and K^+ gradients at their maximum values (mimicking the time point immediately after an injury). Because of the damage, the node fires spontaneously, and as seen, the gradients start to deplete at a steeper pace. When depletion exceeds a certain point, the ectopic APs cease but, stimulated by the left-shifted window current (see Fig. 6B; the leftmost "shoulder" is critical here), the node produces subthreshold oscillations (STOs). Nav channel kinetics are responsible for STO frequency, but since they are associated with a reduced I_{Na} compared to actual APs, pumps are slowly able to restore the ion gradients. As this occurs, STOs decrease in amplitude. However, the interplay of Nav leak and pump current is such that the "leaky" window current eventually causes STOs to grow again and then a new ectopic AP burst occurs.

7 Dynamical Analysis of Ectopic Bursting

Performing a dynamical analysis of a Nav-CLS damaged node with mild damage and well-functioning pumps allows one to examine how the slow relentless activity of the pump interacts with the high speed non-linear kinetic properties of the AP producing channels.

Given the set of variables used for the injured system, a particular state of excitability is associated with each pair of E_{Na}, E_K values (see Fig. 5D in Yu et al. 2012). A bifurcation analysis (Fig. 7) shows how, during one ectopic burst cycle, the node goes into a spontaneous firing state, then becomes bistable, a state from which the node will switch (via a Hopf bifurcation) to a lower amplitude oscillation, then finally enter a quiescent state. It will then return via a similar pathway to the spontaneous firing state.

8 Saltatory Propagation in Axons with Mildly Damaged Nodes

The discussion thus far has concerned a single node's excitability, but damaged nodes also affect saltatory propagation of APs as shown by Boucher et al. (2012). With an emphasis largely on mild damage, Lachance et al. (2014) delved further into how CLS affects propagation, this time including the pump. For saltatory propagation each internode is represented as a single ohmic conductance κ linking two consecutive nodes of Ranvier (Ochab-Marcinek et al. 2009), on the assumption that myelinated internodes have negligible capacitance. The value of κ is chosen to ensure one-to-one propagation across the internode. Using a 10-node myelinated axon model, Lachance et al. (2014) found that quiescent, fully restabilized nodes with mild CLS damage eventually start firing ectopically if repeatedly stimulated

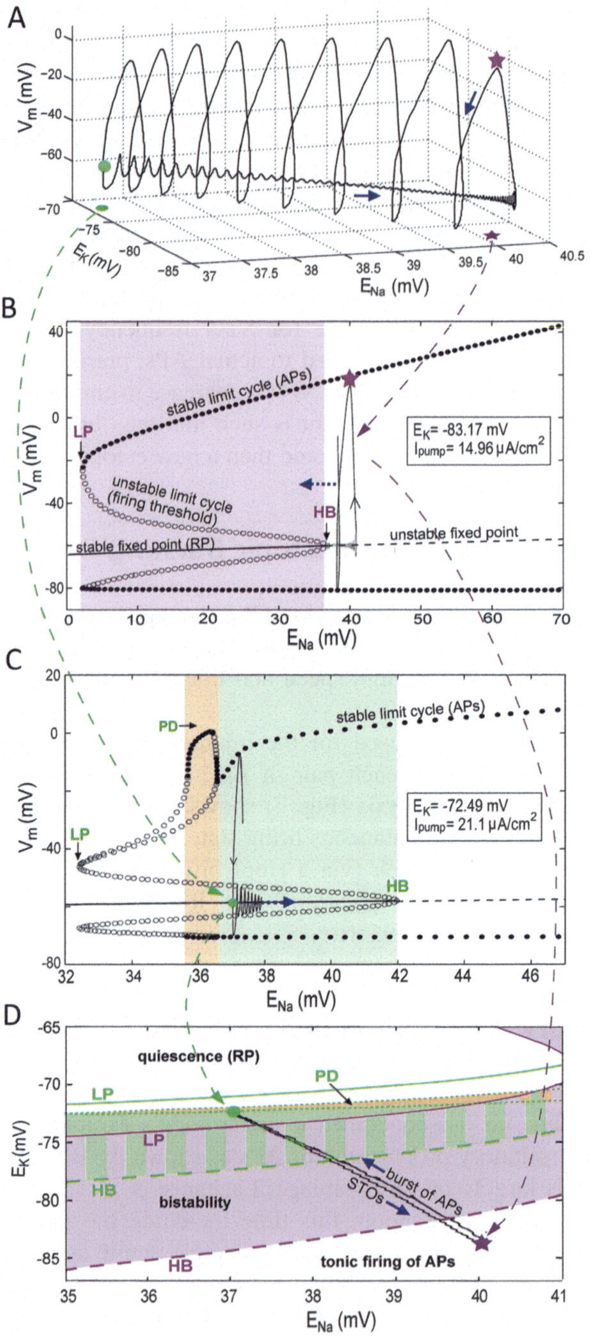

Fig. 7 Burst dynamics explained with three-dimensional V_m trajectories and bifurcation diagrams (for bursts in Fig. 6). Note: X-axis scales are different in each plot. (**A**) The first burst of APs,

by incoming APs (see Fig. 1 in Lachance et al. 2014). This outcome suggests a straightforward explanation for the well-known need to provide a relatively long low-input recovery time (i.e., stimulation of damaged tissue kept to a minimum) for those who have suffered mild traumatic brain injury.

Saltatory propagation simulations revealed another interesting phenomenon. APs can propagate with good fidelity through ectopically firing nodes, provided the frequency of the incoming APs exceeds the intrinsic frequency of the damaged node (i.e., the ectopic firing frequency of the damaged node in the absence of a stimulus). Under these conditions, the ectopic node locks its output to the frequency of the incoming stream of APs. The window of reliable propagation is robust in that it functions even when there are several consecutive ectopic nodes and even with a high level of temporal jitter in the incoming train of APs.

9 Sick Excitable Cells and Nav-CLS in Other Modeling Contexts

Various aspects of neuropathological excitability linked to neuropathic pain, epilepsy, and so on have been addressed computationally (e.g., Rotstein et al. 2006; Cressman et al. 2009; Barreto and Cressman 2011; Prescott et al. 2006; Coggan et al. 2010, 2011; Käger et al. 2000; Kovalsky et al. 2009 for dorsal root ganglion neuronal dysfunction; Choi and Waxman 2011) and discussed previously (Boucher et al. 2012; Yu et al. 2012). These works focus on simulating particular

Fig. 7 (continued) plotted in Fig. 6A as a function of time is plotted here in 3-D as a function of E_{Na} and E_{K} (blue arrow: direction of V_m trajectory). (**B**) The dynamical analysis of the system's transition (a bifurcation) from bursting to STOs is visualized as a diagram of the membrane potential excursions as a function of E_{Na} with fixed E_{K} and I_{pump} values measured at the pink star in Fig. 6A (the beginning of bursting). Such a diagram is known as a bifurcation diagram. At the pink star value the only stable solution is a periodic orbit. The pink oval loop corresponds to one cycle or one AP. Although in reality E_{K} also depolarizes, the graph shows how during a burst the E_{Na} decline shifts orbits leftward into the bistability regime (pink area) where two stable solutions exist. The point where a periodic orbit appears or disappears through a local change in the stability properties of a steady point is known as a Hopf bifurcation (HB). To get a more accurate representation of what happens at the transition, the analysis is repeated in (**C**) with the E_{K} at the green dot. (**C**) Bifurcation diagram for the fixed E_{K} and I_{pump} values at the green dot in (**A**). For E_{Na} at the green dot value (37.27 mV), the system (large green dot) is within the bistability regime (green area). V_m, attracted by this fixed point, has STOs until E_{Na} through the action of the pumps increases beyond the HB point and superthreshold-oscillations (APs) return. The PD region corresponds to period-doubling bifurcation (not attained during the bursting cycle). (**D**) Two-parameter bifurcation diagram for E_{K} and E_{Na}. Pink solid and dashed curves represent LP (saddle-node bifurcation) and HB, respectively, when I_{pump} is fixed as in (**B**). The green solid, dashed, and dash-dotted curves represent LP, HB, and PD, respectively, when I_{pump} is fixed as in *C*. With varying I_{pump} the bistability regime shifts from the pink to the green area (the zone with both colors is the overlap of these two areas). The gray area between two green dash-dotted curves is a zone with PD bifurcations. The black loop shows E_{K} and E_{Na} orbits during a burst (from Yu et al. 2012).

dysfunctions. As such they would not necessarily provide a more general nor cell biophysically grounded theory for dysfunctional excitability in sick excitable cells. Nav-CLS makes this connection by virtue of the commonality of bleb damage.

Jérusalem et al. (2013) incorporated Nav-CLS plus left shift of Kv activation into a multiscale model of injury to myelinated axons. Pumps were not included but in recognition of impaired ion homeostasis, E_K and E_{Na} values are moved toward 0 mV.

The Nav-*CLS* model has been applied in several contexts related to traumatic brain injury by Volman and Ng (2013, 2014, 2016). Most recently, with Ng et al. (2017) they have embedded CLS in the "mechanistic end-to-end concussion model that translates head kinematics to neurologic injury."

The involvement of neural, glial, and extracellular volume changes in dysfunctional excitability as it relates to spreading depression has been addressed computationally (Hübel and Dahlem 2014; Hübel and Ullah 2016). For cells like glia with a substantial anion conductance, including anion fluxes is mandatory. CLS, having been designed around neurons whose resting conductances are not anion based, lacks this feature, but some excitable cells have a resting anion conductances that would certainly need to be taken into account. Skeletal muscle cells and the electrocytes of various electric fish are cases in point. Ma et al. (2017) recently elaborated on dynamical issues emerging from Nav-CLS injury in excitable cells. Since their dominant interest is skeletal muscle malfunctions such as myotonia, their system could be a good platform for linking Nav-CLS to the volume regulatory and electrophysiological consequences that arise when anion fluxes are critical to the homeostatic processes.

10 The CLS Model Within NEURON, the Simulation Environment

As a simple tractable model of node damage, the *CLS* model is suitable for studying the impact of bleb damage beyond a node or series of nodes. A widely used computer environment to study neurons and networks of neurons is NEURON (www.neuron.yale.edu). NEURON was used to produce Fig. 8 which completes the regime diagram of Fig. 5B. Unlike for 5B, the responses to stimulation were tested for the part of the system corresponding to a quiescent membrane. The blue, orange, and yellow regimes represent ectopic spontaneous activity.

Nav CLS could also be adapted in the NEURON environment for regions of excitable membrane that express multiple kinetic types of Nav channels. Examples would include axon initial segments (Duménieu et al. 2017) and nociceptive nerve endings (Choi and Waxman 2011).

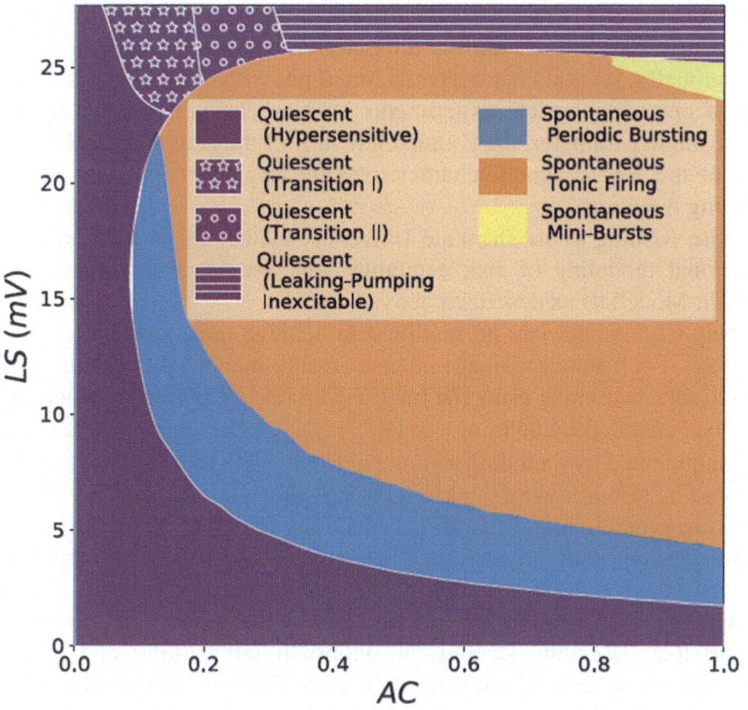

Fig. 8 Regimes of CLS obtained using NEURON Regimes of CLS induced activity for a system in which the pump remains active (damage notwithstanding) calculated for the same parameters as in Fig. 5B but using the NEURON computing environment (ModelDB). For the Quiescent regimes, excitability was tested (not shown) using a constant depolarizing I_{stim} in the same manner as in Fig. 5A (the no-pump situation). Responses to an I_{stim} were not checked for the Spontaneous regimes. In the Quiescent regimes linking the Hypersensitive and Inexcitable zones transitions I and II does not yield normal trains of APs in response to I_{stim} (the former yields complex patterns of damped voltage oscillations and quiescence and the latter exhibits highly damped small oscillations before returning to quiescence).

11 Conclusion

Inappropriately active or "leaky" Nav channels are treacherous. For clinical use, Nav channel antagonists are typically designed to target channels over-active in their slow gating mode, but the clinical efficacy of such drugs for ameliorating sick excitable cell dysfunction has been disappointing (Morris et al. 2012a, b; Morris and Joós 2016). It is well known that when a slow mode Nav leak (often referred to as excess "persistent current") develops, I_{Na} near 0 mV increases. In sick excitable cells however, we suspect that *CLS* underlies "leaky Nav channels" and if so, then the leaky channels are gating in fast, not slow mode. As Fig. 4 shows, where *CLS*-based "Nav leak" develops, a decrease (NOT an increase) in steady-state I_{Na} near

0 mV is predicted. For neuropharmacologists, this electrophysiological feature is an extremely important and straightforward practical tool arising from the *CLS* model.

More globally, several characteristics stand out with the *CLS* model. First is its foundation upon biophysical experiments on Nav channels (Wang et al. 2009). Second, a single parameter, the coupled left shift of the activation and inactivation voltages of the Nav channels, characterizes the degree of injury to the affected Nav-bearing membrane.

For those wishing to use the Yale University mathematical tool, NEURON for computational modeling of sick excitable cells, we have now made Nav-CLS available in ModelDB as described above.

The Nav-CLS model can be modified to achieve more realistic depictions in various ways. For instance, damage intensity can be made to differ in different parts of the excitable membrane using the fraction variable *f*. Complex damage situations can then be generated by defining a set $\{f_i, \mathrm{LS}_i\}_{i=1}^{N}$. The underlying conductances could be augmented by including a slow mode Nav with the ability to left shift with damage and by also allowing the gK(V) to left shift with damage.

A striking general outcome from our modeling to date is that mild *CLS* damage has multi-faceted pathophysiologically realistic effects on excitability. *CLS* (a rigorously observed phenomenon at the channel biophysics level (Wang et al. 2009)), has been deployed in systems that incorporate pumps, that exhibit current noise, and that take into account the abnormal intracellular and extracellular volumes that frequently accompany sick excitable cell conditions. With these characteristics included, *CLS* is sufficient to robustly explain a plethora of known but previously baffling sick excitable cell pathologies. *CLS* predicts hypersensitivity, ectopic tonic and burst firing, depolarizing block, complex subthreshold oscillations, paroxysmal bursts of activity. There is no need to invoke the expression of new channels or to suggest that cell-mediated kinetic modulation of channels must be occurring. CLS provides the only general theory we are aware of to explain dysfunctional excitability in diverse sick excitable cells. Its power, we suspect depends on the fact that *CLS* is not simply a kinetic theory, but one based on observations about the physics of bleb damage.

In summary, bleb damage to Nav-bearing membranes is a widely recognized but insufficiently studied feature of sick excitable cells that, in conjunction with Nav-CLS theory, can explain much of the pathological excitability observed in these cells.

References

Banderali U, Juranka PF, Clark RB, Giles WR, Morris CE (2010) Impaired stretch modulation in potentially lethal cardiac sodium channel mutants. Channels 4:12–21

Barreto E, Cressman JR (2011) Ion concentration dynamics as a mechanism for neuronal bursting. J Biol Phys 37:361–373

Boucher PA, Joós B, Morris CE (2012) Coupled left-shift of Nav channels: modeling the Na+-loading and dysfunctional excitability of damaged axons. J Comput Neurosci 33:301–319

Choi JS, Waxman SG (2011) Physiological interactions between Na(v)1.7 and Na(v)1.8 sodium channels: a computer simulation study. J Neurophysiol 106(6):3173–3184

Coggan JS, Prescott SA, Bartol TM, Sejnowski TJ (2010) Imbalance of ionic conductances contributes to diverse symptoms of demyelination. Proc Natl Acad Sci U S A 107:20602–20609

Coggan JS, Ocker G, Sejnowski TJ, Prescott SA (2011) Explaining pathological changes in axonal excitability through dynamical analysis of conductance-based models. J Neural Eng 8:065002

Cressman JR Jr, Ullah G, Ziburkus J, Schiff SJ, Barreto E (2009) The influence of sodium and potassium dynamics on excitability, seizures, and the stability of persistent states: I. Single neuron dynamics. J Comput Neurosci 26:159–170. Erratum in (2011) 30: 781

Duménieu M, Oulé M, Kreutz MR, Lopez-Rojas J (2017) The segregated expression of voltage-gated potassium and sodium channels in neuronal membranes: functional implications and regulatory mechanisms. Front Cell Neurosci 11:115

Finol-Urdaneta RK, McArthur JR, Juranka PF, French RJ, Morris CE (2010) Modulation of KvAP unitary conductance and gating by 1-alkanols and other surface active agents. Biophys J 98:762–772

Hodgkin AL, Huxley AF (1952) A quantitative description of membrane current and its application to conduction and excitation in nerve. J Physiol 117:500–544

Hübel N, Dahlem MA (2014) Dynamics from seconds to hours in hodgkin-huxley model with time-dependent ion concentrations and buffer reservoirs. PLoS Comput Biol 10(12):e1003941

Hübel N, Ullah G (2016) Anions govern cell volume: a case study of relative astrocytic and neuronal swelling in spreading depolarization. PLoS One 11(3):e0147060

Jérusalem A, García-Grajales JA, Merchán-Pérez A, Peña JM (2013) A computational model coupling mechanics and electrophysiology in spinal cord injury. Biomech Model Mechanobiol 14:1–14

Käger H, Wadman W, Somjen G (2000) Simulated seizures and spreading depression in a neuron model incorporating interstitial space and ion concentrations. J Neurophysiol 84:495–512

Kovalsky Y, Amir R, Devor M (2009) Simulation in sensory neurons reveals a key role for delayed Na+ current in subthreshold oscillations and ectopic discharge: implications for neuropathic pain. J Neurophysiol 102:1430–1442.

Lachance M, Longtin A, Morris CE, Yu N, Joós B (2014) Stimulation-induced ectopicity and propagation windows in model damaged axons. J Comput Neurosci 37:523–531

Läuger P (1991) Electrogenic ion pumps, distinguished lecture series of the society of general physiologists. Sinauer Associates, Sunderland, MA, p 313

Ma QX, Arneodo A, Ding GH, Argoul F (2017) Dynamical study of Nav channel excitability under mechanical stress. Biol Cybern 111(2):129–148

Morris CE, Joós B (2016) Channels in damaged membranes. In: French RJ, Noskov SY (eds) Na channels from phyla to function. Currents topics in membranes, vol 18. Elsevier, Amsterdam, pp 561–597

Morris CE, Juranka PF (2007) Nav channel mechanosensitivity: activation and inactivation accelerate reversibly with stretch. Biophys J 93:822–833

Morris CE, Boucher PA, Joós B (2012a) Left-shifted Nav channels in injured bilayer: primary targets for neuroprotective Nav antagonists? Front Pharmacol 3:19

Morris CE, Juranka PF, Joós B (2012b) Perturbed voltage-gated channel activity in perturbed bilayers: implications for ectopic arrhythmias arising from damaged membrane. Prog Biophys Mol Biol 110:245–256

Ng LJ, Volman V, Gibbons MM, Phohomsiri P, Cui J, Swenson DJ, Stuhmiller JH (2017) A mechanistic end-to-end concussion model that translates head kinematics to neurologic injury. Front Neurol 8:269

Ochab-Marcinek A, Schmid G, Goychuk I, Hänggi P (2009) Noise-assisted spike propagation in myelinated neurons. Phys Rev. E 79(1):011904

Prescott SA, Sejnowski TJ, De Koninck Y (2006) Reduction of anion reversal potential subverts the inhibitory control of firing rate in spinal lamina I neurons: towards a biophysical basis for neuropathic pain. Mol Pain 2:32–51

Rotstein HG, Oppermann T, White JA, Kopell N (2006) The dynamic structure underlying subthreshold oscillatory activity and the onset of spikes in a model of medial entorhinal cortex stellate cells. J Comput Neurosci 21:271–292

Shcherbatko A, Ono F, Mandel G, Brehm P (1999) Voltage-dependent sodium channel function is regulated through membrane mechanics. Biophys J 77:1945–1959

Sheetz MP, Sable JE, Döbereiner HG (2006) Continuous membrane cytoskeleton adhesion requires continuous accommodation to lipid and cytoskeleton dynamics. Annu Rev. Biophys Biomol Struct 35:417–434

Sterratt D, Graham B, Gillies A, Willshaw D (2011) Principles of computational modelling in neuroscience. Cambridge Univ Press, Cambridge

Tabarean IV, Juranka P, Morris CE (1999) Membrane stretch affects gating modes of a skeletal muscle sodium channel. Biophys J 77:758–774

Volman V, Ng LJ (2013) Computer modeling of mild axonal injury: implications for axonal signal transmission. Neural Comput 25:2646–2681

Volman V, Ng LJ (2014) Primary paranode demyelination modulates slowly developing axonal depolarization in a model of axonal injury. J Comput Neurosci 37:439–457

Volman V, Ng LJ (2016) Perinodal glial swelling mitigates axonal degradation in a model of axonal injury. J Neurophysiol 115:1003–1017

Wang JA, Lin W, Morris T, Banderali U, Juranka PF, Morris CE (2009) Membrane trauma and Na + leak from Nav1.6 channels. Am J Physiol Cell Physiol 297:C823–C834

Yu N, Morris CE, Joós B, Longtin A (2012) Spontaneous excitation patterns computed for axons with injury-like impairments of sodium channels and Na/K pumps. PLoS Comput Biol 8: e1002664

Zou S, Chisholm R, Tauskela JS, Mealing GA, Johnston LJ, Morris CE (2013) Force spectroscopy measurements show that cortical neurons exposed to excitotoxic agonists stiffen before showing evidence of bleb damage. PLoS One 8:e73499

Voltage-Gated Sodium Channel β Subunits and Their Related Diseases

Alexandra A. Bouza and Lori L. Isom

Contents

Abstract

Voltage-gated sodium channels are protein complexes comprised of one pore forming α subunit and two, non-pore forming, β subunits. The voltage-gated sodium channel β subunits were originally identified to function as auxiliary subunits, which modulate the gating, kinetics, and localization of the ion channel pore. Since that time, the five β subunits have been shown to play crucial roles as multifunctional signaling molecules involved in cell adhesion, cell migration,

A. A. Bouza
Department of Pharmacology, University of Michigan Medical School, 2200 MSRBIII, 1150 W. Medical Center Dr., Ann Arbor, MI 48109-5632, USA

L. L. Isom (✉)
Department of Pharmacology, University of Michigan Medical School, 2301 MSRB III, 1150 W. Medical Center Dr., Ann Arbor, MI 48109-5632, USA
e-mail: lisom@umich.edu

M. Chahine (ed.), *Voltage-gated Sodium Channels: Structure, Function and Channelopathies*, Handbook of Experimental Pharmacology 246,
https://doi.org/10.1007/164_2017_48

neuronal pathfinding, fasciculation, and neurite outgrowth. Here, we provide an overview of the evidence implicating the β subunits in their conducting and non-conducting roles. Mutations in the β subunit genes (*SCN1B–SCN4B*) have been linked to a variety of diseases. These include cancer, epilepsy, cardiac arrhythmias, sudden infant death syndrome/sudden unexpected death in epilepsy, neuropathic pain, and multiple neurodegenerative disorders. β subunits thus provide novel therapeutic targets for future drug discovery.

Keywords
α subunit · β subunit · Cell adhesion · Channelopathy · Dravet syndrome · Neuronal pathfinding · Sudden unexpected death in epilepsy · Voltage-gated sodium channel

1 The Basics of the Voltage-Gated Sodium Channel β Subunits

There are five voltage-gated sodium channel (VGSC) β subunits, which are encoded by four genes, *SCN1B–SCN4B* (O'Malley and Isom 2015). *SCN1B* encodes the β1 subunit and the developmentally regulated splice variant, β1B, while the β2, β3, and β4 subunits are encoded by *SCN2B–SCN4B*, respectively (Table 1) (Isom et al. 1992, 1995a; Kazen-Gillespie et al. 2000; Morgan et al. 2000; Patino et al. 2011; Qin et al. 2003; Yu et al. 2003). β subunits each contain a large, extracellular V-set immunoglobulin (Ig) domain, making them part of the Ig superfamily of cell adhesion molecules (CAMs) (Brackenbury and Isom 2011; O'Malley and Isom 2015). β1B differs from the other subunits in that it is the only one that is not a type I transmembrane protein, but rather, a soluble, secreted CAM expressed during embryonic development in brain and throughout development, into adulthood, in heart. The C-terminal domain of β1B is encoded by a retained intron, resulting in a unique polypeptide sequence that does not contain a transmembrane segment (Patino et al. 2011). β subunit Ig domains are stabilized by two completely conserved cysteine residues in the extracellular portion, maintaining the β-sheet structure, as

Table 1 VGSC genes and their encoded proteins

VGSC α subunits		VGSC β subunits	
Gene	Protein	Gene	Protein
SCN1A	$Na_v1.1$	*SCN1B*	β1
SCN2A	$Na_v1.2$	*SCN1B*	β1B
SCN3A	$Na_v1.3$	*SCN2B*	β2
SCN4A	$Na_v1.4$	*SCN3B*	β3
SCN5A	$Na_v1.5$	*SCN4B*	β4
SCN8A	$Na_v1.6$		
SCN9A	$Na_v1.7$		
SCN10A	$Na_v1.8$		
SCN11A	$Na_v1.9$		

established by the X-ray crystal structures of β3 and β4 (Gilchrist et al. 2013; Namadurai et al. 2014).

VGSCs are comprised of one pore-forming α subunit and two different β subunits, either a β1 or β3 and a β2 or β4 (Fig. 1) (O'Malley and Isom 2015). β1 and β3 non-covalently associate with α, while β2 and β4 associate with α by cysteine disulfide bonds, Cys-26 and Cys-58, respectively, both of which are located in the extracellular Ig domain (Chen et al. 2012; Gilchrist et al. 2013; McCormick et al. 1998; Meadows et al. 2001; Spampanato et al. 2004). While there is no biochemical evidence to show that β1B associates with α subunits by co-immunoprecipitation, co-expression of β1B and $Na_v1.5$ in heterologous systems results in plasma membrane retention of β1B and increased sodium current density, implicating association with α (Patino et al. 2011; Watanabe et al. 2008). Heterologous co-expression of β1B with $Na_v1.2$ results in changes in current activation and inactivation, while co-expression with $Na_v1.3$ results in subtle alternation of current properties, again suggesting a functional association (Kazen-Gillespie et al. 2000; Patino et al. 2011).

VGSC β subunits are expressed in many tissues and cell types, not all of which are excitable (Brackenbury and Isom 2011). The expression of each specific β subunit is also developmentally regulated. In rodent brain, β1B and β3 are most highly expressed during embryonic development and early life. This differs from heart, in which β1B and β3 expression continues into adult life (Kazen-Gillespie et al. 2000; Patino et al. 2011; Shah et al. 2001). β1 and β2 display peak expression in brain during adulthood (Isom et al. 1992, 1995a). The developmental regulation

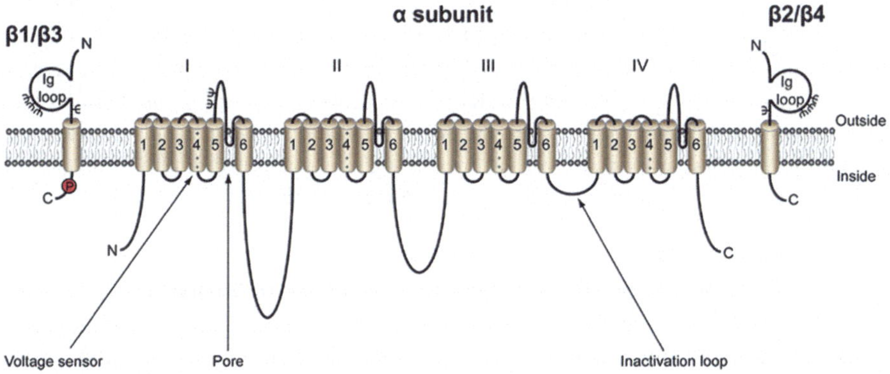

Fig. 1 Cartoon diagram of the VGSC. VGSCs are comprised of one pore-forming, or α subunit, and one or two non-pore forming β subunits. The α subunit is made up four domains each of which contain six transmembrane segments. The voltage sensor is located in transmembrane segment four of each domain (Catterall 2000). There are five β subunits, β1–β4, and the developmentally regulated β1B. β1–β4 all contain an intracellular C-terminal domain, a single transmembrane domain, and a large extracellular immunoglobulin (Ig) domain (Isom et al. 1994). β1B also possesses an Ig domain, but does not contain an intracellular or transmembrane domain, resulting in a soluble, secreted protein (Patino et al. 2011). β1 and β3 are non-covalently linked to the α subunit, while β2 and β4 are linked by disulfide bonds. Each β subunit is heavily glycosylated, denoted by Ψ, and β1 also contains an intracellular phosphorylation site at tyrosine 181 (Isom and Catterall 1996; Malhotra et al. 2004). Figure reproduced from Brackenbury and Isom (2011)

of β4 subunit expression is yet to be determined. β subunits are localized to a variety of specific sub-cellular compartments. In the brain and peripheral nervous system, β subunits are highly enriched the axon initial segment and nodes of Ranvier (Buffington and Rasband 2013; Chen et al. 2002, 2004; Dhar Malhotra et al. 2001; O'Malley et al. 2009). These sites are important in the initiation and propagation of action potentials in neurons and have a high density of VGSC α subunit expression (Brackenbury et al. 2010). In heterologous cells, the β1 C-terminal domain interacts with the scaffolding protein, ankryin-G, in a tyrosine phosphorylation-dependent manner and a similar mechanism is proposed at the axon initial segment and nodes of Ranvier (Malhotra et al. 2000). In cardiomyocytes, the phosphorylation of β1 may regulate its sub-cellular localization. Tyrosine phosphorylated β1 subunits are localized to the intercalated disks, while non-phosphorylated β1 subunits are localized to t-tubules (Malhotra et al. 2004).

In addition to phosphorylation, the β subunits are post-translationally modified by glycosylation and proteolytic cleavage. All five β subunits are highly N-linked glycosylated (Isom et al. 1992). This heavy glycosylation of mature β subunits accounts for about 12 kilodaltons (kDa) of the ~36 kDa total molecular weight. β subunit glycosylation impacts their surface expression and channel modulatory properties (Johnson et al. 2004). Lastly, the transmembrane β subunits are also substrates for sequential cleavage by the β-site amyloid precursor protein cleaving enzyme-1 (BACE1) and γ-secretase (Wong et al. 2005). Initially, BACE1 cleaves β subunits on the extracellular side of the membrane, shedding the Ig domain, which may function as a soluble CAM, similar to β1B. Subsequently, γ-secretase cleaves the β subunits in the lumen of the membrane, generating an intracellular domain (ICD) (Fig. 2) (Haapasalo and Kovacs 2011; Wong et al. 2005). Evidence shows that the β2 subunit ICD translocates to the nucleus where it increases expression of the $Na_v1.1$ α subunit (Kim et al. 2007). A similar mechanism has been proposed, but not shown, for the other β subunits. Sequential cleavage of β subunits may play important roles in mediating neurite outgrowth, migration, and cell adhesion (Brackenbury and Isom 2011; Kim et al. 2005).

1.1 Modulation of the Ion Channel Pore by β Subunits

VGSC β subunits are traditionally known for their functions in modulating the gating and kinetics of the VGSC pore (Calhoun and Isom 2014). In *Xenopus oocytes*, expression of $Na_v1.2$ mRNA alone results in sodium currents that are activated and inactivated much slower than those recorded in neurons. Co-injection of β1 or β2 mRNA altered the sodium current parameters (Isom et al. 1992). Co-expression β1 with $Na_v1.2$ increased peak sodium current density, shifted the voltage-dependence of inactivation negatively, and accelerated activation and inactivation in comparison to $Na_v1.2$ alone (Isom et al. 1992). Co-expression of β2 with $Na_v1.2$ also resulted in increased peak sodium current density and accelerated current inactivation. Expression of the three subunits together, β1, β2, and $Na_v1.2$, yielded the largest peak sodium currents and the most rapidly

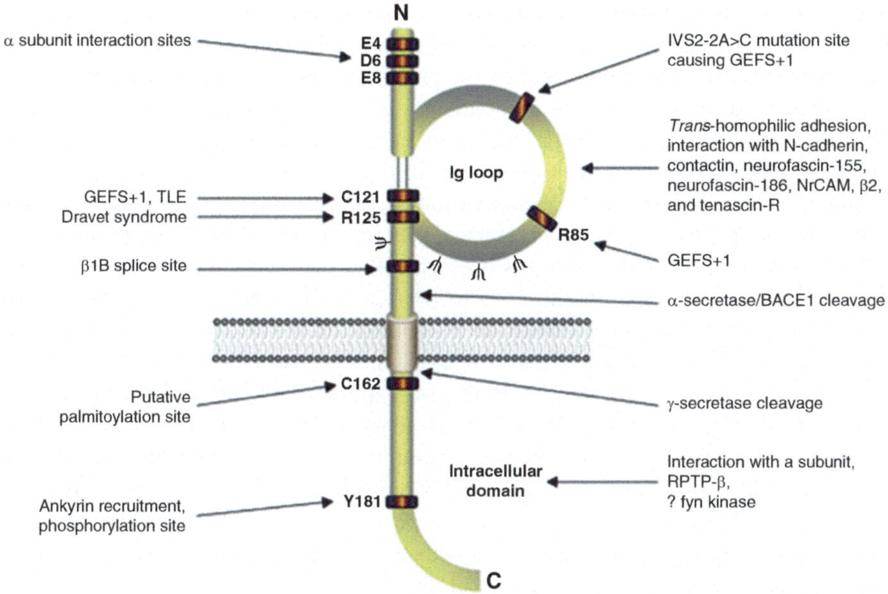

Fig. 2 Cartoon diagram of β1/β1B topology. Both intracellular and extracellular residues on β1 are important for interacting with the α subunit (McCormick et al. 1998; Spampanato et al. 2004). Epilepsy-linked mutation sites are clustered in the Ig domain (Audenaert et al. 2003; Meadows et al. 2002; Patino et al. 2009; Scheffer et al. 2007; Wallace et al. 1998, 2002). The alternative splice site for β1B, ankyrin interaction site (Kazen-Gillespie et al. 2000; Patino et al. 2011; Qin et al. 2003), α-secretase/BACE1/γ-secretase cleavage sites (Wong et al. 2005), N-glycosylation sites (*Ψ*) (McCormick et al. 1998), tyrosine phosphorylation site (Malhotra et al. 2004), and the putative palmitoylation and *fyn* kinase interaction site are designated (Brackenbury et al. 2008; McEwen and Isom 2004). Figure reproduced from Brackenbury and Isom (2011)

inactivating channels, most closely mimicking that observed in neurons (Isom et al. 1995a).

In addition to modulating VGSC activity, β1 can also act on Kv1.1, Kv1.2, Kv1.3, Kv1.6, Kv4.2, Kv4.3, and Kv7.2 in *Xenopus* oocytes or heterologous cells (Deschenes and Tomaselli 2002; Nguyen et al. 2012). Interestingly, β1 and Kv4.2 co-immunoprecipitate from mouse brain. Kv4.2 is a major contributor to A-type potassium current. Knockdown of β1 decreases A-type potassium current and prolongs action potential waveforms in cultured cortical neurons. β1 expression also increases the stability of Kv4.2 in HEK293 cells, leading to increased total and cell surface expression of Kv4.2 (Marionneau et al. 2012).

In addition to oocytes, heterologous mammalian cell lines have been utilized to study VGSC modulation by the β subunits. Although these models are more physiologically relevant, they cannot fully replicate VGSC activity in native excitable cells, such as neurons and cardiomyocytes. In Chinese Hamster Lung (CHL) cells, co-expression of $Na_V1.2$ with β1 increases peak current density and causes a negative shift in the voltage-dependence of activation and inactivation, although to a lesser extent than observed in *Xenopus* oocytes (Isom et al. 1992, 1995b; Patino et al. 2009). Co-expression of $Na_v1.2$ and β2 in CHL cells does not recapitulate the results observed in *Xenopus* oocytes, instead resulting in sodium currents that are unchanged or reduced in comparison to the expression of $Na_v1.2$ alone (Kazarinova-Noyes et al. 2001; McEwen et al. 2004). In addition to altering kinetics of the ion channel pore, co-expression of β1 and/or β2 with $Na_v1.2$ affects α subunit surface expression. Co-expression of β1 with $Na_v1.2$ increases α subunit cell surface expression (Isom et al. 1995a). When β2 is added to the experiment, the cell surface expression of α subunits is even further increased, even though β2 cannot generate this effect in the absence of β1 (Kazarinova-Noyes et al. 2001). The modulatory and localization effects of β subunits on α subunits are impacted by the presence of other Ig-superfamily CAMs. In the case of the CAM, contactin, co-expression with $Na_v1.2$ and β1 increases α subunit cell surface expression and sodium current density approximately fourfold over that observed with $Na_v1.2$ plus β1. This is also displayed with NF186, although to a lesser extent than the effects observed with contactin. β1B co-expression with $Na_v1.2$ in CHL cells also increases α subunit surface expression and peak sodium current density, although this combination only has a modest effect on channel activation and inactivation (Kazen-Gillespie et al. 2000; Patino et al. 2011).

The effects of the β subunits on a variety of α subunits have also been studied in Chinese Hamster Ovary (CHO) cells and Human Embryonic Kidney (HEK) cells. β1 or β3 co-expression with $Na_v1.3$ in CHO cells results in a negative shift in the voltage dependence of inactivation, but does not influence the rate of inactivation. In this same system, co-expression of β2 with $Na_v1.3$ had no effect on the gating or kinetics of the ion channel pore (Meadows et al. 2002). Co-expression of β3 with $Na_v1.5$ in CHO cells results in a negative shift in the voltage dependence of inactivation, but decreases the rate of inactivation (Ko et al. 2005). Co-expression of β1B with $Na_v1.3$ in CHO cells has no effect on $Na_v1.3$ cell surface expression or sodium current density, different from the large effect of β1B observed in CHL cells (Kazen-Gillespie et al. 2000; Patino et al. 2011). In HEK cells, co-expression of $Na_v1.5$ and β4 results in a negative shift in the voltage dependence of inactivation in comparison to expression of $Na_v1.5$ alone (Medeiros-Domingo et al. 2007). The β4 subunit, when expressed with $Na_v1.2$ or $Na_v1.4$, induces a negative shift in the voltage dependence of activation (Yu et al. 2003). This is also the case for co-expression of β4 with $Na_v1.1$, although this results in increased levels of non-inactivating current (Aman et al. 2009). Also in HEK cells, β1 and β3 subunits each modulate activity, cell surface expression, and glycosylation state of $Na_v1.7$. β1 or β3 co-expression with $Na_v1.7$ resulted in shifted activation and inactivation and increased sodium current density. Co-expression of β1 also resulted in alternative glycosylation of $Na_v1.7$, while co-expression with β3 led to increased expression of fully glycosylated $Na_v1.7$ (Laedermann Cé et al. 2013). Overall, studies on β subunit modulation of VGSCs in heterologous systems have revealed cell type, β subunit, and α subunit specific effects.

The most physiologically relevant method to study VGSC modulation by the β subunits is to utilize primary cells, e.g. neurons or cardiomyocytes. In these native cells, β subunit effects are, in general, more modest than observed in heterologous over-expression systems. *Scn1b*-null mice, lacking both β1 and β1B, model the epileptic encephalopathy Dravet syndrome, and exhibit spontaneous seizures, ataxia, and premature death around post-natal day (P)19 (Chen et al. 2004). Acutely isolated P10–P18 *Scn1b*-null pyramidal and bipolar hippocampal neurons show no differences in VGSC activity compared to age-matched wild-type animals (Chen et al. 2004; Patino et al. 2009). However, slice recordings from this age range revealed hyperexcitability in the *Scn1b*-null CA3 hippocampal region as well as epileptiform activity in the hippocampus and cortex, suggesting altered VGSC activity in axons or dendrites (Patino et al. 2009). There are altered sodium currents and decreased excitability in cultured *Scn1b*-null cerebellar granule neurons (CGNs) (Brackenbury et al. 2010). In contrast, acutely isolated *Scn1b*-null dorsal root ganglion (DRG) neurons are hyperexcitable (Brackenbury et al. 2010; Lopez-Santiago et al. 2011). These results suggest that the effects of β1 and β1B in brain are neuronal cell-type specific, consistent with that observed in heterologous cells. Similar to that observed in heterologous systems, β1/β1B expression in vivo affects the expression of α subunits, especially $Na_v1.1$ and $Na_v1.3$. In the *Scn1b*-null hippocampal CA3 region, $Na_v1.1$ expression is decreased, while $Na_v1.3$ expression is increased (Chen et al. 2004). β2 also modulates VGSC gating and kinetics in vivo. Acutely isolated *Scn2b*-null hippocampal neurons display a negative shift in the voltage dependence of inactivation in comparison to neurons from age-matched, wild-type mice (Chen et al. 2002). Acutely isolated *Scn2b*-null small-fast DRG neurons have decreased sodium current density and decreased rates of TTX-sensitive sodium current activation and inactivation (Lopez-Santiago et al. 2006). Importantly, the β4 intracellular domain is postulated to play a role in resurgent sodium current, or the influx of sodium ions through the ion channel pore during repolarization. β4 knockdown in mouse CGNs showed reduced resurgent sodium current and repetitive firing (Bant and Raman 2010). Furthermore, expression of a β4 intracellular domain peptide in CA3 neurons, which do not endogenously express β4 subunits, generates resurgent sodium current (Grieco et al. 2005). This activity is particularly important in high-frequency firing neurons. *Scn4b*-null mice have defects in sodium current modulation. *Scn4b*-null mice have reduced resurgent sodium current and repetitive firing in medium spiny neurons of the striatum, as well as increased failure rates of inhibitory postsynaptic currents with repetitive stimulation (Miyazaki et al. 2014). β1 and β1B are also implicated in regulating resurgent sodium current in the cerebellum, as *Scn1b*-null CGNs have normal transient sodium current, but decreased resurgent sodium current, even though the overall protein expression of β4 is unchanged (Brackenbury et al. 2010). Together, these data indicate that modulation of sodium current by the β subunits in vivo is cell-type-, subcellular domain-, β subunit-, and α subunit-specific.

VGSC β subunits are also important regulators of excitability in the heart. In ventricular cardiomyocytes isolated from *Scn1b*-null mice, transient and persistent sodium currents are increased due to increased *Scn5a* and $Na_v1.5$ expression, resulting in prolongation of action potential repolarization and the QT interval (Lin et al. 2014; Lopez-Santiago et al. 2007). Furthermore, *Scn1b*-null mice display increased susceptibility to polymorphic ventricular arrhythmias. *Scn1b*-null ventricular cardiomyocytes also have increased tetrodotoxin (TTX)-sensitive sodium current, increased $Na_v1.3$ mRNA levels, increased incidence of delayed after-depolarizations, delayed Ca^{2+} transients, and frequent spontaneous Ca^{2+} release. Addition of TTX prevented the majority of changes in Ca^{2+} handling, indicating mutations in *Scn1b* may result in disrupted intracellular Ca^{2+} homeostasis in ventricular myocytes (Lin et al. 2014). *Scn2b* deletion in mice leads to atrial and ventricular arrhythmias and increased levels of atrial fibrosis. These animals exhibit region-specific effects in heart. *Scn2b*-null ventricular myocytes show reduced sodium and potassium currents, with conduction slowing in the right ventricle compared to wild-type. *Scn2b*-null atria had normal levels of sodium current compared to wild-type (Bao et al. 2016). *Scn3b*-null mice also show abnormal cardiac excitability, with ventricular tachycardia from electrical stimulation that is not observed in wild-type mice. *Scn3b*-null hearts also demonstrate atrial tachycardia during atrial burst pacing (Hakim et al. 2008).

1.2 The β Subunits as Cell Adhesion Molecules

All five β subunits have an extracellular Ig domain and belong to the Ig superfamily of CAMs (Isom and Catterall 1996). Importantly, β subunits have also been shown experimentally to function as CAMs (Isom 2002). An especially large body of work in this area has been completed on the β1 subunit. In *Drosophila* S2 cells expressing either β1 or β2, large aggregates form, suggesting these molecules can participate in *trans* homophilic cell adhesion in vitro (Malhotra et al. 2000). Upon β1–β1 *trans* homophilic cell adhesion in *Drosophila* S2 cells, ankyrin is recruited to the point of cell–cell contact (Meadows et al. 2001). Ankyrin binds to the β1 subunit via the intracellular C-terminal domain in a tyrosine phosphorylation-dependent manner. When residue Y181 of β1 is phosphorylated, ankyrin is unbound, while when Y181 is not phosphorylated, ankyrin binds to β1, indicating that downstream signaling events occur in response to cell–cell adhesion (Malhotra et al. 2002). In addition, β1 subunits can form heterophilic interactions with other CAMs, including contactin, N-cadherin, NrCAM, neurofascin-155, neurofascin-186, and the VGSC β2 subunit as well as the extracellular matrix protein, tenascin-R (Fig. 3) (McEwen and Isom 2004; Xiao et al. 1999). β2 subunits can also participate in heterophilic adhesion in vitro with both tenascin-R and tenascin-C (Srinivasan et al. 1998; Xiao et al. 1999). *Drosophila* S2 cells expressing β3 subunits do not aggregate, suggesting that β3 does not participate in *trans* homophilic adhesion (McEwen et al. 2009). In contrast, β3 subunits expressed in HEK cells participate in *trans* heterophilic adhesion with other CAMs, although this does not result in β3-ankyrin binding (McEwen et al. 2009; McEwen and Isom 2004; Ratcliffe et al. 2001). The function of β4 in cell adhesion remains more poorly understood (Brackenbury and Isom 2011). Insights from crystallographic, mutagenic, and photo-crosslinking studies have revealed the structural importance of an antiparallel interface between β4 subunits in *trans* homophilic adhesion (Shimizu et al. 2016). Recent evidence shows that β4 Ig domains interact in a parallel manner involving a disulfide bond

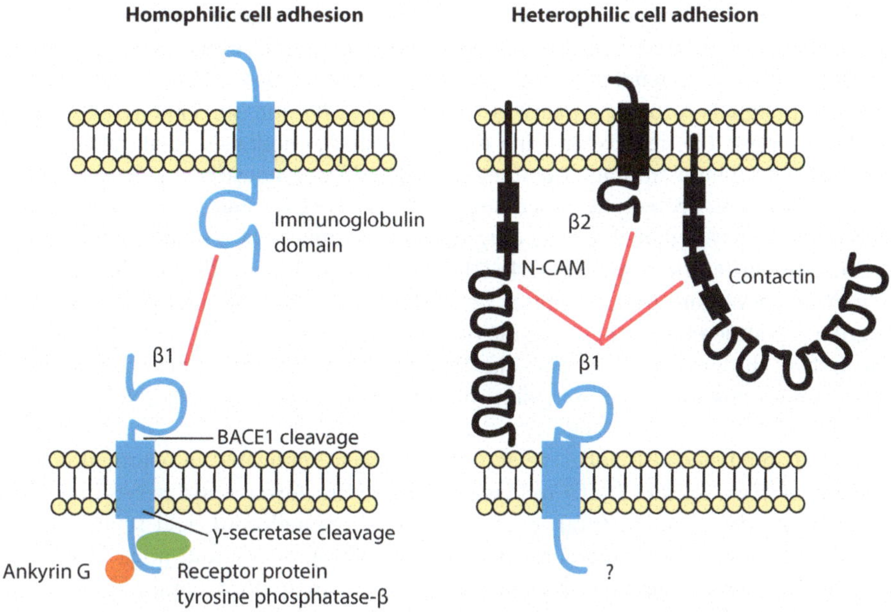

Fig. 3 β1 participates in homophilic and heterophilic cell adhesion (Malhotra et al. 2000). *Left* Schematic of β1–β1 homophilic cell adhesion and its downstream signaling. At points of cell–cell contact β1 binds to ankyrin in a phosphorylation-dependent manner. When β1 is not phosphorylated it is bound to ankyrin, while when tyrosine 181 is phosphorylated it is not bound to ankyrin (Malhotra et al. 2000, 2002). In rat brain β1 interacts with receptor protein tyrosine phosphatase-β which may contribute to regulating the β1 phosphorylation state (Ratcliffe et al. 2000). *Right* β1 participates in heterophilic cell adhesion with N-CAM, VGSC β2 subunits, and contactin

between cysteine 58 and hydrophobic and hydrogen bonding interactions between residues 30 through 35. Deletion of the β4 N-terminal domain led to decreased cell adhesion and increased association with the α subunit, revealing the importance of β4 *cis* dimerization (Shimizu et al. 2017).

Consistent with the role of β1 and β2 in cell adhesion, these molecules have been identified to mediate neurite outgrowth in CGNs (Davis et al. 2004). In this series of experiments, CGNs were grown on a monolayer of CHL cells that either did, or did not, express β subunit proteins. When β1–β1 *trans* homophilic cell adhesion occurred between the CGN and the monolayer expressing β1, neurite length was longer than when it did not. In contrast, β2-mediated homophilic adhesion resulted in decreased neurite length while β4 had no effect on this biological output. These data suggest that β1–β1 *trans* homophilic cell adhesion initiates a signal transduction cascade to drive neurite outgrowth in vitro while β2-mediated signaling may be inhibitory. Cell adhesion-mediated neurite outgrowth has been shown to occur through two downstream pathways: either via an epidermal growth factor receptor (EGFR) or fibroblast growth factor receptor (FGFR) mediated signal transduction cascade, or through the *fyn* kinase pathway (Brackenbury et al. 2008). Inhibitors of

FGFR and EGFR had no effect on β1-mediated neurite outgrowth in CGNs. In contrast, CGNs isolated from *fyn*-null mice grown on a CHL monolayer expressing β1 did not show extended neurite length, suggesting that β1-mediated neurite outgrowth signals through a pathway that involves *fyn* kinase. This hypothesis is further supported by results showing that β1 subunit peptides associate with *fyn* in detergent-resistant membrane fractions solubilized from mouse brain (Brackenbury et al. 2008). The proteolytic processing of β1 by BACE1 and γ-secretase is also important for β1-mediated neurite outgrowth, as inhibitors of γ-secretase block β1-mediated neurite outgrowth (Fig. 4) (Brackenbury and Isom 2011). The secreted *Scn1b* splice variant, β1B, increases neurite outgrowth to a similar extent as full-length β1 (Patino et al. 2011). Outside of the central nervous system, the β1 subunit can induce the growth of neurite-like features from cultured breast cancer cells, suggesting a possible developmental role for β1 in other cell-types (Nelson et al. 2014).

β1/β1B-mediated cell adhesive activity has been implicated in neuronal development in vivo. In the *Scn1b* null mouse, there are fewer optic nerve nodes of Ranvier. At the ultrastructural level, optic nerve, spinal cord, and sciatic nerve nodes have abnormal architecture (Chen et al. 2004). *Scn1b*-null mice also have defects in neuronal pathfinding and fasciculation in multiple brain regions. In normal cerebellum, CGN axons project from the granule layer to the molecular layer, where they form parallel fibers. In the *Scn1b* null mouse, CGN axons are defasiculated, forming a disrupted molecular layer. Abnormal pathfinding and defasiculation are also observed in the *Scn1b*-null corticospinal tract and hippocampus. In a related model, dendritic arborization of pyramidal neurons in subiculum is reduced in *Scn1b*-C121W mutant animals (Reid et al. 2014). The *Scn2b*- and *Scn3b*-null mouse models do not have an apparent neurological phenotype, although *Scn2b*-null mice have increased seizure susceptibility and altered

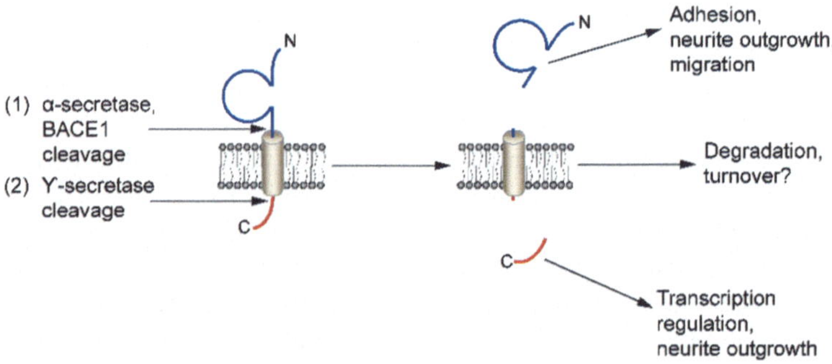

Fig. 4 β subunits are sequentially cleaved by α-secretase and/or the β-site amyloid precursor protein-cleaving enzyme 1 (BACE1) and subsequently by γ-secretase in the lumen of the membrane. Sequential cleavage generates a soluble extracellular N-terminal domain and intracellular C-terminal domain (Kim et al. 2005; Wong et al. 2005). The soluble N-terminal domain participates in cell adhesion and migration, while the intracellular domain modulates neurite outgrowth and, in the case of β2, regulates VGSC gene transcription in vitro (Davis et al. 2004; Kim et al. 2005, 2007; Miyazaki et al. 2007). Figure reproduced from Brackenbury and Isom (2011)

pain sensation (Chen et al. 2002; Hakim et al. 2008, 2010; Lopez-Santiago et al. 2006; O'Malley and Isom 2015). CNS abnormalities in the *Scn4b*-null mouse model were recently described. *Scn4b*-null mice display deficits in balance and motor coordination and resurgent sodium current in null Purkinje neurons was reduced by approximately 50%. This was further validated using in vivo short hairpin RNA knockdown of β4 in adult Purkinje neurons (Ransdell et al. 2017). Overexpression of the β4 subunit in Neuro2a cells results in increased neurite outgrowth, dendrite formation, and filopodia-like protrusions, suggesting a role for β4 in neuronal pathfinding and migration (Oyama et al. 2006).

2 The Role of β Subunits in Pathophysiology

2.1 Cancer

VGSC β subunits are expressed in prostate, breast, lung, and cervical cancers. This expression is subunit specific and varies by cancer type (Brackenbury 2012). β1 is found in breast, prostate, and cervical cancers, while β2 has been detected in breast and prostate cancers, and β3 in prostate and lung cancers (Chioni et al. 2009; Diss et al. 2008; Hernandez-Plata et al. 2012; Jansson et al. 2012; Roger et al. 2007).

β1 and β2 expression levels correspond with metastatic potential in prostate cancer cells (Chioni et al. 2009; Jansson et al. 2012). Experiments performed in vitro with breast cancer cells have shown that β1 expression enhances cell–cell and cell–substrate adhesion and decreases cell migration (Chioni et al. 2009). On the other hand, data suggest that β1 contributes to cell invasion during metastasis in breast cancer cells (Chioni et al. 2009). Overexpression of β1 increases vascular endothelial growth factor secretion and angiogenesis, and decreases apoptosis in endothelial cells (Andrikopoulos et al. 2011). β1 overexpression in an orthotopic mouse model of breast cancer increases tumor growth and metastasis (Nelson et al. 2014). In the well-defined prostate cancer cell line, LNCaP, β2 overexpression increases cell length, but reduces cell volume, which may result in increased cellular motility and invasion. In a wound healing assay, cells overexpressing β2 migrate farther than controls. To the contrary, over-expression of β2 decreases tumor formation and growth after tumor implantation into nude mice. Furthermore, β2 over-expression enhances invasion and growth on laminin (Jansson et al. 2012).

Unlike β1 and β2, β3 is postulated to function as a tumor suppressor because its amino acid sequence contains two *p53* response elements. In *p53*-null mouse embryo fibroblasts, *Scn3b* is increased after adriamycin treatment and β3 expression induces *p53*-dependent apoptosis (Adachi et al. 2004). Less is known about the expression of β4 in cancer, although β4 expression levels are lower in cervical and prostate cancer cells in comparison to noncancerous cells (Diss et al. 2008; Hernandez-Plata et al. 2012). β4 co-expression with $Na_V1.5$ has also been shown to play a role in $CD4^+$ T cell development (Lo et al. 2012). These data suggesting roles for VGSC β subunits in cancer indicate that these molecules are important to the functioning of non-excitable, in addition to excitable, cells.

2.2 Cardiac Arrhythmia

The VGSC β subunits are expressed in the human heart and conduction system. Here, *SCN1B* is expressed at the highest levels in atria and endocardium, while *SCN2B* and *SCN3B* are expressed throughout the human heart (Gaborit et al. 2007). In mouse ventricular cardiomyocytes, β2, β4, and tyrosine-phosphorylated β1 subunits are expressed at the intercalated disc along with $Na_v1.5$, the predominant heart α subunit (Maier et al. 2004; Malhotra et al. 2004). At t-tubules of ventricular cardiomyocytes, β2, β3, and non-phosphorylated β1 are co-expressed with $Na_v1.1$, $Na_v1.3$, and $Na_v1.6$ α subunits (Dhar Malhotra et al. 2001; Maier et al. 2004; Malhotra et al. 2004). Cardiac VGSC β subunits are critical for action potential upstroke, conduction velocity, and excitation-contraction coupling, suggesting that abnormal expression of β subunits may contribute to cardiac disease states (Remme and Bezzina 2010).

Mutations in genes encoding VGSC β subunits are linked to multiple types of cardiac disease (Bao and Isom 2014), including long QT syndrome (LQTS) (Medeiros-Domingo et al. 2007; Riuro et al. 2014), a ventricular arrhythmia in which there is delayed action potential repolarization, resulting in prolongation of the QT interval on the electrocardiogram. LQTS causes an increased risk of ventricular fibrillation (VF) and sudden cardiac death (Alders and Christiaans 1993). There is now an extensive list of LQTS mutations, including mutations in ion channel genes (Nakano and Shimizu 2016; Tester and Ackerman 2014). Two mutations, resulting in gain-of-function activity, have been identified in *SCN1B* and *SCN4B*, respectively (Medeiros-Domingo et al. 2007; Nakano and Shimizu 2016; Riuro et al. 2014), including β1B p.P213T, which results in increased late sodium current and action potential duration, shifted window current, and decreased rate of slow inactivation, and β4 p.L179F, which results in a positive shift in sodium current inactivation causing abnormal action potential repolarization (Medeiros-Domingo et al. 2007; Riuro et al. 2014).

Multiple mutations in *SCN1B* have also been linked to Brugada syndrome (BrS) (Holst et al. 2012; Hu et al. 2012; Watanabe et al. 2008; Yuan et al. 2014). BrS patients have an increased risk of sudden cardiac death due to VF (Watanabe et al. 2008). *SCN1B* mutations are associated with reductions in $Na_v1.5$-generated sodium current density, hyperpolarized voltage-dependence of sodium current inactivation, and/or alterations in the rate of recovery from inactivation (Watanabe et al. 2008). A missense mutation in *SCN2B*, p.D211G, has been linked to BrS and results in reduced sodium current density by decreasing $Na_v1.5$ cell surface expression (Riuro et al. 2013). Mutations in all four of the VGSC β subunit genes are linked to atrial fibrillation (AF) (Li et al. 2013; Olesen et al. 2011; Wang et al. 2010; Watanabe et al. 2009).

Mouse models lacking individual β subunits show the important roles of these subunits in cardiac function. Cardiac function in *Scn1b*-null mice is altered, even after blocking autonomic input. These animals exhibit action potential depolarization and prolonged QT intervals, suggesting a LQTS phenotype. *Scn1b*-null ventricular myocytes have increased transient and persistent sodium current in comparison to wild-type animals, and an increase in $Na_V1.5$ transcript and protein levels (Lopez-Santiago et al. 2007). *Scn1b*-null mice also show increased TTX-sensitive sodium current in the ventricular myocyte midsection, concurrent with increased *Scn3a* mRNA levels. Cardiac-specific *Scn1b*-null mice also display increased *Scn3a* mRNA, lengthened action potential repolarization, delayed after repolarizations and Ca^{2+} transients, and frequent spontaneous release of Ca^{2+}. Alterations in Ca^{2+} levels were blocked by TTX (Lin et al. 2014). *Scn2b*-null mice exhibit a mixed, Brugada-atrial fibrillation like phenotype. *Scn2b*-null ventricular myocytes have alterations in sodium and potassium current densities, particularly in the right ventricular outflow tract. Similar to *Scn2b*-null brain, total levels of $Na_V1.5$ protein were found to be similar to those from wild-type animals, supporting the hypothesis that a main function of β2 in the ventricle is to chaperone VGSC α subunits to the cell surface without changing overall channel expression. In contract, *Scn2b* null atria had normal levels of sodium and potassium currents but increased levels of fibrosis. Lastly, *Scn2b*-null hearts display increased susceptibility to atrial fibrillation and repolarization dispersion compared to wild-type animals (Bao et al. 2016). *Scn3b*-null mice also show cardiac dysfunction. In both atria and ventricles, *Scn3b*-null mice display an increased susceptibility to arrhythmia, reduced peak sodium current, conduction abnormalities that are similar to Brugada syndrome models, bradycardia, AV block, and deficits in sinoatrial node recovery (Hakim et al. 2008). The role of β4 in cardiac function has yet to be reported using the null mouse model.

2.3 Epilepsy

Many mutations in VGSC genes have been linked to epilepsies, including *SCN1B* (Kaplan et al. 2016). There has as yet been no explicit neurological phenotype associated with *SCN3B* and no epilepsy phenotype linked to *SCN4B*. The mutation *SCN1B* p.C121W, identified in a patient with Generalized Epilepsy with Febrile Seizures plus (GEFS+), was one of the first epilepsy mutations ever identified (Wallace et al. 1998). GEFS+ patients initially experience febrile seizures, which then progress to persistent afebrile seizures (Wallace et al. 1998). The heterozygous p.C121W knock-in mouse has been shown to model the GEFS+ phenotype (Wimmer et al. 2010). The p.C121W mutation disrupts a key disulfide bond in the Ig loop (McCormick et al. 1998; Wallace et al. 1998). Although p.C121W traffics appropriately to the plasma membrane and its co-expression increases VGSC α subunit cell surface levels in culture, it is unable to participate in *trans*-homophilic cell adhesion or modulate sodium current in vitro (Meadows et al. 2002). Studies of p.C121W subcellular localization in cultured neurons showed that, unlike wildtype β1, mutant subunits do not traffic to specialized axonal subdomains including the AIS and nodes of Ranvier. Phenotypically, p.C121W homozygous mice model Dravet syndrome, displaying brain-region specific hyper-excitability, reduced dendritic arborization of pyramidal neurons in the subiculum, and an increased susceptibility to febrile and spontaneous seizures (Wimmer et al.

2010). Animals that are heterozygous for this mutation are more susceptible to hyperthermia-induced seizures than *Scn1b*$^{+/-}$ or *Scn1b*$^{+/+}$ animals. Even though β1-C121W is localized to the cell surface of neurons in vivo, they are incompletely glycosylated and do not interact with α subunits (Kruger et al. 2016). Additional GEFS+ mutations in *SCN1B*, p.R85C, and p.R85H, have also been studied in heterologous cells in vitro (Xu et al. 2007). Both mutants have decreased expression compared to wild-type and are unable to modulate α subunits. Although of the two, only p.R85H has been shown to reach the plasma membrane (Patino et al. 2009; Xu et al. 2007).

The *Scn1b*-null mouse line is a model of Dravet Syndrome (DS), and mutations in *Scn1b* are linked to DS (Chen et al. 2004; Patino et al. 2009), a severe and intractable pediatric epileptic encephalopathy that typically presents within the first year of life with myoclonic seizures that can change etiology over time. DS patients also suffer from a variety of comorbidities including ataxia, behavioral and developmental delay, and a high risk of sudden unexpected death in epilepsy, or SUDEP (Gataullina and Dulac 2017). DS mutations in *SCN1B* are homozygous recessive. The first DS mutation identified in *SCN1B* was p.R125C. This mutation has abnormal trafficking and does not reach the cell surface in vitro, resulting in a functional null phenotype (Patino et al. 2009). An additional *SCN1B* DS mutations, p.I106F, was later identified, although the mechanism underlying this mutation remains unknown (Ogiwara et al. 2012). *Scn1b*-null mice further validate the role of β1 in DS. These animals have frequent spontaneous seizures and abnormal neuronal excitability and development, consistent with that observed in DS patients (Chen et al. 2004). In addition, *Scn1b* null mice die at ~P21, and are thus a SUDEP model.

Heterozygous mutations in *SCN1B* have been linked to a variety of other epilepsies. These include p.R85C, p.R85H, p.R125L, and an in-frame deletion mutation (Fendri-Kriaa et al. 2011; Scheffer et al. 2007; Wallace et al. 1998). One mutation that is specific to the developmentally regulated splice variant, β1B, has also been identified, p.G257R, and is linked to idiopathic epilepsy in multiple pedigrees. In vitro, this mutation also has defects in membrane trafficking (Patino et al. 2011). Except for this mutation specific to β1B, all epilepsy-linked mutations in *SCN1B* code for amino acids in the Ig loop domain, suggesting the clinical relevance of cell adhesion in the pathogenesis of epilepsy.

Scn2b-null mice express approximately half of normal levels of cell surface TTX-sensitive VGSCs in brain. These animals are also more prone to pharmacologically induced seizures compared to wild-type animals (Chen et al. 2002). Additionally, a polymorphism in *SCN2B* (rs2298771) has been associated with idiopathic epilepsy (Baum et al. 2014). In conclusion, the β1 and β2 subunits play critical roles in epilepsy.

2.4 Neurodegenerative Disorders

β subunits have been implicated in neurodegenerative disorders including amyotrophic lateral sclerosis (ALS), Alzheimer's disease (AD), Huntington's disease (HD), Multiple sclerosis (MS), and Parkinson's disease (PD) (Calhoun and Isom 2014; O'Malley and Isom 2015). ALS is characterized by the degeneration of motor neurons in the spinal cord, motor cortex, and brainstem (Al-Chalabi et al. 2017). Differential gene expression of *Scn1b* and *Scn3b* has been observed in the *Sod1* mouse model of ALS. *Scn1b* mRNA and protein are decreased, while there is increased *Scn3b* mRNA and protein in ventral dorsal horn. Neuronal hyperexcitability is found in ALS, thus, alterations in the expression of *Scn1b* and *Scn3b*, as well as the changes reported in the expression of $Na_V1.6$, may explain hyperexcitability in ALS patients (Nutini et al. 2011).

Like that of the Amyloid Precursor Protein (APP), most famously known for its potential implications in AD, β subunits are substrates for sequential cleavage by β-site APP cleaving enzyme-1 (BACE1) and γ-secretase, potentially linking the β subunits to AD (Wong et al. 2005). In AD pathology, APP is initially cleaved on the extracellular portion of the membrane by BACE1 and then subsequently cleaved in the lumen of the membrane by γ-secretase, generating the amyloid β (Aβ) peptide. Aβ then accumulates and forms amyloid plaques. BACE1 is ubiquitously expressed throughout the body, but is expressed at highest levels in the pancreas and brain (Cole and Vassar 2007). The expression of BACE1 increases with age in the cortex of AD patients (Evin et al. 2010). Interestingly, AD patients are at increased risk of seizures, further supporting a potential role of VGSCs in AD (Pandis and Scarmeas 2012). BACE1 cleavage of β2 reverses normal β2 modulation of VGSC β subunits. In BACE1-null mice, decreased cleavage of β2 (or possibly other BACE1 substrates, including other VGSC β subunits) may contribute to the increased neuronal excitability observed in AD patients (Kim et al. 2011). In addition, *SCN3B* mRNA is lower in AD brains with neurofibrillary tangles (NFTs), another pathological issue displayed in some AD cases, suggesting β subunits may be implicated in the formation of NFTs and hyperexcitability in AD (Dunckley et al. 2006).

SCN2B and *SCN4B* have been linked to HD, a genetic, neurodegenerative disease that affects motor coordination and mental ability. Ultimately, many of these patients lose their ability to walk and/or talk. In HD patient postmortem brain samples, *SCN4B* is downregulated in the striatum. This is mimicked in mouse models, where it has been shown to occur prior to loss of motor coordination. In vitro, β4 overexpression is implicated in neuronal development, suggesting that in HD, β4 dysregulation may contribute to neural degeneration. A decrease in β2 expression is also observed in the same mouse model of HD, but later in the pathogenesis of disease than observed for *Scn4b* (Oyama et al. 2006).

Although *Scn2b*-null mice have normal myelination, at least in the optic nerve, deletion of *Scn2b* is neuroprotective in the Experimental Allergic Encephalomyelitis (EAE) model of MS. Interestingly, *Scn2b* deletion in the EAE mouse model leads to decreased axonal degeneration, fewer demyelinated and dysmyelinated axons, reduced phenotypic severity, and increased survival (O'Malley et al. 2009). β2 may also be implicated in MS through sequential cleavage by BACE1 and γ-secretase. In cerebrospinal fluid from MS patients, there is decreased BACE1 activity and this biomarker in MS is linked to a more severe and prolonged disease state. Throughout MS progression, BACE1 expression continues to decline (Mattsson et al. 2009).

VGSC β1 subunits are also implicated in maintaining normal myelination. *Scn1b*-null mice phenotypically display abnormal optic nerve myelination, spinal cord dysmyelination, increased axonal degeneration, fewer optic nerve nodes of Ranvier, and defects in nodal ultrastructure in both the central and peripheral nervous systems. Loss of β1 expression, and thus adhesion, at nodes of Ranvier leads to abnormalities in the formation of paranodal junctions, suggesting β1 contributes to myelination (Chen et al. 2004).

Increased expression and glycosylation of the VGSC β4 subunit compared to wild-type animals has been identified in a mouse model of PD. Studies of neurite outgrowth in response to expression of WT vs. mutant β4 that could not be glycosylated showed that neurite outgrowth was accelerated, with an increased level of filopdia-like protrusions. Thus, the glycosylation state of β4 may be critical for neuronal morphology and may be involved in PD pathogenesis (Zhou et al. 2012). Overall, β subunits contribute to myelination and neurodegenerative disease states through a variety of mechanisms.

2.5 Neuropathic Pain

A variety of factors can cause neuropathic pain, including genetic mutations and nerve injury. This leads to defects in nociception, the neuronal pathways implicated in sensing noxious stimuli. The VGSC β subunits are expressed in dorsal root ganglion (DRG) neurons and peripheral nerves, suggesting potential roles for these proteins in neuropathic pain (Lopez-Santiago et al. 2006). Behavioral pain phenotypes are difficult, if not impossible, to study in *Scn1b*-null mice due to their severe seizures and early post-natal death (Chen et al. 2004). In spite of this, *Scn1b*-null DRG neurons are hyperexcitable, suggesting that these mice may have some form of allodynia (Lopez-Santiago et al. 2011). On the other hand, *Scn1b* mRNA levels are increased in a model of chronic constrictive nerve injury, complicating the interpretation of the role of β1 in neuropathic pain (Blackburn-Munro and Fleetwood-Walker 1999).

Studies examining the role of β2 in neuropathic pain have also led to conflicting results. While *Scn2b*-null mice are less sensitive than wild-type littermates in models of inflammatory and neuropathic pain, β2 protein levels are increased in injured and non-injured wild-type neurons in spared nerve injury and spinal nerve ligation models in rat (Lopez-Santiago et al. 2006; Pertin et al. 2005). The latter occurs without a corresponding increase in mRNA levels (Pertin et al. 2005). Lastly, *Scn2b* mRNA levels are downregulated in cervical sensory ganglia after avulsion injury, but increased in a model of chronic constrictive nerve injury (Blackburn-Munro and Fleetwood-Walker 1999; Coward et al. 2001).

Scn3b mRNA expression is increased in multiple pain models, including in small C-fibers, in a chronic constrictive injury model in rats, in Aδ fibers in the streptozotocin model of diabetic neuropathy in rat, in the small and medium fibers in the sciatic nerve transection model, and finally, in the spared nerve injury model of neuropathic pain suggesting *Scn3b* may play a role in modulating pain (Shah et al. 2000, 2001; Takahashi et al. 2003).

Although there are little data to directly implicate β4 in pain, the C-terminal portion of β4 plays a role in generating resurgent sodium current in DRG neurons (Grieco et al. 2005). Paroxysmal Extreme Pain Disorder (PEPD) is an inherited neuropathic pain syndrome linked to gain-of-function mutations in *SCN9A*, encoding $Na_v1.7$. When PEPD-linked $Na_v1.7$ mutants are co-expressed with the C-terminal β4 peptide, differential enhancement of resurgent current is observed, suggesting a potential role for β4 in pain (Theile et al. 2011). In all, the β subunits contribute to pain phenotypes in a cell-type and subunit-specific manner.

2.6 Sudden Infant Death Syndrome (SIDS) and Sudden Unexpected Death in Epilepsy (SUDEP)

Sudden Infant Death Syndrome, or SIDS, is the unexpected death of a child up to 1 year of age where a clear cause of death cannot be identified via autopsy (Krous et al. 2004). The mechanism of SIDS remains to be elucidated, but one out of ten cases is associated with cardiac ion channel gene mutations, including in genes encoding the β subunits (Van Norstrand and Ackerman 2009). p.V36M and p.V54G mutations in *SCN3B* and p.S206L in *SCN4B* have been linked to SIDS (Tan et al. 2010). Importantly, p.V36M in *SCN3B* has also been linked to idiopathic ventricular fibrillation, a potential fatal cardiac arrhythmia, and p.S206L in *SCN4B* also leads to abnormal excitability in rat ventricular myocytes (Tan et al. 2010; Valdivia et al. 2010). There has been one reported instance of SIDS in a child with a p. R214Q in β1B, which has also been associated with Brugada Syndrome (Hu et al. 2012). β1B modulates $Na_v1.5$ function, potentially providing an underlying mechanism for *SCN1B* linked cardiac dysfunction (Patino et al. 2011). To date, no mutations in *SCN2B* have been linked to SIDS.

Some ion channel genes that have been linked to SIDS have also been linked to Sudden Unexpected Death in EPilepsy (SUDEP). SUDEP is defined as the sudden and unexpected death of a person with epilepsy without any identifiable cause of death during autopsy (Nashef et al. 2012). SUDEP occurs in up to 17% of epileptic patients and those diagnosed with Dravet syndrome (DS) are at an especially high risk for SUDEP (Ficker et al. 1998). Seizures that are difficult to treat by pharmacological intervention are also associated with increased SUDEP risk (Hesdorffer et al. 2012). Currently, there are no reliable biomarkers for SUDEP, but it is likely that death is initiated by dysfunction in multiple organ systems, including autonomic dysfunction, cardiac arrhythmia, central or obstructive apnea, hypoventilation, and pulmonary edema (Surges and Sander 2012). Several types of cardiac events are known to occur during or after seizure activity in epilepsy patients. These include asystole, atrial fibrillation, bradycardia, tachycardia, and T-wave alterations (Jansen and Lagae 2010). Epileptic activity may affect the autonomic nervous system, which is known to be a critical regulator of cardiac function. Dysregulation of the autonomic nervous system and spreading depression to brain stem centers during an epileptic event can result in fatal cardiac abnormalities (Jansen and Lagae 2010; Massey et al. 2014; Surges and Sander 2012).

Multiple DS and epilepsy animal models also serve as models for SUDEP, including the previously discussed *Scn1b*-null mouse line (Chen et al. 2004). Additional models include *Scn1a*$^{+/-}$ mice, which model the haploinsufficiency observed in most DS patients, and the *Kcna1* null mouse line, which deletes the voltage-gated potassium channel Kv1.1 (Glasscock et al. 2010; Oakley et al. 2011). Intriguingly, each of these SUDEP models presents with different cardiac alterations that may mechanistically contribute to SUDEP. *Scn1b*-null mice display increased cardiac sodium current and prolonged QT and RR intervals. *Scn1b*-null mice treated with atropine or propranolol do not show differences in QT interval compared to vehicle treated animals, indicating the cardiac phenotype may not be a result of an abnormal autonomic activity (Lopez-Santiago et al. 2007). *Scn1a*$^{+/-}$ mice also display increased cardiac sodium current, but additionally have bradycardia, focal discharges, a variable RR interval, and bundle branch block (Auerbach et al. 2013; Kalume et al. 2013). *Kcna1*-null mice display a cardiac phenotype as well, including atrioventricular (AV) block, bradycardia, premature ventricular contractions and altered heart rate variability (Glasscock et al. 2010). Contrary to *Scn1b*-null mice, treatment of *Scn1a*$^{+/-}$ and *Kcna1*-null animals with atropine reverses AV block, suggesting parasympathetic hyperexcitability in these models (Glasscock et al. 2010; Kalume et al. 2013). In summary, studies with animal models and patient mutations provide evidence that β subunits are likely key regulators in the pathogenesis of SIDS and SUDEP, although additional work must be completed to further understand and ultimately prevent SIDS and SUDEP events.

3 Conclusion

In conclusion, VGSC β subunits play critical roles in modulating the gating, localization, and kinetics of the VGSC pore as well as modulate the activities of some potassium channels. In addition, these non-pore-forming proteins function as CAMs and signaling molecules in both excitable and non-excitable cell types. Their importance as CAMs is implicated in neurite outgrowth, axonal pathfinding and fasciculation, and migration in cancerous cells. Sequential β subunit cleavage by BACE and γ-secretase also affects the expression of other genes. Mutations in the genes encoding β subunits are linked to a variety of devastating diseases, including epilepsy, SIDS and SUDEP, cancer, neuropathic pain, and some of the major neurodegenerative disorders (Fig. 5). Additional research needs to be completed in order to further understand the biology of these critical proteins and their potential as novel therapeutic targets for a wide variety of disease states.

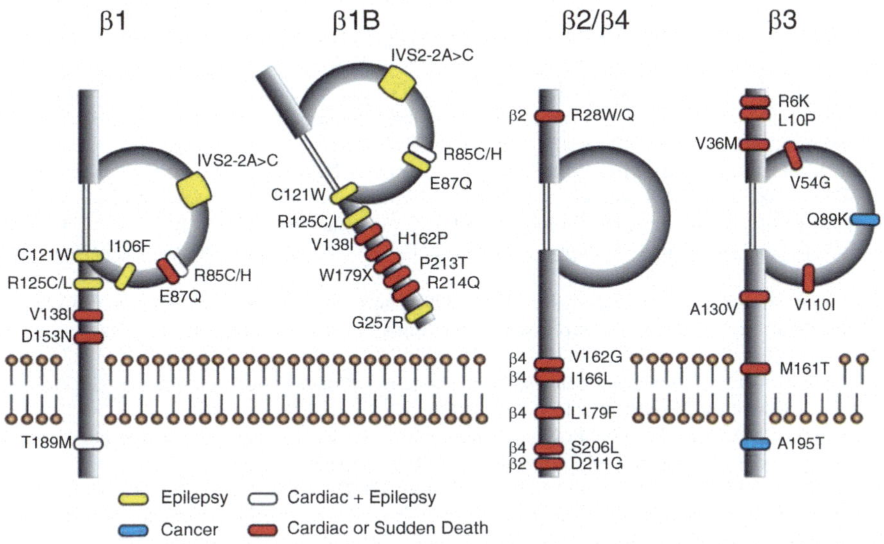

Fig. 5 Disease-linked β subunit mutations. Figure reproduced from O'Malley and Isom (2015)

References

Adachi K, Toyota M, Sasaki Y, Yamashita T, Ishida S, Ohe-Toyota M, Maruyama R, Hinoda Y, Saito T, Imai K, Kudo R, Tokino T (2004) Identification of SCN3B as a novel p53-inducible proapoptotic gene. Oncogene 23:7791–7798. https://doi.org/10.1038/sj.onc.1208067

Al-Chalabi A, van den Berg LH, Veldink J (2017) Gene discovery in amyotrophic lateral sclerosis: implications for clinical management. Nat Rev Neurol 13:96–104. https://doi.org/10.1038/nrneurol.2016.182

Alders M, Christiaans I (1993) Long QT syndrome. In: Pagon RA, Adam MP, Ardinger HH, Wallace SE, Amemiya A, Bean LJH, Bird TD, Ledbetter N, Mefford HC, Smith RJH, Stephens K (eds) GeneReviews®. University of Washington, Seattle, Seattle

Aman TK, Grieco-Calub TM, Chen C, Rusconi R, Slat EA, Isom LL, Raman IM (2009) Regulation of persistent Na current by interactions between beta subunits of voltage-gated Na channels. J Neurosci 29:2027–2042. https://doi.org/10.1523/jneurosci.4531-08.2009

Andrikopoulos P, Fraser SP, Patterson L, Ahmad Z, Burcu H, Ottaviani D, Diss JK, Box C, Eccles SA, Djamgoz MB (2011) Angiogenic functions of voltage-gated Na^+ channels in human endothelial cells: modulation of vascular endothelial growth factor (VEGF) signaling. J Biol Chem 286:16846–16860. https://doi.org/10.1074/jbc.M110.187559

Audenaert D, Claes L, Ceulemans B, Lofgren A, Van Broeckhoven C, De Jonghe P (2003) A deletion in SCN1B is associated with febrile seizures and early-onset absence epilepsy. Neurology 61:854–856

Auerbach DS, Jones J, Clawson BC, Offord J, Lenk GM, Ogiwara I, Yamakawa K, Meisler MH, Parent JM, Isom LL (2013) Altered cardiac electrophysiology and SUDEP in a model of Dravet syndrome. PLoS One 8:e77843. https://doi.org/10.1371/journal.pone.0077843

Bant JS, Raman IM (2010) Control of transient, resurgent, and persistent current by open-channel block by Na channel beta4 in cultured cerebellar granule neurons. Proc Natl Acad Sci U S A 107:12357–12362. https://doi.org/10.1073/pnas.1005633107

Bao Y, Isom LL (2014) $Na_V1.5$ and regulatory β subunits in cardiac sodium channelopathies. Card Electrophysiol Clin 6:679–694. https://doi.org/10.1016/j.ccep.2014.07.002

Bao Y, Willis BC, Frasier CR, Lopez-Santiago LF, Lin X, Ramos-Mondragon R, Auerbach DS, Chen C, Wang Z, Anumonwo J, Valdivia HH, Delmar M, Jalife J, Isom LL (2016) Scn2b deletion in mice results in ventricular and atrial arrhythmias. Circ Arrhythm Electrophysiol 9: e003923. https://doi.org/10.1161/circep.116.003923

Baum L, Haerian BS, Ng HK, Wong VC, Ng PW, Lui CH, Sin NC, Zhang C, Tomlinson B, Wong GW, Tan HJ, Raymond AA, Mohamed Z, Kwan P (2014) Case-control association study of polymorphisms in the voltage-gated sodium channel genes SCN1A, SCN2A, SCN3A, SCN1B, and SCN2B and epilepsy. Hum Genet 133:651–659. https://doi.org/10.1007/s00439-013-1405-1

Blackburn-Munro G, Fleetwood-Walker SM (1999) The sodium channel auxiliary subunits beta1 and beta2 are differentially expressed in the spinal cord of neuropathic rats. Neuroscience 90:153–164

Brackenbury WJ (2012) Voltage-gated sodium channels and metastatic disease. Channels (Austin) 6:352–361. https://doi.org/10.4161/chan.21910

Brackenbury WJ, Isom LL (2011) Na channel beta subunits: overachievers of the ion channel family. Front Pharmacol 2:53. https://doi.org/10.3389/fphar.2011.00053

Brackenbury WJ, Davis TH, Chen C, Slat EA, Detrow MJ, Dickendesher TL, Ranscht B, Isom LL (2008) Voltage-gated Na^+ channel beta1 subunit-mediated neurite outgrowth requires Fyn kinase and contributes to postnatal CNS development in vivo. J Neurosci 28:3246–3256. https://doi.org/10.1523/jneurosci.5446-07.2008

Brackenbury WJ, Calhoun JD, Chen C, Miyazaki H, Nukina N, Oyama F, Ranscht B, Isom LL (2010) Functional reciprocity between Na^+ channel $Na_v1.6$ and beta1 subunits in the coordinated regulation of excitability and neurite outgrowth. Proc Natl Acad Sci U S A 107:2283–2288. https://doi.org/10.1073/pnas.0909434107

Buffington SA, Rasband MN (2013) Na^+ channel-dependent recruitment of Na_vbeta4 to axon initial segments and nodes of Ranvier. J Neurosci 33:6191–6202. https://doi.org/10.1523/jneurosci.4051-12.2013

Calhoun JD, Isom LL (2014) The role of non-pore-forming beta subunits in physiology and pathophysiology of voltage-gated sodium channels. Handb Exp Pharmacol 221:51–89. https://doi.org/10.1007/978-3-642-41588-3_4

Catterall WA (2000) From ionic currents to molecular mechanisms. Neuron 26:13–25. https://doi.org/10.1016/S0896-6273(00)81133-2

Chen C, Bharucha V, Chen Y, Westenbroek RE, Brown A, Malhotra JD, Jones D, Avery C, Gillespie PJ 3rd, Kazen-Gillespie KA, Kazarinova-Noyes K, Shrager P, Saunders TL, Macdonald RL, Ransom BR, Scheuer T, Catterall WA, Isom LL (2002) Reduced sodium channel density, altered voltage dependence of inactivation, and increased susceptibility to seizures in mice lacking sodium channel beta 2-subunits. Proc Natl Acad Sci U S A 99:17072–17077. https://doi.org/10.1073/pnas.212638099

Chen C, Westenbroek RE, Xu X, Edwards CA, Sorenson DR, Chen Y, McEwen DP, O'Malley HA, Bharucha V, Meadows LS, Knudsen GA, Vilaythong A, Noebels JL, Saunders TL, Scheuer T, Shrager P, Catterall WA, Isom LL (2004) Mice lacking sodium channel beta1 subunits display defects in neuronal excitability, sodium channel expression, and nodal architecture. J Neurosci 24:4030–4042. https://doi.org/10.1523/jneurosci.4139-03.2004

Chen C, Calhoun JD, Zhang Y, Lopez-Santiago L, Zhou N, Davis TH, Salzer JL, Isom LL (2012) Identification of the cysteine residue responsible for disulfide linkage of Na+ channel alpha and beta2 subunits. J Biol Chem 287:39061–39069. https://doi.org/10.1074/jbc.M112.397646

Chioni AM, Brackenbury WJ, Calhoun JD, Isom LL, Djamgoz MB (2009) A novel adhesion molecule in human breast cancer cells: voltage-gated Na^+ channel beta1 subunit. Int J Biochem Cell Biol 41:1216–1227. https://doi.org/10.1016/j.biocel.2008.11.001

Cole SL, Vassar R (2007) The Alzheimer's disease beta-secretase enzyme, BACE1. Mol Neurodegener 2:22. https://doi.org/10.1186/1750-1326-2-22

Coward K, Jowett A, Plumpton C, Powell A, Birch R, Tate S, Bountra C, Anand P (2001) Sodium channel beta1 and beta2 subunits parallel SNS/PN3 alpha-subunit changes in injured human sensory neurons. Neuroreport 12:483–488

Davis TH, Chen C, Isom LL (2004) Sodium channel beta1 subunits promote neurite outgrowth in cerebellar granule neurons. J Biol Chem 279:51424–51432. https://doi.org/10.1074/jbc.M410830200

Deschenes I, Tomaselli GF (2002) Modulation of Kv4.3 current by accessory subunits. FEBS Lett 528:183–188

Dhar Malhotra J, Chen C, Rivolta I, Abriel H, Malhotra R, Mattei LN, Brosius FC, Kass RS, Isom LL (2001) Characterization of sodium channel alpha- and beta-subunits in rat and mouse cardiac myocytes. Circulation 103:1303–1310

Diss JK, Fraser SP, Walker MM, Patel A, Latchman DS, Djamgoz MB (2008) Beta-subunits of voltage-gated sodium channels in human prostate cancer: quantitative in vitro and in vivo analyses of mRNA expression. Prostate Cancer Prostatic Dis 11:325–333. https://doi.org/10.1038/sj.pcan.4501012

Dunckley T, Beach TG, Ramsey KE, Grover A, Mastroeni D, Walker DG, LaFleur BJ, Coon KD, Brown KM, Caselli R, Kukull W, Higdon R, McKeel D, Morris JC, Hulette C, Schmechel D, Reiman EM, Rogers J, Stephan DA (2006) Gene expression correlates of neurofibrillary tangles in Alzheimer's disease. Neurobiol Aging 27:1359–1371. https://doi.org/10.1016/j.neurobiolaging.2005.08.013

Evin G, Barakat A, Masters CL (2010) BACE: therapeutic target and potential biomarker for Alzheimer's disease. Int J Biochem Cell Biol 42:1923–1926. https://doi.org/10.1016/j.biocel.2010.08.017

Fendri-Kriaa N, Kammoun F, Salem IH, Kifagi C, Mkaouar-Rebai E, Hsairi I, Rebai A, Triki C, Fakhfakh F (2011) New mutation c.374C>T and a putative disease-associated haplotype within SCN1B gene in Tunisian families with febrile seizures. Eur J Neurol 18:695–702. https://doi.org/10.1111/j.1468-1331.2010.03216.x

Ficker DM, So EL, Shen WK, Annegers JF, O'Brien PC, Cascino GD, Belau PG (1998) Population-based study of the incidence of sudden unexplained death in epilepsy. Neurology 51:1270–1274

Gaborit N, Le Bouter S, Szuts V, Varro A, Escande D, Nattel S, Demolombe S (2007) Regional and tissue specific transcript signatures of ion channel genes in the non-diseased human heart. J Physiol 582:675–693. https://doi.org/10.1113/jphysiol.2006.126714

Gataullina S, Dulac O (2017) From genotype to phenotype in Dravet disease. Seizure 44:58–64. https://doi.org/10.1016/j.seizure.2016.10.014

Gilchrist J, Das S, Van Petegem F, Bosmans F (2013) Crystallographic insights into sodium-channel modulation by the beta4 subunit. Proc Natl Acad Sci U S A 110:E5016–E5024. https://doi.org/10.1073/pnas.1314557110

Glasscock E, Yoo JW, Chen TT, Klassen TL, Noebels JL (2010) Kv1.1 potassium channel deficiency reveals brain-driven cardiac dysfunction as a candidate mechanism for sudden unexplained death in epilepsy. J Neurosci 30:5167–5175. https://doi.org/10.1523/jneurosci.5591-09.2010

Grieco TM, Malhotra JD, Chen C, Isom LL, Raman IM (2005) Open-channel block by the cytoplasmic tail of sodium channel beta4 as a mechanism for resurgent sodium current. Neuron 45:233–244. https://doi.org/10.1016/j.neuron.2004.12.035

Haapasalo A, Kovacs DM (2011) The many substrates of presenilin/gamma-secretase. J Alzheimers Dis 25:3–28. https://doi.org/10.3233/jad-2011-101065

Hakim P, Gurung IS, Pedersen TH, Thresher R, Brice N, Lawrence J, Grace AA, Huang CL (2008) Scn3b knockout mice exhibit abnormal ventricular electrophysiological properties. Prog Biophys Mol Biol 98:251–266. https://doi.org/10.1016/j.pbiomolbio.2009.01.005

Hakim P, Brice N, Thresher R, Lawrence J, Zhang Y, Jackson AP, Grace AA, Huang CL (2010) Scn3b knockout mice exhibit abnormal sino-atrial and cardiac conduction properties. Acta Physiol (Oxf) 198:47–59. https://doi.org/10.1111/j.1748-1716.2009.02048.x

Hernandez-Plata E, Ortiz CS, Marquina-Castillo B, Medina-Martinez I, Alfaro A, Berumen J, Rivera M, Gomora JC (2012) Overexpression of $Na_V1.6$ channels is associated with the invasion capacity of human cervical cancer. Int J Cancer 130:2013–2023. https://doi.org/10.1002/ijc.26210

Hesdorffer DC, Tomson T, Benn E, Sander JW, Nilsson L, Langan Y, Walczak TS, Beghi E, Brodie MJ, Hauser WA (2012) Do antiepileptic drugs or generalized tonic-clonic seizure frequency increase SUDEP risk? A combined analysis. Epilepsia 53:249–252. https://doi.org/10.1111/j.1528-1167.2011.03354.x

Holst AG, Saber S, Houshmand M, Zaklyazminskaya EV, Wang Y, Jensen HK, Refsgaard L, Haunso S, Svendsen JH, Olesen MS, Tfelt-Hansen J (2012) Sodium current and potassium transient outward current genes in Brugada syndrome: screening and bioinformatics. Can J Cardiol 28:196–200. https://doi.org/10.1016/j.cjca.2011.11.011

Hu D, Barajas-Martinez H, Medeiros-Domingo A, Crotti L, Veltmann C, Schimpf R, Urrutia J, Alday A, Casis O, Pfeiffer R, Burashnikov E, Caceres G, Tester DJ, Wolpert C, Borggrefe M, Schwartz P, Ackerman MJ, Antzelevitch C (2012) A novel rare variant in SCN1Bb linked to Brugada syndrome and SIDS by combined modulation of $Na_{(v)}1.5$ and $K_{(v)}4.3$ channel currents. Heart Rhythm 9:760–769. https://doi.org/10.1016/j.hrthm.2011.12.006

Isom LL (2002) The role of sodium channels in cell adhesion. Front Biosci 7:12–23

Isom LL, Catterall WA (1996) Na^+ channel subunits and Ig domains. Nature 383:307–308. https://doi.org/10.1038/383307b0

Isom LL, De Jongh KS, Patton DE, Reber BF, Offord J, Charbonneau H, Walsh K, Goldin AL, Catterall WA (1992) Primary structure and functional expression of the beta 1 subunit of the rat brain sodium channel. Science 256:839–842

Isom LL, De Jongh KS, Catterall WA (1994) Auxiliary subunits of voltage-gated ion channels. Neuron 12:1183–1194

Isom LL, Ragsdale DS, De Jongh KS, Westenbroek RE, Reber BF, Scheuer T, Catterall WA (1995a) Structure and function of the beta 2 subunit of brain sodium channels, a transmembrane glycoprotein with a CAM motif. Cell 83:433–442

Isom LL, Scheuer T, Brownstein AB, Ragsdale DS, Murphy BJ, Catterall WA (1995b) Functional co-expression of the 1 and type IIA subunits of sodium channels in a mammalian cell line. J Biol Chem 270:3306–3312. https://doi.org/10.1074/jbc.270.7.3306

Jansen K, Lagae L (2010) Cardiac changes in epilepsy. Seizure 19:455–460. https://doi.org/10.1016/j.seizure.2010.07.008

Jansson KH, Lynch JE, Lepori-Bui N, Czymmek KJ, Duncan RL, Sikes RA (2012) Overexpression of the VSSC-associated CAM, beta-2, enhances LNCaP cell metastasis associated behavior. Prostate 72:1080–1092. https://doi.org/10.1002/pros.21512

Johnson D, Montpetit ML, Stocker PJ, Bennett ES (2004) The sialic acid component of the beta1 subunit modulates voltage-gated sodium channel function. J Biol Chem 279:44303–44310. https://doi.org/10.1074/jbc.M408900200

Kalume F, Westenbroek RE, Cheah CS, FH Y, Oakley JC, Scheuer T, Catterall WA (2013) Sudden unexpected death in a mouse model of Dravet syndrome. J Clin Invest 123:1798–1808. https://doi.org/10.1172/JCI66220

Kaplan DI, Isom LL, Petrou S (2016) Role of sodium channels in epilepsy. Cold Spring Harb Perspect Med 6. https://doi.org/10.1101/cshperspect.a022814

Kazarinova-Noyes K, Malhotra JD, McEwen DP, Mattei LN, Berglund EO, Ranscht B, Levinson SR, Schachner M, Shrager P, Isom LL, Xiao ZC (2001) Contactin associates with Na^+ channels and increases their functional expression. J Neurosci 21:7517–7525

Kazen-Gillespie KA, Ragsdale DS, D'Andrea MR, Mattei LN, Rogers KE, Isom LL (2000) Cloning, localization, and functional expression of sodium channel beta1A subunits. J Biol Chem 275:1079–1088

Kim DY, Ingano LA, Carey BW, Pettingell WH, Kovacs DM (2005) Presenilin/gamma-secretase-mediated cleavage of the voltage-gated sodium channel beta2-subunit regulates cell adhesion and migration. J Biol Chem 280:23251–23261. https://doi.org/10.1074/jbc.M412938200

Kim DY, Carey BW, Wang H, Ingano LA, Binshtok AM, Wertz MH, Pettingell WH, He P, Lee VM, Woolf CJ, Kovacs DM (2007) BACE1 regulates voltage-gated sodium channels and neuronal activity. Nat Cell Biol 9:755–764. https://doi.org/10.1038/ncb1602

Kim DY, Gersbacher MT, Inquimbert P, Kovacs DM (2011) Reduced sodium channel $Na_{(v)}1.1$ levels in BACE1-null mice. J Biol Chem 286:8106–8116. https://doi.org/10.1074/jbc.M110.134692

Ko SH, Lenkowski PW, Lee HC, Mounsey JP, Patel MK (2005) Modulation of $Na_{(v)}1.5$ by beta1- and beta3-subunit co-expression in mammalian cells. Pflugers Arch 449:403–412. https://doi.org/10.1007/s00424-004-1348-4

Krous HF, Beckwith JB, Byard RW, Rognum TO, Bajanowski T, Corey T, Cutz E, Hanzlick R, Keens TG, Mitchell EA (2004) Sudden infant death syndrome and unclassified sudden infant deaths: a definitional and diagnostic approach. Pediatrics 114:234–238

Kruger LC, O'Malley HA, Hull JM, Kleeman A, Patino GA, Isom LL (2016) Beta1-C121W is down but not out: epilepsy-associated Scn1b-C121W results in a deleterious gain-of-function. J Neurosci 36:6213–6224. https://doi.org/10.1523/jneurosci.0405-16.2016

Laedermann Cé J, Syam N, Pertin M, Decosterd I, Abriel H (2013) β1- and β3-voltage-gated sodium channel subunits modulate cell surface expression and glycosylation of $Na_{(v)}1.7$ in HEK293 cells. Front Cell Neurosci 7:137. https://doi.org/10.3389/fncel.2013.00137

Li RG, Wang Q, YJ X, Zhang M, XK Q, Liu X, Fang WY, Yang YQ (2013) Mutations of the SCN4B-encoded sodium channel beta4 subunit in familial atrial fibrillation. Int J Mol Med 32:144–150. https://doi.org/10.3892/ijmm.2013.1355

Lin X, O'Malley H, Chen C, Auerbach D, Foster M, Shekhar A, Zhang M, Coetzee W, Jalife J, Fishman GI, Isom L, Delmar M (2014) Scn1b deletion leads to increased tetrodotoxin-sensitive sodium current, altered intracellular calcium homeostasis and arrhythmias in murine hearts. J Physiol 593:1389. https://doi.org/10.1113/jphysiol.2014.277699

Lo WL, Donermeyer DL, Allen PM (2012) A voltage-gated sodium channel is essential for the positive selection of $CD^{4(+)}$ T cells. Nat Immunol 13:880–887. https://doi.org/10.1038/ni.2379

Lopez-Santiago LF, Pertin M, Morisod X, Chen C, Hong S, Wiley J, Decosterd I, Isom LL (2006) Sodium channel beta2 subunits regulate tetrodotoxin-sensitive sodium channels in small dorsal root ganglion neurons and modulate the response to pain. J Neurosci 26:7984–7994. https://doi.org/10.1523/jneurosci.2211-06.2006

Lopez-Santiago LF, Meadows LS, Ernst SJ, Chen C, Malhotra JD, McEwen DP, Speelman A, Noebels JL, Maier SK, Lopatin AN, Isom LL (2007) Sodium channel Scn1b null mice exhibit prolonged QT and RR intervals. J Mol Cell Cardiol 43:636–647. https://doi.org/10.1016/j.yjmcc.2007.07.062

Lopez-Santiago LF, Brackenbury WJ, Chen C, Isom LL (2011) Na^+ channel Scn1b gene regulates dorsal root ganglion nociceptor excitability in vivo. J Biol Chem 286:22913–22923. https://doi.org/10.1074/jbc.M111.242370

Maier SK, Westenbroek RE, McCormick KA, Curtis R, Scheuer T, Catterall WA (2004) Distinct subcellular localization of different sodium channel alpha and beta subunits in single ventricular myocytes from mouse heart. Circulation 109:1421–1427. https://doi.org/10.1161/01.cir.0000121421.61896.24

Malhotra JD, Kazen-Gillespie K, Hortsch M, Isom LL (2000) Sodium channel beta subunits mediate homophilic cell adhesion and recruit ankyrin to points of cell-cell contact. J Biol Chem 275:11383–11388

Malhotra JD, Koopmann MC, Kazen-Gillespie KA, Fettman N, Hortsch M, Isom LL (2002) Structural requirements for interaction of sodium channel beta1 subunits with ankyrin. J Biol Chem 277:26681–26688. https://doi.org/10.1074/jbc.M202354200

Malhotra JD, Thyagarajan V, Chen C, Isom LL (2004) Tyrosine-phosphorylated and nonphosphorylated sodium channel beta1 subunits are differentially localized in cardiac myocytes. J Biol Chem 279:40748–40754. https://doi.org/10.1074/jbc.M407243200

Marionneau C, Carrasquillo Y, Norris AJ, Townsend RR, Isom LL, Link AJ, Nerbonne JM (2012) The sodium channel accessory subunit Navβ1 regulates neuronal excitability through

modulation of repolarizing voltage-gated $K^{(+)}$ channels. J Neurosci 32:5716–5727. https://doi.org/10.1523/jneurosci.6450-11.2012
Massey CA, Sowers LP, Dlouhy BJ, Richerson GB (2014) SUDEP mechanisms: the pathway to prevention. Nat Rev Neurol 10:271–282. https://doi.org/10.1038/nrneurol.2014.64
Mattsson N, Axelsson M, Haghighi S, Malmestrom C, Wu G, Anckarsater R, Sankaranarayanan S, Andreasson U, Fredrikson S, Gundersen A, Johnsen L, Fladby T, Tarkowski A, Trysberg E, Wallin A, Anckarsater H, Lycke J, Andersen O, Simon AJ, Blennow K, Zetterberg H (2009) Reduced cerebrospinal fluid BACE1 activity in multiple sclerosis. Mult Scler 15:448–454. https://doi.org/10.1177/1352458508100031
McCormick KA, Isom LL, Ragsdale D, Smith D, Scheuer T, Catterall WA (1998) Molecular determinants of Na^+ channel function in the extracellular domain of the beta1 subunit. J Biol Chem 273:3954–3962
McEwen DP, Isom LL (2004) Heterophilic interactions of sodium channel beta1 subunits with axonal and glial cell adhesion molecules. J Biol Chem 279:52744–52752. https://doi.org/10.1074/jbc.M405990200
McEwen DP, Meadows LS, Chen C, Thyagarajan V, Isom LL (2004) Sodium channel beta1 subunit-mediated modulation of $Na_v1.2$ currents and cell surface density is dependent on interactions with contactin and ankyrin. J Biol Chem 279:16044–16049. https://doi.org/10.1074/jbc.M400856200
McEwen DP, Chen C, Meadows LS, Lopez-Santiago L, Isom LL (2009) The voltage-gated Na^+ channel beta3 subunit does not mediate trans homophilic cell adhesion or associate with the cell adhesion molecule contactin. Neurosci Lett 462:272–275. https://doi.org/10.1016/j.neulet.2009.07.020
Meadows L, Malhotra JD, Stetzer A, Isom LL, Ragsdale DS (2001) The intracellular segment of the sodium channel beta1 subunit is required for its efficient association with the channel alpha subunit. J Neurochem 76:1871–1878
Meadows LS, Malhotra J, Loukas A, Thyagarajan V, Kazen-Gillespie KA, Koopman MC, Kriegler S, Isom LL, Ragsdale DS (2002) Functional and biochemical analysis of a sodium channel beta1 subunit mutation responsible for generalized epilepsy with febrile seizures plus type 1. J Neurosci 22:10699–10709
Medeiros-Domingo A, Kaku T, Tester DJ, Iturralde-Torres P, Itty A, Ye B, Valdivia C, Ueda K, Canizales-Quinteros S, Tusie-Luna MT, Makielski JC, Ackerman MJ (2007) SCN4B-encoded sodium channel beta4 subunit in congenital long-QT syndrome. Circulation 116:134–142. https://doi.org/10.1161/circulationaha.106.659086
Miyazaki H, Oyama F, Wong HK, Kaneko K, Sakurai T, Tamaoka A, Nukina N (2007) BACE1 modulates filopodia-like protrusions induced by sodium channel beta4 subunit. Biochem Biophys Res Commun 361:43–48. https://doi.org/10.1016/j.bbrc.2007.06.170
Miyazaki H, Oyama F, Inoue R, Aosaki T, Abe T, Kiyonari H, Kino Y, Kurosawa M, Shimizu J, Ogiwara I, Yamakawa K, Koshimizu Y, Fujiyama F, Kaneko T, Shimizu H, Nagatomo K, Yamada K, Shimogori T, Hattori N, Miura M, Nukina N (2014) Singular localization of sodium channel beta4 subunit in unmyelinated fibres and its role in the striatum. Nat Commun 5:5525. https://doi.org/10.1038/ncomms6525
Morgan K, Stevens EB, Shah B, Cox PJ, Dixon AK, Lee K, Pinnock RD, Hughes J, Richardson PJ, Mizuguchi K, Jackson AP (2000) Beta 3: an additional auxiliary subunit of the voltage-sensitive sodium channel that modulates channel gating with distinct kinetics. Proc Natl Acad Sci U S A 97:2308–2313. https://doi.org/10.1073/pnas.030362197
Nakano Y, Shimizu W (2016) Genetics of long-QT syndrome. J Hum Genet 61:51–55. https://doi.org/10.1038/jhg.2015.74
Namadurai S, Balasuriya D, Rajappa R, Wiemhofer M, Stott K, Klingauf J, Edwardson JM, Chirgadze DY, Jackson AP (2014) Crystal structure and molecular imaging of the Na_v channel beta3 subunit indicates a trimeric assembly. J Biol Chem 289:10797–10811. https://doi.org/10.1074/jbc.M113.527994

Nashef L, So EL, Ryvlin P, Tomson T (2012) Unifying the definitions of sudden unexpected death in epilepsy. Epilepsia 53:227–233. https://doi.org/10.1111/j.1528-1167.2011.03358.x

Nelson M, Millican-Slater R, Forrest LC, Brackenbury WJ (2014) The sodium channel beta1 subunit mediates outgrowth of neurite-like processes on breast cancer cells and promotes tumour growth and metastasis. Int J Cancer 135:2338–2351. https://doi.org/10.1002/ijc.28890

Nguyen HM, Miyazaki H, Hoshi N, Smith BJ, Nukina N, Goldin AL, Chandy KG (2012) Modulation of voltage-gated K^+ channels by the sodium channel beta1 subunit. Proc Natl Acad Sci U S A 109:18577–18582. https://doi.org/10.1073/pnas.1209142109

Nutini M, Spalloni A, Florenzano F, Westenbroek RE, Marini C, Catterall WA, Bernardi G, Longone P (2011) Increased expression of the beta3 subunit of voltage-gated Na^+ channels in the spinal cord of the SOD1G93A mouse. Mol Cell Neurosci 47:108–118. https://doi.org/10.1016/j.mcn.2011.03.005

O'Malley HA, Isom LL (2015) Sodium channel beta subunits: emerging targets in channelopathies. Annu Rev Physiol 77:481–504. https://doi.org/10.1146/annurev-physiol-021014-071846

O'Malley HA, Shreiner AB, Chen GH, Huffnagle GB, Isom LL (2009) Loss of Na^+ channel beta2 subunits is neuroprotective in a mouse model of multiple sclerosis. Mol Cell Neurosci 40:143–155. https://doi.org/10.1016/j.mcn.2008.10.001

Oakley JC, Kalume F, Catterall WA (2011) Insights into pathophysiology and therapy from a mouse model of Dravet syndrome. Epilepsia 52(Suppl 2):59–61. https://doi.org/10.1111/j.1528-1167.2011.03004.x

Ogiwara I, Nakayama T, Yamagata T, Ohtani H, Mazaki E, Tsuchiya S, Inoue Y, Yamakawa K (2012) A homozygous mutation of voltage-gated sodium channel beta(I) gene SCN1B in a patient with Dravet syndrome. Epilepsia 53:e200–e203. https://doi.org/10.1111/epi.12040

Olesen MS, Jespersen T, Nielsen JB, Liang B, Moller DV, Hedley P, Christiansen M, Varro A, Olesen SP, Haunso S, Schmitt N, Svendsen JH (2011) Mutations in sodium channel beta-subunit SCN3B are associated with early-onset lone atrial fibrillation. Cardiovasc Res 89:786–793. https://doi.org/10.1093/cvr/cvq348

Oyama F, Miyazaki H, Sakamoto N, Becquet C, Machida Y, Kaneko K, Uchikawa C, Suzuki T, Kurosawa M, Ikeda T, Tamaoka A, Sakurai T, Nukina N (2006) Sodium channel beta4 subunit: down-regulation and possible involvement in neuritic degeneration in Huntington's disease transgenic mice. J Neurochem 98:518–529. https://doi.org/10.1111/j.1471-4159.2006.03893.x

Pandis D, Scarmeas N (2012) Seizures in Alzheimer disease: clinical and epidemiological data. Epilepsy Curr 12:184–187. https://doi.org/10.5698/1535-7511-12.5.184

Patino GA, Claes LR, Lopez-Santiago LF, Slat EA, Dondeti RS, Chen C, O'Malley HA, Gray CB, Miyazaki H, Nukina N, Oyama F, De Jonghe P, Isom LL (2009) A functional null mutation of SCN1B in a patient with Dravet syndrome. J Neurosci 29:10764–10778. https://doi.org/10.1523/jneurosci.2475-09.2009

Patino GA, Brackenbury WJ, Bao Y, Lopez-Santiago LF, O'Malley HA, Chen C, Calhoun JD, Lafreniere RG, Cossette P, Rouleau GA, Isom LL (2011) Voltage-gated Na^+ channel beta1B: a secreted cell adhesion molecule involved in human epilepsy. J Neurosci 31:14577–14591. https://doi.org/10.1523/jneurosci.0361-11.2011

Pertin M, Ji RR, Berta T, Powell AJ, Karchewski L, Tate SN, Isom LL, Woolf CJ, Gilliard N, Spahn DR, Decosterd I (2005) Upregulation of the voltage-gated sodium channel beta2 subunit in neuropathic pain models: characterization of expression in injured and non-injured primary sensory neurons. J Neurosci 25:10970–10980. https://doi.org/10.1523/jneurosci.3066-05.2005

Qin N, D'Andrea MR, Lubin ML, Shafaee N, Codd EE, Correa AM (2003) Molecular cloning and functional expression of the human sodium channel beta1B subunit, a novel splicing variant of the beta1 subunit. Eur J Biochem 270:4762–4770

Ransdell JL, Dranoff E, Lau B, Lo WL, Donermeyer DL, Allen PM, Nerbonne JM (2017) Loss of Navbeta4-mediated regulation of sodium currents in adult Purkinje neurons disrupts firing and impairs motor coordination and balance. Cell Rep 19:532–544. https://doi.org/10.1016/j.celrep.2017.03.068

Ratcliffe CF, Qu Y, McCormick KA, Tibbs VC, Dixon JE, Scheuer T, Catterall WA (2000) A sodium channel signaling complex: modulation by associated receptor protein tyrosine phosphatase beta. Nat Neurosci 3:437–444. https://doi.org/10.1038/74805

Ratcliffe CF, Westenbroek RE, Curtis R, Catterall WA (2001) Sodium channel beta1 and beta3 subunits associate with neurofascin through their extracellular immunoglobulin-like domain. J Cell Biol 154:427–434

Reid CA, Leaw B, Richards KL, Richardson R, Wimmer V, Yu C, Hill-Yardin EL, Lerche H, Scheffer IE, Berkovic SF, Petrou S (2014) Reduced dendritic arborization and hyperexcitability of pyramidal neurons in a Scn1b-based model of Dravet syndrome. Brain 137:1701–1715. https://doi.org/10.1093/brain/awu077

Remme CA, Bezzina CR (2010) Sodium channel (dys)function and cardiac arrhythmias. Cardiovasc Ther 28:287–294. https://doi.org/10.1111/j.1755-5922.2010.00210.x

Riuro H, Beltran-Alvarez P, Tarradas A, Selga E, Campuzano O, Verges M, Pagans S, Iglesias A, Brugada J, Brugada P, Vazquez FM, Perez GJ, Scornik FS, Brugada R (2013) A missense mutation in the sodium channel beta2 subunit reveals SCN2B as a new candidate gene for Brugada syndrome. Hum Mutat 34:961–966. https://doi.org/10.1002/humu.22328

Riuro H, Campuzano O, Arbelo E, Iglesias A, Batlle M, Perez-Villa F, Brugada J, Perez GJ, Scornik FS, Brugada R (2014) A missense mutation in the sodium channel beta1b subunit reveals SCN1B as a susceptibility gene underlying long QT syndrome. Heart Rhythm 11:1202–1209. https://doi.org/10.1016/j.hrthm.2014.03.044

Roger S, Rollin J, Barascu A, Besson P, Raynal PI, Iochmann S, Lei M, Bougnoux P, Gruel Y, Le Guennec JY (2007) Voltage-gated sodium channels potentiate the invasive capacities of human non-small-cell lung cancer cell lines. Int J Biochem Cell Biol 39:774–786. https://doi.org/10.1016/j.biocel.2006.12.007

Scheffer IE, Harkin LA, Grinton BE, Dibbens LM, Turner SJ, Zielinski MA, Xu R, Jackson G, Adams J, Connellan M, Petrou S, Wellard RM, Briellmann RS, Wallace RH, Mulley JC, Berkovic SF (2007) Temporal lobe epilepsy and GEFS+ phenotypes associated with SCN1B mutations. Brain 130:100–109. https://doi.org/10.1093/brain/awl272

Shah BS, Stevens EB, Gonzalez MI, Bramwell S, Pinnock RD, Lee K, Dixon AK (2000) Beta3, a novel auxiliary subunit for the voltage-gated sodium channel, is expressed preferentially in sensory neurons and is upregulated in the chronic constriction injury model of neuropathic pain. Eur J Neurosci 12:3985–3990

Shah BS, Stevens EB, Pinnock RD, Dixon AK, Lee K (2001) Developmental expression of the novel voltage-gated sodium channel auxiliary subunit beta3, in rat CNS. J Physiol 534:763–776

Shimizu H, Miyazaki H, Ohsawa N, Shoji S, Ishizuka-Katsura Y, Tosaki A, Oyama F, Terada T, Sakamoto K, Shirouzu M, Sekine S, Nukina N, Yokoyama S (2016) Structure-based site-directed photo-crosslinking analyses of multimeric cell-adhesive interactions of voltage-gated sodium channel beta subunits. Sci Rep 6:26618. https://doi.org/10.1038/srep26618

Shimizu H, Tosaki A, Ohsawa N, Ishizuka-Katsura Y, Shoji S, Miyazaki H, Oyama F, Terada T, Shirouzu M, Sekine SI, Nukina N, Yokoyama S (2017) Parallel homodimer structures of the extracellular domains of the voltage-gated sodium channel beta4 subunit explain its role in cell-cell adhesion. J Biol Chem 292(32):13428–13440. https://doi.org/10.1074/jbc.M117.786509

Spampanato J, Kearney JA, de Haan G, McEwen DP, Escayg A, Aradi I, MacDonald BT, Levin SI, Soltesz I, Benna P, Montalenti E, Isom LL, Goldin AL, Meisler MH (2004) A novel epilepsy mutation in the sodium channel SCN1A identifies a cytoplasmic domain for beta subunit interaction. J Neurosci 24:10022–10034. https://doi.org/10.1523/jneurosci.2034-04.2004

Srinivasan J, Schachner M, Catterall WA (1998) Interaction of voltage-gated sodium channels with the extracellular matrix molecules tenascin-C and tenascin-R. Proc Natl Acad Sci U S A 95:15753–15757. https://doi.org/10.1073/pnas.95.26.15753

Surges R, Sander JW (2012) Sudden unexpected death in epilepsy: mechanisms, prevalence, and prevention. Curr Opin Neurol 25:201–207. https://doi.org/10.1097/WCO.0b013e3283506714

Takahashi N, Kikuchi S, Dai Y, Kobayashi K, Fukuoka T, Noguchi K (2003) Expression of auxiliary beta subunits of sodium channels in primary afferent neurons and the effect of nerve injury. Neuroscience 121:441–450

Tan BH, Pundi KN, Van Norstrand DW, Valdivia CR, Tester DJ, Medeiros-Domingo A, Makielski JC, Ackerman MJ (2010) Sudden infant death syndrome-associated mutations in the sodium channel beta subunits. Heart Rhythm 7:771–778. https://doi.org/10.1016/j.hrthm.2010.01.032

Tester DJ, Ackerman MJ (2014) Genetics of long QT syndrome. Methodist Debakey Cardiovasc J 10:29–33

Theile JW, Jarecki BW, Piekarz AD, Cummins TR (2011) $Na_v1.7$ mutations associated with paroxysmal extreme pain disorder, but not erythromelalgia, enhance Na_vbeta4 peptide-mediated resurgent sodium currents. J Physiol 589:597–608. https://doi.org/10.1113/jphysiol.2010.200915

Valdivia CR, Medeiros-Domingo A, Ye B, Shen WK, Algiers TJ, Ackerman MJ, Makielski JC (2010) Loss-of-function mutation of the SCN3B-encoded sodium channel {beta}3 subunit associated with a case of idiopathic ventricular fibrillation. Cardiovasc Res 86:392–400. https://doi.org/10.1093/cvr/cvp417

Van Norstrand DW, Ackerman MJ (2009) Sudden infant death syndrome: do ion channels play a role? Heart Rhythm 6:272–278. https://doi.org/10.1016/j.hrthm.2008.07.028

Wallace RH, Wang DW, Singh R, Scheffer IE, George AL, Phillips HA, Saar K, Reis A, Johnson EW, Sutherland GR, Berkovic SF, Mulley JC (1998) Febrile seizures and generalized epilepsy associated with a mutation in the Na^+-channel sz1 subunit gene SCN1B. Nat Genet 19:366–370

Wallace RH, Scheffer IE, Parasivam G, Barnett S, Wallace GB, Sutherland GR, Berkovic SF, Mulley JC (2002) Generalized epilepsy with febrile seizures plus: mutation of the sodium channel subunit SCN1B. Neurology 58:1426–1429

Wang P, Yang Q, Wu X, Yang Y, Shi L, Wang C, Wu G, Xia Y, Yang B, Zhang R, Xu C, Cheng X, Li S, Zhao Y, Fu F, Liao Y, Fang F, Chen Q, Tu X, Wang QK (2010) Functional dominant-negative mutation of sodium channel subunit gene SCN3B associated with atrial fibrillation in a Chinese GeneID population. Biochem Biophys Res Commun 398:98–104. https://doi.org/10.1016/j.bbrc.2010.06.042

Watanabe H, Koopmann TT, Le Scouarnec S, Yang T, Ingram CR, Schott JJ, Demolombe S, Probst V, Anselme F, Escande D, Wiesfeld AC, Pfeufer A, Kaab S, Wichmann HE, Hasdemir C, Aizawa Y, Wilde AA, Roden DM, Bezzina CR (2008) Sodium channel beta1 subunit mutations associated with Brugada syndrome and cardiac conduction disease in humans. J Clin Invest 118:2260–2268. https://doi.org/10.1172/JCI33891

Watanabe H, Darbar D, Kaiser DW, Jiramongkolchai K, Chopra S, Donahue BS, Kannankeril PJ, Roden DM (2009) Mutations in sodium channel beta1- and beta2-subunits associated with atrial fibrillation. Circ Arrhythm Electrophysiol 2:268–275. https://doi.org/10.1161/circep.108.779181

Wimmer VC, Reid CA, Mitchell S, Richards KL, Scaf BB, Leaw BT, Hill EL, Royeck M, Horstmann MT, Cromer BA, Davies PJ, Xu R, Lerche H, Berkovic SF, Beck H, Petrou S (2010) Axon initial segment dysfunction in a mouse model of genetic epilepsy with febrile seizures plus. J Clin Invest 120:2661–2671. https://doi.org/10.1172/jci42219

Wong HK, Sakurai T, Oyama F, Kaneko K, Wada K, Miyazaki H, Kurosawa M, De Strooper B, Saftig P, Nukina N (2005) Beta subunits of voltage-gated sodium channels are novel substrates of beta-site amyloid precursor protein-cleaving enzyme (BACE1) and gamma-secretase. J Biol Chem 280:23009–23017. https://doi.org/10.1074/jbc.M414648200

Xiao ZC, Ragsdale DS, Malhotra JD, Mattei LN, Braun PE, Schachner M, Isom LL (1999) Tenascin-R is a functional modulator of sodium channel beta subunits. J Biol Chem 274:26511–26517

Xu R, Thomas EA, Gazina EV, Richards KL, Quick M, Wallace RH, Harkin LA, Heron SE, Berkovic SF, Scheffer IE, Mulley JC, Petrou S (2007) Generalized epilepsy with febrile seizures plus-associated sodium channel beta1 subunit mutations severely reduce beta subunit-mediated modulation of sodium channel function. Neuroscience 148:164–174. https://doi.org/10.1016/j.neuroscience.2007.05.038

Yu FH, Westenbroek RE, Silos-Santiago I, McCormick KA, Lawson D, Ge P, Ferriera H, Lilly J, DiStefano PS, Catterall WA, Scheuer T, Curtis R (2003) Sodium channel beta4, a new disulfide-linked auxiliary subunit with similarity to beta2. J Neurosci 23:7577–7585

Yuan L, Koivumaki JT, Liang B, Lorentzen LG, Tang C, Andersen MN, Svendsen JH, Tfelt-Hansen J, Maleckar M, Schmitt N, Olesen MS, Jespersen T (2014) Investigations of the Na_vbeta1b sodium channel subunit in human ventricle; functional characterization of the H162P Brugada syndrome mutant. Am J Physiol Heart Circ Physiol 306:H1204–H1212. https://doi.org/10.1152/ajpheart.00405.2013

Zhou TT, Zhang ZW, Liu J, Zhang JP, Jiao BH (2012) Glycosylation of the sodium channel beta4 subunit is developmentally regulated and involves in neuritic degeneration. Int J Biol Sci 8:630–639. https://doi.org/10.7150/ijbs.3684

The manufacturer's authorised representative in the EU is Springer Nature Customer Service Centre GmbH, Europaplatz 3, 69115 Heidelberg, Germany. If you have any concerns regarding our products, please contact ProductSafety@springernature.com

Printed and bound by CPI Group (UK) Ltd, Croydon, CR0 4YY
30/09/2026
02225654-0015